M.A. Levitan · Yu. A. Lavrushin

Sedimentation History in the Arctic Ocean and Subarctic Seas for the Last 130 kyr

Springer

Dr. M. A. Levitan
Russian Academy of Sciences
Vernadsky Inst. Geochemistry
Analytical Chemistry
Kosygina Str. 19
Moscow
Russia 119991
m-levitan@mail.ru

Dr. Yu. A. Lavrushin
Russian Academy of Sciences
Inst. Geology
Pyzhevsky per. 7
Moscow
Russia 119017
lavrushin@ginras.ru

ISSN 0930-0317
ISBN 978-3-642-00287-8 e-ISBN 978-3-642-00288-5
DOI 10.1007/978-3-642-00288-5
Springer Dordrecht Heidelberg London New York

Library of Congress Control Number: 2009926693

Cover design: Bauer, Thomas

Printed on acid-free paper

Springer is part of Springer Science+Business Media (www.springer.com)

Contents

Part I Geological and Paleoecological Events of the Late Pleistocene and Holocene in Northern Eurasia

1 **Geological and Paleoecological Events of the Late Pleistocene along Eurasian Coastal Areas of the Arctic Ocean** 3
General Upper Pleistocene Stratigraphic Scheme for Northern Eurasia ... 3
Duration of the Mikulino Interglaciation 5
Correlation of the Natural Events Correlative with MIS 5d–5a in Northern West Europe and Northwestern Russia 6

2 **Late Pleistocene Geologic-Paleoecological Events in the North of European Russia** 11
Relationship between Land and Sea Areas during the Mikulino Interglacial in Northern Eurasia 11
Genetic Types of Continental Sediments 14
Marine Sediments of the Boreal Transgression in the North of European Russia 15

3 **Main Geologic-Paleoecological Events of the Late Pleistocene in the North of Western Siberia** 31

4 **Geologic-Paleoecological Events of the Late Pleistocene in the Northern-Siberian Lowland and Taimyr Peninsula** 37

5 **The Late Glacial Time and Holocene of Northern Eurasia** 43

6 **Outlines of the Late Pleistocene and Holocene History of the East Arctic Seas** 47

7 **The Deglaciation Time and Holocene of Northern Eurasia** 57

Part II Marine Sedimentation in the Arctic Ocean and Subarctic Seas

8 The Seas of West Subarctic Region ... 61
Geologic and Oceanographic Setting ... 61
History of Sedimentation ... 68
History of Sedimentation Rates ... 68
History of Sedimentation on the Vøring Plateau During the Last 25 ka 88
History of Sedimentation at the Continental Margin of Eastern and South-Eastern Greenland During the Last 130 ka ... 107

9 The Arctic Ocean ... 113
Recent Environment ... 113
Morphostructure, Oceanographic and Sea-Ice Setting, Recent Sediments and Their Mineral Composition ... 113
Facies Variations of Holocene Sediments on the Yermak Plateau (According to Study Data of >63 mkm Fraction) ... 124
History of Sedimentation ... 136
History of Sedimentation Rates During the Last 130 ka ... 136
History of Sedimentation on the Yermak Plateau During the Last 190 ka ... 148
Organic-Geochemical Sediment Studies of the Eastern Part of the Central Arctic ... 173

10 The Western Arctic Seas ... 177
Recent Sedimentation Environment ... 177
The Barents Sea ... 177
The Kara Sea ... 178
Surface Sediments of the Pechora Sea ... 179
Surface Sediments of St. Anna Trough ... 186
Facies Zonality of Surface Sediments in the Eastern Kara Sea ... 194
History of Sedimentation ... 210
Late- and Post-Glacial History of Sedimentation in the Eastern Part of the Barents Sea ... 210
Holocene Sedimentation History in the Southern Novaya Zemlya Trough ... 224
History of Sedimentation in the Pechora Sea During the Late Pleistocene and Holocene ... 241
Light Fraction Mineralogy of the Upper Quaternary Sediments from the Saint Anna Trough and Its Paleoceanographic Interpretation ... 247
Holocene History of Yenisei River Discharge ... 256
Holocene History of Ob River Discharge ... 272

11 Eastern Arctic Seas 289
Recent Sedimentation Environment 289
The Laptev Sea 289
The East Siberian Sea 290
The Chukchi Sea 291
History of Sedimentation 291
History of Sedimentation in the Laptev Sea During the Late Weichselian to Holocene by Geophysical and Geochemical Data 292
Holocene History of the Lena and Other Rivers Discharge in the Laptev Sea 295
Organic Geochemical Data About Sedimentation History Along the Continental Slope of the East Siberian Sea During the Last Climatic Cycle 297
Preliminary Data About Accumulation of Diatom-Bearing Clayey Silts at the Chukchi Sea Shelf 298

12 Seas of the Eastern Subarctic 301
Recent Sedimentation Environment 301
History of Sedimentation 307
History of Sedimentation in the Deep-Water Part of the Shirshov Ridge (Bering Sea) During the Last Three Marine-Isotope Stages 307
History of Sedimentation in the Northern Sea of Okhotsk During the Last 1.1 Ma 310

Part III The Late Pleistocene Paleogeographic Events of Northern Eurasia and History of Sedimentation in the Subarctic Seas and the Arctic Ocean in Relation to the Northern Hemisphere Glaciation during the Last Climatic Cycle

13 Characteristic Features of the Mikulino Landscapes 333

14 Results of Paleoclimate Studies 337

15 Particularities of Sedimentation Processes Within the Continental Blocks and Marine Basins 345
Deglaciation Peculiarities 345
Facies Variability during Glaciations, Deglaciations, Interglacials 349
Geological History of the Arctic Ocean Sea Ice during the Last 60 ka 350
Intercoupling of Atmo-, Hydro-, Cryo-, Bio-, and Lithospheres 353

References 357

Index 381

List of Figures

1.1 Oxygen-isotope records of the lower parts of the GRIP and GISP 2 Greenland ice cores (Johnsen et al., 1995) 6

1.2 Climatostratigraphic scheme of main natural events of the Late Pleistocene (by Yu A. Lavrushin, based on (Behre, 1989; Spiridonova, 1983; Mangerud, 1989). Palynological, geological and geomorphological data are used qualitatively 7

2.1 Schematic distribution map of the Karelian Interglacial sea (Biske, 1959). 1 – coast lines; 2 – sea; 3 – locations of interglacial sedimentary sections with paleontological characteristics; 4 – locations of interglacial sedimentary sections without paleontological characteristics 12

2.2 Distribution of the interglacial boreal transgression (Lavrova and Troitsky, 1960). 1 – marine boreal transgression waters; 2 – freshened bays; 3 – waters of Eemian transgression 13

2.3 Boreal sea on the Russian North (according to S.L. Troitsky (1964)). 1 – territories covered by boreal sea during the maximal transgression; 2 – Baltic Basin during the transgression maximum; 3 – freshened areas; 4–8 – boundaries of distribution of zoogeographic groups and individual species of mollusks and barnacles: 4 – Lusitanian (south-boreal) species; 5 – *Mactra elliptica,* 6 – *Cardium edule,* 7 – *Pholas crispatus,* 8 – *Cyprina rina islandica;* 9 – basins (I – White Sea, II – Pechora, III – West Siberia, IV – Taimyr) 14

2.4 Extension of the Late Valday Ice Sheet during the deglaciation (Lavrov and Potapenko, 2005), simplified. *A* – Barents Sea-Novaya Zemlya-Kara Sea Ice Sheet. I–IX – lobes: I – Kozhvinskaya; II – Lyzhskaya; III – Laisko-Izhemskaya; IV – Pechorskaya; V – Kolvinskaya; VI – Malozemel'skaya; VII – Bol'shezemel'skaya; VIII – Kuloisko-Mezenskaya; IX – Chöshskaya. *B* – Scandinavian Ice Sheet. I–V – lobes: I – Verkhnemezenskaya; II – Vashskaya; III – Pinezhskaya; IV – Severodvinskaya; V – Vazhskaya 28

3.1 Paleogeographic scheme of glaciation and marine transgression boundaries in northern West Siberia and the North Siberian Lowland during the Late Pleistocene (Arkhipov, 2000). 1 – Kazantzev Sea (MIS 5e); 2 – Kargin Sea (MIS 3.1 and 3.3); 3 – location of sediments: *a* – well studied Kazantzev sediments; *b* – Kargin sediments; 4 – boundary of Sartan Glaciation (MIS 2); 5 – Lokhpodgort motions (MIS 3.2); 6 – boundary of Yermak Glaciation (MIS 4) . . . 32

4.1 Main elements of the Taimyr glacial morpho-sculpture (Anthropogen of Taimyr, 1982). (Muruktinian Glaciation, North Siberian Stage): 1 – moraine swells: *a* – pressure moraines; *b* – bulkload moraines; 2–3 – deglaciation time for Muructinian Ice Cover: 2 – fracture-kame swells and elevated massifs; 3 – kame terraces; Severokokorskaya Stage: 4 – pressure-bulkload marginal features; 5 – sag-and-swell topography of the “dead” ice fields; 6 – kame terraces; 7 – rises of interlobe massifs; 8 – inter-tongue massifs; 9–12 – Sartanian Glaciation, Karaulian Phase: 9 – pressure-bulkload marginal features; 10 – rises of interlobe massifs; 11 – inter-tongue massifs; 12 – kame terraces; 13 – N’yapanskaya Phase: *a* – marginal pressure-bulkload and bulkload features, *b* – intertongue massifs; 14 – Noril’skaya Phase: bulkload marginal features in the foothill of Putoran Plateau; 15–21 – elements of glacial morpho-sculpture deciphered at radiolocation images on the Northern Taimyr: 15 – proposed marginal glacial features on the north of Taimyr; 16 – crests of proposed end-moraine swells; 17 – areas of hilly topography; 18 – areas covered by thick sequence of questionable glacial deposits; 19 – areas with traces of glacial exaration, partly covered by questionable glacial deposits; 20 – areas not covered by loose deposits but with clear traces of exaration treatment; 21 – areas without clear exaration traces but with erratics on the surface; 22 – crests of end-moraine swells; 23 – troughs; 24 – rises of North Siberian Lowland composed from hard rock and possibly were not covered by ice; 25 – north-western Taimyr with only valley-net glaciation in the Late Pleistocene; 26 – bottom of glacial depressions with marine, lacustrine and partially fluvial sediments; 27 – profile lines; 28 – glacier swells: I – Urdakhskaya; II – Sakshsinskaya; III – Severokokorskaya; IV – Dzhangodskaya; V – Syntabul’skaya; VI – Bai-kuranerskaya; VII – Mokorittskie; VIII – Upper Taimyr; IX – North Taimyr; X – N’yapanskaya . . . 38

4.2 Marginal glacial features and source petrographic provinces (Anthropogen of Taimyr, 1982). 1 – boundaries of pre-Cenozoic petrographic provinces; 2–9 – petrographic provinces: 2 – North Taimyr (Proterozoic and Archean granites, gneisses, greenstone shales, limestones, dolomites), 3 – Byrranga (Upper Paleozoic sandstones, siltstones (P) and limestones (C), 4 – Yenisei-Khatanga (Mesozoic sands and sandstones (K), 5 – Putoran-Vilyui (Mesozoic trapps), 6 – Kotui (Cambrian-Silurian limestones and

dolomites), 7, 8 – Anabar: 7 – Archean granitoids, 8 – proterozoic quartz-feldspar sandstones of Mukun Serie, 9 – intrusions of subalcaline basaltoids; 10, 11 – large features of glacial topography of marginal accumulation zone: 10 – sag-and-swell and swell topography, swells and massifs, 11 – small terminal moraine swells; 12–13 – boundaries: 12 – maximal development of Zyryan Glaciation, 13 – the studie area of composition of glacial deposits, clastic matter and products of its rewashing 40
8.1 Scheme of the Nordic Seas bottom topography (Vogt, 1986) 62
8.2 Surface water masses, main surface currents, and sea-ice margins in the Nordic Seas (Hald, 2001). 1–4 – surface waters: 1 – Coastal, 2 – Arctic, 3 – Atlantic, 4 – Polar; 5, 6 – sea-ice margins: 5 – in summer, 6 – in winter; 7, 8 – surface currents: 7 – warm, 8 – cold 63
8.3 Lithology of surface sediments in the Nordic Seas (Vogt, 1986). 1 – marly mud without IRD; 2 – sandy clay; 3 – clayey mud 64
8.4 Amount of IRD (*a*), size of its largest fragments (*b*), and its lithological variability (*c*) in the Holocene and Upper Pleistocene sediments of the Nordic Seas (Bischof, 2000) 66
8.5 Maps of smectite distribution in the surface sediments of the Nordic Seas (compiled by M.A. Levitan, based on data from (Berner, 1991)). 1 – bottom currents, 2 – surface currents 67
8.6 Location of sediment cores with established stratigraphy in the Nordic Seas 69
8.7 Recent sedimentation rates (SR) in the Nordic Seas (cm/ky) (Paetsch et al., 1992), modified by M.A. Levitan 75
8.8 MIS 1 sedimentation rates (SR) in the Nordic Seas (cm/ky) (for data source see text) 76
8.9 MIS 2 sedimentation rates (SR) in the Nordic Seas (cm/ky) (for data source see text) 77
8.10 MIS 3 sedimentation rates (SR) in the Nordic Seas (cm/ky) (for data source see text) 78
8.11 MIS 4 sedimentation rates (SR) in the Nordic Seas (cm/ky) (for data source see text) 79
8.12 MIS 5 sedimentation rates (SR) in the Nordic Seas (cm/ky) (for data source see text) 80
8.13 MIS 5a-d sedimentation rates (SR) in the Nordic Seas (cm/ky) (for data source see text) 81
8.14 MIS 5e sedimentation rates (SR) in the Nordic Seas (cm/ky) (for data source see text) 82
8.15 Ratio of ice volume in the Northern Hemisphere (dark) (Loutre, Berger, 2003) and average sedimentation rates in the Nordic Seas (light) for the last five MIS (conventional units) (Levitan et al., 2007a) 86
8.16 Location of studied sediment cores in the Vøring Plateau area (Levitan et al., 2005b) 89

8.17 Water-mass stratification in the Vøring Plateau area (Levitan et al., 2005b). 1 – Bottom Water; 2 – Deep Water; 3 – Atlantic Water; 4 – Norwegian Coastal Water 90
8.18 Lithology, TOC and $CaCO_3$ amount, and grain-size distribution in sediment core ASV 1372 (Levitan et al., 2005b). 1–8 – lithology: 1 – clayey mud, 2 – hardground, 3 – sandy-silty clay, 4 – silty clay, 5 – sandy-silty-clayey mictite, 6 – marly mud (10–30% $CaCO_3$), 7 – fragments of dense clay, 8 – IRD; 9 – erosion boundary; 10–16 – color: 10 – brown, 11 – brownish-gray, 12 – olive-gray, 13 – greenish-gray, 14 – light gray, 15 – dark gray, 16 – light yellowish-gray; 17–19 – texture: 17 – homogenous, 18 – laminated, 19 – bioturbated. ^{14}C dates concern iceberg sediments from Marion Dufresne and Jan Mayen sediment cores (their location is shown at Fig. 8.16) 91
8.19 Examples of grain-size distribution (histograms) from sediment core ASV 1372 (Levitan et al., 2005b). *a–d* – core depths of samples, cm: *a* – 0 – 3, *b* – 27 – 29, *c* – 82 – 84, *d* – 122 – 124 92
8.20 Concentration of mineral grains and foraminifers in the >0.1 mm fraction (Levitan et al., 2005b). 1 – mineral grains, 2 – benthic foraminifers, 3 – planktonic foraminifers 93
8.21 Composition of benthic foraminifer assemblages in sediment core ASV 1372 (Levitan et al., 2005b). 1 – *Islandiella spp.;* 2 – *Fissurina/Parafissurina;* 3 – *Elphidium clavatum;* 4 – *Cibicides wuellerstorfi;* 5 – *Epistominella exigua;* 6 – *Epistominella pusilla;* 7 – *Melonis barleeanus;* 8 – *Oridorsalis umbonatus;* 9 – *Cassidulina reniforme;* 10 – *Cassidulina teretis*; 11 – other species 94
8.22 Reconstruction of annual SST based on planktonic foraminifers (Levitan et al., 2005b) 95
8.23 Compilation of data on mass accumulation rates and source provinces of terrigenous matter as well as advection of Atlantic Water in the area of sediment core ASV 1372 for the time interval 9–14 cal. kyrs. BP (Levitan et al., 2005b) 102
8.24 Lithostratigraphic correlation of sediment cores along the Vøring Plateau latitudinal transect (see Fig. 8.16 for location of sediment cores) (Levitan et al., 2005b). 1 – iceberg deposits, 2 – "high productivity" sediments enriched in $CaCO_3$, 3 – turbidites. For letter designations see text. In parentheses – water depths, m 106
8.25 Comparison of main paleoceanographic events in the Nordic Seas, glaciation history of Svalbard and western Scandinavia, insolation change at 78° N, and oxygen-isotope record of the Greenland Ice Core during the last climatic cycle (Hald, 2001) 107
8.26 Location of studied sediment cores at the East Greenland continental margin off Scoresby Sund (Nam et al., 1995) 108
8.27 Comparison of paleoproductivity and IRD in the North Atlantic during the last 60 ka (Sarnthein et al., 2001) 110

8.28 Oxygen-isotope record of the GRIP Ice Core (Dansgaard et al., 1993), IRD content and magnetic susceptibility in sediment core PS1726 (from (Stein et al., 1996)). Numbers in column for ice core: number of Dansgaard-Oeschger interstadials (warming periods) 110
8.29 Variations of sedimentation rates (cm/ky) in selected sediment cores from the East Greenland continental margin (Nam and Stein, 1999). In parentheses – water depths, m 111
8.30 Variability of mass accumulation rates (g/cm^2/ky) in selected sediment cores from the East Greenland continental margin (Nam and Stein, 1999) 112
9.1 Elements of the Arctic Ocean morphostructure (Johnson, 1990) 114
9.2 Distribution of winter and summer sea ice in the Arctic Ocean (Macdonald et al., 2003). 1 – compact ice; 2 – non-compact ice (>15 %); 3 – open sea surface 115
9.3 Stratification of the Arctic Ocean water column (Aagaard and Carmack, 1989). PSW – Polar Surface Water; AIW – Arctic Intermediate Water; CBDW – Canada Basin Deep Water; UAODW – Upper Arctic Ocean Deep Water; LAODW – Lower Arctic Ocean Deep Water 116
9.4 Arctic Ocean surface-water circulation (Aagaard and Carmack, 1989) 117
9.5 Arctic Ocean Intermediate-Water circulation (Rudels et al., 1994) 118
9.6 Heavy-mineral distribution in Arctic Ocean surface sediments (Belov and Lapina, 1961). 1– 8– heavy-mineral associations: 1 – epidot-pyroxene, 2 – black ore-epidot-pyroxene, 3 – chloritoid, 4 – dolomite, 5 – garnet, 6 – black ore minerals, 7 – pyroxene, 8 – pyroxene-amphibole 120
9.7 Smectite distribution in eastern and central Arctic Ocean surface sediments (Wahsner et al., 1999) 123
9.8 Illite distribution in eastern and central Arctic Ocean surface sediments (Wahsner et al., 1999) 124
9.9 Chlorite distribution in eastern and central Arctic Ocean surface sediments (Wahsner et al., 1999) 125
9.10 Kaolinite distribution in eastern and central Arctic Ocean surface sediments (Wahsner et al., 1999) 126
9.11 Location of studied sediment cores at Yermak Plateau and elements of surface circulation (Levitan et al., 2000b). Isobaths are given in meters. 1 – West Spitsbergen Current; 2 – East Greenland Current; 3 – Arctic Ocean Deep Water currents; 4 – Bottom Water of the Norwegian Sea (N) and the Arctic Ocean (A); 5 – sea-ice margin in summer of 1997; 6 – RV *Polarstern* sediment cores. I – Barents Sea; II – Spitsbergen; III – Sophia Deep; IV – Yermak Plateau; V – Fram Strait; VI – Nansen Basin; VII – Greenland; VIII – Lena Deep 127
9.12 Bathymetric profiles across the Yermak Plateau (Stein and Fahl, 1997) 131

9.13 Lithological profiles across the Yermak Plateau from box core records; for location of cores see Fig. 9.12 (Levitan et al., 2000b). 1 – unrounded coarse rock fragments, gravel, and pebble; 2 – sand; 3 – silts and silty clay; 4 – hemipelagic mud; 5 – lenses; 6 – hardground 131
9.14 Distribution of biogenic remains in the >0.063 mm fraction of surface sediments (Levitan et al., 2000b). 1 – polychaets; 2 – planktonic foraminifers (number in parentheses – amount in fraction 2.0–0.063 mm, %); 3 – benthic foraminifers including *Biloculina* and *Triloculina*; 4 – agglutinated benthic foraminifers; 5 – silica sponges; 6 – bivalves 133
9.15 Location of studied sediment cores in the Arctic Ocean 143
9.16 MIS 5 sedimentation rates (SR) (cm/ky) in the central Arctic Ocean (Levitan and Stein, 2007) 144
9.17 MIS 4 sedimentation rates (SR) (cm/ky) in the central Arctic Ocean (Levitan and Stein, 2007) 145
9.18 MIS 3 sedimentation rates (SR) (cm/ky) in the central Arctic Ocean (Levitan and Stein, 2007) 146
9.19 MIS 2 sedimentation rates (SR) (cm/ky) in the central Arctic Ocean (Levitan and Stein, 2007) 147
9.20 MIS 1 sedimentation rates (SR) (cm/ky) in the central Arctic Ocean (Levitan and Stein, 2007) 148
9.21 Histograms of average SR (cm/ky) distribution for glacial (black) and non-glacial (light) continental margins of the Arctic Ocean for the last five marine isotope stages (MIS) (Levitan and Stein, 2007) ... 149
9.22 Histograms of average SR (cm/ky) distribution for deep-sea basins (black) and ridges (light) of the Arctic Ocean for the last five marine isotope stages (MIS) (Levitan and Stein, 2007) 151
9.23 Lithology of sediment core PS2837 (Levitan et al., 2002a). 1 – clayey mud; 2 – silty clay; 3 – fine silt; 4 – coarse silt; 5 – mictite (sandy-silty clay); 6 – mictites with IRD; 7 – silty clay with sand admixture; 8 – clay with silt admixture; 9 – silty clay with carbonate admixture; 10 – hydrotroilite; 11 – planktonic foraminifers. H1-H6 – Heinrich events (Thiede and Tiedemann, 2001) 152
9.24 Lithology of sediment core PS2834 (Levitan et al., 2002a). 1 – clayey mud; 2 – hardground; 3 – silty clay; 4 – silty clay with carbonate admixture; 5 – clayey silt; 6 – mictite; 7 – clayey mud with carbonate admixture; 8 – IRD; 9 – bivalves shells 155
9.25 Lithology of sediment core PS2859 (Levitan et al., 2002a). 1 – clayey mud; 2 –silty clay with sand admixture; 3 – sandy silt; 4 – hardground; 5 – silty clay; 6 – clayey mud with carbonate admixture; 7 – silt with sand admixture; 8 – silty clay; 9 – sandy-silty clay; 10 – sandy silt with carbonate admixture; 11 – silty clay with carbonate admixture; 12 – clayey mud with carbonate admixture; 13 – turbidites 159

9.26 Lithostratigraphic correlation of sedimentary sections along to the northern transect at Yermak Plateau (Levitan et al., 2002b). Location of profiles is shown at Fig. 9.11. 1 – clayey mud; 2 – sand; 3 – silt; 4 – silty clay; 5 – hardground; 6 – clayey mud with carbonate admixture; 7 – silt with carbonate admixture; 8 – IRD; 9 – turbidites 162
9.27 Lithostratigraphic correlation of sedimentary sections along to the southern transect at Yermak Plateau (Levitan et al., 2002b). Location of profiles is shown at Fig. 9.11. 1 – clayey mud; 2 – sand; 3 – silt; 4 – silty clay; 5 – clayey mud with foraminifers admixture; 6 – silty clay with foraminifers admixture; 7 – interlayering of clayey mud and sandy silt; 8 – hardground; 9 – clayey mud with carbonate admixture; 10 – interlayering of silt and clayey mud; 11 – silt with carbonate admixture; 12 – hydrotroilite; 13 – IRD; 14 – admixture of foraminifers; 15 – bivalves shells; 16 – turbidites 163
9.28 Distribution of IRD in sedimentary sections from Yermak Plateau (Levitan et al., 2002b). For core locations see Fig. 9.11 167
9.29 Evolution scheme of some paleoceanographic and paleogeographic parameters in the Arctic (Levitan et al., 2000b) 171
9.30 TOC, hydrogen index (HI) values, and C/N ratios in organic matter of Arctic Ocean sediments representing the last about 30 ka (Stein et al., 2004b). Arrows show ^{14}C dates; light gray color matches MIS 1 175
9.31 TOC, hydrogen index (HI) values, and C/N ratios in organic matter of the Amundsen Basin core PS2174-5 possibly representing the last 7–8 MIS (Stein et al., 2004b) 176
10.1 Scheme of the Pechora Sea lithodynamics (Tarasov, 1996). 1 – sandy-gravely-pebbly sediment; 2 – sand; 3 – silty sand; 4 – sandy silt; 5 – clayey silt; 6 – silty clay; 7 – sandy-silty-clayey sediment; 8 – areas of bottom erosion of Pleistocene deposits; 9 – areas of bottom erosion of Paleozoic rock; 10 – transportation directions of sand and silt; 11 – lithological boundaries 180
10.2 Lithology of surface sediments of Varandey Site (Levitan et al., 2003b). 1 – clayey-silty sand; 2 – small-fine sand; 3 – fine-small sand; 4 – small-grained sand 180
10.3 Terrigenous-mineralogical provinces of the Pechora Sea sediments (Levitan et al., 2003b). 1 – Vaigach; 2 – South Novaya Zemlya; 3 – East Pechora; 4 – Central Pechora; 5 – West Pechora 185
10.4 Light-mineral assemblages in surface sediments from the St. Anna Trough (Kukina et al., 1999). 1 – quartz, feldspars, rock fragments; 2 – quartz and feldspars; 3 – feldspars, rock fragments, quartz; 4 – isobaths, m; 5 – geological stations 187
10.5 Heavy-mineral assemblages in surface sediments from the St. Anna Trough (Levitan et al., 1999). 1 – black ore-epidote-clinopyroxene; 2 – epidote-clinopyroxene-black ore; 3 – epidote-black ore-clinopyroxene; 4 – black ore-amphibol-epidote-cliopyroxene;

5 – clinopyroxene-epidote-black ore; 6 – garnet-epidote-clinopyroxene-black ore; 7 – isobaths, m; 8 – geological stations . . . 191
10.6 Clay-mineral assemblages in surface sediments from the St. Anna Trough (Levitan et al., 1999). 1 – kaolinite-chlorite-illite-smectite; 2 – kaolinite-smectite-chlorite-illite; 3 – chlorite-illite-smectite-kaolinite; 4 – kaolinite-chlorite-smectite-illite; 5 – isobaths, m; 6 – geological stations . . . 192
10.7 Facies zonality and location of geological stations in the East Kara Sea (Levitan et al., 2005a). Cruises: BP – RV *Academik Boris Petrov*; DM – RV *Dmitry Mendeleev*. A–D – facies codes . . . 196
10.8 Locations of sediment cores and lithological profiles of RV *Academik Sergey Vavilov* Expedition (Murdmaa et al., 2006) . . . 212
10.9 Lithostratigraphic correlation of sediment cores along the western (40–41° E) transect. Location of sediment cores is shown on the water depth transect; for location see also Fig. 10.8 (Murdmaa et al., 2006) . . . 213
10.10 Lithostratigraphic correlation of sediment cores along the northern (79°45′–79°57′ N) transect; for location see also Fig. 10.8 (Murdmaa et al., 2006) . . . 214
10.11 Lithostratigraphic correlation of sediment cores along the eastern (60° E) transect; for location see also Fig. 10.8 (Murdmaa et al., 2006) . . . 216
10.12 Summary diagram of results of sediment core ASV880 (Murdmaa et al., 2006) . . . 217
10.13 Summary diagram of results of sediment core ASV1183 (Murdmaa et al., 2006). See legend in Fig. 10.12 . . . 218
10.14 Summary diagram of results of sediment core ASV1200 (Murdmaa et al., 2006). See legend in Fig. 10.12 . . . 220
10.15 Location of sediment core ASV 1157 in the South Novaya Zemlya Trough (Levitan et al., 2004). 1 – location of sedimentary section with peat layer with an age of 15310 years (Serebryanny and Malyasova, 1998); 2 – southern part of moraine of Admiralty complex (Gataullin and Polyak, 1977) . . . 225
10.16 Lithology and amount of the >0.05 mm fraction (in %) in sediment core ASV 1157 (Levitan et al., 2004). 1 – silty clay, 2 – silty clay with sand admixture, 3 – boundaries of layers with different textures, 4 – hydrotroilite, 5 – bivalves shells, 6 – polychaets tubes; 7–10 – sediment color: 7 – yellowish-gray, 8 – olive-gray, 9 – olive, 10 – black; 11–14 – texture: 11 – massive, 12 – with weak and intermediate bioturbation, 13 – with intensive bioturbation, 14 – laminated. *Arrows* show calendar age of mollusks shells . . . 227
10.17 Distribution of heavy minerals (in %) and their ratios in sediment core ASV 1157 (Levitan et al., 2004) . . . 229

10.18 Distribution of TOC, $CaCO_3$, benthic foraminifers, and oxygen and carbon isotope in *N. labradoricum* tests of the sediment core ASV 1157 (Levitan et al., 2004). 1–7 – species of benthic foraminifers: 1 – *Spiroplectammina biformis,* 2 – *Nonion labradoricum,* 3 – *Elphidium clavatum,* 4 – *Cassidulina reniforme,* 5 – *Haynesina orbiculare,* 6 – *Pyrgo williamsoni,* 7 – other species 231
10.19 Changes of some sedimentation parameters in the South Novaya Zemlya Trough during Holocene times (Levitan et al., 2004) 240
10.20 Location of geological stations in the Pechora Sea 242
10.21 Geological model (profile) of Quaternary deposits in the south-eastern Pechora Sea (Levitan et al., 2003b). 1 – sand; 2 – "black clay"; 3 – "gray clay"; 4 – diamicton 247
10.22 Lithology of sediment core PL94-60 (Levitan and Kukina, 2002). 1 – silty clay; 2 – silt; 3 – sand; 4 – hydrotroilite; 5 – IRD........... 250
10.23 Lithology of sediment core PL94-08 (Levitan and Kukina, 2002). 1 – silty clay; 2 – silt; 3 – sand; 4 – IRD 250
10.24 Lithology of sediment core PL94-64 (Levitan and Kukina, 2002). See legend at Fig. 10.23 252
10.25 Location of Yenisei Transect and geological stations (Levitan et al., 2004). 1 – marginal filter (*a* – distal part, *b* – depocentre, *c* – proximal part), 2 – delta 258
10.26 Lithology and mineral composition of sediment core BP00-23/7 (Levitan et al., 2004). I–IV – lithostratigraphic horizons. *Arrows* show radiocarbon dates. 1 – silty clay; 2 – sandy-silty clay; 3 – clayey-sandy silt and silty sand; 4 – shells of bivalve mollusks ... 259
10.27 Chemical composition of sediment core BP00-23/7 (Levitan et al., 2004) 261
10.28 Age model and environmental evolution of sediment core BP00-23/7 (Levitan et al., 2004)... 263
10.29 Lithology and mineral composition of sediment core BP00-07/6 (Levitan et al., 2004). I–II – lithostratigraphic horizons. Radiocarbon dates are shown by single arrows for sediment core BP00-07/6 and double arrows for sediment core BP00-07/7 (Stein et al., 2002). See legend at Fig. 10.26 266
10.30 Chemical composition of sediment core BP00-07/6 (Levitan et al., 2004) 267
10.31 Age distribution of clinopyroxenes, Fe-hydroxides, and clinopyroxene/epidote ratio in sediment core BP00-07/6 (Levitan et al., 2004)... 269
10.32 Map of the study area of the "*Academik Boris Petrov*" Expedition 2001 in the inner Kara Sea and Ob and Yenisei estuaries, presenting locations of sampling sites and sediment echograph profiles (Stein and Stepanets, 2002). Based on the profiling results, the extent of the southeastern margin of the Last Glacial Maximum (LGM) ice sheet and the LGM paleo-river channels were reconstructed (from

(Stein et al., 2002)). Further sediment cores from the Ob Transect investigated in this study, were added. 1 – geophysical lines; 2, 3 – sediment cores: 2 – with radiocarbon dates, 3 – without age determination; 4 – paleo-Ob valley; 5– 7 – land: 5 – 11 ka ago, 6 – 9 ka ago, 7 – recent . . . 273
10.33 Lithology and age of sediment core BP03-19 (Levitan et al., 2007c). 1 – clayey mud; 2 – silt; 3 – silty clay; 4 – clay with silt admixture; 5 – sandy clays; 6 – bioturbation; 7 – bivalve mollusk shells . . . 274
10.34 Distribution of chemical elements in sediment core BP03-19 (Levitan et al., 2007c) . . . 276
10.35 Age distribution of quartz and K_2O/Al_2O_3 ratio in sediment core BP03-19 (Levitan et al., 2007c) . . . 280
10.36 Lithology and age of sediment core BP03-07 (Levitan et al., 2007c). 1 – silty clay; 2 – clay with silt admixture; 3 – sandy clays; 4 – bioturbation; 5 – ikaite; 6 – bivalve mollusk shells . . . 282
10.37 Distribution of chemical elements in sediment core BP03-07 (Levitan et al., 2007c) . . . 283
10.38 Age distribution of quartz and K_2O/Al_2O_3 ratio in sediment core BP03-07 (Levitan et al., 2007c) . . . 284
10.39 Correlation diagram of HM-TA (see text for explanation) for sediments of the Ob Transect which allows to separate sediments belonging to different facies (Levitan et al., 2007c). 1 – upper unit of core BP03-19; 2 – lower unit of core BP03-19; 3 – core BP03-07 . . 285
11.1 Distribution of biogenic opal (wt. %) in surface sediments of the Chukchi Sea (Pavlidis, 1982). 1 – <1; 2 – 1–5; 3 – 5–10; 4 – >10 . . . 292
11.2 TOC distribution in sediment cores from the Laptev Sea continental slope (Stein and Fahl, 2004) . . . 294
11.3 Location of studied sediment cores in the Laptev Sea (Bauch et al., 2001). 1 – transform fault (Explanatory note to tectonic map of the Barents and Kara Seas, 1998); 2 – boundary between complete and uncomplete cross-sections of the Holocene . . . 295
12.1 Lithology of the Navarin Basin Subatlantic sediments (the Bering Sea north-western shelf) and distribution of TOC (Marina et al., 1985). *a* – lithology: 1 – gravel and pebble, 2 – sand, 3 – silt, 4 – clayey mud, 5 – outcrops of hard rock, 6 – geological stations; *b* – TOC content (wt. %): 1 – >1.7, 2 – 1.7–1.5, 3 – 1.5–1.3, 4 – 1.3–1.1, 5 – 1.1–0.9, 6–0.9–0.7, 7 – 0.7–0.5, 8 – <0.5, 9 – geological stations . . . 302
12.2 Lithological transect through the Navarin Basin (*a*) and TOC distribution (*b, c*) (Marina et al., 1985). *a*: 1–5 – stratigraphic horizons: 1 – Subatlantic, 2 – Subboreal, 3 – Atlantic, 4 – Boreal, 5 – Preboreal; 6–12 – lithology: 6 – sand, 7 – silty sand, 8 – silt, 9 – silty clay, 10 – marly mud, 11 – clayey mud, 12 – diatom mud, 13 – pebble, 14 – shell, 15 – pyrite. *b*: 1–8 – TOC (wt. %) distribution: 1 – 0–0.5, 2 – 0.5–0.7, 3 – 0.7–0.9, 4 – 0.9–1.1,

5 – 1.1–1.3, 6 – 1.3–1.5, 7 – 1.5–1.7, 8 – >1.7, 9 – TOC values; 10 – boundaries of stratigraphic horizons, 11–15 – stratigraphic horizons. (see Fig. 12.2a) *c*: 1–6 – TOC mass accumulation rates ($g/cm^2/ka$): 1 – <0.25, 2 – 0.25–0.50, 3 – 0.50–0.75, 4 – 0.75–1.00, 5 – 1.00–1.25, 6 – >1.25; 7 – lines of equal mass accumulation rates; 8 – boundaries of stratigraphic horizons 304

12.3 Scheme of surface-water circulation (Chernyavsky et al., 1993) and location of RV *Marion Dufresne* sediment cores in the Sea of Okhotsk 305

12.4 Distribution of selected geochemical parameters in sediment core DM 2594 (the Bering Sea) (Gorbarenko, 1996) 309

12.5 Lithology and stratigraphy of sediment core MD01-2415 (Levitan et al., 2007b). 1 – sand; 2 – silty clay; 3 – clay; 4 – diatom ooze; 5 – diatom clay; 6 – IRD; 7 – volcanic ash lenses; 8 – volcanic ash layers 312

12.6 Histograms of main lithologies and grain-size distribution (Levitan et al., 2007b). *a* – diatom ooze, *b* – diatom clay, *c* – silty clay, *d* – silty-clayey sand, *e* – vitric volcanic ash, *f* – host sediment for lenses of volcanic ash 313

12.7 Distribution of Na, Ba, Sr, and Sc in sediment core MD01-2415 (cofm – recalculated for carbonate-opal-free matter) (Levitan et al., 2007b) 318

12.8 Distribution of Th/Sc ratios in sediment core MD01-2415 (Levitan et al., 2007b) 319

12.9 REE spectra of sediments (chondrite normalized) (*a*), and distribution of Eu anomaly in sediment core MD01-2415 (*b*) (Levitan et al., 2007b). Eu anomaly for all samples was calculated according to formula "(Eu sample/Eu chondrite)/(Sm sample/Sm chondrite)". Those REE are shown which were determined by INAA method; other element concentrations are calculated. Numbers in Fig. 12.9*a:* sampling core depth (cm below sea floor) 320

12.10 Distribution of smectite/illite ratios in sediment core MD01-2415 (Levitan et al., 2007b) 323

12.11 Schematic summary compilation showing the sedimentation history of the Northern Sea of Okhotsk for the last 1.1 Ma (Levitan et al., 2007b). MAR – mass accumulation rates ($g/cm^2/ky$) (Nürnberg and Tiedemann, 2004), with change for terrigenous matter MAR. Paleoproductivity downcore up to level 130 ka is given according to (Nürnberg and Tiedemann, 2004), below – on our data 326

12.12 Changes of oxygene-isotope composition and concentrations of greenhouse gases in ice cores of greenland and Antartic (Brook et al., 2006) for the last 90 ka **(a)** and for the last 750 ka **(b)** in comparison to the global benthic oxygen isotope stack from (Liciecki and Raymo, 2005). **(b)** (Levitan et al., 2007b) 328

14.1 Surface currents during, the Eemian Interglaciation (Haake and Pflaumann, 1989). *a* – currents of MIS 5e beginning; *b* – currents of MIS 5e end. Area of distribution of pack sea ice in the Vøring Plateau area is hatched 338
14.2 Stadial-interstadial events of the Late Pleistocene in the West Scandinavia and their reflections in sediments of the adjacent shelf (cal. age scale) (Baumann et al., 1995) 340
14.3 Climate changes in the West Arctic between 15 000 and 11 000 years ago (by Yu.A. Lavrushin) 341
14.4 Structure of coastal late glacial terraces (Lavrov and Potapenko, 2005). *a* – location of main sedimentary sections; *b* – terrace structure: 1 – peat, 2 – peat under moraine deposits, 3 – sand, 4 – sand with gravel and pebble, 5 – boulder clays, 6 – flow till, 7 – sand and silt ("kame" sediments), 8 – limestone, dolomite, 9 – clay, siltstone 343
15.1 Sedimentation zones and ecological conditions during glacial expansion (**A**) and waning (**B**) (fjord variant) 346
15.2 Barents sea level changes during post-glacial time 347
15.3 Movements of the Barents and White Seas coastal lines (Lavrov and Potapenko, 2005). 1 – Kola Peninsula (Koshechkin, 1975) – highest (1*a*) and lowest (1*b*) position; 2 – Malozemel'skaya and Bol'shezemel'skaya tundra (Arslanov et al., 2005); 3, 4 – sea level: 3 – according to (Fairbanks, 1989), 4 – according to (Shepard, 1973) 350

List of Tables

2.1	Composition and the distribution of mollusks, barnacles and brachiopods in the interglacial sediments of the Kola Peninsula [*Gudina, Euzerov*, 1973]	18
2.2	Chronostratigraphy and vegetation of the North-Eastern European part of Russia in the late Neopleistocene (according to [*Lavrov, Potapenko*, 2005])	23
3.1	Chronstratigraphy of the late Pleistocene geological events of the northern West Siberia plain (according to [*Arkhipov*, 1997])	34
3.2	The event-related chronostratigraphy of Kargin – Middle Valday stage (according to [*Arkhipov*, 2000])	35
5.1	Basic glaciological and morphogenetic events of last glaciation during its degradation stage	44
6.1	Radiocarbon dates, calibrated calendar age and calculated averaged sedimentation rates [*Bauch et al.*,2001]	49
8.1	Sedimentation rates (cm/ky) in the Norwegian-Greenland Basin during the last five marine oxygen isotope stages (MIS) (compiled by M.A. Levitan)	69
8.2	Average values of SR ratio for MIS in the studied regions (by M.A. Levitan)	83
8.3	Physical properties in sediment core ASV 1372 [*Levitan et al.*, 2005]	95
8.4	Clay mineral composition in sediment core ASV 1372 [*Levitan et al.*, 2005]	96
8.5	Sedimentation rates and mass accumulation rates of major sediment components in sediment core ASV 1372 [*Levitan et al.*, 2005]	99
9.1	Average composition (%) of clay mineral associations in fraction <2 mkm of surface layer sediments at the continental margins of the Arctic Ocean [*Levitan et al.*, 1995b], with changes	121
9.2	Average composition (%) of clay mineral associations in the fraction <2 mkm of surface layer sediments in the deep-water areas of the Arctic Ocean [*Levitan et al.*, 1995b], with changes	122

9.3 Average composition (%) of clay mineral associations in surface layer sediments of the main oceans [*Levitan et al.*, 1995b], with changes 122
9.4 Temperature and salinity of bottom waters [*Levitan et al.*, 2000b] 130
9.5 Major zones of Holocene sedimentation [*Levitan et al.*, 2000b], with changes 134
9.6 Age difference between Heinrich events in the North Atlantic and iceberg sediments related to the Arctic part of the Laurentian Ice Sheet [*Darby et al.*, 2002], simplified 137
9.7 Sedimentation rates (cm/ky) in the Arctic Ocean during the last five marine oxygen isotope stages (MIS) (compiled by R. Stein and M.A. Levitan) 139
9.8 Spatial and temporal distribution of sedimentation rates (cm/ky) and its ratio in the Arctic Ocean during the last five MIS (by M.A. Levitan) 150
9.9 Sedimentation rates (cm/ky) on the Yermak Plateau 151
9.10 Epidote/pyroxene ratio in sediments of different marine isotope stages [*Levitan et al.*, 2002b] 169
9.11 Total content of mica and amphiboles (rel. %) in sediments of the Yermak Plateau calculated for IRD-free matter [*Levitan et al.*, 2002b] 169
9.12 Location of the studied stations [*Stein et al.*, 1994a] 174
10.1 Content of grain-size fractions (wt. %) in the surface layer sediments of the Pechora Sea [*Levitan et al.*, 2003b] 181
10.2 Calculated grain-size parameters [*Levitan et al.*, 2003b] 182
10.3 Content of main heavy minerals (rel. %) in the surface sediment layer of the Pechora Sea [*Levitan et al.*, 2003b] 184
10.4 Location of geological stations of RV *Polarstern* in the eastern Kara Sea 185
10.5 Average grain-size composition (wt. %) of the recent facies [*Levitan et al.*, 2005a] 202
10.6 Average content of major clay minerals in different facies (fraction <2 mkm) (rel. %) [*Levitan et al.*, 2005a] 203
10.7 Average content of major light minerals in different facies (fraction 0.1–0.05 mm)(rel. %) [*Levitan et al.*, 2005a] 203
10.8 Average content of major heavy minerals in different facies (fraction 0.1–0.05 mm) (rel. %) [*Levitan et al.*, 2005a] 204
10.9 Average chemical composition of different facies [*Levitan et al.*, 2005a] 205
10.10 Characteristics of the studied sediment cores from the central and eastern parts of the Barents Sea [*Murdmaa et al.*, 2006] 211
10.11 Radiocarbon dates of samples from sediment cores ASV 880, 1200 and 1183 [*Murdmaa et al.*, 2006] 215
10.12 Content of clay minerals in the fraction of less than 0.001 mm in the eastern Barents Sea sediments (rel. %) [*Murdmaa et al.*, 2006] 221

10.13 Location of sediment cores and thickness of the individual units (by I.O Murdmaa and M.A. Levitan) 224
10.14 Age of bivalve mollusk shells from sediment core ASV 1157 [*Levitan et al.*, 2003b] 233
10.15 Calendar age (cal. yrs. BP) of Holocene stages by Blitt-Sernander scale [*Merkt*, 1999] 233
10.16 Radiocarbon age of bivalve mollusk shells *Astarta borealis* and *Macoma calcarea* from the "blanket" sand of the Pechora Sea studied in 13-th cruise of RV *Academik Sergey Vavilov* [*Levitan et al.*, 2000], with changes 243
10.17 Grain-size composition (wt. %) of the Pechora Sea Quaternary sediments [*Levitan et al.*, 2003b] 244
10.18 Arithmetic mean values of TOC and $CaCO_3$, and the physical properties values of the basic Quaternary sedimentary units [*Levitan et al.*, 2003a] 245
10.19 Degree of sediment sorting according to normalized entropy [*Romanovsky*, 1977] 248
10.20 Radiocarbon dating of the sediment sample from the St. Anna Trough 249
10.21 Radiocarbon age of the bivalve *Portlandia* shell samples [*Levitan et al.*, 2004] 257
10.22 REE content (ppm) in sediments of the Yenisei marginal filter [*Levitan et al.*, 2004] 263
10.23 Location of studied sediment cores of the Ob River transect 272
10.24 Results of the radiocarbon analysis of mollusk shells 273
10.25 Correlation matrix of sediment core BP03-19 (upper horizon) 277
10.26 Correlation matrix of sediment core BP03-19 (lower horizon) 278
10.27 Correlation matrix of sediment core BP03-07 279
10.28 Chemical composition (wt. %) of sediment cores of Ob River transect 286
11.1 Sedimentation rate (cm/ky) and bulk mass accumulation rate ($g/cm^2/ky$) at the continental slope of the East Siberian Sea [*Stein et al.*, 2001] 298
11.2 Grain-size (wt. %) and chemical composition (wt. %) of diatom-bearing silty clays in sediment core PSh 40 at the Chukchi Sea shelf [*Aksenov et al.*, 1987] 299
12.1 Thickness of Holocene diatom oozes in the Bering Sea according to M.A. Levitan (29-th cruise of RV *Dmitry Mendeleev*) 304
12.2 Distribution of chemical elements in the major lithotypes of sediment core MD01–2415 315
12.3 Average concentration of elements in the Okhotomorsk (I) and Vanin (II) sequences 317

12.4 Average value of Eu anomaly in different geologic objects (composed by M.A. Levitan) 321
15.1 Geochronological correlations of high productive events, Heinrich events (Bonds et al.1993; Shaw, 1995, Dokken and Hald, 1999; Hald, 2001) and the Middle Valday Interstadials of theRussian Plain (by Yu.A. Lavrushin) 352

Introduction

The establishment of relationships between sediment composition and climatic environment in the sediment basin and subsequent evolution of climate relates to the classical problems of fundamental sedimentology. The widely known publications by the Russian academicians N.M. Strakhov, A.B. Ronov, and A.P. Lisitsin are dedicated to different aspects of this problem. In particular, the monograph published by A.P. Lisitsin "Sea-ice and iceberg sedimentation in the Ocean: recent and past" (Lisitsin, 2002) closely corresponds to the issues examined in this book. This monograph discusses in detail the environments and means of accumulations of recent marine and oceanic sediments in the ice zone of sedimentation of the Ocean, however, much less attention is given to the history of ice sedimentation, especially to high-resolution paleoceanography.

In the present work the authors accepted the following basic principles:

1. Study not only of the Arctic, but also of the Subarctic, especially of those regions, where there were conducted the original studies by the authors.
2. Study of climatic history influence (first of all, – the glaciation evolution of Northern Hemisphere) on sedimentation for the last 130 ka (MIS 5e – MIS 1) not only in the *marine periglacial* environment (term of G.G. Matishov), but also in the deep water areas and on the adjacent continental blocks.
3. Imperative description of recent sedimentation environment for subsequent application of the comparative-lithological method.
4. Detailed consideration of accessible stratigraphic and geo-chronometric data for partition and correlation of various sedimentary facies.

Some of the above-mentioned principles require further explanation. First of all, stratigraphy of the recent marine and oceanic sediments of the Arctic ice zone was essentially undeveloped compared to the stratigraphy of terrestrial Quaternary sediments. This circumstance was due to the predominantly terrigenous composition of marine sediments, which, as a rule, do not contain sufficient amount of biogenic calcite for dating by the traditional methods of radiocarbon analysis. As it was shown by numerous studies, in these sediments the scattered organic matter is mainly terrigenous and redeposited. Therefore, it is not suitable for radiocarbon dating. Just at the beginning of the 80's of the last century the new method of radiocarbon dating appeared which was based on the accelerating mass-spectrometry

(AMS ^{14}C), which made it possible to date sediments using the very small (first milligrams) hinges of calcite.

Only then, the active determination of age of the Arctic and Subarctic siliciclastic sediments has become widely available within the age limits of radiocarbon method (approximately within 50 ka). Unfortunately, the AMS ^{14}C facility does not exist in Russia. The presented data either are taken from the reference or determined in the course of joint projects with the associates from Germany, USA, France. Lately some funding has become available for payment of the corresponding studies abroad, which gives hope for the wider use of AMS ^{14}C method in Russia.

Secondly, the next important stratigraphic method for the Arctic sediments is oxygen-isotope study. Development of new generation of mass spectrometers and special prefixes for the automatic mass measurement of samples allowed obtaining appropriate data literally on several species of the carbonate foraminifer planktonic and secretional tests.

Thirdly, the actively developing procedures of magnetostratigraphy, which investigate magnetic excurses, played an important role in the stratigraphy of the Arctic sediments. Such works have already led to a whole series of important discoveries, in particular, to the "rejuvenation" of many geological cross-sections and the proof of the higher sedimentation rates in the pelagic regions of the Arctic Ocean, than it was considered earlier.

Finally, a wide application of optically simulated luminescence (OSL) method and dating of exposure age on the earth's surface on ^{10}Be is very important for the stratigraphy of Quaternary sediments and understanding of the glacial history of the continental blocks, which adjoin the Arctic Ocean.

In the first part of the book a prime attention is given to geology and paleoecology of the terrestrial Upper Quaternary and Holocene sediments at the territory of North Eurasia since the authors mainly conducted their own studies in these regions, or in the surrounding seas (for example, the longest chapter is dedicated to the West Arctic seas). However, there is much information regarding the history of glaciation of Greenland and North America in the chapters devoted to the West Subarctic seas and the Arctic Ocean. It should be noted that the studies are presented here starting from the Mikulino Interglaciation (MIS 5e), in order to have the comparative material for the analysis of the last interglacial period – the Holocene.

Again, the authors would like to emphasize the comprehensive character of the conducted investigations. There was studied the influence of the most varied glaciation features on evolution of sedimentation, in particular: continental and marine ice sheets, mountain glaciations, permafrost, sea ice and icebergs. The influence of these features on the Earth climate and its change is completely obvious.

Is is important to note two other main components of the Arctic climate: the influence of water exchange of the Arctic Ocean with the warm and saline waters of the Atlantic Ocean and, to a lesser degree, with the Pacific waters, and the abundant liquid and solid riverine discharge of Eurasia and North America into the Arctic Ocean. The above influence of both factors on accumulation, primary production, thermohaline circulation, water chemistry, temperature fields in the water layer and the atmosphere is very significant. These factors have been studied to a different degree in various regions. Some data are presented for the first time.

In order to solve the specified fundamental problem not only the "direct" task is imperative, for example as the climatic features influence sedimentation, but also the "reverse" task, as according to sedimentology proxies (grain-size, mineralogical, geochemical, and micropaleontological data), to determine climate environment and its changes. Here the authors proposed a whole set of indicators, which are valid within the regional framework of its application. At the first place, the book deals with the heavy, light and clay mineral data, and results of geochemical (organic and inorganic) studies.

Introduction is written by Prof. Dr. M.A. Levitan (V.I. Vernadsky Institute for Geochemistry and Analytical Chemistry, Russian Academy of Sciences, Moscow, Russia) with participation of Prof. Dr. Yu.A. Lavrushin (Geological Institute, Russian Academy of Sciences, Moscow, Russia), the second part is composed by M.A. Levitan (with one exception: Section "Organic-Geochemical Sediment Studies of the Eastern Part of the Central Arctic" is written by Prof. Dr. R. Stein, Alfred-Wegener Institute for Polar and Marine Research, Bremerhaven, Germany). The author of the first and the third parts is Yu.A. Lavrushin (with the participation of M.A. Levitan in the third part). The authors of the book have been studying the above scientific issues for a long period. It is worthy to note that the first monograph of Yu.A. Lavrushin, dedicated to the Quaternary sediments of Spitsbergen, was published in 1969. The first cruise of M.A. Levitan into the Sea of Okhotsk happened in 1976, and it did in 1982 into the Bering Sea. For the last 15 years almost all publications of M.A. Levitan were dedicated to the Arctic and the Subarctic regions.

It makes sense to briefly mention the title of this book. It is necessary to note that the sedimentation basins of the extensive Arctic region have been studied unevenly (for example, it is difficult to compare the Barents and East Siberian Sea degree of research). At present, there are still very few datings and not enough bore holes. For the most part, sedimentation history on the shelves could be described only for the latest 10–15 ka according to the research results for the shelf seas of North Eurasia. There are only a few paragraphs that discuss sedimentation history of much longer time periods (up to 1.1 million years).

The severe ice conditions still do not allow conducting scientific studies on board of the common research vessels in many Arctic regions, for example, in the Kara and East Siberian Sea. Thus far, the continental margin of the Arctic Ocean has been poorly studied. In this book the authors have not considered the applied scientific aspects, such as oil and gas exploration, geological engineering, geo-ecology, placer deposits, fishery, etc., but the importance of the book for these studies is obvious.

To summarize, the authors have chosen this particular book title and proposed it to the readers because the necessity for scientific generalization of the Arctic studies had already appeared that, in particular, emphasized by numerous publications dedicated to the studies of other aspects of marine geology of the Arctic region [(Svitoch, 2003; Romankevich and Vetrov, 2001; Stein and Macdonald, 2004) and other].

The international studies of the high-latitude regions, in particular, the Arctic and the Subarctic are of particular importance. The authors participated in many international programs, mainly in joint cooperation with the German associates from the AWI (Bremerhaven) and GEOMAR (Kiel) scientific institutions. Besides the joint papers, there were published four collections, in which the authors participated in composition and editing (Stein et al., 1996; 1999; Schoster and Levitan, 2003, 2004). M.A. Levitan had an unforgettable impression during the expeditions on board of the German research ice-breaker Polarstern (1997) in the area of the Yermak Plateau and on board of the French research vessel Marion Dufresne (2001) in the Japan Sea, Sea of Okhotsk and the northwestern part of the Pacific Ocean. The authors are deeply grateful to all associates and crews of the above scientific expeditions.

Finally, it is necessary to focus attention on one more crucial point. Several international expeditions have been working for the last 10–15 years in the Subarctic regions of the Russian part of Eurasia and the adjacent seas. As a result of these studies, many new interesting materials have been obtained, which made it possible to formulate a number of the fundamental conclusions, presented in many publications. The overwhelming majority of these publications are known to the authors of this book, but the references are absent in the book due to the following reason. It is caused by significant controversy and insufficient authenticity of the part of the conclusions, based on geochronological methods. In particular, the datings, obtained by modern modification of thermoluminescent method. As far as TL-method is concerned, the absolutization of the results obtained, causes doubt in many Russian and foreign researchers.

With respect to some conclusions, based on the datings of AMS ^{14}C method, it seems that the sampling procedure did not take into account the local features of sedimentation particularities. In particular, the discussion deals with the fact that in Pleistocene the alluvial deposits of the north of Eurasia contained plant detrite of unequal-age, which accumulation is respectively connected with the wash-out of the peat bogs of different age.

Despite the above considerations, obtaining the large set of geochronological data, although of various degree of authenticity, by itself is the important result of the international research, since previously only single datings have been known. These findings made it possible to date many important paleogeographic events of the Subarctic. Some of the obtained results have been taken into account, and are being discussed at the I and III parts of the proposed book.

Authors used different approaches: M.A. Levitan described mainly results of his regional studies while Yu.A. Lavrushin concentrated on important results obtained by Russian researchers in the Arctic/Subarctic region. This still makes the book to be an important contribution in Arctic Ocean climate research because it will present a lot of new data and interpretation which are not available in the open literature for western readers yet.

Funding of the executed work was achieved due to the Grants of Russian Foundation for Basic Research (RFFI), RFFI-CNRS (France), RFFI-GFEN (China), DFG (Germany), NRL (USA), FTSP "World Ocean", Presidium RAS (Russian Academy of Sciences) (program # 16), ONZ-14. The authors are grateful to the funding agencies and organizations, which contributed to the creation of this monograph. The authors hope that the proposed book will find its readers among the specialists in the areas of sedimentology, Quaternary and marine geology, paleoceanography, marine geoecology, oceanology, climatology, and also among students of the enumerated specialties.

The translation of slightly modified version of book by (Levitan et al., 2007a) is performed by E.S. Shelekhova, Ph. D. (Moscow, Russia). Editing of translation was done by M.A. Levitan (whole manuscript), R. Stein (Part II), and T.N. Golodnyuk (Parts I and III).

Part I
Geological and Paleoecological Events of the Late Pleistocene and Holocene in Northern Eurasia

Chapter 1
Geological and Paleoecological Events of the Late Pleistocene along Eurasian Coastal Areas of the Arctic Ocean

In the Late Pleistocene and Holocene a number of important geological-paleoecological events took place that were related to significant global climatic changes, relationships of land and sea areas, specific terrestrial sedimentation, and different types of natural phenomena some which had a catastrophic character and, accordingly, geoecological effect.

General Upper Pleistocene Stratigraphic Scheme for Northern Eurasia

In contrast to the lower units of the Pleistocene, stratigraphy of the upper one has been studied for a long time and developed in detail. The most comprehensive investigations were carried out in the Eurasian temperate zone of Northern Hemisphere. Therefore, it seems expedient to begin describing the Upper Pleistocene stratigraphy from this zone. In spite of numerous debatable problems, application of physical methods and usage of results of paleomagnetic, biostratigraphic, paleoclimatic and lithological investigations enabled approaching still inadequately substantiated but essentially new scheme of Late Pleistocene event stratigraphy (Lavrushin et al., 2002). It should be stressed that such a scheme can be developed on the base of multidisciplinary studies only. The application of physical methods to different (continental and marine) natural environments helps to compile a calendar of significant global events, which occurred both in the subaqual and subaerial conditions, including subglacial ones, and provides an important tool of reliable correlations of many natural phenomena.

After the preliminary brief notes some general trends in the Late Pleistocene environmental history should be outlined. Now this time interval is estimated to be approximately 130 ka.

Traditionally, the Russian researchers distinguish a single interglacial of the early Late Pleistocene, which is called Mikulino in the European part of Russia and Kazantsev in the Asian part and correlative to the Eemian Interglacial of Western Europe. Two glacials and two interstadials of the later time were established. The glacials correspond to the Kalinin and Ostashkov glaciations (correlated with MIS 4 and MIS 2 respectively) in European Russia and to the Zyryansk and Sartan glaciations in Siberia. The interstadials embrace a larger (post-Eemian) part of MIS 5

M.A. Levitan, Y.A. Lavrushin, *Sedimentation History in the Arctic Ocean and Subarctic Seas for the Last 130 kyr*, Lecture Notes in Earth Sciences 118,
DOI 10.1007/978-3-642-00288-5_1, © Springer-Verlag Berlin Heidelberg 2009

(d–a) and MIS 3. The latter interstadial is called the Middle Valday. The interstadials were characterized by climatic instability with periodic alternation of warming and cooling phases. According to integrated data (mainly palynological), the warming phases were colder than the present-day climate. Landscapes close to the periglacial ones with some cryogenic horizons dominated in the cooling phases. Shifts of Scandinavia and Spitsbergen mountain glaciers to the shelves were fixed at MIS 5 d.

The mentioned events reflect a complex climatic history of the Late Pleistocene. Since this book describes processes both in the Arctic seas and adjacent land areas, the noticeable fluctuations of the World Ocean level are worth noting. During the last glaciation the sea level fell by 110–120 m and a large part of the Arctic shelves was dried. Consequently, since the Kalinin Glaciation is supposed to be the most significant one, the dried shelf area was wider at that time.

Some important facts related to the glaciations should be also mentioned. The glacier degradation caused a very rapid rise of sea level. The available data indicates a catastrophic rate of post-glacial transgression and avalanche sedimentation. As shown below, there was a rapid rearrangement of sea currents, which reflected the high dynamics of water mass and its complex structure. Periodic changes of marine biota (the cold Arctic fauna was replaced by the warmer boreal and Arctic-boreal ones and vice versa) occurred even during the interstadials, which may serve as indicators of paleoclimatic conditions on the adjacent land areas. In this case we deal with paleodynamics of different-type water masses rather than the ocean level fluctuations. Particularly, the cooling could be caused by deep penetration of high-Arctic water masses into the shelf seas, while the warming could be brought about by the warm currents, e.g., a branch of the paleo-Gulf Stream. A role of atmospheric processes must not be neglected. According to available data, vegetation of European Russia included some Siberian floral elements in the post-Eemian phases of MIS 5d–5a and some West European elements at MIS 3.

Finally, one more important fact is worth noting. Changes in the sea current system occurred not only in the Late Glacial phases, but also in the relatively warm intervals, particularly, in the interglacials. The high changeability of sea currents in the early Eemian Interglacial (Lavrushin and Alekseev, 1999) was found out by the German researchers (Haake and Pflaumann, 1989) who studied foraminifers of the Vøring Plateau of the Norwegian Sea.

The cold East Spitsbergen current was originated in the Holocene (Subboreal) in the northwestern Barents Sea, between the Bear Island and Spitsbergen (Lavrushin et al., 1990). This current brought about an increase of ice volume in the Barents Sea and climatic deterioration in European Russia. Certainly, its influence should not be overestimated, because this area experienced a reduced effect of the Atlantic air masses and an increased action of the Siberian air masses (Spiridonova and Lavrushin, 1997).

Thus, the climatic conditions were formed under the influence of both highly dynamical water masses and genetically different air masses.

The presented trends in the Late Pleistocene environmental development are considered below in more detail.

Duration of the Mikulino Interglaciation

As it was noted, the important markers of the early Late Pleistocene are the Eemian Interglaciation in Western Europe and the Mikulino and Kazantsev Interglaciations in Russia. At present there are two opinions about duration of the Mikulino Interglaciation. Some researchers interprete this event to be a short-term episode 15–20 ka long. Others consider it lasted for approximately 70 ka. So, according to the EPR dating, the interglaciation in Northwestern Russia occurred 140–70 ka ago (Molodkov and Raukas, 1998c). The duration of the Eemian Interglaciation is also differently estimated by West European researchers. The lowermost Eemian deposits of the Netherlands (the Amsterdam region) dated by the U/Th method have an age of 117 ka, and the entire duration of the Eemian interglaciation is about 19–20 ka according to the uranium method (136.2–117.2 ka ago) (Kruk et al., 1998). According to the U/Th method, the Eemian in the Lower Saxony (Germany) ranged 149–115 ka ago (Urban, 1998). The U/Th dating for corals of Hawaii, California, Florida, and Bahamas yielded the interval of 132–115 ka ago, which corresponds to MIS 5e and, possibly, a part of MIS 5d (Muhs, 1998). It should be noted that the previous correlation of the Eemian to MIS 5 is supported now by very few researchers. The common opinion is that the Eemian Interglaciation corresponds to MIS 5e and lasted approximately 15–20 ka. Importantly, according to the oxygen-isotope records from the lower part of the Greenland ice sheet, the Eemian duration is also limited to MIS 5e comprising several phase-type units (Johnsen et al., 1995) reflecting short-term warmings and coolings (Fig. 1.1).

The results of the oxygen-isotope investigations of the ice sheet (the GRIP borehole) appeared to be essentially inconsistent with paleoclimatic data for the land areas. These data revealed a semisinusoidal-type trend, which can be briefly described as follows: gradual warming → climatic optimum→ gradual cooling. No phases complicating the trend have been distinguished.

Thus, the available date of 135 ka for the Early Eemian and the Holocene time of 10 ka suggest the overall length of the Late Pleistocene as 125–130 ka.

However, some debatable problems appeared recently. First of all, three, but not two, glaciations can be established in the Late Pleistocene. This idea was forwarded by the Scandinavian researchers, who dated the glaciations as 90–80, 60–50 and 20–10 ka ago respectively. But dates for the Late Pleistocene glacial events vary in different regions. This fact seems to stem from two reasons. The first reason is the application of different methods of dating, some of which are still far from being reliable. The age determinations of the Mikulino Interglaciation can serve as an example. Second, the dates obtained by the OSL and U/Th methods for the same strata are quite different. Attention should be paid to one more methodological point. Ages of the glacial events of the last, third part of the Late Pleistocene (the Late Visla Glacial) are usually shown both in calibrated and radiocarbon dates, while dates for older stadials, interstadials, and glacials are not sufficiently reliable.

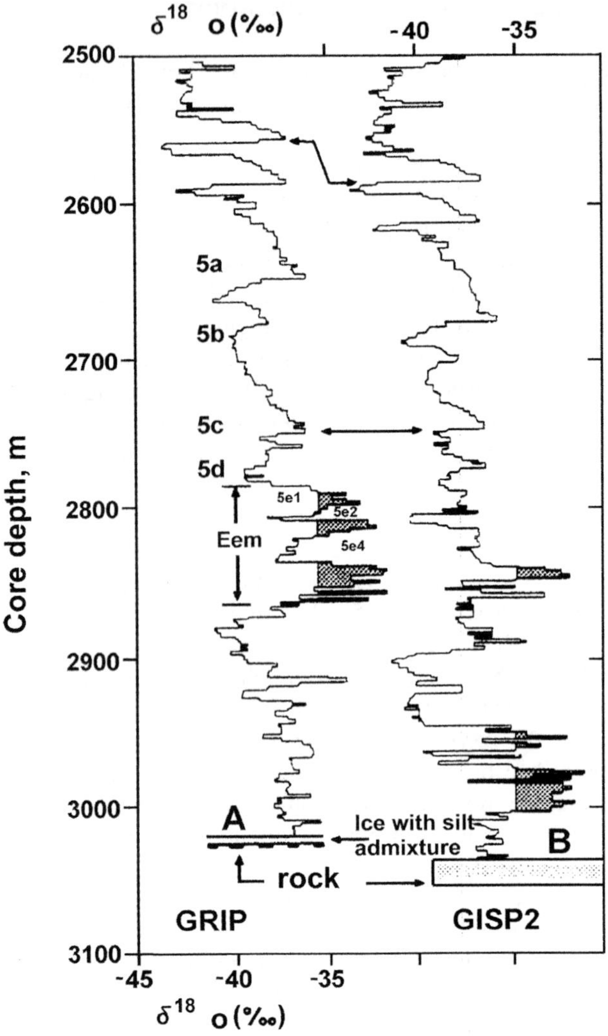

Fig. 1.1 Oxygen-isotope records of the lower parts of the GRIP and GISP 2 Greenland ice cores (Johnsen et al., 1995)

Correlation of the Natural Events Correlative with MIS 5d–5a in Northern West Europe and Northwestern Russia

As noted above, the natural events correlative with MIS 5 (substages 5d–5a) were characterized by considerable climatic instability, which was pronounced in alternation of relatively warm and cold phases similar to interstadials and stadials (Fig. 1.2). In the northwestern part of European Russia the following events are known: Kurgolov Stadial, Upper Volga Interstadial, Kileshinsky Stadial, Tosnensky Interstadial, Bologovsky Stadial, Berezaysky Interstadial, and Yedrovsky Stadial. In West-

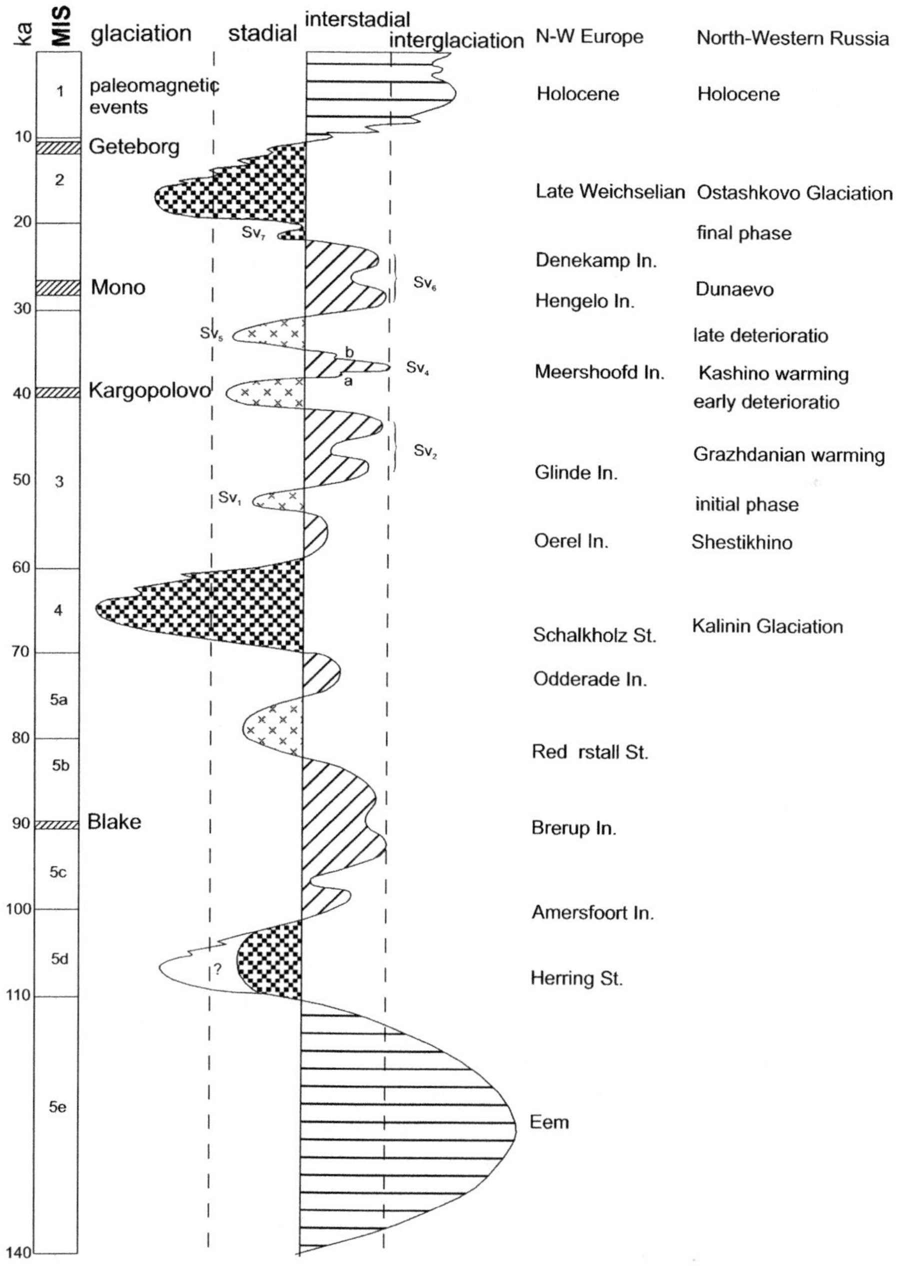

Fig. 1.2 Climatostratigraphic scheme of main natural events of the Late Pleistocene (by Yu A. Lavrushin, based on (Behre, 1989; Spiridonova, 1983; Mangerud, 1989). Palynological, geological and geomorphological data are used qualitatively

ern Europe there were Herring Stadial, Amersfoort and Brerup Interstadials separated by insignificant cooling (they are correlated with the Upper Volga Interstadial), Rederstall Stadial, and Odderade Interstadial. The latter is correlated to MIS 5a and dated as 72–61 ka ago. Traditionally this time interval is assigned to the Early Valday. But since the West European events were correlated to the isotopic substages, it seems reasonable to establish this interval as a unit of the Late Neo-Pleistocene. It is about 40 ka long and precedes the Kalinin Glacial. This unit can be named Late Eemian or eo-Early Valday (eo-Visla). The latter name is probably preferable for the stratigraphic scheme of European Russia. Besides, the Blake paleomagnetic event occurred approximately 90 ka ago, which coincided with the Amersfoort–Brerup Interstadial.

The next important event is the Kalinin Glaciation, which corresponds to MIS 4 and assigned to the Early Valday. According to the available data, this glaciation was very short-term and lasted about 5–10 ka. If this is real, this event should be considered as a serious natural catastrophe.

The Middle Valday (MIS 3) is characterized by essential climate changes reflected in repeated episodes of cooling and warming. The Kashin, Grazhdanskoe and Dunabe warmings can be distinguished. The following successive warmings were established in West Europe: Oeerel (58–54 ka ago), Glinde (51–48 ka ago), Meershoofd (46–44 ka ago), Hengelo (39–36 ka ago), and Denekamp (32–28 ka ago). Many important natural and archaeological events took place in East Europe during the Middle Valday. The most dramatic events are rapid landscape changes, alterations of the Late Paleolithic material cultures, changes in paleozoological assemblages, ashfalls dated as 38 and 33 ka ago and probably indicating a contact zone of two (Atlantic and Turkestan) air masses, reoccurrence of slope processes, abrupt changes in the river hydrological regime reflected in specific alluvium structure. Two paleomagnetic events were recorded in the Middle Valday: Lashamp (42 ka ago) and Mono (25 ka ago). In general, the Middle Valday is similar to the eo-Early Valday in the internal structure. The both intervals can be considered to be pre-glacial time of sharp climatic oscillations leading to essential environmental changes. The general trend similar to the glacial-type one, however, is variously pronounced in different regions, which assumes a nonlinear character of the processes.

The Late Valday (MIS 2) is the time of the Ostashkov Glaciation. Short-term stadials and interstadials at the end of this glaciation were succeeded by a sharp warming at the Late Neopleistocene–Holocene boundary. In Greenland a temperature rose by 10°C. Such climatic change can be attributed to a crisis of the Late Glacial. The end of the last glaciation was characterized by the following sedimentation processes: avalanche fluvioglacial sedimentation, peculiar hydrological regime of river flows without floods, intensive sediment accumulation on slopes, which dammed small river flows. The Geteborg paleomagnetic event occurred about 12 ka ago. The Late Glacial is marked by the Late Paleolithic cultures, disappearance of many large mammals of the Late Paleolithic fauna, radical landscape changes, etc.

During the Late Pleistocene glaciations, the glaciosedimentary processes in the glaciated regions were accompanied by sedimentation in the periglacial zones, which was proceeded in three phases. The slope sedimentation in wet climatic conditions dominated in the early phase. The middle phase was characterized by arid climate, intensive loess accumulation, cryogenic process, and formation of structures and deposits typical of a cryozone. The final phase (Late Glacial) was described above in detail.

Certainly, the dates used herein (Behre, 1989; Behre and Plicht, 1992) are being refined or should be refined in future. Nevertheless, they can provide a possibility to develop an event-based stratigraphic scale for the Late Pleistocene.

The described general trends in the Late Pleistocene environmental evolution have not been completely fixed in the Arctic regions, especially in the shelves and the ocean.

The picture presented actually demonstrates the level of studies to be approached by researchers of the Arctic regions. As shown below, some works are in progress, but the Upper Pleistocene stratigraphy of the Arctic regions is far from being so refined as that of southern Russia or West Europe. This fact should be taken in account when considering materials on the Arctic and sub-Arctic areas.

There is one more point to be paid attention. Geological heterogeneity of the Arctic seas determines specific features of their geological history and hampers reliable correlations of geological events of different types.

The southern Arctic seas adjacent to Eurasia can be classified into two groups. The first group includes the glacial shelf seas, i.e., the sea areas, which were repeatedly dried out and covered by continental ice during the Quaternary. The second group of the Arctic seas, the Laptev Sea and eastward, was located in the non-glaciation region, but in the large cryozone of very severe climate and with permanent sea ice cover. These seas differ also in geological structure, as it will be shown below.

Chapter 2
Late Pleistocene Geologic-Paleoecological Events in the North of European Russia

Relationship between Land and Sea Areas during the Mikulino Interglacial in Northern Eurasia

The main object to be considered is sediments of the Eemian (Mginsk) or Boreal, transgression. Some general trends of their distribution and biostratigraphic evidence are presented. These sediments have been long studied by many researchers, and the aim of this chapter is to summarize the results.

According to (Lavrova and Troitsky, 1960), the Boreal basin was mostly shallow. Coastal sandy facies contained a rich molluscan fauna with dominating boreal species (*Cyprina islandica, Astarte borealis, Mactra elliptica* and others), Lusitanian elements (*Anomia striata, Cardium edule*), which suggests the high water salinity, positive winter temperatures, and active aeration of the basin. Of sublittoral diatoms the warm-water benthic and semibenthic forms are especially abundant (Polyakova, 1997). The climate of the Boreal transgression was one of the warmest in the second half of the Pleistocene. Coniferous forests dominated the coastal vegetation, and broad-leaved species grew on the southern shore of the White Sea. Glaciers disappeared from the Scandinavian mountains and the Kola Peninsula (Geology of Quaternary deposits of the north-western European part of the USSR, 1967).

The results of investigations and published data allow conclusion that the sediments of the Boreal transgression spread continuously from the Gulf of Finland to the White Sea. Therefore the transgression can be considered to be a significant event, which changed relationship between the land and sea areas. The sea transgression spread along valleys of large rivers (Severnaya Dvina, Mezen′, and Pechora). The White Sea was connected with the Gulf of Finland by a strait between the Baltic crystalline shield and the Russian Plate. For a short time Scandinavia became a huge island (Fig. 2.1) (Lavrova and Troitsky, 1960; Biske, 1959; Znamenskaya and Cheremisinova, 1962; Geomorphology and Quaternary deposits of the north-western European part of the USSR, 1969; Lavrushin and Spiridonova, 1995). According to the available geological data, the oldest non-glacial Quaternary deposits of the White Sea coast are the marine sediments of the Mikulino time. This fact allows suggestion that the White Sea was originated at that time (Lavrushin

M.A. Levitan, Y.A. Lavrushin, *Sedimentation History in the Arctic Ocean and Subarctic Seas for the Last 130 kyr*, Lecture Notes in Earth Sciences 118,
DOI 10.1007/978-3-642-00288-5_2, © Springer-Verlag Berlin Heidelberg 2009

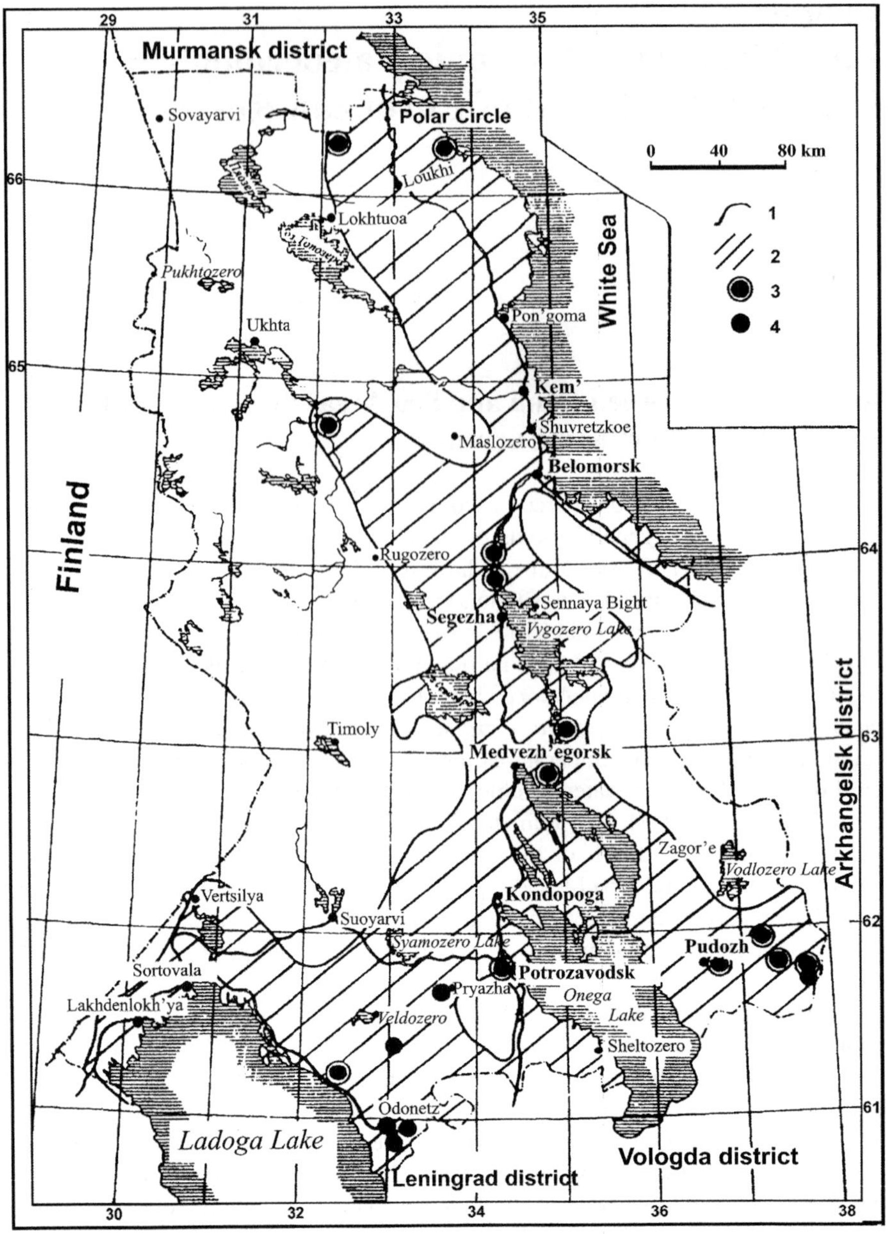

Fig. 2.1 Schematic distribution map of the Karelian Interglacial sea (Biske, 1959). 1 – coast lines; 2 – sea; 3 – locations of interglacial sedimentary sections with paleontological characteristics; 4 – locations of interglacial sedimentary sections without paleontological characteristics

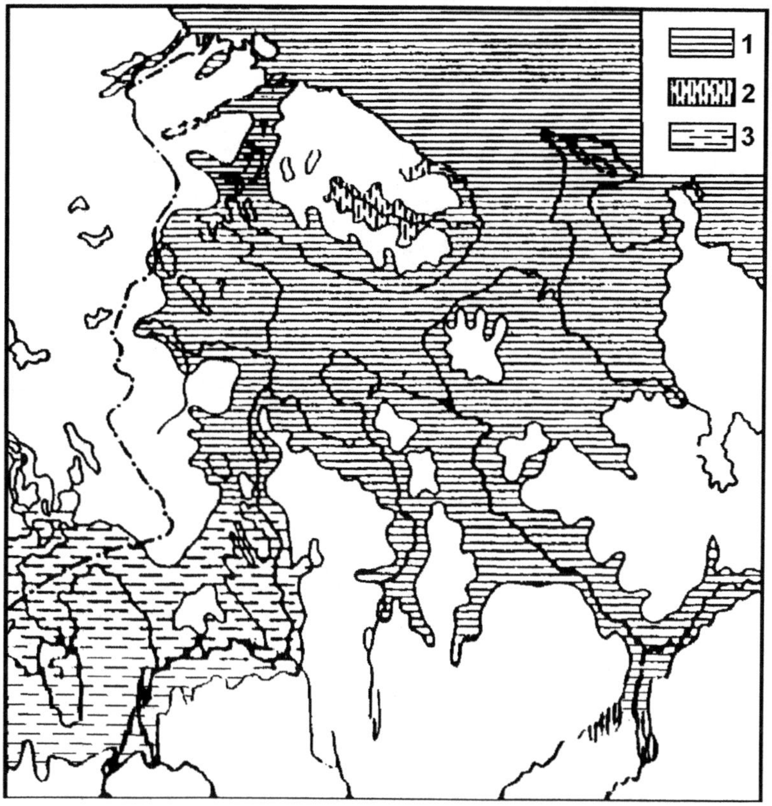

Fig. 2.2 Distribution of the interglacial boreal transgression (Lavrova and Troitsky, 1960). 1 – marine boreal transgression waters; 2 – freshened bays; 3 – waters of Eemian transgression

and Spiridonova, 1995). The Mikulino marine sediments are also the oldest ones in grabens filled by the Ladoga and Onega lakes and the grabens may be formed through the early Mikulino (early Eemian) neotectonic processes (Garetsky et al., 1999).

In the south of the East European Platform the Mikulino time is marked by liman sedimentation during the Late Karangat transgression of the Black Sea and the Late Khazaria transgression of the Caspian Sea.

Thus, two interrelated events took place in the Mikulino time. The intensive neotectonic graben formation occurred at the contact of the Russian Plate and the Baltic crystalline shield in the north and northwest of European Russia. The origination of the recent tectonic structures coincided with the Eemian (Mginsk) sea transgression, which essentially effected the relationship of land and sea areas (Figs. 2.1, 2.2, 2.3).

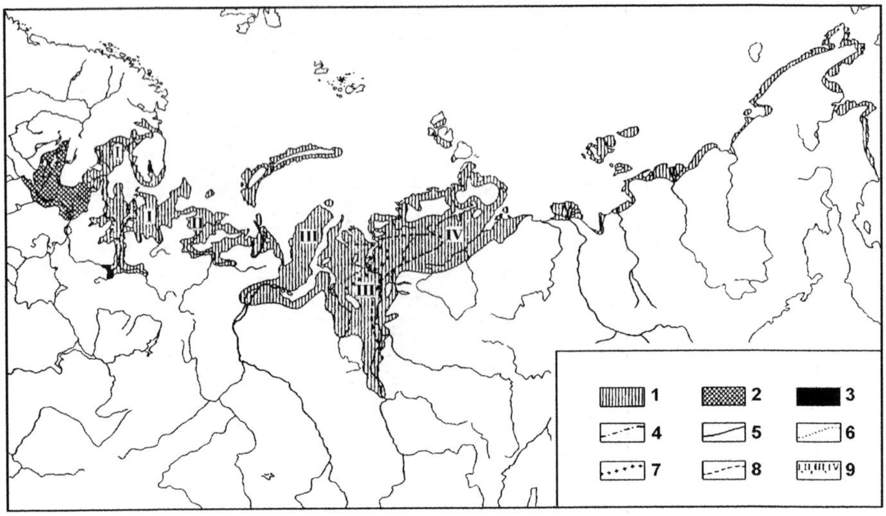

Fig. 2.3 Boreal sea on the Russian North (according to S.L. Troitsky (1964)). 1 – territories covered by boreal sea during the maximal transgression; 2 – Baltic Basin during the transgression maximum; 3 – freshened areas; 4–8 – boundaries of distribution of zoogeographic groups and individual species of mollusks and barnacles: 4 – Lusitanian (south-boreal) species; 5 – *Mactra elliptica*, 6 – *Cardium edule*, 7 – *Pholas crispatus*, 8 – *Cyprina rina islandica;* 9 – basins (I – White Sea, II – Pechora, III – West Siberia, IV – Taimyr)

Genetic Types of Continental Sediments

Besides the marine sediments, many continental sequences of the Mikulino Interglacial were discovered in European Russia. The most widespread are lacustrine and boggy deposits; buried soils are abundant in the south. Alluvial and proluvial sediments are less common.

The lacustrine-boggy sediments are characteristic of the northern and central regions of European Russia (to the south from the sea basin). This suggests a high level of ground waters, which might be caused by the sea transgression. Southwards, in the present zone of forest-steppe and steppe, there are numerous horizons of buried soils, which evidence of weak slope denudation, and local proluvial formations. Less is known about specific features of alluvial sediments. A deltaic sequence of the Severnaya Dvina is 70 m thick. Accumulation of so thick sequence during the short-time Boreal Eemian (Mginsk) transgression speaks of an avalanche sedimentation. However, at some distance from the sea the plain river alluvium formed sequences of usual thickness and facies relationship.

Thus, the distribution of the continental sediment types shows that the specific feature of the Mikulino sedimentation stemmed from the high level of ground waters in the land area was a wide bog formation.

Marine Sediments of the Boreal Transgression in the North of European Russia

The sediments of the Boreal transgression correlative with the Eemian one, were distinguished by M.A. Lavrova and S.L. Troitsky (1960). They have a number of characteristic features. The most important features are as follows.

1. The Boreal transgression was produced by two different (Arctic and Boreal (Eemian)) water masses penetrating far into the land area. This was reflected in mixed composition of marine biota.
2. Changes of the genetically different biotic assemblages upward the sequences suggest an alternation of dominating water masses in the Boreal basin. Some strata contain predominantly the Arctic fauna with single boreal forms, but others enclose an increased amount of the latter. So, the penetration of the Arctic waters into the land may occur in colder climatic conditions similar to the stadial ones. The noticeable increase of the boreal forms may evidence of the interstadial climate in the adjacent land. The predominance of the Lusitanian (boreal) fauna indicates the interglacial environment as confirmed later by palynological studies.
3. The study of some boreal marine sequences revealed that they experienced significant glaciotectonic action. Certainly, the glacio-dislocations considerably hamper reconstruction of the boreal sedimentation.
4. The Boreal sea sequences demonstrate distinct signs of sea level fluctuations, which point to complicated dynamics of the transgression.

Below some sections of the Boreal transgression are described whose lithological and biostratigraphical characteristics reflect one or several mentioned features.

The Pas'va and Koleshki sections at the Vaga River were studied by many researchers (Devyatova, 1982; Zhuze and Poretsky, 1937; Liyvrand, 1981; Loseva, 1978; Lavrushin and Spiridonova, 1995; and others). The sediments of the Boreal transgression lie between thin peat beds and represented by greenish-gray clay with plant remains and shells of marine mollusks. Thickness of the sediments is 3.5–4.0 m.

According to palynological data of E.D. Lyivrand and E.A. Spiridonova, the top of the marine strata and the overlapping peat bed correspond to a climatic optimum of the Mikulino Interglaciation, which is reflected by the following palynological zones (Grichuk, 1961):

M4 – zone of oak, elm and fir; the beginning of maximal abundance of alder-tree and filbert;

M5 – zone of oak and elm with an admixture of linden and hornbeam; the end of the maxima of filbert and alder-tree; fir maximum;

M6 – zone of fir and hornbeam.

M7 and M8 zones were established by E.A. Spiridonova (Lavrushin and Spiridonova, 1995) above the climatic optimum.

Thus, the presence of the peat beds under and above the marine sediments suggests a high level of ground waters in the adjacent land during both the transgressive and regressive phases of the Boreal sea.

In the Koleshki section the eroded upper peat bed was overlain by a thin sandy member containing molluscan shells. Its presence can evidence of repeated sea level changes, i.e., the next transgressive phase.

The palynological zones of V_1, V_2, V_3, V_4, V_5, and V_6 were distinguished in the Lower Valday strata. The overlaying Middle Valday sediments correspond to Svd_{1-7} zones (Spiridonova, 1983).

In these sections the Lower and Middle Valday deposits are represented by fluvial sediments with many signs of washing out and located in the river valley, which cuts almost across the terminal moraine ridge of the Kalinin glaciation (Map of Quaternary deposits of European part of the USSR and adjacent territories, 1971). Nevertheless, above the Mikulino deposits of the Pas'va section there are sediments of the Late Volga Interstadial (V_2) reflecting a predominance of thin birch-pine forest; the birch forests became prevailing in the Late Brerup Interstadial (V_4) (Lavrushin and Spiridonova, 1995).

E.I. Devyatova (1982) reconstructed paleolandscapes of the Onega River valley at the Boreal sea coast. The succession of vegetation phases are as follows: periglacial vegetation (grass, bushes); dominating birch forests; coniferous forests (fir, pine); mixed forests (combination of pollen maxima of broad-leaved species and those of birch and alder) during the climatic optimum; coniferous-birch forests; fir forests; periglacial vegetation (grass, bushes, birch) at the end of the interglaciation.

In the east, the sediments of the Boreal transgression increased in thickness up to 60–70 m. They were recovered by drilling at the lower course of the Severnaya Dvina River. Palynological analysis of the sequence (about 40 m) of Hole 15 near the Lake Izhmy (Pleshivtzeva, 1972) showed the typical pattern of appearance and culmination of the broad-leaved plants (oak – elm – filbert – hornbeam) during the optimum of the Mikulino Interglaciation, in spite of some differences from the southern regions. The broad-leaved species occurred as insignificant admixture in the forest of firs and birches. The hypo-Arctic species *Betula nana* was associated with some thermophilic elements. Significantly, the climatic optimum of the Mikulino Interglaciation was recorded at the depth of 90.0–76.1 m, whereas the base of the Boreal transgression sequence was at the depth of 91.0 m.

The initial short-term phase of the Boreal transgression may proceed in cold climatic conditions, which were probably formed by the cold Arctic waters penetrating into the land.

The similar succession of vegetational phases was discovered in sections of some other holes drilled at the lower course of the Severnaya Dvina River. The lower part of the Boreal strata (up to 14 m) contains shells of *Portlandia arctica* (Gray), *Yoldia arctica* Gray, *Yoldia hyperborea* (Loven), *Macoma calcarea* (Chemn.). Sediments of the climatic optimum recovered at the lower course of Severnaya Dvina and in the Mezen and Pezy river valleys enclose Arctic-boreal (to 50%) and boreal (to 30%) molluscan species, Arctic forms are single. The prevailing boreal species are *Carbula gibba* Olivi, *Cardium echinatum* L., *Cardium paucicostatum* Beck, *Mactra elliptica* B., *Cyptina islandica* L., *Panopea norvegica* Speng., *Buccinum undatum* L., *Littorina littorea* (L.), *Nucella lapillus* (L.).The most frequent boreal-Lusitanian element is *Cardium edule* L. Analogous but thicker walled shells were found in the coastal sandy-gravel-pebble deposits.

The described assemblages reflect only one of the regressive phases of the Boreal sea. The high-Arctic thick-walled species *Serripes grönlandicus* (Chemn.) was found in brownish-gray sandy loams of the Peza River outcrop (near the Bychie Village), which overlie coastal-marine sediments enclosing mollusks of the climatic optimum. Occurrence of this form most likely indicates a short-term penetration of the Arctic waters into the land and related cooling. Palynological data indicates a predominance of northern taiga in the coasts at that time (Geology of Quaternary deposits of the north-western European part of the USSR, 1967). Special investigations and geological mapping in the Mezen and Peza valleys revealed that the maximal level of the Boreal transgression approached the present-day hypsometric mark of approximately 100 m. The lithofacies composition reflects repeated sea level fluctuations and changes of sedimentation regime.

To the northwest, in the Kola Peninsula, the sediments of the Boreal transgression called Ponoy Layers were recognized in outcrops at the Varzuga, Chapoma and Panoy rivers, and in cliffs of the Svyatonossk Bay.

The Ponoy Layers are represented by loams and clays (up to 9.5 m) varying in color from black to light gray and containing abundant marine mollusks (Table 2.1) as well as foraminifers and diatoms (Gudina and Yevzerov, 1973; Cheremisinova, 1952; Lavrov, 1960).

According to V.I. Gudina, the foraminiferal assemblage is composed by "Lusitanian, boreal, Arctic-boreal, boreal-Arctic and Arctic forms. Boreal and Arctic-boreal species are prevailing. Lusitanian and boreal-Lusitanian forms: *Lenticulina orbicularis, Amphicoryna scalaris f. compacta, Globulina inaequalis, Guttulina lactea, Sigmomorphina undulosa, Fissurina latistoma, Discorbis punctulatus, Gavelinopsis praegeri, Rosalina globularis, Hyalinea balthica, Elphidium excavatum, Trifarina angulosa* (12 species or 15% of the assemblage); boreal forms: *Pyrgo williamsoni, Lagena gracillima, L. semilineata, L. sulcata, Polymorphina novangliae, Oolina hexagona, Buccella troitzkyi, Paromalina bilateralis, Elphidium boreale, Bulimina aculeata, Bolivina pseudoplicata, Cassidulina laevigata,* etc. (23 species or 28.5%); Arctic-boreal forms: *Quinqueloculina arctica, Pateoris hauerinoides, Lagena apiopleura, Oolina globosa, Fissurina laevigata, F. latistoma, F. marginata, F. serrata, Eponides wrightii, Cibicides rotundatus, Astrononion gallowayi, Nonionella auricula, Bolivina pseudopunctata,* etc. (18 species or 22%). Warm-water species (Lusitanian, boreal-Lusitanian, boreal and Arctic-boreal) compose 65.5% of the assemblage (53 species). Cold-water elements (Arctic and boreal-Arctic species) are in subordinate amount (26%). The boreal-Arctic species: *Quinqueloculina borea, Tappanella arctica, Buccella inusitata, Nonionellina labradorica, Cribroelphidium granatum, Cr. goesi, Elphidiella arctica, Globigerina pachyderma, Cassidulina subacuta, Islandiella islandica,* etc. (11 species or 14%); the Arctic species: *Dentalina baggi, D. frobisherensis, D. ittai, Globulina glacialis, Buccella hannai arctica, Patellina corrugata, Cribrononion obscurus,*

Table 2.1 Composition and the distribution of mollusks, barnacles and brachiopods in the interglacial sediments of the Kola Peninsula [*Gudina, Euzerov*, 1973]

Species	Location and number of cross-section																	
	Varzuga River, 1		Chavanaga River , 2	Kamenka River, 3	Strelna River, 4	Chanoma River, 5		Pyalitsa River, 7		Big Kumzhevaya River 9	Ludyanoy Stream, 10	Ponoy River, 13a	Branch of Pakhtenny Stream 13б	Ponoy River, 13в	Popov Stream , 13г	Small Kachkovka River, 14	Top of the St.Nose Bay, 15		
																	Ponoy Layers		
	Ponoy Layers	Loam lens				Ponoy Layers	Strelna Layers										ower part	Middle part	Strelna Layers
1	2	3	4	5	6	7	8	9	10	11	12	13	14	15	16	17	18	19	20
Boreal–Lusitanian																			
Cardium edule L.												+		+	+	+			
Boreal																			
Acmaea virginea Müll.																+			
Anomia ephippium L.	+													+					
Buccinum undatum L.													+		+	+			
Cardium elegantulum Beck.						+													
C. echinatum L.												+		+		+			
C. fasciatum Mont.	+	+										+		+	+	+	+		
Arctica(-Cyprina) islandica (L.)	+		+			+		+				+	+	+	+	+	+		+
Dentalium entalis L.		+	+			+						+		+			+		
Hydrobia ulvae Pennant.																+			
Littorina littorea L.														+					

Table 2.1 (continued)

1	2	3	4	5	6	7	8	9	10	11	12	13	14	15	16	17	18	19	20
Lucina borealis L.												+		+			+		
Mactra elliplica Brown.											+	+	+		+	+	+		
Mya arenaria L.														+					
Panopaea norvegica Spengl.	+	+			+	+											+		
Balanus balanoides L.		+																	
B. hameri Asc.		+	+			+											+		
Mainly Boreal																			
Anomia squamula L. var. *aculeata* Müll.												+		+					
Astarte borealis Chenm. var. *arctica* Gray					+							+	+			+			
Lacuna divaricata Fabr.																+	+		
Lora trevelliana (Turt.)		+																	
Modiolus modiolus (L.)	+											+	+	+	+	+	+		
Mytilus edulis L.	+													+	+				
Neptunea despecta L.		+				+						+		+	+				
Arctic-Boreal																			
Acribia (Amauropsis) islandica (Gmel.)	+														+	+			
Anomia squamula L.						+					+	+	+	+	+	+	+		
Astarte elliptica Brown.	+	+	+		+	+	+				+	+	+	+	+		+		
Boreonatica clausa (Brod. et Sow.)	+										+				+	+	+		
Hiatella arctica (L.)	+		+			+					+	+	+		+	+	+	+	
Leda pernula (Müll.)	+	+				+											+	+	+

Table 2.1 (continued)

1	2	3	4	5	6	7	8	9	10	11	12	13	14	15	16	17	18	19	20
L. minuta (Müll.)						+						+		+					
Lepeta coeca (Müll.)	+																+		
Macoma calcarea Gmel.	+	+				+		+				+		+			+	+	+
Margarites cinereus (Couth.)															+	+			
M. groenlandicus (Chemn.)	+														+	+			
Musculus laevigatus (Gray)												+							
Mya truncata L.	+	+			+	+		+			+	+	+		+	+	+		
M. truncata var. *uddevalensis* Hanc.														+	+		+		
Nucula tenuis (Mont.)	+																	+	+
Chlamys islandicus (Müll.)	+	+		+		+				+	+	+	+	+	+	+	+	+	
Euspira pallida (Brod. et Sow.)						+						+		+		+	+		
Puncturella noachina (L.)						+								+		+	+		
Sipho islandicus (Chemn.) (?)																	+	+	
Boreotrophon truncatus (Ström.)																+			
Velutina undata Brown.																+			
Balanus balanus L.		+									+				+				
B. crenatus Brug.		+				+						+							
Mainly Arctic																			
Admete viridula Fabr.	+					+													
Astarte borealis Chemn.	+	+			+	+		+		+	+	+	+	+	+	+	+		
A. crenata (Gray) var *crebricostata* Andr. et Forb.	+	+	+	+		+											+		

Table 2.1 (continued)

1	2	3	4	5	6	7	8	9	10	11	12	13	14	15	16	17	18	19	20
A. montagui (Dillw.)	+	+				+				+	+	+	+	+	+		+		
A. montagui Dillw. var. *striata* Leach.	+					+								+			+		
A. montagui (Dillw.) var. *warhami* Leche											+	+							
Lora harpularia (Couth.)												+		+					
L. scalaris (Möll.)																	+		
Margarites groenlandicus (Chenm.) var. *umbilicalic* Brod. et Sow.															+				
Boreotrophon clathratus L.															+				
Trichotropis borealis Brod. et Sow.																	+		
Arctic																			
Acmaea rubella Fabr.												+			+	+			+
Bathyarca glacialis Gray										+							+		
Astarte borealis (Chemn.) var. *placenta* Morch.					+		+										+		
A. crenata (Gray)	+	+	+	+		+		+		+	+				+		+		+
Cardium ciliatum Fabr.	+	+								+						+			
Yoldiella lenticula (Möll.)																		+	
Propeamussium groenlandicum Sow.																		+	+
Hemithiris psiltacea (Gmelin)	+					+													
Serripes groenlandicus (Chemn.)	+				+	+										+	+		

Protelphidium orbiculare, etc. (10 species or 12%). Unidentified species amount to about 10%. Thus, the assemblage is mainly constituted by moderate-cold-water species (50.5% of boreal and Arctic-boreal forms) with the significant portion of warm-water elements (15% of Lusitanian and boreal-Lusitanian forms). Such composition indicates a boreal character of the assemblage. The enclosing sediments were accumulated predominantly in the lower sublittoral zone of the open sea characterized by normal oceanic salinity and positive bottom temperatures" (Gudina and Yevzerov, 1973; pp. 52–54).

Now the general situation in the eastern regions, in particular, at the lower course of the Pechora River, should be considered. The section called Vas'tyansky Kon' was studied by many researchers and described in literature (Gol'bert et al., 1973; Lavrova, 1949a; Lavrova and Troitsky, 1960; Popov, 1963; Troitsky, 1964, 1966; Epstein, 1990; and others). This section is of interest due to specific features of structure and morphology. First, it is characterized by glacial dislocations. Second, along with the Mikulino marine sediments, the Middle Valday alluvial and alluvial-deltaic deposits are exposed above the Pechora shoreline. Third, the section cuts the terminal moraine of the last glaciation along its strike.

S.L. Troitsky suggested that the alluvial sandy sequence of the middle part of the section had been transformed by glacial thrusts into blocks separated by moraine loam interbeds (Gol'bert et al., 1973). The lower marine sequence of dark gray clay seems to experience essential glaciotectonic action as well. This action was pronounced in breccia-like structure of the clays with gliding planes and, as it will be shown below, glacial thrusts. A.V. Gol'bert with co-authors (1973) subdivided the clayey sequence into three members. At the base of Member 1 there is a block of till 1 × 1.5 m in size; its basal interbed 3–4 cm thick includes abundant thin-walled shells. Granulometric changes are observed upward Member I due to an increase of silt material. The clays (about 8 m) contain a cold-water boreal-Arctic assemblage of marine bivalves. According to V.I. Gudina, the mollusks were associated with foraminifers; the Arctic species *Elphidium subclavatum* being most abundant (64%). The foraminiferal assemblage was the Arctic deep water (depth of 150–200 m) one, which lived at temperatures $0 \pm +1°C$. A boreal-Arctic foraminiferal assemblage was found higher up the section.

Member 2 consists of four thin interbeds of fine-grained sands enclosing sandy-clayey balls. Signs of local gaps and erosion of the top are observed. The authors (Gol'bert et al., 1973) united Members 1 and 2 into a single sedimentation cycle. This interpretation does not seem confident. Member 2 contains no molluscan shells, but a noticeable amount of foraminifers, which form, according to V.I. Gudina, a boreal-Arctic assemblage including euryhaline forms, elfides and nonionides. The presence of warm-water and shallow-water *discorbis, noniones, buccella* is explained by a small depth of the basin and proximity of the delta. According to the microfaunal data, depth of the basin was less than 50 m, salinity did not exceed 30–32%, and temperature of bottom waters was +2 or +3°C.

Overlying Member 3 was subdivided into three parts (3a, 3b, 3c) (Gol'bert et al., 1973).

Table 2.2 Chronostratigraphy and vegetation of the North-Eastern European part of Russia in the late Neopleistocene (according to [*Lavrov, Potapenko*, 2005])

Age, ka	Cronostratigraphy			Vegetation: Lower Pechora	Vegetation: Mezen-Vychegda Basin
	Interstadial	Warming		Sparse growth of spruce-birch	
30, 32		Deep cooling	3352+470(ЛУ-513A) 34540+1570 (ЛУ-513B)	Periglacial tundra	
34, 36	Interstadial	Tyrybeiskoe warming	37990+440(ЛУ-609) 38320+1630(ЛУ-518) 38400+640(ЛУ-517A) 38670+870(ЛУ-512A) 39840+570(ЛУ-512B)	Sparse growth of spruce-birch, association of wormwood, cereals, sedges	Pines and sparse growth of spruce-birch, association of cereals and motley grass
38		Cooling		Sparse growth of birches, tundra-steppe	
40, 42		Urdynzhskoe warming	39170+440(ЛУ-588) 40650+7(ЛУ-550) 40680+1180(ЛУ-1113) 42040+1900(ЛУ-1119) 42660+970(ЛУ-519) 42810+1200(ЛУ-533)	Sparse growth of birch and spruce	Pine and birch-spruce forests
44		Deep cooling	43240+1040(ЛУ-394) 4580+1200(ЛУ-515B)	Pereglacial tundra and forest-tundra	
46		Shapkinskoe warming		Sparse growth of birch and spruce	Birch-spruce forest
	Interstadial	Warming	46880+2090(ЛУ-624)	Forest-tundra	Sparse growth of pine and spruce
48, 50		Anorginskoe warming	47210+1270(ЛУ-674) 47520+1000(ЛУ-566) 49930+1420(ЛУ-1051)	Birch-spruce forests	Pine-spruce forests
				Forest-tundra	
52		Deep cooling	>50150(ЛУ-671) >53000(KE-1050)	Tundra; Pereglacial forest-tundra	
	Mikulino Interglaciation				

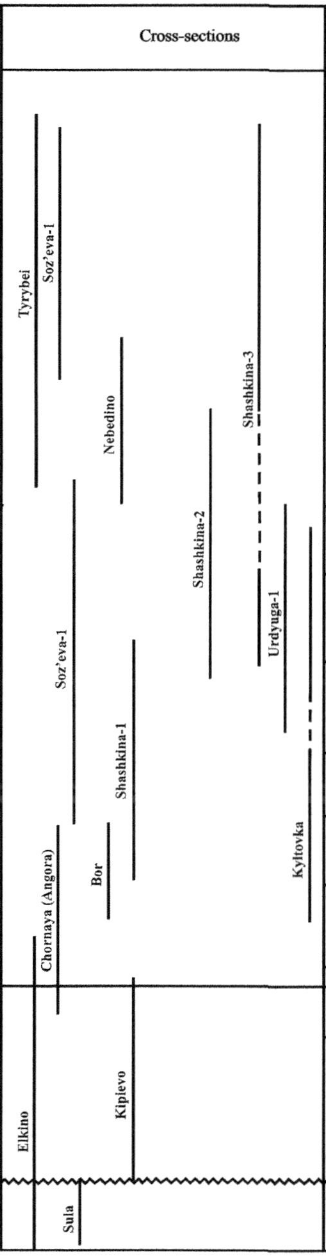

Member 3a is composed by coarse-grained sand with gravel, pebbles, small boulders, and lenses of coquina. Among the mollusk shells S.L. Troitsky identified boreal, predominantly boreal, Arctic-boreal and Arctic species. A.V. Gol'bert et al. (1973) noted on allochthonous occurrence of the thanatocenosis, which combines biotic elements from different depths and sedimentation environments. They explain this combination by sedimentation processes but it may be also caused by glaciotectonic events.

Member 3a yielded four new species of foraminifers, three of which are boreal and one (*E. subclavatum*) is Arctic (Gol'bert et al., 1973). The first appearance of *Quinqueloculina oviformis* known from the Kazantsev sediments of West Siberia indicates the Late Pleistocene age of this member. As a whole, the foraminiferal assemblage can be described as Arctic-boreal. Depth of the basin was less than 20 m, salinity 30–32‰, and temperature of bottom waters +2 or +3°C. Thickness of Member 3a is approximately 0.2 m.

Member 3b is represented by unsorted loam similar to till. Its thickness is about 0.2 m. There are no mollusks. Foraminiferal assemblage includes Arctic-boreal species from Members 3a and 3c (Gol'bert et al., 1973).

Member 3c is gray clay. A molluscan assemblage characteristic of the middle sublittoral zone is succeeded by biocenosis of middle and lower sublittoral zones and then by a deep-water assemblage. A foraminiferal assemblage also indicates the basinal deepening down to 150–200 m. The assemblage has some specific features. According to V.I. Gudina (Gol'bert et al., 1973), elfidides are prevailing forms (60–80%), and *Cassidulina subacata* occurs most frequently in the middle part of the member. Arctic-boreal species show maximal abundance. In general, the assemblage suggests a considerable increase of water and air temperature, which confirms a conclusion on the Mikulino age and boreal origin of the foraminifers. Thickness of Member 3c is approximately 7–8 m.

Let us summarize the data on the marine clayey sequence subdivided by interbeds of predominantly sandy material (Members 2 and 3a) into two units. It is a common opinion of all those, who studied the Vast'yansky Kon' section, that the units correspond to the regressive stage of the Boreal sea. In other words, their accumulation was attributed to the sedimentation factor related to sea level fluctuations. Our investigations revealed that the two-unit structure of the sequence was mainly resulted from glaciotectonic activities. In the Vast'yansky Kon' section the clayey sequence is well exposed in the frontal part of an embankment facing the Pechora course. In the lateral part of the embankment sandy Members 2 and 3a are dipping at 45° to the north-north-east (dipping azimuth is 10°). This allows supposition that the sandy (or sandy gravel) sediments with marine shells were brought by glacial thrust into the middle part of the sequence. Initially they lay under the clays and were a coastal-marine facies. Thus, the clayey sequence acquired the two-unit structure because of the glaciotectonic activities. The breccia-like texture of the clays and gliding planes can be also explained by the same activities.

The marine clayey deposits are overlain by the thick sequence of predominantly sandy alluvial sediments. At its base there is a unit of sand enclosing shells of boreal mollusks, which was previously considered by many (including us) to be accumulated during the regressive phase of the Boreal transgression (Zarkhidze et al., 1970). However, the occurrence of reindeer bones in the unit has not been taken into consideration. Ferruginous sandy pebbled sediments of the pra-Pechora midstream alluviual facies were described from the base of the unit. In general, the entire alluvial sequence is subdivided by numerous glacial thrusts into separate scales, thus complicating the structure of the terminal moraine ridge. Therefore it can be assumed that the sandy unit with the marine mollusks represents also the coastal-marine sediments of the transgressive phase of the Boreal transgression, which were brought by steeply dipping glacial thrusts, as convincingly shown by A.V. Gol'bert et al., (1973). Our investigations confirmed S.L. Troitsky's opinion that the scales appeared as a result of subdividing the sandy sequence by glacial thrusts, which brought the till interbeds into the bases of blocks (Gol'bert et al., 1973). Later slightly differing ideas on the glaciotectonic origin of the Vast'yansky Kon' Ridge were published by O.G. Epstein (1990), A.S. Lavrov and L.M. Potapenko (2005). The total thickness of the deformed alluvial formation varies in outcrops from 35 to 50 m.

Granulometric analysis of the sandy sequence showed that sands were well sorted, predominantly fine-grained (fraction of 0.1–0.25 mm constitutes 80%); amount of coarse silt varies from 1.2 to 30% at the absence of clay and fine silt fractions. More silt was discovered near the alluvial peat interbeds.

According to A.V. Gol'bert et al. (1973), mineral composition of the sands is persistently uniform: ilmenite and magnetite – 30–33%, amphiboles – 8–10%, garnet – 12%, tourmaline – 1–2%, zircon – 4–5%.

The mineral composition of the sandy sequences differs from that of the underlying marine clays. The latter contain less ilmenite, magnetite, zircon, garnet, rutile, but more amphiboles and minerals of epidote-zoisite group.

A specific feature of the sandy sequence is the presence of allochthonous sandy-peaty lenses, which allow the radiocarbon dating. Previously the lenses were dated by V.L. Yakhimovich by the radiocarbon method. Five dates have been obtained. The oldest date (11 m above the lower boundary of the sandy sequence) is 29470±450 (Bash GI-7), and the youngest one (30 m above the alluvium base) is 24790±500 (Bash GI-8). Later the radiocarbon dating of several peat samples yielded the out-of-limit ages: ≥47910 (LU-1491), ≥48040 (LU-1487), ≥48050 (LU-1484), ≥52120 (LU-1481), ≥53690 (LU-372), although one date (42560±1080 – LU-1489) was final. There are different but final estimations of age of the mammoth incisor from the sandy sequence 32440±850 (LU-3973) and 39000±850 (T-13200). The data published by J. Mangerud et al. (1999) cannot provide accurate dating of the sequence. The express mass spectrometry yielded out-of-limits dates for majority of samples, whereas final radiocarbon dates were from 31.6 to 43.1 ka.

The TL and OSL methods provided considerably different dates for the sandy sequence: 30000±3000 (TL); 26000±3000, 25000±2000, 27000±3000 and 66000±7000 (OSL) (Tveranger et al., 1995). Dates obtained by the OSL method in another laboratory appeared to be considerably older (63±5, 62±5 и 58±7 ka) (Mangerud et al., 1999). These dates were close to one of those published by

J. Tveranger. This, however, did not make the obtained age more reliable. Nevertheless, basing on a number of final radiocarbon dates A.S. Lavrov and L.M. Potapenko (2005) referred these sediments to the early Middle Valday (Table 2.2). They explained inaccurate radiocarbon dates for the plant remains by allochtonous occurrence of the latter as a result of reworking of different-aged peats during the accumulation of the sandy sequence. Therefore taking into consideration the final dates (except for evidently younger dates obtained in the Bashkirian laboratory), we can agree with A.S. Lavrov and L.M. Potapenko (2005), who inferred the early Middle Valday age of the sandy sequence from careful studies of sections in the north of European Russia, their correlations and geological mapping. It is necessary to keep in mind that the youngest date (31.6 ka) as well as the visible part of the alluvium sequence does not reflect the final phase of the deposition. Undoubtedly, the upper part was cut off and assimilated by the overlying Upper Pleistocene moraine, as evidenced by the presence of alluvial outliers in it. The latter is of significance because it definitely indicates shifts of the glacial sheet edge during the formation of the Vast'yansky Kon' terminal moraine. This conclusion is supported by the presence of till brought with glacial thrusts and by the mentioned analogous glaciotectonic features of the Boreal sea sequences.

Summing up the preliminary notes on the Boreal transgressive deposits, it should be added that no thorough palynological studies have been conducted in the Pechora Lowland in contrast to the western areas and therefore climate variations during the Boreal transgression have not been studied well. In this respect, the mentioned microfaunistic studies of V.I. Gudina seem to be very promising. They show small changes in the foraminiferal assemblages under an alternating effect of the high Arctic and boreal waters. Certainly, the water changes also influenced climatic conditions, but they cannot be directly correlated with the Upper Pleistocene stadials and interstadials.

In the stratigraphic scheme of the northeast of European Russia the Boreal transgressive sediments and contemporaneous continental deposits are united into the Sulina Horizon (Andreicheva, 2002). The Sulina alluvium reflects one climatic optimum and two maxima of coniferous species. During the climatic optimum there was the southern taiga where *Picea, Pinus silvestris, Betula sect. Albae, Alnus* were associated with single *Quereus, Ulmus, Corylus, Tilia, Carpinus* (Duryagina and Konovalenko, 1993).

It should be noted that most of the radiocarbon dates for the Vast'yanskiy Kon' alluvium indicate its intensive redeposition. Therefore results of previous palynological studies are ambiguous. Of interest is another Late Pleistocene chronostratigraphic scheme for the northeast of European Russia (Lavrov and Potapenko, 2005), which shows climatic and vegetation changes during the interval of 52–30 ka ago (Table 2.2). The scheme presents a complex dynamics of landscapes and glacial events during the Middle Valday, although being inconsistent with some available dates for individual sections. This fact is not of critical importance at the present stage of investigations. An attempt to produce such generalization is very useful and can serve as a base for developing a refined scale of the Late Pleistocene events in the north of European Russia.

Noteworthy is that the Kalinin glaciation in this region was not described in the summarizations (Lavrov and Potapenko, 2005; Andreicheva, 2002), although considerable cooling was reported by these authors. L.N. Andreicheva (2002) described the periglacial sediments of the Kalinin glaciation without indicating distribution of the glacial sediments.

A moraine of the Early Valday (Kalinin) Glacial (MIS 5b) was recovered in quarries of the Kola Peninsula and core sections of bore holes in the northern piedmont of the Lovozero Mountains. Unfortunately, there is no reliable data to outline even approximately the area of this glaciation.

Some signs of the Early Valday (MIS 4) glacial sheet were found in the Sokli Massif of northern Finland near the Russian boundary, but its limits are still unknown.

The scheme of L.N. Andreicheva (2002) shows seven climatic phases in the Middle Valday (the Byzov Stage, 48–33 ka ago), which correspond to variations in the landscape development (Duryagina and Konovalenko, 1993). At the initial phase (Bz_I) dominating hypo-Arctic mesophytes typical of the present-time tundra were associated with the Arctic and Arctic-Alpine species as well as steppe xerophytes. There were also dark-coniferous-taiga associations and elements of periglacial vegetation. The climate was cold and arid. The radiocarbon date for this phase is about 48000 (Andreicheva, 2002).

The phase of early warming (Bz_{II}) was characterized by dark-coniferous-taiga and boron vegetation in combination with birch and fir woods. A landscape was of northern taiga type, with fir and birch sparse growths. Sedges and cereals prevailed in the grassy cover. The time range of this phase is estimated as 47–45 ka ago.

The phase of early cooling (Bz_{III}) was similar in vegetation to phase Bz_I. Shrub and grass associations dominated, but boggy-tundra associations with xerophytic communities were spread widely. The climate was more humid.

In the middle warming phase (Bz_{IV}) northern taiga forests prevailed in the north of the region and fir and pine forests in the south. Many cereals and different grasses.

The phase of late cooling (Bz_V) is similar to phase Bz_{III}.

The phase of late warming (Bz_{VI}) was characterized by prevailence of northern taiga forests. It is estimated to range 42500–38000 (Andreicheva, 2002).

In the terminal phase of sharp cooling (Bz_{VII}) there were open landscapes with elements of periglacial flora and small islands of rare birch forests.

These climatic rhythms reflect an attempt to develop the Middle Valday climatostratigraphy of the northeast of European Russia. It is quite clear that so called terminal cooling phase of the Byzov time is not the last one. In the range from 38000 to 23000 there were several more climatic rhythms which appeared to be of global significance. However, the above publications mentioned nothing about them. This means that either the climatostratigraphic units do not reflect a complete pattern of climatic variations or the radiocarbon dates for some units are in disagreement with palynological data.

In general, the Middle Valday scheme of A.S. Lavrov, L.M. Potapenko is not well agreed with that of L.N. Andreicheva. Nevertheless, a direction of further investigations has been outlined.

In the north of European Russia a marine transgression of the Middle Valday terminated at about 35 ka (Lavrushin and Epstein, 2001). A cliff of the Madakha River (the Kanin Peninsula) exposes two coastal-marine sequences separated by a peat interbed. The radiocarbon age of the latter is 42 ka. This structure fixed a short-term regression of the Middle Valday sea and, at least, two transgressive phases of the Barents Sea. The second phase was about 7 ka long. In the Kolguyev Island sediments of the same transgression are overlain by a moraine of the last glaciation. They are represented by deltaic deposits in the coasts of the Kolakolkov Inlet and

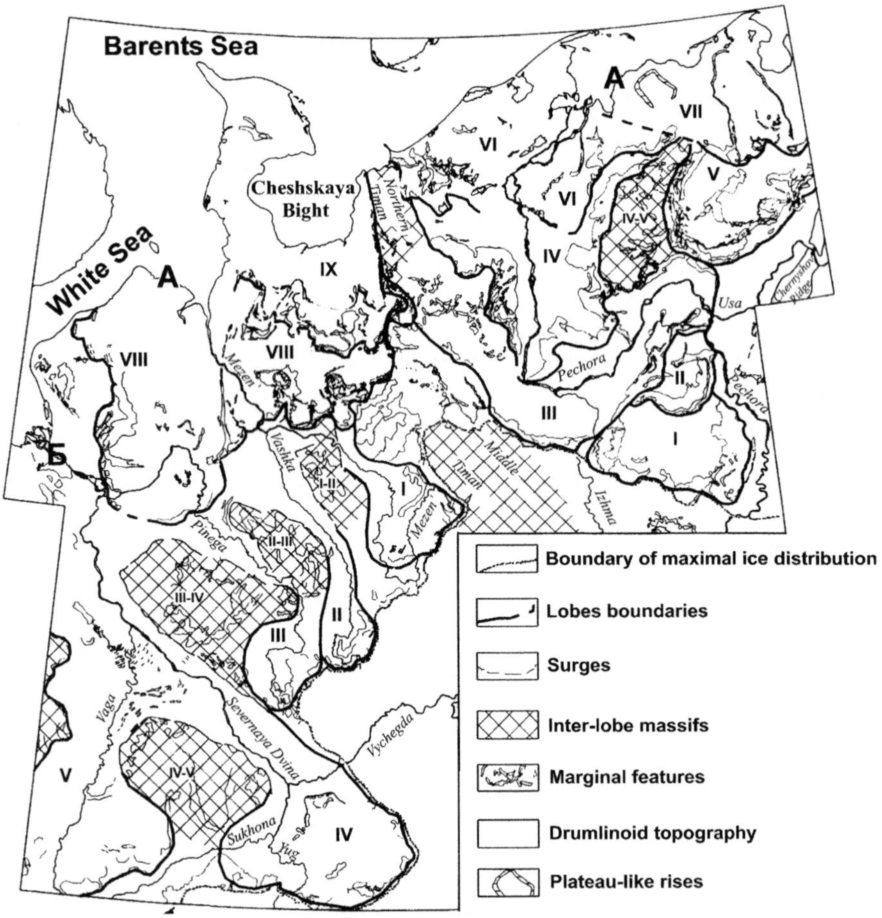

Fig. 2.4 Extension of the Late Valday Ice Sheet during the deglaciation (Lavrov and Potapenko, 2005), simplified. *A* – Barents Sea-Novaya Zemlya-Kara Sea Ice Sheet. I–IX – lobes: I – Kozhvinskaya; II – Lyzhskaya; III – Laisko-Izhemskaya; IV – Pechorskaya; V – Kolvinskaya; VI – Malozemel'skaya; VII – Bol'shezemel'skaya; VIII – Kuloisko-Mezenskaya; IX – Chöshskaya. *B* – Scandinavian Ice Sheet. I–V – lobes: I – Verkhnemezenskaya; II – Vashskaya; III – Pinezhskaya; IV – Severodvinskaya; V – Vazhskaya

the Bolvansky Nose Cape and by alluvial and alluvial-lacustrine sediments upstream the Pechora River (Lavrushin and Epstein, 2001).

Finally, some short-time geological events of the terminal Late Pleistocene (Ostashkov) Glacial should be mentioned. The QUEEN project results cast doubts upon previous views on limits of the last glaciation in the north of European Russia. The doubts appeared from numerous dates obtained by different methods for fossiliferous sediments underlying the relief-forming moraine. Analysis of these dates by A.S. Lavrov and L.M. Potapenko (2005) convincingly showed that there were no reasons to revise the existing views on the last glaciation limits (Fig. 2.4). So, it is unlikely expedient to discuss this problem herein. Another moment is of greater importance. During the last glaciation, as well as the previous ones, the Arctic shelf basins disappeared almost entirely because of sea level fall. In the remaining deepest parts of the basins there may be some bays "sealed up" by the glacier ice. The relationship between land and sea areas essentially changed due to a noticeable enlargement of the former. Large ice-dammed basins appeared in valleys of the north-flowing rivers. At the final phase of the last glaciation (Late Glacial Maximum) the accumulated water masses run off from the basins. An intensive glacioeustatic transgression contributed to the process by destroying the ice cover and greatly accelerating the deglaciation. These are main differences in the process of deglaciation between the glacial paleoshelves and adjacent land.

One more problem of the Late Pleistocene glacial history in the north of European Russia remains unsolved. It concerns the Kalinin Glaciation correlative to the Zyryan Glaciation of West Siberia.

Chapter 3
Main Geologic-Paleoecological Events of the Late Pleistocene in the North of Western Siberia

The major geological and paleoecological events of the Late Pleistocene in the north of West Siberia are two glaciations (Yermak (Zyryan) and Sartan) and three transgressions (Kazantsev, Kargin and post-glacial) (Arkhipov, 2000) (Fig. 3.1). As stated above, the Kazantsev transgression is correlated with the Eemian transgression of West Europe and MIS 5e (Troitsky, 1966; Gudina, 1976; Arkhipov, 1997, 2000). The Kazantsev Horizon lies on erosional surface of the Sanchug marine sequence and the Tazov moraine (Troitsky, 1969). It is composed of alternating sand, silt and clays (40–50 m) (Arkhipov et al., 1992).

The horizon encloses Arctic-boreal and boreal assemblages of foraminifers and mollusks. The foraminiferal assemblage includes warm-water, boreal and Arctic-boreal (56%) and cold-water, boreal-Arctic and Arctic (over 27%) forms (Gudina, 1976; Levchuk, 1984). The dominating warm-water species are *Retroelphidium atlanticum, R. boreale, R. propinquum, Haynesina magna, H. orbicularis, H. asterotuberculata, Cassandra helenae, Cassidulina reniformis, Cribroelphidium granatum.* The former four species are characteristic of the Kazantsev beds, but *Retroelphidium boreale, R. hyalinum, R. williamsoni, Elphidiella arctica, Discorbis deplanatus, Quinqueloqulina agglutinata, Q. oviformis, Q. deplanata, Pyrulina cylindroides* were found nowhere else.

Distribution of the Kazantsev glaciation and other important geological–paleoecological processes in the north of West Siberia and in the North Siberian Lowland is shown in Fig. 2.3. A fragment of the Late Pleistocene stratigraphic scale for the northern regions of West Siberia is presented in Table 3.1.

As seen in this table, the Kazantsev Horizon is overlain by the Yermak Horizon. In the Salekhard area, the Yermak moraine (correlated with MIS 5d) lie above the Kazantsev deposits and under the lower Kargin (Kharsoim) beds. At the lower course of the Ob' River a key hole recovered the Kormuzhikhan glacial and aqueoglacial sediments of the Yermak Horizon. Higher up the section, the Bogdashin Beds were recognized; their stratotype being in the southern margin of the Belogorskaya Upland. The Bogdashin Beds are represented by lacustrine loamy-silt sediments with three peat interbeds. Two upper interbeds are united into so called Upper Peat up to 2 m thick. The Lower Peat is also up to 2 m thick. The Lower and Upper Peats are separated by a silt interbed 0.9–1.1 m thick. According to palynological data, climatic conditions of that time were similar to the present-day ones. This allowed the peat interbeds to be correlated to the Ammersfort and Odderade

M.A. Levitan, Y.A. Lavrushin, *Sedimentation History in the Arctic Ocean and Subarctic Seas for the Last 130 kyr*, Lecture Notes in Earth Sciences 118,
DOI 10.1007/978-3-642-00288-5_3, © Springer-Verlag Berlin Heidelberg 2009

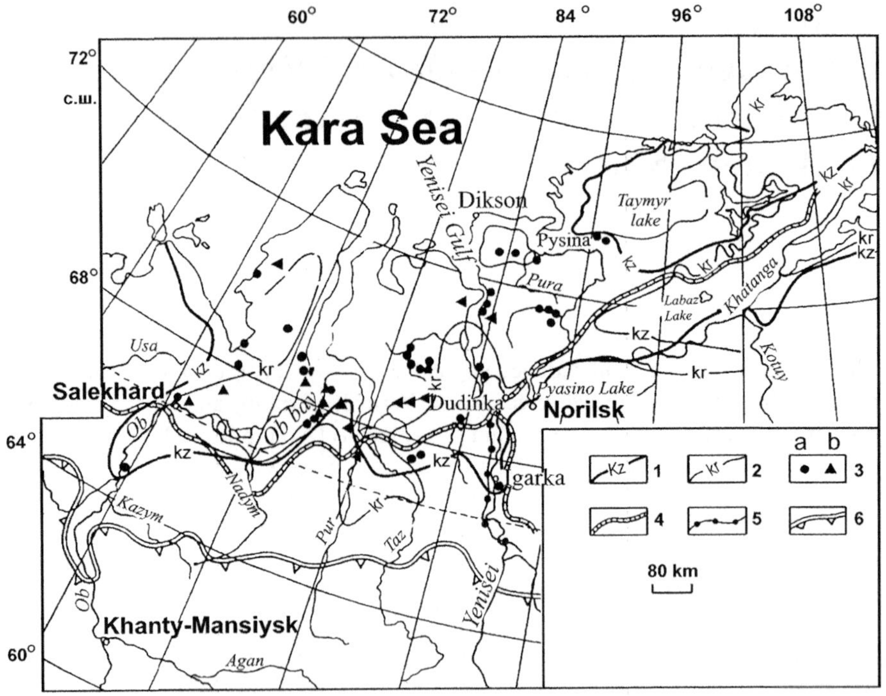

Fig. 3.1 Paleogeographic scheme of glaciation and marine transgression boundaries in northern West Siberia and the North Siberian Lowland during the Late Pleistocene (Arkhipov, 2000). 1 – Kazantzev Sea (MIS 5e); 2 – Kargin Sea (MIS 3.1 and 3.3); 3 – location of sediments: *a* – well studied Kazantzev sediments; *b* – Kargin sediments; 4 – boundary of Sartan Glaciation (MIS 2); 5 – Lokhpodgort motions (MIS 3.2); 6 – boundary of Yermak Glaciation (MIS 4)

Interstadials. The Bogdashin Beds are overlain by the Khoshgort glacial sediments, which are also exposed in the valley of the lower Ob River and the Salekhard lakes area. The Khoshgor Subhorizon consists of a moraine and varved clays.

At the lower course of the Yenisei River as well as at the Ob' lower course higher the Polar Circle the Yermak glacial deposits form undivided sequences. To the south of the both areas two stadial moraines can be observed (Arkhipov, 1997).

In conclusion of description of the Yermak (lower Zyryan) Horizon two debatable points should be noted. First, the geochronology of the events is based on unreliable TL dates. The unreliability of the TL method is aggravated by absence of statistics and usage of single dates.

Second, correlations of the glacial deposits to the isotopic stages and substages of the oceanic scale. The correlation of the Kormuzhikhan moraine to MIS 5d and the Khoshgort moraine to MIS 4 seems incorrect, and hence, the assignment of the latter moraine to the Yermak Horizon is unproved. Sometimes MIS 5d is correlated to the Kalinin glaciation of European Russia and MIS 4 is usually correlated by the West European researchers (in terms of the Russian stratigraphic terminology) to a glaciation in the interval between the Early and Middle Valday (in our opinion, the Kalinin glaciation corresponds to MIS 4). If these correlations are correct, the sequences of these two glaciations cannot be united into a single horizon. In this case the Yermak Horizon should be ranked as a superhorizon and the Kormuzhikhan and Khoshgort glacial complexes as horizons. However, if the existing correlations are true, these complexes should be considered to represent two independent glaciations of different age. The presence of the single moraine to the north of the Arctic Circle cannot serve as a confident argument because no genetical-structural studies have been performed. We can only refer to three successive moraines exposed at the town of Uglich and differing from each other in color, long axes orientation of boulders, and partially lithological composition. A detailed analysis of moraine structure is difficult without special investigations even by using enough core output.

So, a conclusion is apparent that the lower part of the post-Kazantsev sequence has not been adequately studied.

An age of the Kargin sediments is estimated within the interval from 55 (50?) to 23–22 ka ago. In terms of climatostratigraphy, the Kargin time is considered to be an interglacial or mega-interstadial. At that time marine, alluvial, and lacustrine-glacial sediments were accumulated (Table 3.2).

The Kargin Horizon is subdivided into three subhorizons represented by (1) the lower Kargin (61.4(56.7)–47.8(45.5) ka) and Kharsoim (>40–36.4±08 ka) beds; (2) the Lokhpodgort–Kyzym–Konoshchel glacial and lacustrine-glacial sediments; (3) the upper Kargin marine and alluvial sediments (28.6–25.9 ka). Thus, in West Siberia the Kargin sediments also reflect two transgressive phases, which were separated by a regression marked by significant cooling of the Lokhpodgort Stadial. Two transgressive phases of the Kargin basin were recorded in a single section of the Kanin Peninsula, the north of European Russia, in opposite to several sections of West Siberia (see Fig. 2.3).

L.K. Levchuk (1984) found three assemblages of foraminifers in the Kargin marine sediments. The Early Kargin (boreal-Arctic) assemblage corresponds to the beginning of the transgression. The Malokhet (boreal in the deep-water facies and Arctic-boreal in the shoals) assemblage relates to the maximum of the transgression, which preceded the Lokhpodgort glacier retreat. The Late Kargin (Lipov-Novoselov) assemblage corresponds to the termination of the transgression. V.I. Gudina (1976) correlated it with the Kharsoim and Kanin assemblages.

The Kharsoim (Lower Kargin) Subhorizon is a sequence of clay, silt, and sand (5–7 to 30 m) with Arctic and boreal-Arctic foraminifers *Protelphidium orbiculare* (Brady), *Cribroelphidium goesi* (Stsaedr.), *C. granata* (Gud.), *Elphidiella arctica* (Cushm), *E. gröenlandica* (Cushm) (Gudina, 1976). Allochthonous plant fossils of the Kharsoim clays yielded two radiocarbon dates: 36.4±8 ka and older than 40 ka.

A similar foraminiferal assemblage was discovered in local exposures at the Nadoyakha River (Yamal Peninsula, dated as >47170 (LU-1045)), along the eastern coast of the Ob' Inlet, and at the Messoyakha River (Gydan Peninsula), where they are overlain by glacial-type sediments: sandy loams with pebbles, gravel and shell fragments (Levchuk, 1984).

Table 3.1 Chronostratigraphy of the late Pleistocene geological events of the northern West Siberia plain (according to [*Arkhipov*, 1997])

Chronostratigraphic frame work				Chronology of geological events				
Stage	Horizon	Geological age, ka	Isotope stages	Northern zone of marine transgressions and glaciations				
				Al. sedim.	Terraces 10.2–9.5			
Upper Pleistocene	Sartan (Upper Zyryan)	0.2–10.3	2	Polar-Ural gl.-lgl.	m. - al. sediments Transit 55(60)–40(30) m terrace	Norilsk gl.-lgl. 11.5–10.3	Val'kov lgl.-lim. 19.9+0.5 – 10.7+0.2	Ingression m. 19200+200
Upper Pleistocene	Sartan (Upper Zyryan)	12	2					
Upper Pleistocene	Sartan (Upper Zyryan)	13	2	Sopkey gl.-lgl.	III - II terraces of Lower Ob 12.2+0.17	N'yapan gl.-lgl. 15–13		
Upper Pleistocene	Sartan (Upper Zyryan)	14	2			Tiutniy lim. 15.3–0.2		
Upper Pleistocene	Sartan (Upper Zyryan)	17(16)	2	Salekhard-Ural gl.-lgl.	Lgm. - lim. sedim. of ice-dammed 80(100)–60м. terrace	Gydan gl.-lgl.		
Upper Pleistocene	Kargin	23–22	3	Upper Kargin al. and m. (Gydan-Messoyakha) 28.6–25.9				
Upper Pleistocene	Kargin	30	3	gl.-lgl.	Lohpodgor-Kazym-Konoshchel lim. 33(34)–30(29)	lim. 33–30		
Upper Pleistocene	Kargin	33	3	Horsoim m. >40–36.4+0.8 Lower Kargin 61.4(56.7)–47.8(45.5) mean. 52,27		Zolotomyss al. >45–38 Kargin (Small Kheta) > 45–35		
Upper Pleistocene	Yermakovsk (Lower Zyryan)	60–55	4	Hoshtogor gl.-lgl.		Yermak gl.-lgl.		
Upper Pleistocene	Yermakovsk (Lower Zyryan)	70	a-b-c	Bogdashkin lim. 80+11(13)–65+8				
Upper Pleistocene	Yermakovsk (Lower Zyryan)	90(80)	d 5	Kormuzhihan gl.-lgl. 110(100) –25(31)				
Upper Pleistocene	Kazantsev	110(100)	e	Kazantsev m.-al. 150=10(15)–120=16				
Upper Pleistocene	Kazantsev	130–150						

Table 3.2 The event-related chronostratigraphy of Kargin – Middle Valday stage (according to [*Arkhipov*, 2000])

<table>
<tr><th colspan="2" rowspan="2">Horizons, subhorizons</th><th colspan="3">Lower course of the Ob River</th><th colspan="3">Lower course of the Yenisei River</th><th colspan="3">North of European Russia</th></tr>
<tr><th>Foraminifer complex</th><th colspan="2">Climatic rhythms</th><th colspan="3">Transgression stages, foraminifer associations</th><th colspan="2">Geochronological scale</th><th>MIS</th></tr>
<tr><td colspan="2">Sartan 23(22)–10</td><td colspan="6">Glaciation period</td><td>Late Valday</td><td>25.0–16.0</td><td>2</td></tr>
<tr><td colspan="2">Upper Kargin 30(31)–22 (23)</td><td>?</td><td>Upper Kargin Warming 30(31)–23(22)</td><td>Lipovsk-Novoselov Warming 30–22</td><td colspan="2">Completed transgression Lipovsk-Novoselovsk Arctic – boreal</td><td rowspan="6">Marine sediments</td><td rowspan="6">Middle Valday</td><td>Dunaev (Bryansk) Interglacial 32.0–25.0</td><td>3.1</td></tr>
<tr><td rowspan="5">Kargin</td><td>Lokhpodgort 35–30(31)</td><td colspan="2">Lokhpodgort cooling 35–30(31)</td><td>Konoshchel Cooling 33–30</td><td colspan="2">Regression Hiatus?</td><td>Leyamstsinsk (Mikhailov) Cooling 36.0–32.0</td><td>3.2</td></tr>
<tr><td rowspan="4">Kharsoim 55(60?)–50</td><td rowspan="4">Marine sediments</td><td rowspan="2">Zolotomys Warming 42–35(33)</td><td rowspan="2">Malokhet Warming 43–33</td><td colspan="2">Maximum of transgression Malokhet</td><td rowspan="2">Grazhdansky Prospect Interglacial 42.5–36.0</td><td rowspan="2">3.3</td></tr>
<tr><td>Deep sea Boreal</td><td>Shallow waters Arctic boreal</td></tr>
<tr><td>Cooling 44(45)–42</td><td>Early Cooling</td><td colspan="2">Regression Hiatus?</td><td>Shapkin Cooling 45.0–42.5</td><td>3.4</td></tr>
<tr><td>Shuryshkar Warming</td><td>Early Warming 50–45</td><td colspan="2">Beginning of transgression Early Kargin Boreal-Arctic often with domination of the Arctic species 56.7–47.8 61.4–47.8 average 52.27</td><td>Krasnogorsk (Rokay) Interglacial 32.0–25.0</td><td>3.5</td></tr>
<tr><td colspan="2">Yermak 110+/–27–55(60?)</td><td colspan="5">Glaciation period</td><td></td><td>Early Valday</td><td>72.0–58.0</td><td>4</td></tr>
</table>

Age of marine molluscan shells from the Lower Kargin beds (with the synonymous assemblage of foraminifers) of the Pasadochnaya River valley (Taimyr) was determined by the EPR method at the Cologne University of Germany. It appeared to be 72.3 ka (at the base) and 49.6 ka (at the top). Mollusks from the Kanin Peninsula (the Barents Sea) were dated as 54.7, 51.3, 47.4 and 46.5 ka (the specimens were given by V.S. Zarkhidze). Shells from the October Revolution Islands of the Arctic were dated as 52 ka (Molodkov et al., 1992). The age determinations of these specimens were also performed by the same method in the Geological Institute of Tallinn (Estonia). The specimens from one section were dated as 56±4.2 ka and 68±7 ka (according to ^{14}C dating, 51540±430 ka), whereas those from the other section were dated as follows: the dates of 52±4, 66±6 and 65±8.7 ka (by the EPR method) were obtained for the mollusks from the basal marine beds and 45550±230 ka (by the ^{14}C studies) for the forms from the top of these beds. According to the ^{14}C dating, the Lower Kargin marine beds correspond to the early Middle Valday.

At the lower course of the Yenisey River the Kargin marine sediments were recognized only near the Vorontsovo Settlement, where they are represented by sands and silts with the Malokhet assemblage of foraminifers. They are overlain by the Sartan moraine and fluvioglacial sediments.

The data obtained recently enabled S.A. Arkhipov (2000) to establish a chronological succession of geological events during the Kargin interval of the Late Pleistocene (Table 3.2).

Owing to the refined subdivision of the Kargin sequence five substages of MIS 3 were distinguished. The substages are numbered from 3a to 3e from the top downwards. Substage 3a corresponds to the Late Kargin–Lipovsk–Novoselovsk–Late Lobanov warming; Substage 3b is related to the Lokhpodgort–Konoshchel Glaciation (cooling); Substage 3c is correlated with the Middle Kargin warming; Substage 3d is correlated with a short-term cooling; and Substage 3e corresponds to the Early Kargin warming (Arkhipov, 2000). Although the correlations with the oceanic oxygen-isotope scale are formal and insufficiently substantiated, they can provide a base of future studies in this direction.

Formation of the glacial Sartan Horizon is the latest of the main geological and paleoecological events of the Late Pleistocene. The Sartan Horizon consists of the glacial sequence of the same name, lacustrine-glacial, dammed lacustrine marine ingressive sediments, and river terraces. All the listed formations are indicators of the geological and paleoecological events. So, the Gydan (Salekhard-Ural) terminal moraine ridge (20–18 ka ago) marks the maximal limit of the Sartan glaciation. The Nyapan (Sopkey) (15–13 ka ago) and Norilsk (the Polar Urals) (11.5–10.4 ka ago) moraines are those of recession. The sediments of dammed lakes suggest that the north-flowing rivers were closed by the ice sheets. The ingressive marine sediments reflect the Late Glacial transgression, sediments which include banks with *Portlandia arctica* and river terraces formed during of its regressive phase (Arkhipov, 2000).

Chapter 4
Geologic-Paleoecological Events of the Late Pleistocene in the Northern-Siberian Lowland and Taimyr Peninsula

As in the other parts of Russia, the Upper Pleistocene succession of the North Siberian Lowland begins with the Kazantsev deposits which are represented there by different facies of marine and lacustrine-alluvial sediments. These are sands, silts, clays showing cross bedding of different type, frequently rhythmical one (Anthropogen of Taymyr, 1982). The Kazantsev sediments are 40 m thick and overlain by a moraine of the Early Zyryan (Yermak) Glaciation. Different glacial dislocations are frequently observed.

Marine mollusks occur in abundance. Their assemblage includes 41 Arctic, boreal and boreal-Arctic species approximately in equal proportions. The boreal species, such as *Mytilus edulis* Linné, *Macoma baltica* (Linné), *Cyrtodaria jenisseae* (Aachs), *Astarte invicata* Merclin et Petrov, *Astarte leffingwelli* Dall, are index-species. *Hiatella arctica* (Linné), *Macoma calcarea* (Gmelin), *Astarte montagui* (Dill.), *A. montagui striata* Leach., *A. borealis* (Schum.), *Mya truncata truncata* Linné are widespread.

The molluscan analysis shows that the Kazantsev sea was an open basin of small depth (within the upper-middle sublittoral zone) and probably lower salinity. From the northeast to the southwest the fauna became more diverse owing to increasing influence of the western warm waters.

The foraminiferal assemblage from the Kazantsev sediments of the North Siberian Lowland (Levchuk, 1984) was briefly described in the previous chapter.

Palynological studies of the marine and continental Kazantsev sediments of the North Siberian Lowland suggest that the climate was warmer than in nowadays (Anthropogen of Taymyr, 1982).

The Lower Zyryan (Yermak) glacial deposits of the North Siberian Lowland are called the Muruktin sediments. They include the North Siberian Beds of the first stage of the glaciation (the beginning and maximal advance), the Yantardakh Beds of the beginning of deglaciation, and the North Kokor Beds representing the basal moraine and the marginal deposits of the second stage of the glaciation. The distribution of the listed beds is shown in Fig. 4.1.

The North Siberian Beds are a basal moraine composed of boulder loams and sandy loams up to 40 m thick. Differences in the boulder composition allowed establishing the Taimyr, Putoran, and Anabar petrographical provinces in the eastern part

M.A. Levitan, Y.A. Lavrushin, *Sedimentation History in the Arctic Ocean and Subarctic Seas for the Last 130 kyr*, Lecture Notes in Earth Sciences 118,
DOI 10.1007/978-3-642-00288-5_4, © Springer-Verlag Berlin Heidelberg 2009

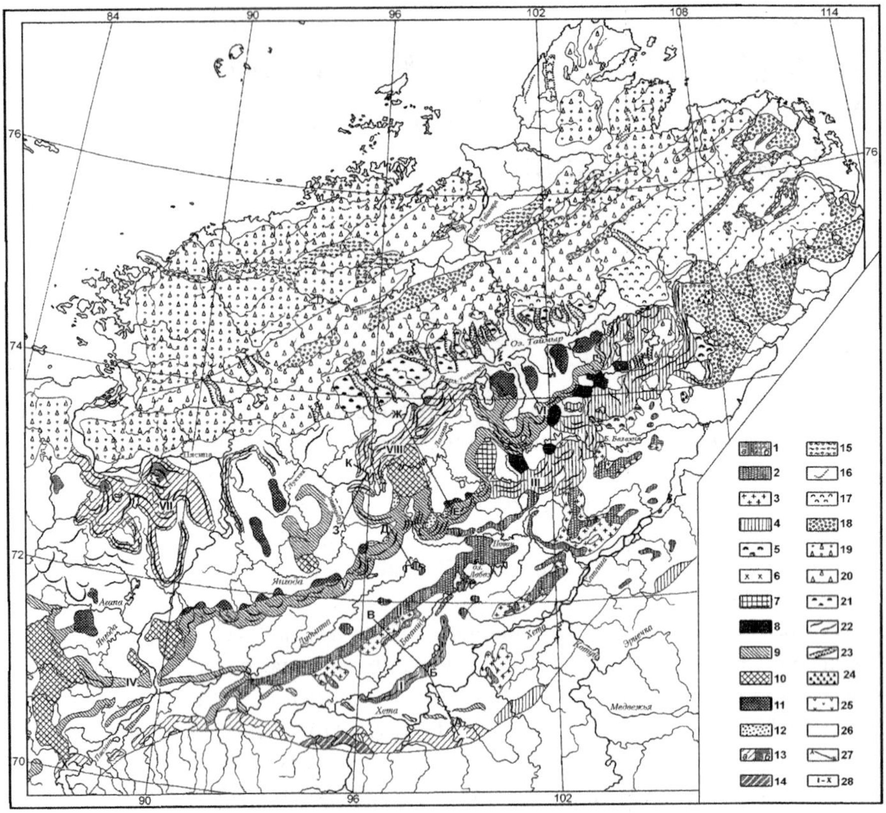

Fig. 4.1 Main elements of the Taimyr glacial morpho-sculpture (Anthropogen of Taimyr, 1982). (Muruktinian Glaciation, North Siberian Stage): 1 – moraine swells: *a* – pressure moraines; *b* – bulkload moraines; 2–3 – deglaciation time for Muructinian Ice Cover: 2 – fracture-kame swells and elevated massifs; 3 – kame terraces; Severokokorskaya Stage: 4 – pressure-bulkload marginal features; 5 – sag-and-swell topography of the "dead" ice fields; 6 – kame terraces; 7 – rises of interlobe massifs; 8 – inter-tongue massifs; 9–12 – Sartanian Glaciation, Karaulian Phase: 9 – pressure-bulkload marginal features; 10 – rises of interlobe massifs; 11 – inter-tongue massifs; 12 – kame terraces; 13 – N'yapanskaya Phase: *a* – marginal pressure-bulkload and bulkload features, *b* – intertongue massifs; 14 – Noril'skaya Phase: bulkload marginal features in the foothill of Putoran Plateau; 15–21 – elements of glacial morpho-sculpture deciphered at radiolocation images on the Northern Taimyr: 15 – proposed marginal glacial features on the north of Taimyr; 16 – crests of proposed end-moraine swells; 17 – areas of hilly topography; 18 – areas covered by thick sequence of questionable glacial deposits; 19 – areas with traces of glacial exaration, partly covered by questionable glacial deposits; 20 – areas not covered by loose deposits but with clear traces of exaration treatment; 21 – areas without clear exaration traces but with erratics on the surface; 22 – crests of end-moraine swells; 23 – troughs; 24 – rises of North Siberian Lowland composed from hard rock and possibly were not covered by ice; 25 – north-western Taimyr with only valley-net glaciation in the Late Pleistocene; 26 – bottom of glacial depressions with marine, lacustrine and partially fluvial sediments; 27 – profile lines; 28 – glacier swells: I – Urdakhskaya; II – Sakshsinskaya; III – Severokokorskaya; IV – Dzhangodskaya; V – Syntabul'skaya; VI – Bai-kuranerskaya; VII – Mokorittskie; VIII – Upper Taimyr; IX – North Taimyr; X – N'yapanskaya

of the North Siberian Lowland (Fig. 4.2). Analysis of the clastic material showed that the ice sheet covering the lowland moved from the north and northwest (Anthropogen of Taymyr, 1982).

The moraine usually lies on rhythmically bedded lacustrine-glacial deposits. The rhythmical bedding is due to alternation of varved clays (0.5–1.0 m) with members of thin (0.6–2 m) sand and silt interbeds.

The Yantardakh Beds are mostly sediments of aqueoglacial origin. They include crack-kame sediments of interfluvial ridges, pressure ridges, terminal moraines, and kame terraces. In the east of the lowland (at the lower course of the Khatanga River) the marine beds enclose mollusks and scarce foraminifers *Protoelphidium orbiculary* (Brady), *Retroelphidium* ex gr. *subelavatum* Gud., *Cribrononion insertus* (Williamson), *Miliolinella subrotunda* (Montagui), *Triloculina subtricorinata* Gud., *Tr.* cf. *oblonga, Cribrononion* cf. *incertus* Juv. (identifications of V. I. Gudina). The appearance of marine sediments can be caused by intensive melting of the thin ice cover in the eastern and central parts of the lowland at the proximity of the Laptev Sea. The deglaciation was accompanied by the glacier disintegration into a number of dead ice blocks. As the latter melted, the transgression spread further into the lowland. However, there are some other views. According to V.I. Gudina, the microfauna can be related to the Sanchug sediments of the Middle pleistocene. Another version is that the marine sediments occur in outliers inside the moraine. A lot of versions concerning genesis and age of the sediments with foraminifers do not allow any unambiguous conclusion.

The North Kokor Beds are aqueoglacial deposits of the second stage of the Muruktin Glaciation as well as marine sediments (see Fig. 4.1). The marine sediments were discovered at the lower courses of the Khatanga, Malaya and Bol'shaya Balakhnya rivers, i.e., in the area adjacent to the Khatanga Bay. Mostly, they are typical coastal and upper sublittoral sediments. They enclose scarce foraminifers dominated by the Arctic and boreal-Arctic species. Age of tills was defined by their position in the section. They are underlain by the Kazantsev sediments and overlain by the Kargin marine and continental deposits. Many radiocarbon dates obtained for the latter deposits range from 50 to 26 ka ago.

The Kargin sediments of the North Siberian Lowland are both marine and continental ones. The marine sedimentation underwent two stages: at the early stage (50–39 ka ago) the sediments spread widely and at the later stage (38–26 ka ago) the area of the transgression was considerably reduced.

The Lower Kargin sediments are commonly shallow-water facies of alternating clays, silts, sands with pebble lenses and interbeds. Marine mollusks are represented by Arctic-boreal, Arctic and boreal species characteristic of a shallow basin. Boreal species *Mytilus edulis* Linné, *Macoma baltica* (Linné) and Arctic taxa *Portlandia arctica* (Gray), *P. arctica aestuariorum* Mossev, *P. arctica siliqua* (Reeve) occur in abundance.

The radiocarbon date of 37000±1000 (SOAN-834) for plant detritus from the lacustrine deposits lying on the Lower Kargin marine sediments is supposed to indicate a termination of the first stage of the transgression.

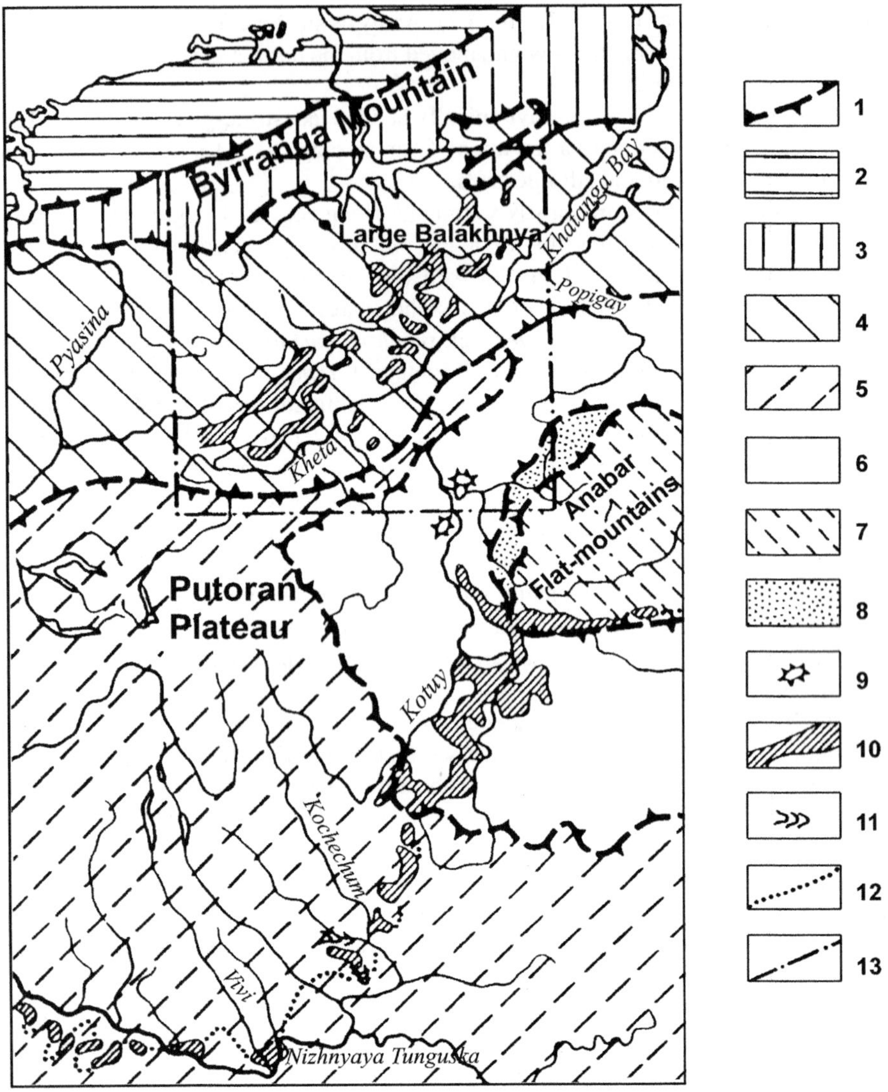

Fig. 4.2 Marginal glacial features and source petrographic provinces (Anthropogen of Taimyr, 1982). 1 – boundaries of pre-Cenozoic petrographic provinces; 2–9 – petrographic provinces: 2 – North Taimyr (Proterozoic and Archean granites, gneisses, greenstone shales, limestones, dolomites), 3 – Byrranga (Upper Paleozoic sandstones, siltstones (P) and limestones (C), 4 – Yenisei-Khatanga (Mesozoic sands and sandstones (K), 5 – Putoran-Vilyui (Mesozoic trapps), 6 – Kotui (Cambrian-Silurian limestones and dolomites), 7, 8 – Anabar: 7 – Archean granitoids, 8 – proterozoic quartz-feldspar sandstones of Mukun Serie, 9 – intrusions of subalcaline basaltoids; 10, 11 – large features of glacial topography of marginal accumulation zone: 10 – sag-and-swell and swell topography, swells and massifs, 11 – small terminal moraine swells; 12–13 – boundaries: 12 – maximal development of Zyryan Glaciation, 13 – the studie area of composition of glacial deposits, clastic matter and products of its rewashing

The Upper Kargin marine sediments were accumulated in the interval of 33000–26000 ago (according to the radiocarbon dating). They are represented by dark gray clays with silt interbeds or by fine-grained sands. Occurrence of marine mollusks *Macoma baltica* (L.), *Mytilus edulis* L., *Portlandia arctica* (Gray), *P. arctica siliqua* (Reeve) evidences for a shallow basin. A foraminiferal assemblage is dominated by the Arctic-boreal species *Retroelphidium hiallinum* Brodn., *Protoelphidium parvum* Gudina, *Protoelphidium orbiculare* (Brady), *Cribroelphidium goesi* (Stshedrina), *Elphidiella groenlandica* (Gushman) as well as nonionides, miliolides, and globulins (Anthropogen of Taymyr, 1982).

The Kargin continental sediments are the lacustrine-alluvial ones occurring in the central part of the North Siberian Lowland. They reflect three climatic warmings and two coolings. Most frequently these sediments fill large glaciogenic depressions and compose lake terraces of small thermokarst sinks. They are represented by bedded sands and gray silts tinged locally with brown and interbeds of alluvial plant detritus. The characteristic feature of the sediments is a high content of Fe.

The Sartan sediments of the North Siberian Lowland and the Byrranga Mountains are glacial, aqueoglacial, lacustrine and alluvial ones formed predominantly during the active degradation of the glacial covers and discharge from glacial lakes. Deposits of the Karaul, Nyapan and Norilsk stages were distinguished.

In the Sartan time the substantial part of the Taimyr Peninsula was covered by ice. Three belts of marginal sequences were discovered by geological mapping. They correspond to three (Syntabul, Upper Taimyr and North Taimyr) stages of the glacier degradation.

Nevertheless, the limits of the Sartan glaciation in Taimyr and the North Siberian Lowland are still under discussion. There is an opinion that the ice covers existed only in the northern and northwestern parts of the Taimyr Peninsula whereas mountain-valley glaciers occurred in separate areas of the Byrranga Mountains. This view assumes that the Kara shelf was the center of the glaciation in the western part of the peninsula. However, this paleoglaciological problem has not been solved yet.

Chapter 5
The Late Glacial Time and Holocene of Northern Eurasia

A significant event took place in the Late Glacial time both on the shelf and adjacent land areas. This is a glacioeustatic transgression related to intensive degradation of the ice cover. A time range of the transgression is estimated from the end of Bølling to Younger Dryas, i.e., from 12.5 to 10.5 ka ago (Lavrov and Potapenko, 2005). Geomorphologically this transgression is manifested in so called coastal terrace (the term of A.S. Lavrov and L.M. Potapenko) extending between the Czech and Khaypudyrka inlets. The maximum height of the terrace in the joint area reaches 25–30 m. The terrace is composed by heterogranular sands with lenses of poorly rounded coarse clastics. The upper part of the terrace exposed along the Barents Sea coast includes lenses and short interbeds of autochthonic peat not more than 0.2 m thick. Thus, the clastic sequence of the terrace is subdivided by the peat lenses into several horizons.

Radiocarbon dates for the buried peat suggest repeated short-term fluctuations of sea level during the transgressive phase. Episodes of sea level lowering may occur at 11.7; 10.5–10.3; 10.0; 9.7; 9.2−9.0 ka ago. Some close dates for the base of the subaerial peat beds were obtained at different hypsometric marks. In these cases the inherited overlaying of the younger peat on the older one may took place. Because of poorly studied structure of suraerial peat beds (4–5 m thick) a time of the superposition remains unclear. Besides, little is known about the sea level lowering during the short-term regressions. The peat beds were recovered by drilling in the shelf area near the Varandey Settlement (Veynsberg et al., 1995) under the Holocene marine sediments at the absolute mark of –15.8 m. Radiocarbon date for the peat beds is 10300±130. According to A.S. Lavrov and L.M. Potapenko (2005), the short-time sea level fall by 5 m occurred about 9.2 ka ago, according to the radiocarbon dating of alluvium in the cut-off meander of the Pra-Tanyuy River.

Thus, the glacioeustatic transgression was characterized by a complex sea level dynamics. Far from all sea level changes have been documented. Further studies are needed in this direction.

The terminal phase of the Early Khvalynian transgression of the Caspian Sea (the maximal one of the Pleistocene) proceeded at the same time as the glacioeustatic transgression of the Barents Sea. The initial short-term phase of the Khvalynian transgression only was related to melt waters of the last glacier. It cannot be

M.A. Levitan, Y.A. Lavrushin, *Sedimentation History in the Arctic Ocean and Subarctic Seas for the Last 130 kyr*, Lecture Notes in Earth Sciences 118,
DOI 10.1007/978-3-642-00288-5_5, © Springer-Verlag Berlin Heidelberg 2009

Table 5.1 Basic glaciological and morphogenetic events of last glaciation during its degradation stage

0	utf8

Dynamics of ice sheet and lithosphere reaction							Basic events and the morphogenetic types of the relief of the deglaciation period in the limb zone of the Late Valday Ice Sheet	Azonal sediments
Ice thickness	Dynamic state	Deglaciation character	Isostatic rebound of lithosphere	Glaciodynamic stages				
Decrease	Stage of local motion-pulsations (surges)	Areal deglaciation	Regional (arch) uplift Activation of differentiated motions of lithosphere blocks	Deglaciation	Final Stage	Late–postglacial	Relief and surge sediments (local moraines). Local motions of glacial edge (surges). Okunev (estuarine) terrace, the coastal terrace of the Barents Sea. Ingression of the Barents sea into the valley of the Pechora River.	Basic moraine, accumulative boundary moraines, shallows of melting waters and glacial lake waters
	Curtailment of ice motion in the limb zone				Third Stage		De Geer's Moraines. Marginal kame and kame fields. Thaw limnokames, large kames. Subglacial compression moraines (moraine of under-ice extrusion)	
	Destruction of the source province	Dissecting deglaciation			Second Stage		*Revival of the Pechora River (Okunev and Denisov alluvium} and Vychegda River.* Lymnokames. Terraces: Lower-Pechora, Gamsk, epigenetic Formation of the Lower-Pechora intraglacial lake — the estuary of the Barents Sea and Gamsky Basin. Voysk terrace. Descent of the Upper-Pechora and Oz"yagsk near-glacial lakes.	
	Decrease of dynamics. Development of lobes	Frontal and areal deglaciation	Beginning stage of the arch uplift and fault activation		First Stage		Pressure-accumulative moraines of the ledges of glacioeustatic plateaus. Pressure-accumulative and accumulative-pressure boundary moraines. Accumulation of the glacial-lake sediments of glacial dammed lakes.	
Stabilization	Stabilization		Development of the regional glacioeustatic depression	Stabilization of the ice edge			Drainage of dammed glacial lakes into the Caspian Region. Pressure-accumulative and accumulative-pressure final moraines. Drainage of glacial lakes into the Caspian Region through the Mylvin, Kel'tmen and Kama canyons	
Increase	Advance			Active state of the ice edge			Formation of the Upper-Pechora and Oz"yagsk near-glacial and dammed lakes. Formation of the moraine ice. Deposition of the basic moraine.	

excluded that similar sea level changes in the remote basins were due to some global hydrogeological processes of unclear nature.

In the coastal part of West Siberia (the Yamal, Gydan and Tazov peninsulas) there is a terrace of similar type and two altitude levels of 40–55 and 20–35 m, which is also brought about by the Late Glacial glacioeustatic transgression. The terrace was called "transitional" and dated as 12260±170. Signs of this sea ingression were discovered also at the Taimyr Peninsula, at the lower courses of the Agapa and Pyasina rivers (Arkhipov, 1997).

In the north of European Russia the Late Glacial period was marked by the formation of an intraglacial terrace at the lower course of Pechora, a terrace of glacier contacts and the Okunev terrace (Lavrov and Potapenko, 2005).

In the opinion of these authors, deposits of the Lower Pechora intraglacial terrace are sediments of the synonymous estuary of the Barents Sea in the Late Glacial, formation of which began 14.5–14 ka ago.

Deposits of so called terrace of glacier contacts should be referred to the outwash-plain apron assigned conventionally to the Late Glacial (Lavrov and Potapenko, 2005).

Finally, the Okunev terrace was formed during the synonymous ingression into the Pechora valley, which proceeded, according to the radiocarbon dating (Lavrov and Potapenko, 2005), in the interval of 12.7–10.8 ka ago. The ingression spread along the Pechora valley to the present-time altitude of 25 m correlative with the mentioned terrace of the West Siberian coast. The dates indicate that the ice dam across the Pechora River was destructed somewhat before the ingression and the modern drainage network appeared about 12.7–13.0 ka ago. Noteworthy is that the ice dam was destructed through actions of two water masses: waters of ice-dammed lake from the south and waters of the Late Glacial transgression from the north. The main geological events of the Late Glacial period in the north of European Russia are shown in Table 5.1 (after Lavrov and Potapenko, 2005).

Chapter 6
Outlines of the Late Pleistocene and Holocene History of the East Arctic Seas

In general, the East Arctic (Laptev, East Siberian and Chukchi) seas and adjacent land areas have not received adequate studies. It is impossible therefore to present any comprehensive concept of the Late Pleistocene and Holocene history of the seas. Herein their geological history is presented in fragments.

As noted above, manifestations of the Boreal transgression of the early Late Pleistocene (the Kazantsev-Mikulino time) were found to the west of this region. No reliable data is available on the geological structure of the Pleistocene sediments in the Laptev Sea between the Severnaya Zemlya and Novosibirskie Islands. The Kazantsev sediments were discovered in the North Siberian Lowland, as stated in the previous chapter. But no marine sediments of this age occur in the coasts of Anabar and Olenek Inlets as well as in the Lena delta, where the Upper Pleistocene is represented by terrestrial deposits.

The lower Upper Pleistocene sediments referred to the Kazantsev Horizon were described from the western coast of the Kotelny Island in the Laptev Sea, where they form a marine terrace 8–9 m above the sea level. This terrace is composed by a sand sequence with mollusks *Hiatella arctica* (Linne) , *Astarte (tridonta) borealis* (Schumacher), and *Sipho togatus* Morch. In opinion of M.N. Alekseev and his co-authors, pollen of *Betula sect.*, *Nanae* and *Alnast*er found in the shallow-water sediments are indicative of vegetation characteristic of bushy tundra and climatic conditions warmer than the present-time one. However, the conclusion concerning the climatic situation does not seem to be well substantiated, the marine terrace of the Kotelny Island may be younger.

The radiocarbon date of 36400±500 (GIN-3524) was obtained for plant detritus of another marine terrace in the western coast of the Kotelny Island. This date indicates that the terrace was formed in the Kargin time. According to M.N. Alekseev and co-authors, the high altitude of the Kargin marine sediments of the Kotelny Island is due to neotectonic activity. At the Kargin time the sea level was considerably lower than at present and, hence, there was a large land between the Novosibirskie Islands and the modern coast of the Laptev Sea. The land was inhabited by mammals of the Upper Paleolithic complex, which could penetrate far to the north. Actually, the modern Laptev Sea did not exist even in the Sartanian time. In its present limits this sea appeared in the Younger Dryas and was finally formed in the

M.A. Levitan, Y.A. Lavrushin, *Sedimentation History in the Arctic Ocean and Subarctic Seas for the Last 130 kyr*, Lecture Notes in Earth Sciences 118,
DOI 10.1007/978-3-642-00288-5_6, © Springer-Verlag Berlin Heidelberg 2009

Holocene, as suggested by numerous radiocarbon dates obtained by the international expedition to the Laptev Sea (Bauch et al., 2001).

Analysis of the published dates clarified the following important points.

Coring in the continental slope yielded four sediment cores from the water depths of 983 m (PS 2458-4), 270 m (PS 51/154-11), 122 m (PS 51/118-3) and 114 m (PS 51/118-2) (Table 6.1). The dates obtained showed hiatuses and repeated age inversions. Most likely, the disturbed age succession can be attributed to gravitational processes provoked by the increased seismicity of the region. Insignificant inversions recorded even at the sea depth of 114 m may be also due to imprecise dating. But the other sediment cores show more essential inversions. For example, in the sediment core from the water depth of 122 m the inversions were recorded at 749, 230 and 126 cm from the coretop. Hiatuses may occur between sampling sites KIA 6919 and 6920, and between KIA 6923 and 6922 in the sediment core from the water depth of 270 m. The same sediment core (Sample KIA-6925) yielded the oldest ^{14}C date of 13120±60 correspondent to the calibrated date of 15293 cal. yrs. BP (it is necessary to emphasize that the dated sediment core is from the continental slope). Hiatuses (age inversions) and gravitationally displaced beds yielding close dates were fixed in the sediment core from the water depth of 983 m. These are beds about 70 cm and 20 cm thick in the intervals of 436–369 and 486–467 cm from the coretop respectively. Sedimentation rates in the continental slope were calculated without accounting specific structural features and therefore are doubtful.

Dates obtained for the sediment cores from the outer (water depths of 60 and 70 m), middle (42–51 m) and inner (21–32 m) shelf zones indicate the same structural features. These features can be attributed to the high seismicity of the region and mud beds instability. This also casts doubts on the published rates of sedimentation (Bauch et al., 2001).

These dates also show that the outer shelf began to be formed in the Younger Dryas, and the sea extended into the modern middle shelf at the very end of this interval. This suggests that the initial stage of the Laptev Sea origination was related to the final stage of the Late Glacial transgression. Finally, marine sedimentation within the modern inner shelf was caused by the Flandrian transgression of the Atlantic stage of the Holocene. In the Preboreal and Boreal stages there was a land.

The specific features of the Holocene marine sequences, particularly, many age inversions caused by the increased seismicity of the region, which led to displacement of bottom sediments, are worth special attention (Table 6.1).

The absence of proved marine sediments on the modern southern coast of the Laptev Sea can serve as a convincing argument for appearance of this basin in the terminal Late Pleistocene.

More complete materials on the Late Pleistocene history of the east-central segment of the Russian Arctic have been accumulated for the Severnaya Zemlya Archipelago. They enable estimating how far the Arctic water masses penetrated into the land areas in the Kazantsev and Kargin times forming possibly, an ingression bay in the riftogenic Gakkel Ridge.

In the October Revolution, Pioneer, and Komsomolets islands of the Severnaya Zemlya Archipelago the Kazantsev sediments formed two terraces 80–70 m and

Table 6.1 Radiocarbon dates, calibrated calendar age and calculated averaged sedimentation rates [*Bauch et al.,*2001]

Sediment core number	Core depth, cm	^{14}C years	Cal. yrs. BP	Sedimentation rate, cm/ka
1	2	3	4	5
		Continental slope		
PS2458–4				
AAR–2417	0.5	recent	0	
AAR–3082	252	8830±55	9437	27
AAR–2419	294	9030±100	9785	
AAR–2421	335	9340±120	10073	
AAR–3083	369	10020±70	10963	
AAR–3084	399	10090±65	11102	
AAR–2418	436	10050±170	11086	
AAR–3085	467	10600±75	11815	89
AAR–2420	486	10540±120	11666	
AAR–3086	530	11560±100	13081	
AAR–3087	578	12270±65	13823	
AAR–3088	625	12750±150	14285	
KIA–6113	667	12600±110	14110	
PS51/154–11	0		0	
KIA–6919	31`	1540±45	1136	12
KIA–6920	138	10120±55	11143	
KIA–6921	204	10235±45	11151	
KIA–6922	300	10725±50	12022	
KIA–6923	375	12180±60	13680	92
KIA–6924	440	12525±55	14086	
KIA–6925	518	13120±60	15293	
PS51/118–3				
KIA–6898	5	280±30	0	8
KIA–6899	75	8535±55	8994	
KIA–6900	126	8470±60	8961	
KIA–6901	230	9020±60	9699	
KIA–6902	300	8730±50	9238	
KIA–6903	442	9440±60	10268	
KIA–6904	549	9575±55	10300	402
KIA–6905	640	9970±50	10951	
KIA–6906	749	9660±90	10320	
KIA–6907	765	9760±60	10463	
KIA–6908	834	9760±55	10465	
KIA–6909	866	10010±70	10960	

Table 6.1 (continued)

1	2	3	4	5
PS51/118–2	0		0	2
KIA–6889	20	8495±45	8974	
KIA–6890	56	8470±55	8961	
KIA–6891	122	8775±50	9247	
KIA–6892	206	8905±50	9535	
KIA–6893	31	9100±45	9809	386
KIA–6894	300	9290±50	9907	
KIA–6895	388	9365±50	10124	
KIA–6896	442	9350±50	10104	
KIA–6897	537	9635±60	10314	
Outer shelf				
PS2725–5				
KIA–2747	0	recent	0	
KIA–114	115	8340±60	8891	13
KIA–115	207	9170±90	9828	
KIA–117	392	9280±60	9903	266
KIA–118	430	9340±60	10073	
PS51/150–10	0		0	
KIA–6927	11	845±30	499	11
KIA–6928	56	4980±35	5321	
KIA–6929	90	6305±35	6783	17
KIA–6930	131	8955±40	9609	
KIA–6931	215	9420±50	10262	
KIA–6932	315	9650±45	10317	113
KIA–6933	410	10720±55	12020	
KIA–6934	485	11060±70	12728	
Central shelf				
PS51/135–4	0		0	
KIA–6910	4	recent	0	2
KIA–6911	8	4920±40	5291	
KIA–6912	40	6480±50	6999	31
KIA–6913	80	7100±55	7593	
KIA–6914	143	500±25	145 ?	
KIA–8547	152	715±30	411 ?	13
KIA–6915	266	8945±55	9605	
KIA–6916	403	9580±45	10301	
KIA–6917	456	10187±60	11135	171
KIA–6918	562	10360±55	11331	

Table 6.1 (continued)

1	2	3	4	5
PM9499–2°	0		0	
KIA–1794	24	2140±30	1768	4
KIA–3115	29.5	6510±50	7027	
KIA–1795	46	8990±50	9625	
KIA–1796	77	9470±70	10276	
KIA–1797	94	9590±50	10303	
KIA–3117	116	8870±50	9458	31
KIA–1793	122	8660±50	9047	
KIA–1799	149	10090±50	11102	
KIA–1817	157	10140±50	11120	
PS51/141–2	0		0	
KIA–7932	15	620±35	288	11
KIA–7933	112	9040±70	9789	
KIA–7934	143	9910±60	10804	
KIA–7935	175	9830±70	10612	194
KIA–7936	395	10320±70	11251	
KD9502–14°	0		0	
KIA539	3	recent	0	3
KIA540	23	6420±30	6930	
KIA541	31	6440±50	6954	
KIA542	41	6640±30	7210	
KIA543	59	6790±40	7333	
KIA544	85	6630±50	7203	
KIA545	107	7340±50	7822	101
KIA546	119	7610±70	8104	
KIA547	139	7700±70	8173	
KIA548	157	7900±40	8335	
KIA549	171	8330±50	8885	
KIA550	185	8300±60	8866	
KIA551	225	8420±80	8936	
PS51/092–12	0		0	
KIA–6877	2	590±25	273	
KIA–6878	64	1505±35	1078	41
KIA–6879	160	1680±35	1267	
KIA–6880	300	6725±40	7270	
KIA–6881	402	7280±45	7754	176
KIA–6882	500	7950±55	8408	

Table 6.1 (continued)

1	2	3	4	5
PS51/092–13	0		0	
KIA–6883	13	530±45	243	
KIA–6884	36	615±25	285	
KIA–6885	95	2050±45	1671	70
KIA–6886	172	3735±35	3685	
KIA–6887	269	3860±35	3851	
KIA–6888	356	6520±40	7062	
PM9462				
KIA–3137	5	recent	0	
AAR–2702	27.5	900±55	525	
KIA–3138	90	2090±30	1700	
KIA–3139	120	2680±30	2356	
AAR–2703	158	3235±65	3091	
KIA–3140	195	4050±40	4118	
AAR–2704	241	4620±65	4845	50
AAR–2705	270	5060±70	5455	
KIA–3141	315	5920±40	6353	
AAR–2706	349	6480±75	6999	
AAR–2707	406	7120±75	7610	
KIA–3142	445	8350±50	8897	
PS51/080–13	0		0	
KIA–6873	41.5	1910±25	1503	28
KIA–6874	71.5	1940±25	1522	
KIA–6875	142	4795±30	5097	
KIA–6876	202	5950±35	6393	46

Notes: ^{14}C age according to Fahl and Stein, 1999; Bauch et al., 1999; AAR – Aarhaus University; KIA – University of Kiel

70–60 m high. The first terrace 2 km wide is composed by loams and sands with admixture of pebbles (overall thickness varies from 6 to 40 m). The deposits enclose mollusks *Saxicava arctica f. pholadis* L., *S. arctica f. arctica* L., *Mya truncate var. uddevalonsis* H., *Astarte borealis var. placenta* Morch.

The second terrace 2.5 km wide is composed by brown clays with fine pebble interbeds, which are overlain by greenish-gray clays. Some sand beds also occur. Thickness is 15–20 m. The following mollusks were found: *Pecten islandicus* Mull., *Cardium grønlandicum* Chem., *C. ciliatum* Fabr., *Astarte borealis* Chemn., *A. borealis* var. *placenta* Morch., *A. crenata* (Grey.), *Leda pernula* Mull., *Arca glacialis* Grey., *Mya truncate* H., *Saxicava arctica* L., *Natica grønlandica* Beek., *Margarita cineree* (Goth.).

The described molluscan assemblages are ecologically different. Warmer water fauna characterizes the terrace 60–70 m high. The terraces may be of different age, as somewhat confirmed by the OSL dating.

In the Shokalsky and Neupokoyev islands of the archipelago the Kazantsev sediments are locally up to 50 m thick.

A terrace 30–40 m high is supposed to be formed in the Kargin time. Brown clays, sands and pebbles up to 10 m thick enclose mollusks *Astarte borealis* Chemn., *A. borealis* var. *jeniseae* Sachs, *Astarte compassa* L., *Macoma calcarea* (Chemn.), *May truncate* H., *Saxicava arctica* L.

Few data was published on the Upper Pleistocene and Holocene sediments of the East Siberian Sea coasts. These sediments are best documented in the coast of the Chaun Bay (Veynsberg, 1984; Bezrodnykh and Nazarov, 1984; Valpeter, 1984) where three rhythmic members of similar structure were distinguished in the Quaternary sediments (Bezrodnykh and Nazarov, 1984). The members are composed of marine sediments and subaerial deposits.

The lower rhythmic member is mainly formed by lagoonal sediments represented by thin- and fine-grained sands and silts up to 20 m thick, which are overlain by lacustrine-boggy and alluvial sediments. The sediments compose a terrace 20–25 m high. They are correlated with the Enmakay Formation of the Valkarayskaya Lowland, which has the Eo-Pleistocene age according to paleomagnetic data.

Assignment of the second rhythmic member to the Kazantsev time is conventional and poorly documented. Sediments of the third rhythmic member constitute several terraces from 5 to 13 m high. Basal lagoonal and deltaic deposits are overlain by subaerial ones. Among the latter there are ice-saturated dusty loams and little icy eolian sands and sandy loams. Age of the marine sediments was determined by the radiocarbon dating of drift-wood buried in coastal bars. The dates for the drift-wood found on the western coast of the Nolde Inlet appeared to be 33800±560 and 34600±510. A peat bed of the lagoonal sediments in the 5–10-m terrace of the Valkarayskaya Lowland yielded the dates of 33200±300 and 33700±780. Taking into consideration the sediment types, the dates indicate the final stage of the Kargin sea transgression. The lower part of the loess-ice cover overlying the marine sediments yielded a date of 22980±640 and corresponds to the Sartan Glaciation.

Thus, the dates obtained suggest that there may be a still undiscovered subaerial member between the sediments of the terminal phase of the sea transgression and the subaerial sediments of “edoma” type.

Marine terraces 3.0–3.5 m high were formed during the Holocene Flandrian transgression; the present-day sea level was reached 4–5 ka ago. Some data on thermokarst development is of interest. Short-term episodes of thermokarst formation related to climatic warmings were documented by the radiocarbon dating: 11000, 9400, 8900–8200 (up to six thermokarst manifestations were recorded for every 100–200 years in this interval), 7000 and 5700 (Valpeter, 1984). The dates suggest that the most intensive thermokarst processes took place in the Boreal time.

Lower Pleistocene marine and continental sediments of the Chukchi Peninsula coasts were described in the monograph of O.M. Petrov (1966), who distinguished the Valkatlen, Konergin, Vankarem, Amguema and Iskatel beds.

The marine Valkatlen, alluvial and lacustrine Konergin beds represent interglacial deposits of the early Late Pleistocene.

The Valkatlen deposits constitute marine terraces 30 m high. They were accumulated in the upper sublittoral and coastal-marine zones. The sublittoral deposits are represented by yellowish-gray fine-grained sands with lenses and interbeds of silty sand and silt. The sand contains molluscan shells: *Macoma calcarea, May truncate, M. truncate ovata, Gomphina fluctuosa, Mytilus edulis, Serripes grønlandicus, Astarte borealis, A. alaskensis, Axinoxida orbiculata, Buccinum* cf. *glacialis, Natica clausa, Lepeta concentrica, Buccinum baaeri, Margarites helicinus*. The coastal-marine sediments are usually sand-gravel, pebble, and locally boulder formations. Thickness of the Valkatlensky deposits reaches 20–25 m.

The Konergin deposits are predominantly sands, locally loams, with peat interbeds and lenses and wood fragments. These sediments lie on erosional surface of the marine Valkatlen sediments. Basing on palynological studies of few samples, O.M. Petrov (1966) suggested an existence of forest-tundra landscapes at the Konergin time. Moreover, opinions were raised that this time was marked by the interglaciation optimum. These ideas, however, are poorly substantiated because of scarcity of palynological material. Thickness of the Konerginsky sediments reaches 20 m.

The Vankaremsky beds are glacial and aquaeoglacial sediments widespread in the Chukchi Peninsula, especially in the Vankaremskaya Lowland. They lay on the eroded marine sands of the same age and on the Konerginsky continental deposits.

The Amguema deposits form two terraces. The first one is inserted into the fluvioglacial sediments of the Vankarem Glaciation along the river valleys (Petrov, 1966) and passes into the marine terrace at the river mouths. Sands of the latter terrace contain the following mollusks: *Macoma calcarea, Mya pseudoarenaria, Astarte borealis, A. borealis arctica, A. alaskensis, Serriges grønlandicus, Mytilus edulis, Macoma baltica.* The Amguema beds were referred to the second interglacial period of the Late Pleistocene (Petrov, 1966). Probably the radiocarbon dates of 33350±530 and 29150±110 were obtained for these sediments. Wood fragments from alluvium of the second flood-plain terrace at the Amguema River yielded a date of 9350±230. Further studies are needed to refine its relationships with the marine terrace.

The Iskatel beds are glacial deposits widespread in the upper river courses of mountains and piedmonts. In the areas of their distribution, well pronounced glacial landforms are represented by terminal moraine ridges and by hills of subglacial ablative moraines. In the river valleys the first flood-plain terrace is located inside the Iskatel deposits. O.M. Petrov (1966) related the glaciation of the Kara River valley to the end of the Late Pleistocene, although its Holocene age cannot be excluded. As this author supposed, the initial phase of the Iskatel Glaciation was characterized by the river incision processes and formation of a scarp of the second flood-plain terrace.

The first flood-plain terrace (3–4 m high) and floodplain deposits are commonly related to the Holocene. The Holocene marine sediments form marine and lagoon terraces (5–7 m) and some accumulations of lower level. Formations of higher level are dated in the range of 9–6 ka ago (Svitoch, 2003).

Chapter 7
The Deglaciation Time and Holocene of Northern Eurasia

Numerous palynological studies carried out for the last 50 years allowed reconstruction of Holocene paleolandscapes and paleotemperatures (Grichuk, 1950, 1961; Katz, 1958; Neishtadt, 1957; Khotinsky, 1977; Spiridonova and Lavrushin, 1997; and others). A lot of publications devoted to Northern Eurasia dealt with migrations of plant assemblages, relationships between the forest and tundra zones as a result of climatic changes, atmosphere evolution, sea level changes, dynamics of bog formation, geological history of Anthropogene communities inhabiting Northern Eurasia in the Holocene, etc. All the questions need to be especially analyzed outside the framework of this book. Therefore only several points of the geological history of the Arctic coast, which are directly related to the problem under consideration, are discussed below.

These are, first of all, essential changes in relationships between the land and sea areas in the Holocene of northeastern Russia, particularly, the appearance of the Laptev and East Siberian seas. In this region, thick cryogenic icy deposits of Quaternary age were partly destroyed by thermoabrasion in the course of the glacioeustatic and Flandrian transgressions. They are preserved in the Lyakhov and partly Anzhu islands. Formation of the estuaries, such as the Ob Inlet, gulfs of Yenisei, Khatanga, Yana and Chaun Inlet, were probably brought about by the last transgression. Thick delta fans were formed due to intensive deposition by the Lena, Yana and Indigirka rivers.

In the European north, the situation is somewhat different. The glacial depressions (Throat of the White Sea and Czech Bay), which are occupied by the sea now, can be clearly outlined.

At the lower course of the Pechora River, which is the only large river flowing into the Barents Sea, the picture is more complex. There are the Gulyaevskiye Koshki Islands, which represent washed-out fragments of the Pra-Pechora delta appeared due to a catastrophic breakthrough of the glacier-dammed Pechora Lake. Later this delta was washed away by the Flandrian transgression, which contributed to the formation of the Pechora Inlet. The Khaypudyrskaya and Baydaratskaya inlets can be also considered as glacial depressions. Although the glacial depressions appeared in the terminal Late Pleistocene, the present configuration of the shelf seas were formed in later times, including the Holocene.

M.A. Levitan, Y.A. Lavrushin, *Sedimentation History in the Arctic Ocean and Subarctic Seas for the Last 130 kyr*, Lecture Notes in Earth Sciences 118,
DOI 10.1007/978-3-642-00288-5_7, © Springer-Verlag Berlin Heidelberg 2009

In the north of European Russia, in the northern margin of the Bolshezemelskaya Tundra, just as in that of West Siberia, some beds of buried gletcher ice of the last glaciation were found. Ice melting may produce lake basins near the proximal part of the terminal moraine. Under the favorable conditions the moraines may be broken through by waters of the lake basins. The flows saturated by washed out till become mud-and-gravel streams (sel). Such processes took place in the Markhida mouth area, where the moving sels acquired features of subglacial moraines. Moreover, the high-density currents deform their beds by seizing fragments of frozen rocks, which can be taken for small glacial detached masses. Wood fragments from the mud flow sediments of the Markhida area were dated approximately as 9.3 ka ago (one date is 9.8 ka ago). The dates suggest that the mud flow arose in the Preboreal stage. Sediments of another mud flow of younger age were also discovered in the same area. The first people who studied these formations relate them to surge (Lavrov and Potapenko, 2005). From examination of the Markhida section a conclusion was made that these moraine-like formations containing trunks of trees were similar in many lithogenetic features to glacial sel accumulations. The Early Holocene age and large diameter (up to 15 cm) of tree trunks doubted correctness of the genetic interpretation of the wood-enclosing sediments as the subglacial moraine. A forest (even rarefied) hardly grew in immediate proximity of the glacial edge in the northers latitudes of Russia.

Part II
Marine Sedimentation in the Arctic Ocean and Subarctic Seas

Chapter 8
The Seas of West Subarctic Region

The Western Subarctic region includes Iceland, Norwegian and Greenland seas which are commonly referred to as the Norwegian-Greenland Basin in Russian, and the Nordic Seas in English (Vogt, 1986). The studied region is constrained from south by Iceland, the Denmark Strait and the Faeroe-Iceland sill, from west – by Greenland, from north – by the Fram Strait, from east – by Spitsbergen, the Barents Sea and Scandinavia.

Geologic and Oceanographic Setting

The Norwegian-Greenland Basin (Schäfer et al., 2001) consists of some deep-sea basins (Norwegian, Lofoten, Greenland, and Boreas), divided by system of mid-oceanic ridges (Kolbeinsen, Mohn, and Knipovich) and major fault zones (Greenland and Jan-Mayen) (Fig. 8.1). Depth of the Greenland Deep exceeds 2000 m whereas the Lofoten and Norwegian deeps have a water depth of about 3000 m. The Egir Ridge, an old spreading center, the Iceland Plateau and the micro-continent Jan Mayen are related to this part of the sea floor. Deep-water areas are surrounded by continental margins that consist of shelves as deep as up to 400 m, continental slopes (the slope on a traverse of Central Norway is complicated by the marginal Vøring Plateau) and continental rises with common occurrence of fans in the east of the basin. An extremely wide development of fjords on inner shelves of Norway, Spitsbergen and Greenland should be noted.

Climate of the Norwegian-Greenland Basin is mainly defined by thermohaline circulation (Fig. 8.2) and ice conditions. The northwest is covered by all-year-round ice, while in the eastern part the ice cover is restricted to winter times (Hald, 2001). Such ice conditions are substantially caused by development of East-Greenland current carrying cold Polar surface waters from the Arctic ocean to the south, and Norwegian current bringing warm Atlantic surface waters from the North Atlantic to the north (to the west of Spitsbergen the current is called the West-Spitsbergen current). The Arctic waters are located in a border zone between the above water masses and circulate on the surface as anticyclonic gyre. Interaction between main surface water masses and addition of cold saline and dense waters forming at the

M.A. Levitan, Y.A. Lavrushin, *Sedimentation History in the Arctic Ocean and Subarctic Seas for the Last 130 kyr*, Lecture Notes in Earth Sciences 118,
DOI 10.1007/978-3-642-00288-5_8, © Springer-Verlag Berlin Heidelberg 2009

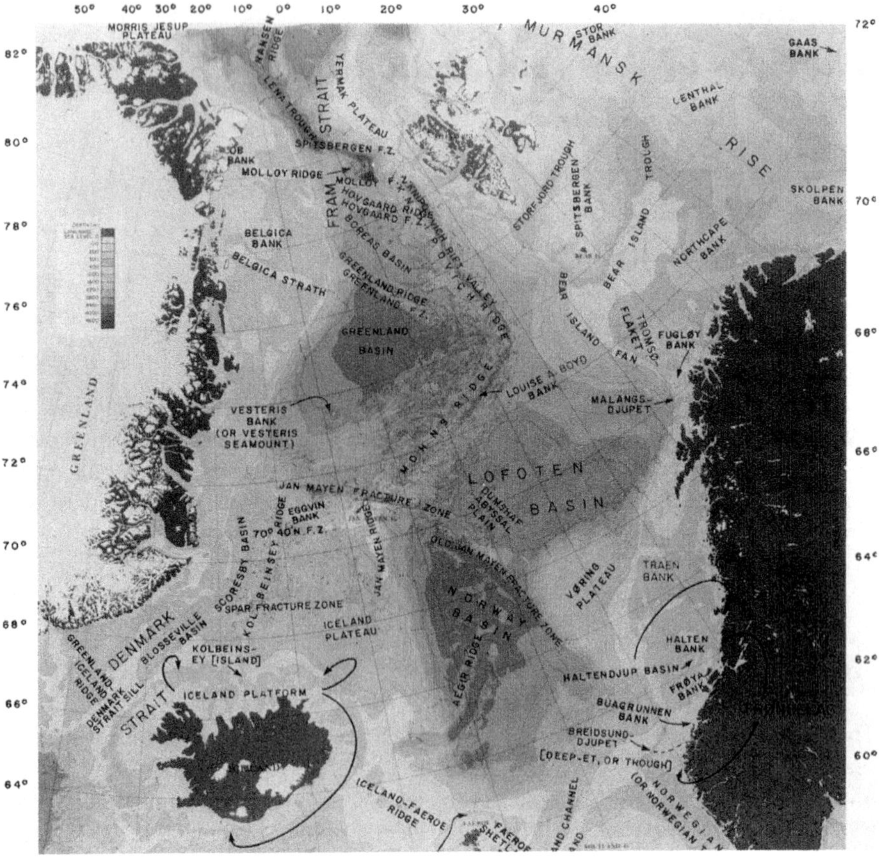

Fig. 8.1 Scheme of the Nordic Seas bottom topography (Vogt, 1986)

surface of open-water areas, on shelves and in the Barents Sea during ice formation, results in active convection and vertical circulation and formation of main water masses of the basin.

The Atlantic water has salinity not less than 35‰, and its temperature shifts from 3 up to 8°C in winter and from 6 up to 10°C in summer (Johannessen, 1986). The Polar water extends from surface down to approximately 150 m. From top to bottom of this water mass its salinity varies from 30 up to 34‰, the temperature changes from 1.5 up to 0°C. In intermediate waters salinity of the Polar water varies between 34 and 34.7‰. Changes of salinity between 34.8 and 35.0‰ are characteristic of the Arctic water. The temperature varies between 0 and 2°C. Deep waters of the Norwegian and Greenland seas have salinity about 34.92‰ and negative temperature. The coastal Norwegian current carries warm fresh-mixed (32–35‰) waters, and its demineralization is related to an active contribution of fresh water from the fjords during spring and summer seasons. The borders between water masses at the

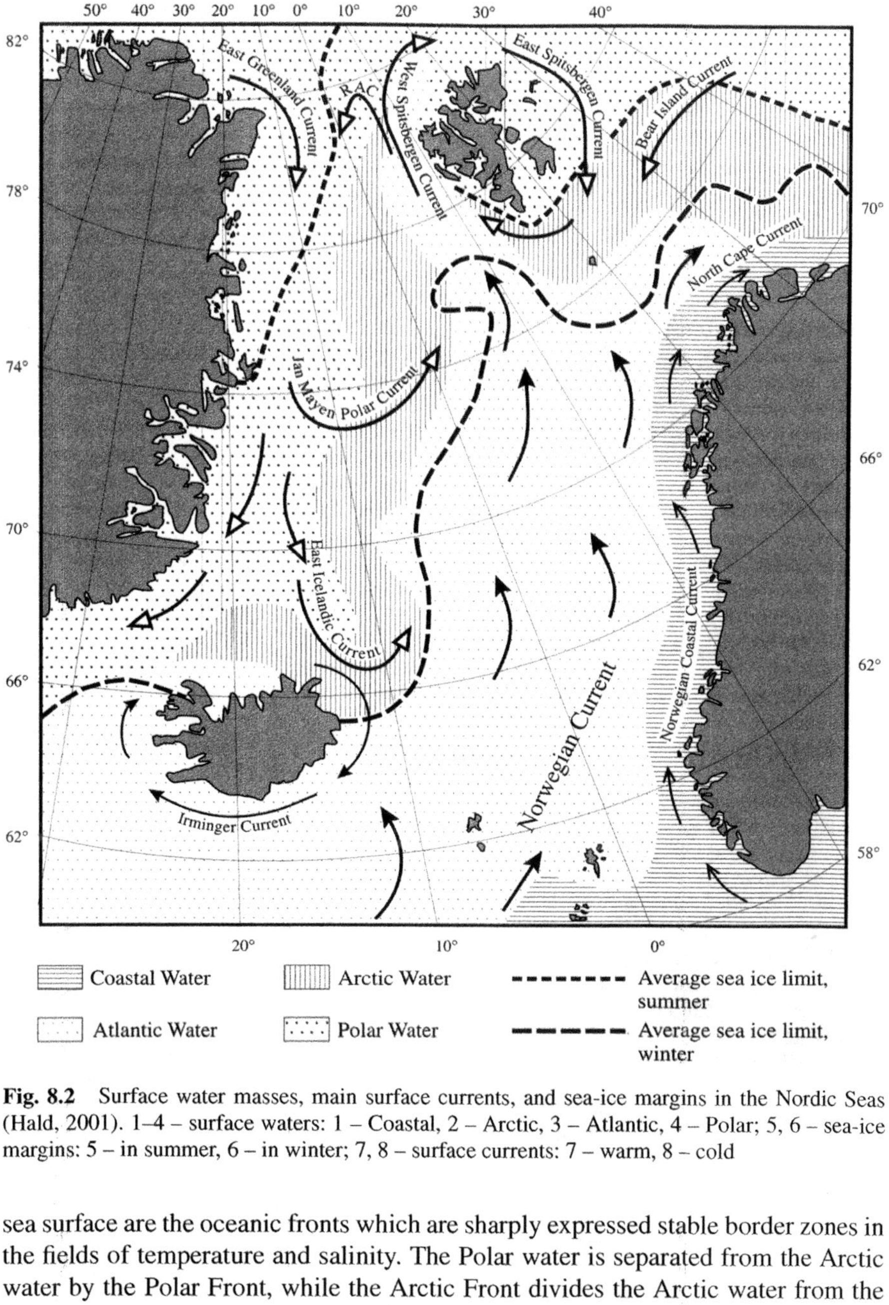

Fig. 8.2 Surface water masses, main surface currents, and sea-ice margins in the Nordic Seas (Hald, 2001). 1–4 – surface waters: 1 – Coastal, 2 – Arctic, 3 – Atlantic, 4 – Polar; 5, 6 – sea-ice margins: 5 – in summer, 6 – in winter; 7, 8 – surface currents: 7 – warm, 8 – cold

sea surface are the oceanic fronts which are sharply expressed stable border zones in the fields of temperature and salinity. The Polar water is separated from the Arctic water by the Polar Front, while the Arctic Front divides the Arctic water from the Atlantic water. It is of interest that East Icelandic current stops in the area of the Arctic Front. It has to be mentioned that the deep waters of the Norwegian-Greenland Basin give rise to deep waters of the North Atlantic and as a whole to global saline conveyor (Broecker, 1991).

The Norwegian-Greenland Basin can clearly be subdivided into two parts: the smaller northwestern (western) part and the larger southeastern (eastern) part. The cold less saline Polar water mass dominates in the west while the warm and more saline Atlantic water occupies the east. This pattern is also reflected in composition of surface sediments (Fig. 8.3). Terrigenous siliciclastic mud dominates in surface sediments in the west. They hardly contain any remnants of carbonate organisms (mainly foraminifers) and include coarse grained ice-rafted debris (IRD). In the west a similar mud is essentially enriched by biogenic carbonate, down to forming marly lithologies (30–50% $CaCO_3$), and there are only minor amounts of IRD (Matthiessen et al., 2001). Detailed studies show that $CaCO_3$ distribution is caused mainly by particularities of production and dilution of this component, but not by dissolution which is expressed poorly (Henrich, 1998). It is of interest that even far to the north, on the Yermak Plateau, already outside of the studied basin, the branches of Atlantic water are reflected in enrichment of surface sediments by rem-

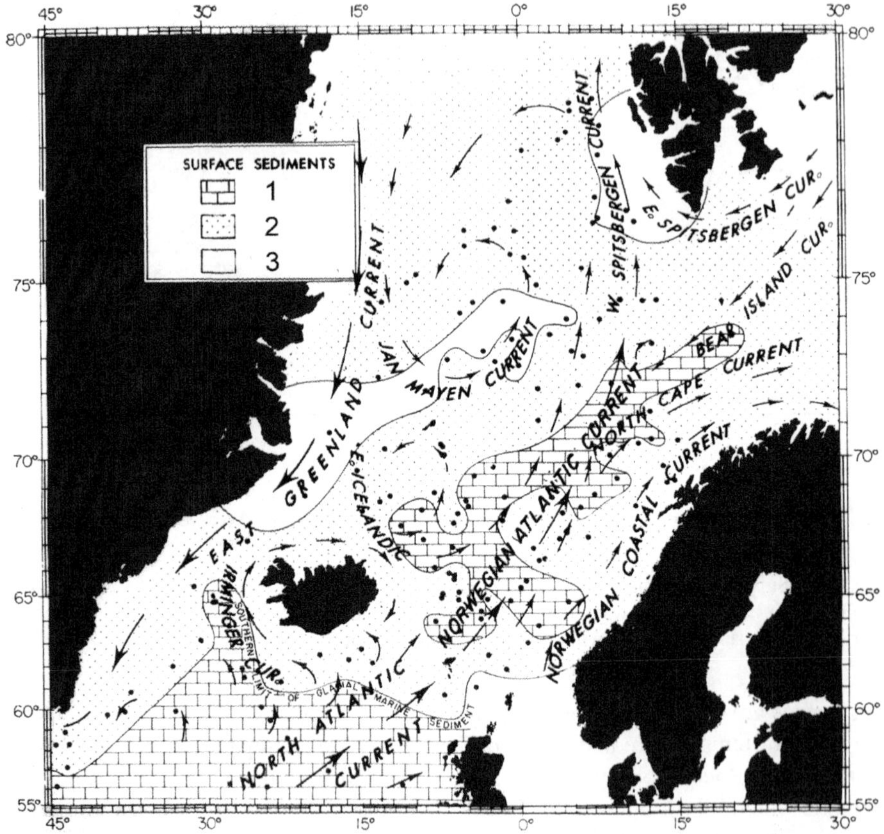

Fig. 8.3 Lithology of surface sediments in the Nordic Seas (Vogt, 1986). 1 – marly mud without IRD; 2 – sandy clay; 3 – clayey mud

nants of planktonic foraminifers (Levitan et al., 2000b). The morphological partition of the deep-water floor pointed above mainly influences the grain-size composition of surface sediments. The more coarse-grained sediments are developed due to enhanced hydrodynamical activity of bottom water above the ridges, plateaus and rises compared to the deep-sea basins. The noticeable impurity of volcanogenic material, even to formation of tuffites, is present in surface sediments of the Iceland Plateau and Kolbeinsen Ridge. Iceland is a main source of volcanogenic matter in the region, and the related ash is found in the Pleistocene-Holocene cross-sections even in the Fram Strait area (Wallrabe-Adams and Lackschewitz, 2003).

The raised primary production is related to the marginal areas of sea-ice fields (Peinert et al., 2001). In sediments formed at perennial sea-ice covered areas, Corg (TOC) content is rather low. Fluxes of organic carbon in Holocene sediments (Taylor et al., 2002b) reveal a clear circumcontinental zonation according to which the raised mass accumulation rates (MAR) of TOC gravitates towards the continents surrounding the basin and the large islands.

Numerous studies are devoted to analysis of various facies features: ice-rafted debris, various minerals, chemical and isotope sediment composition, physical properties, and biogenic remnants. The ice-rafted debris and oxygen and carbon isotopes in foraminifers are mostly studied in details. The results of such studies are subsequently used in analysis of cross-sections of Quaternary sediments for a reconstruction of paleoclimatic and paleoceanographic features of the region.

IRD is determined by researchers differently. For example, it may include grain-size fractions >150 mkm, or >250 mkm, or >400 mkm, or >500 mkm. IRD is studied mostly in the fraction >500 mkm by J. Bischof (2000). For example, his component analysis of 23000 samples (139 728 grains) from Pleistocene-Holocene sediments of the Norwegian Sea (Fig. 8.4) showed that the most grains are sedimentary rock (27.54%), igneous and metamorphic rock (30.86%), quartz (26.18%) and feldspar (12.46%). He studied possible source provinces and transport pathways of IRD in detail.

M.A. Levitan compiled schematic maps distribution of smectite, illite, chlorite and kaolinite in the fraction < 2 mkm in surface sediments of Norwegian-Greenland Basin based on raw data of H. Berner (1991). One can clearly see on the distribution map of smectite (Fig. 8.5) that its main sources are related to Iceland. The Faeroe-Iceland sill, the Jan-Mayen fault zone, rock exposures on the Egir Ridge, and southeastern Greenland play a subordinate role. The listed regions are areas of basaltoid development which were subjected to smectite mineralization. The main sources of kaolinite are Scandinavia, the Barents Sea and Spitsbergen. Chlorite enters approximately from the same sources, and also from east and northeastern parts of Greenland. The illite distribution pattern is the most complex. It enriches sediments of the Iceland Sea, the deep water regions of the Norwegian Sea and continental margin of Spitsbergen.

It is important to pay attention to the large role of oxygen isotope data in the tests of planktonic foraminifers, in particular, for the agreement of temperatures of surface water mass, obtained by interpretation of oxygen isotopes, and by direct measurements (Sarnthein et al., 2003).

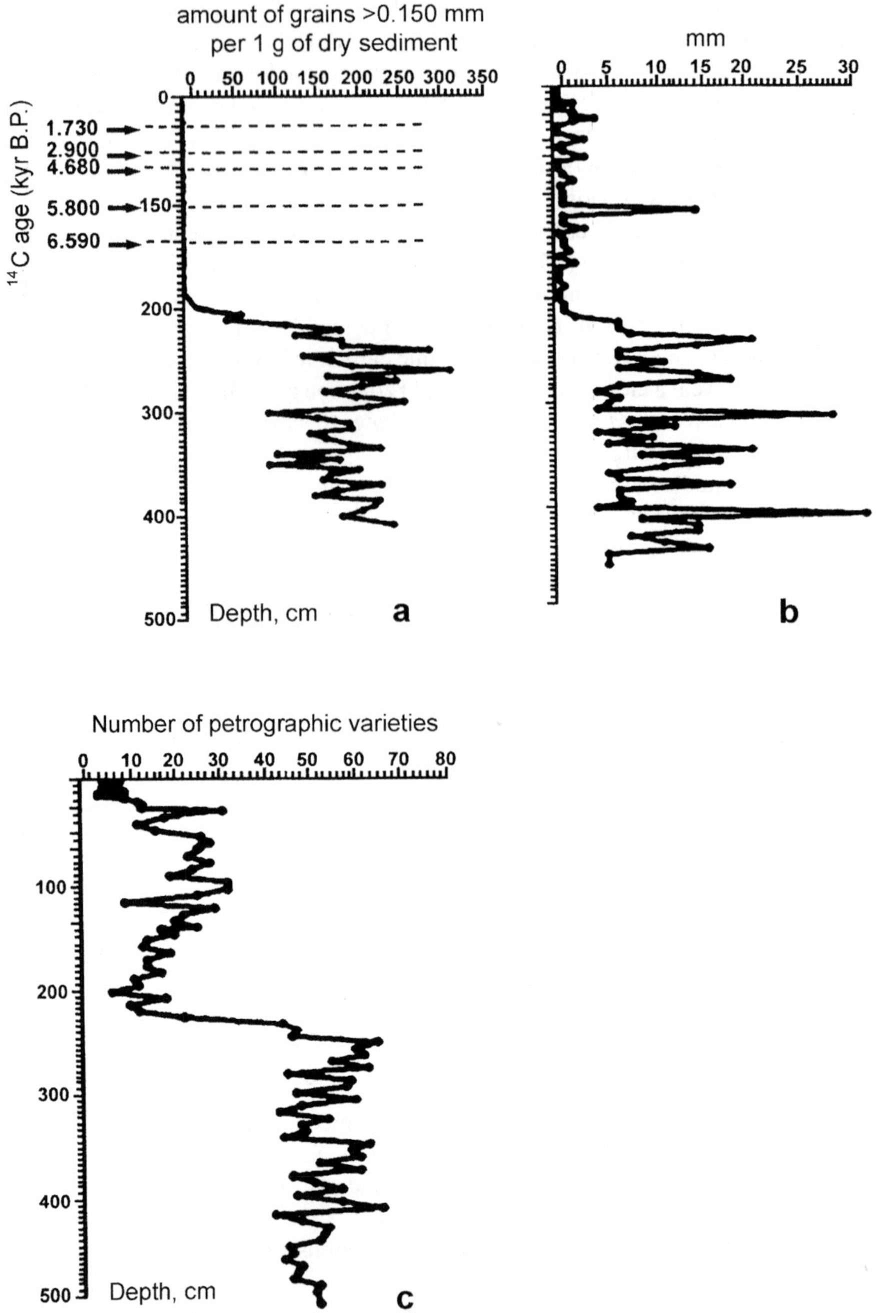

Fig. 8.4 Amount of IRD (*a*), size of its largest fragments (*b*), and its lithological variability (*c*) in the Holocene and Upper Pleistocene sediments of the Nordic Seas (Bischof, 2000)

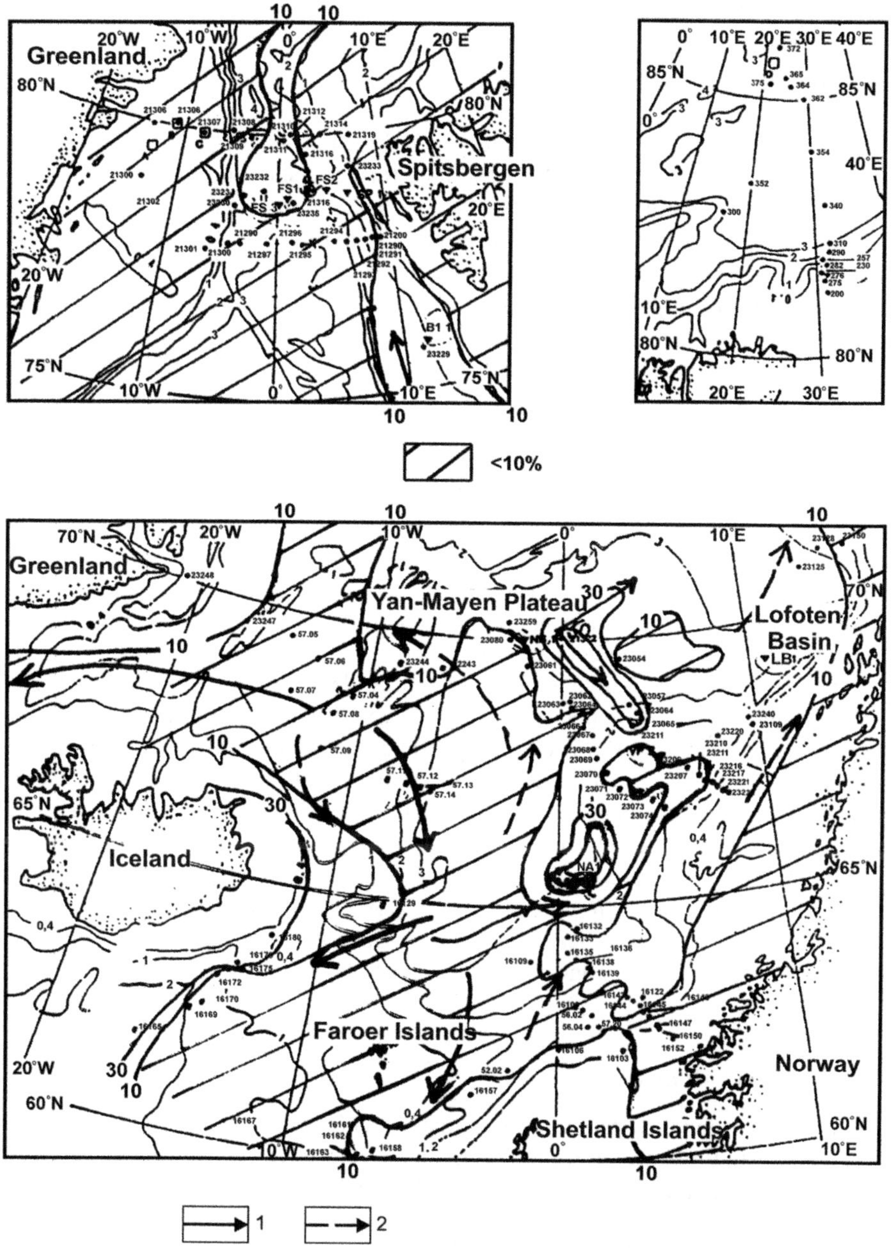

Fig. 8.5 Maps of smectite distribution in the surface sediments of the Nordic Seas (compiled by M.A. Levitan, based on data from (Berner, 1991)). 1 – bottom currents, 2 – surface currents

History of Sedimentation

History of Sedimentation Rates

The value of quantitative parameters such as sedimentation rate (SR) and mass accumulation rate (MAR) is hard to overestimate for understanding the history of sedimentation. The Nordic Seas sediments play an important role as the archive of glacial-interglacial changes, which occurred both in the basin and on the surrounding continents and large islands. Therefore this region has been intensively studied in the course of the past three decades. A number of scientific cruises, including four deep-sea drilling expeditions, have been carried out. As a result, about 200 sediment cores have been stratified by oxygen-isotope method and radiocarbon method (accelerating mass-spectrometry; AMS ^{14}C). In a number of cases, also biostratigraphy, lithostratigraphy, magnetic stratigraphy and ^{10}Be-stratigraphy were used.

Stratigraphy serves as a basis for age model development and determination of sedimentation rates and mass accumulation rates. As a rule, authors of such stratigraphic works attempt to interpret the obtained data for explaining the most general regularities governing the history of sedimentation. However, there is a problem of limited number of sediment cores available for such studies. Actually, some of the described phenomena are of local propagation; quite often the proposed by different authors regularities contradict each other.

Until recently there were practically no works, dedicated to the most general regularities governing sedimentation history of the Norwegian-Greenland basin. The only exception is the work (Taylor et al., 2002a), in which schematic maps of sedimentation rates for the Holocene and Last Glacial Maximum (LGM) were published, using data available in the PANGAEA repository for georeferenced data from Earth system research (http://www.pangaea.de/). However, maps of sedimentation rates for the older epochs of the last glacial-interglacial cycle are still absent. Furthermore, there are differences between large regions such as, for example, the Norwegian Sea, Greenland Sea, Greenland continental margins and others, not studied yet. This gap in knowledge is filled by the study described below.

Materials and Methods

This work includes data on approximately 160 sediment cores, which have a reliable stratigraphic breakdown. Location of cores is shown at Fig. 8.6. Most data is obtained from (Henrich et al., 1989; Vogelsang, 1990; Spielhagen, 1991; Lackschewitz, 1991; Dokken, 1995; Sarnthein et al., 1995, 2003; Nam, 1997; Struck, 1997; Goldschmidt, 1997; Hebbeln et al., 1998; Henrich, 1998; Völker, 1999; Hald et al., 2003) and many others. The data are summarized in Table 8.1.

Using AMS ^{14}C data, we tried to work in a scale of calendar age, where the values are obtained from reservoir radiocarbon datings corrected to the age in accordance with the procedure, described in (Stuiver et al., 1998). Age boundaries

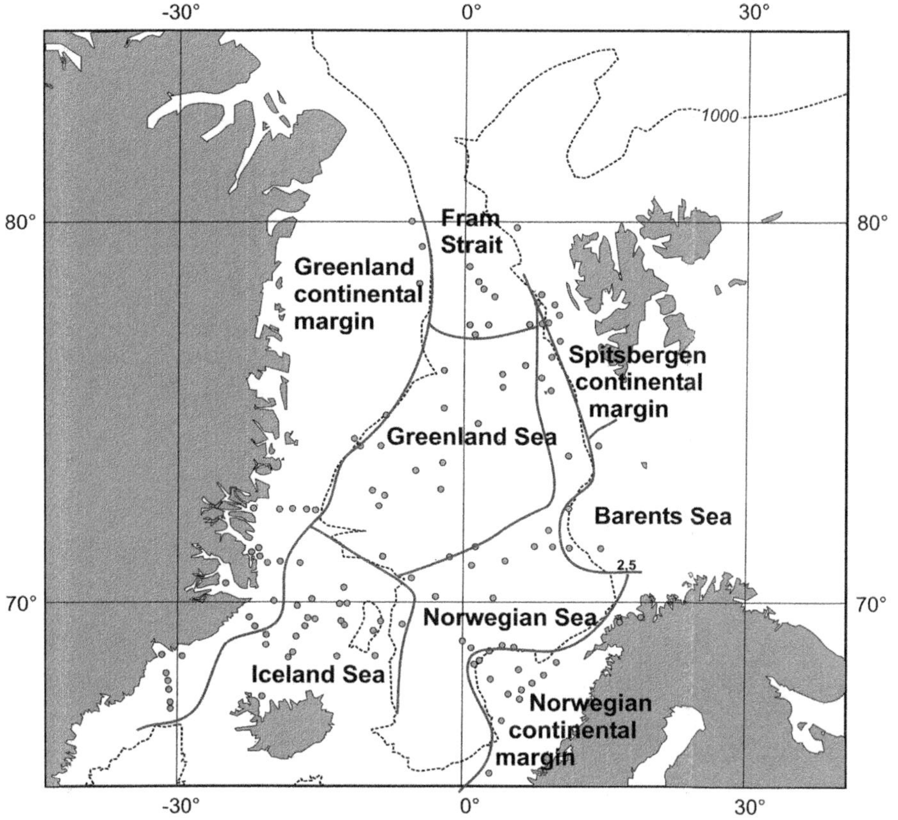

Fig. 8.6 Location of sediment cores with established stratigraphy in the Nordic Seas

between the distinguished marine isotope stages (MIS) are taken from (Martinson et al., 1987). There are several interesting publications, for example, (Sarnthein et al., 2003; Shackleton et al., 2002), in which the authors establish a new age-qualification boundaries between some MIS, which differ from (Martinson et al., 1987). However, we carried out calculations based on the age boundaries according to (Martinson et al., 1987), because authors of the relevant publications used this scale. Secondly, the new age boundaries have not been recognized officially.

Since the gravity corers often destroy near-surface sediments, for SR calculations we tried to use summary core depths (taking into account data of box corers and gravity corers at one station). In accordance with (Backman et al., 2004), reservoir age has possibly reached 1000 years during glaciation epochs, when ventilation of bottom water in the deep-water basins was poor. This means that actual SR for such epochs might be somewhat higher than it follows from our calculations.

Comparison of SR and MAR variations with time is found to be of interest. We expected that real SR values should be slightly higher than the calculated ones

Table 8.1 Sedimentation rates (cm/ky) in the Norwegian-Greenland Basin during the last five marine oxygen isotope stages (MIS) (compiled by M.A. Levitan)

Sediment core	Latitude, N	Longitude	Water depth, m	MIS 1	MIS 2	MIS 3	MIS 4	MIS 5a–5d	MIS 5e	MIS 5	Reference
Spitsbergen continental margin											
NP 90–36	77°37.04'	09°56.18'E	1360	16.67	44.24	–	–	–	–	–	Hald, 2001
NP 90–38	77°31.54'	08°23.04'E	2328	–	–	–	–	4.78	more 14.55	6.7	Hald, 2001
NP 90–12	78°24.47'	09°24.87'E	683	5.5	22.73	–	–	–	–	–	Hald., 1995
PCM 7	78°37'	07°54'E	1073	11.25	31.08	–	–	–	–	–	Lloyd et al., 1996
PCM 5	78°29'	07°43'E	2139	7.5?	7.89?	–	–	–	–	–	Lloyd et al., 1996
NP 90–21	78°11.30'	09°56.56'E	322	53.04	–	–	–	–	–	–	Elverhøi et al., 1995
PS 1290	78°00.8'	08°43.2'E	1522	5.83?	>13	–	–	–	–	–	Elverhøi et al., 1995
Barents Sea continental margin											
M 23258	75°00'	14°E	1768	27.25	35.06	–	–	–	–	–	Sarnthein et al., 2003
MD 95–2012	72°00'	11°E	2200	8.59	44.05	44.64	–	–	–	–	Sarnthein et al., 2003
T–88–2	71°55.3'	14°21.5'E	1500	21.67	79.17	–	–	–	–	–	Koch et al., 1996
Norway continental margin											
JM 98–1–PC	69°29.9'	18°23.6'E	213	210.04	336.59	–	–	–	–	–	Hald et al., 2003
M 23055	68°25.4'	04°01.3'E	2311	3.33	4.23	2.86	2.5	–	–	1.51	Henrich et al., 1989
M 23062	68°43.7'	00°10.1'E	2244	2.08	3	2.66	0.92	–	–	1.54	Henrich et al., 1989
HM 79–6/4	62°58.0'	02°42.0'E	850	20.83	50	–	–	–	–	–	Koç et al., 1996
MD 95–2011	66°58'	07°38'E	1048	59.13	–	–	–	–	–	–	Calvo et al., 2002
JM 96–42/1	67°23.3'	08°46'E	1068	41.67	–	–	–	–	–	–	Laberg et al., 2002
JM 96–45/1	67°52'	09°44'E	945	3.75	59.01	–	–	–	–	–	Laberg et al., 2002
M 23544	67°25.0'	08°32.8'E	1223	45.45	–	–	–	–	–	–	Laberg et al., 2002
M 23545	67°21.2'	08°18.5'E	1158	50.86	–	–	–	–	–	–	Laberg et al., 2002
MD 95–2010	66°41.05'	04°33.97'E	1226	2.9	48.3	13.1	–	–	–	–	Dokken and Jansen, 1999
JM 98–625/1	67°07'	07°09'E	1342	1.7	25.8	5.7	–	–	–	–	Dahlgren and Vorren, 2003
JM 98–626/1	66°24'	05°58'E	733	0.8	55.36	–	–	–	–	–	Dahlgren and Vorren, 2003
JM 98–628/1	66°49'	06°18'E	1068	0.8	34.3	–	–	–	–	–	Dahlgren and Vorren, 2003
T–79–51	69°18.00'	16°23.00'E	505	26.67	–	–	–	–	–	–	Hald et al., 1996
M 23063	68°45.0'	00°00.0'E	2299	1.5	2.08	–	–	–	–	–	Vogelsang, 1990
M 23071	67°05.10'	02°54.50'E	1308	4.17	15.42	6.71	3.33	3.78	1.8	3.39	Vogelsang, 1990
M 23074	66°40.0	04°54.30'E	1157	9.17	17.08	10.43	–	–	–	–	Vogelsang, 1990
M 23199	68°22.53'	05°13.48'E	1968	2.08	8.33	3	–	–	–	–	Vogelsang, 1990
M 23205	67°36.58'	05°45.57'E	1411	3.33	4.17	10	3.75	–	–	2.05	Vogelsang, 1990
ODP 642	67°13.5'	02°55.7'E	1286	–	–	–	–	–	–	2.05	Henrich, 1989
ODP 643	67°42.9'	01°02.0'E	2753	–	–	3.57	1.07	–	–	2.5	Henrich, 1989
M 23068	67°50.0'	01°30.3'E	2230	3.42	6.07	3.84	1.75	–	–	2.39	Henrich et al., 1989
M23411	65°47.9'	03°30.6'W	2849	3.59	–	–	–	–	–	–	Hass et al., 2002
V 27–85	67°30'	4°E	1500?	–	>12.5	3.43	2.14	1.71	6.82	2.41	Henrich, 1998

Table 8.1 (continued)

					Norwegian Sea						
M 17730	72°03'	07°19'E	2769	8.83	6.17	4.62	–	–	–	–	Sarnthein et al., 1995
M 23259	72°02'	09°16'E	2518	6.5	12.25	5.95	–	–	–	–	Sarnthein et al., 1995
HM 94–13	71°38'	01°38'W	1946	2.5	2.33	–	–	–	–	–	Sarnthein et al., 1995
MD95–2012	72°	11°E	2100?	8.63	49.33	–	–	–	–	–	Weinelt et al., 2003
M 17732	71°36.8'	04°12.8'E	3103	8.17	4.75	2.71	1.79	–	–	2.91	Bauch, 1997
M 23359	65°31.7'	04°09.6'E	2820	1.99	1.5	1.58	1.36	0.86	1.91	1.02	Struck, 1997
M 23256	73°10.30'	10°56.60'E	2061	2	8.48	–	–	–	–	–	Nees, 1997
M 23269	71°26.3'	00°40.1'E	2867	–	>4.17	3.66	2.29	2.22	5	2.77	Kuhlemann et al., 1993
MD95–2009	62°	5°W	1500?	–	–	–	–	7.78	8.18	7.86	Rasmussen et al., 2003
M 23044	68°14.10'	02°32.90'E	2387	1.42	2.03	–	–	–	–	–	Vogelsang, 1990
M 23056	68°30.10'	03°50.30'E	2665	1	1.19	–	–	–	–	–	Vogelsang, 1990
NP 90–39	77°15.49'	09°05.58'E	2119	6.22	4.87	>13.69	–	–	–	–	Hald et al., 2003
M 23398	76°26.30'	08°59.60'E	2224	–	4	6.29	2.86	–	–	1.79	Hald et al., 2003
M 23385	74°44.35'	10°52.19'E	2498	3.33	7	6.86	3.57	–	–	–	Hald et al., 2003
M 23059	70°18.3'	03°07.4'E	2281	2.5	3.13	2.44	1.25	–	–	1.05	Henrich et al., 1989
M 23065	68°29.7'	00°49.1'E	2802	1.83	2.87	2.97	1.17	–	–	1.86	Henrich et al., 1989
M23063	68°45'	00°50'W	2500?	1.25	?	?	?	–	–	1.78	Bauch et al., 2001
V 28–56	68°	6°W	2500?	2.5	5.42	1.57	0.71	1.44	2.27	1.61	Henrich, 1998
V 29–210	67°	6°30'W	2000?	2.08	3.75	2.29	1.79	1.67	4.55	2.23	Henrich, 1998
V 27–86	66°45'	1°E	2000?	1.67	5.42	2.43	5	1.56	5.45	1.52	Henrich, 1998
V 27–60	72°30'	8°E	1800?	8.33	7.92	5.43	2.86	2.11	2.73	2.23	Henrich, 1998
V 23–70	71°	6°30'E	3050?	2.92	2.92	3.86	1.07	2.33	2.27	2.32	Henrich, 1998
V 23–60	70°	8°E	3000	2.92	4.17	7.29	2.14	1.78	2.27	1.7	Henrich, 1998
					Iceland Sea						
M23352	70°00.4'	12°25.8'W	1819	3.33	2.92	1.71	2.29	–	–	1.48	Struck, 1997
PS1243	69°22.3'	06°32.1'W	2710	4.17	1.83	1.63	2.14	–	–	1.59	Struck, 1997
PS1244	69°22.0'	08°40.0'W	2162	3.18	0.67	3.13	0.83	–	–	1.53	Struck, 1997
PS1246	69°23.0'	12°55.0'W	1902	2.58	2.75	1.34	2.79	–	–	1.48	Struck, 1997
M23353	70°34.2'	12°43.3'W	1394	6	2.83	3.57	1.86	–	–	1.61	Struck, 1997
M23342	71°38.2'	08°38.2'W	1974	3.42	3.25	3.14	2.29	–	–	1.57	Struck, 1997
PS1852	70°15.7'	15°49.5'W	1117	–	–	1.76	0.86	–	–	1.61	Struck, 1997
B997-323	66°30.0'	21°03.0'W	250	6.13	–	–	–	–	–	–	Andrews and Principato, 2002
P0004/1	68°48.2'	17°42.6'W	1481	3.1	2	–	–	–	–	–	Lackschewitz, 1991
P0006/1	69°11.8'	16°49.2'W	950	4	2.9	–	–	–	–	–	Lackschewitz, 1991
P0020/2	67°59.8'	18°31.9'W	858	45	30	–	–	–	–	–	Lackschewitz, 1991
PS1845	69°27.6'	15°45.3'W	922	4.17	2.5	2	1.67	–	–	0.98	Lackschewitz, 1991
P0017	68°11.2'	18°02.9'W	762	2.67	10.83	3.29	1.67	–	–	–	Lackschewitz, 1991
PS1842	69°27'	16°31'W	982	0.67	3.93	–	–	–	–	–	Sarnthein et al., 1995
HM 71-15	69°59.90'	17°26.00'W	1547	3.17	4.17	2.83	1.29	1	0.24	1.11	Vogelsang, 1990
HM 71-17	70°00.40'	13°01.10'W	1460	2.08	3.58	4.14	0.64	1.18	1	1.14	Vogelsang, 1990
HM 71-19	69°29.0'	09°30.60'W	2210	3	3.67	0.54	0.43	–	–	0.66	Vogelsang, 1990

Table 8.1 (continued)

ODP 907	69°14.989'	12°41.894'W	1801	2.33	1.8			1.1	0.91	1.07	Mc Manus et al., 1996
PS 2644	67°52.0'	21°45.4'W	777	>3.90	>7.83	16.73	6.31	–	–	–	Vöelker, 1999
PS 2646	68°33.4'	21°12.6'W	1113	5.84	7.5	14.43	5.71	–	–	3	Vöelker, 1999
PS 2647	68°46.4'	21°03.3'W	1372	7.91	12.09	10.85	4.64	–	–	2.24	Vöelker, 1999
V 27–46	67°	12°W	1500?	1.67	4.17	2.14	2.5	2.11	3.64	2.41	Henrich, 1998
V 27–47	68°	13°45'W	1500?	2	?	?	?	?	1.82	?	Henrich, 1998
V 23–74	68°	9°20'W	1800?	5	10.42	4	2.5	2.33	2.27	2.32	Henrich, 1998
V 28–35	66°30'	9°W	1400?	1.25	3.33	?	?	?	1.82	?	Henrich, 1998
Greenland Sea											
PS1906-2	76°02.1'	02°09.2'W	2939	1.25	2.92	3.14	0.98	–	–	1.07	Magnus, 2000
PS1745-6	77°46.6'	00°55.3'E	3052	2.08	3.08	1.23	1.43	–	–	1.52	Magnus, 2000
M17728-2	76°31.1'	03°57.3'E	2485	1.25	1.25	1.71	1.07	–	–	1.25	Magnus, 2000
PS1878-3	73°15.33'	09°00.74'W	3048	4.9	5.72	3.17	–	–	–	–	Antonow et al., 1997
PS1882-2	73°35.96'	08°19.29'W	3175	2.24	3.65	3.75	2.21	–	–	1.64	Antonow et al., 1997
PS1892-3	73°44.06'	09°41.17'W	3002	5.39	–	–	2.21	–	–	1.05	Antonow et al., 1997
M23357	70°57.3'	05°32.6'W	1736	1	8.19	1.41	0.71	–	–	0.86	Goldschmidt, 1997
PS1736-3	74°19.7'	05°10.9'W	3460	1.25	2.92	3.14	2.14	–	–	1.7	Jünger, 1994
PS1900-6	74°31.7'	02°20.27'W	3538	1.67	3.33	1.86	1.07	–	–	1.07	Jünger, 1994
HM 94-34	73°46'	02°32'W	3004	2.92	2.73	4.43	–	–	–	–	Samthein et al., 1995
PS 1906	76°55.5'	02°09.0'W	2990	1.79	4.15	5.73	–	–	–	–	Nørgaard-Pedersen et al., 2003
PS 1320	77°59.40'E	00 32.03'E	3104	1	3.75	2.57	1.29	–	–	1.13	Mangerud, 1989
HM 94-25	75°36'	01°18'E	2469	1.67	2.14	–	–	–	–	–	Samthein et al., 1995
M 17728	76°50.0'	03°57.3'E	2485	1.5	1	2	1.33	1.77	0.73	1.32	Hebbeln et al., 1998
M 23454	76°45.1'	08°02.6'E	2144	4	4.33	4.69	6.94	–	–	1.67	Goldschmidt, 1997
M 23456	77°04.0'	06°20.5'E	2182	6.25	5.83	4.81	2.58	–	–	1.58	Goldschmidt, 1997
PS 1295	77°59.20'	02°24.80'E	3112	1.25	3.33	2.29	1.43	–	–	1.16	Mangerud, 1989
K 11	72°	1°40'E	2100?	2.5	3.75	3.86	0.71?	–	–	more 1.43	Henrich, 1998
Greenland continental margin											
M 23230	78°51.5'	04°46.8'W	1235	2	2.6	1.14	2	–	–	0.54	Spielhagen, 1991
PS 2876	81°45.7'	09°24.0'W	1991	4.68	18.83	15.4	–	–	–	–	Nørgaard-Pedersen et al., 2003
PS 2887	79°36.0'	04°36.5'W	1411	1.5	3.67	1.49	–	–	–	–	Nørgaard-Pedersen et al., 2003
PS 1728	70°45'	25°00.0'W	385	15	–	–	–	–	–	–	Dowdeswell et al., 2000
PS 1725	70°06.9'	18°49.9'W	879	>2.3	4.1	2.3	2	–	–	1.1	Nam and Stein, 1999
PS 1726	70°07.0'	18°38.9'W	1174	>3.9	5.2	2.9	2.3	–	–	2.1	Nam and Stein, 1999
PS 1730	70°07.2'	17°42.1'W	1617	6.5	8.6	7.8	2.3	–	–	1.8	Nam and Stein, 1999
PS 1927	71°29.7'	17°07.1'W	1734	>4.3	12.6	7.1	2.1	–	–	–	Nam and Stein, 1999
PS 1951	68°50.5'	20°49.2'W	1481	>4.5	10.6	11.2	2.5	–	–	3	Nam and Stein, 1999
PS 1919	75°12'	11°40'W	1700	0.83	4.33	1.71	2.71	–	–	1.79	Funder et al., 1998
91-K7	68°15.7'	32°05.83'W	862	283.42	–	–	–	–	–	–	Andrews et al., 1996
91-K8	68°07.79'	31°51.77'W	872	472	–	–	–	–	–	–	Andrews et al., 1996

Table 8.1 (continued)

91-K14	68°06.02'	29°35.74'W	459	134.83	–	–	–	–	–	–	Andrews et al., 1996
91-K15	68°06.02'	29°27.16'W	445	17.43	–	–	–	–	–	–	Andrews et al., 1996
91-K5	67°24.59'	31°03.98'W	622	11.71	–	–	–	–	–	–	Andrews et al., 1996
88-5A	67°07.54'	30°54.26'W	707	8.14	–	–	–	–	–	–	Andrews et al., 1996
88-10A	66°12.19'	30°39.29'W	496	2.19	25.76	–	–	–	–	–	Andrews et al., 1996
91-18B	65°57.77'	30°38.0'W	470	8.36	18.25	–	–	–	–	–	Andrews et al., 1996
91-K11A	68°06.94'	31°25.9'W	244	7.66	–	–	–	–	–	–	Andrews et al., 1996
PO 175/15	66°45.83'	30°50.42'W	295	3.23	126.74	–	–	–	–	–	Andrews et al., 1996
PS 2627	73°07.40'	15°40.85'W	2009	2.86	11.53	–	–	–	–	–	Ó Cofaigh et al., 2002
PS 2628	73°09.79'	15°57.98'W	1694	1.92	31.79	–	–	–	–	–	Ó Cofaigh et al., 2002
PS 1923	71°29.7'	20°31.0'W	260	4.17	–	–	–	–	–	–	Nam, 1997
PS 1926	71°29.6'	18°16.5'W	1493	3.7	>25.5	–	–	–	–	–	Nam, 1997
PS 1920	74°59.7'	11°04.3'W	2717	5	7.5	6.86	3.57	–	–	3.04	Stein et al., 1996
PS 1723	70°07.1'	19°59.9'W	283	8.5	–	–	–	–	–	–	Nam, 1997
PS 1724	70°07.2'	19°13.3'W	363	12.9	–	–	–	–	–	–	Nam, 1997
PS 1922	75°00.0'	08°46.3'W	3350	9.17	23.33	–	–	–	–	–	Stein et al., 1996
PS 1924	71°30.0'	19°10.9'W	288	7.2	–	–	–	–	–	–	Nam, 1997
PS 1946	69°36.3'	22°22.6'W	320	5	–	–	–	–	–	–	Nam, 1997
PS 1947	69°16.5'	21°45.6'W	379	5	–	–	–	–	–	–	Nam, 1997
PS 1925	71°30.0'	18°43.5'W	811	>8	>9.5	–	–	–	–	–	Nam, 1997
PS 1948	69°04.4'	21°17.3'W	624	>3.3	>19.9	–	–	–	–	–	Nam, 1997
PS 1949	68°59.1'	21°09.0'W	1106	>4.7	>33	–	–	–	–	–	Nam, 1997
PS 1950	68°53.4'	20°55.7'W	1480	>5.7	>32.5	–	–	–	–	–	Nam, 1997
PS 2631	73°10.7'	22°11.0'W	430	45.4	–	–	–	–	–	–	Evans et al., 2002
PS 2641	73°09.3'	19°28.9'W	469	51.67	–	–	–	–	–	–	Evans et al., 2002
PS 2630	73°09.5'	18°04.1'W	287	?	–	–	–	–	–	–	Evans et al., 2002
PS 2629	73°09.5'	16°29.0'W	850	?	30	–	–	–	–	–	Evans et al., 2002
PS 2423	80°02.3'	05°26.9'W	829	1.25	8.33	8.14?	–	–	–	–	Notholt, 1998
PS 2424	80°02.07'	05.44.57'W	445	30	57.33	–	–	–	–	–	Notholt, 1998
PS 1733	71°51.95'	22°16.55'W	328	40	–	–	–	–	–	–	Notholt, 1998
PS 1734	71°41.4'	21°32.92'W	307	–	1.73	–	–	–	–	–	Notholt, 1998
PS 1735	71°54.97'	21°41.78'W	412	22.82	16.15	–	–	–	–	–	Notholt, 1998
PS 2419	80°27.8'W	13°40.04'W	333	3.57	–	–	–	–	–	–	Notholt, 1998
PS 1894	75°48.8'	08°18.0'W	1975	5.31	18.63	–	–	–	–	–	Nørgaard-Pedersen et al., 2003
M23351	70°21.5'	18°13.2'W	1672	3.09	11.09	27?	–	–	–	–	Vöelker, 1999
Fram Strait and Yermak Plateau											
PS 1535	78°45'	01°51'E	2557	2.08	3.92	2.24	2.36	1.78	4.55	2.32	Sarnthein et al., 1995
ODP 909	78°35.07'	03°04.38'E	2519	3.33	5	3.43	5	–	–	2.68	Hevrøy et al., 1996
M 23235	78°51.55'	01°18.59'E	2456	1.7		3.2		–	–	2.6	Eisenhauer et al., 1994
PS 1235	78°55'	01°30'E	2500		2.5		3.7	–	–	2.44	Nowaczyk and Baumann, 1992
M 21535	78°44.8'	01°52.8'E	2557	1.6	4.6	3.2	1.3	–	–	2.2	Köhler, 1992
MG 123	79°12'	00°30'E	3000	1.67	5.42	1.14	4.64	–	–	2.5	Köhler, 1992
PS 1291	78°00.6'	08°03.1'E	2400	3.5	6.67	5.14	1.43?	–	–	2.23	Spielhagen, 1991
PS 1294	77°59.90'	06°39.90'E	2458	1.5	6.67	4.57	2.86	–	–	1.52	Mangerud et al., 1998
PS 1230	78°51.5'	04°46.5'W	2700	1.33	1.92	1.36	–	–	–	–	Nørgaard-Pedersen et al., 2003

Note: Blank means no data

due to compaction and the difference should increase with the age of sediments. Nevertheless, the direct comparison of SR and MAR for terrigenous sediments of the Greenland continental margin in same cores revealed a high similarity in the behavior of both parameters with time (Stein et al., 1996). Hence, the SR variability governs variations of bulk MAR, and variations in physical properties are not of major importance.

We do not agree with the approach (Taylor et al., 2002b), where a constant dry density of 1.00 g/cm^3 for all sediments of the Norwegian-Greenland Basin is used, which have been accumulating during MIS 1 and MIS 2. Schematic maps of SR, presented in (Taylor et al., 2002a), are not connected with the relief of the sea floor and give much less information than they could possibly do. We produced maps of SR for different MIS in the Norwegian-Greenland Basin on the bathymetric basis (using data bank of GEBKO).

In order to verify the validity of this systematic approach and to compare SR for MIS 1 with the recent SR, a new version of the schematic SR map for surface sediment layer of the studied aquatory was compiled, based on data received by U-Th method (Paetsch et al., 1992) (Fig. 8.4).

Finally, for studying SR distribution in different parts of the investigated basin, we divided it (from 63°N to 80°N) into a number of regions, located both at the continental margins and in the deep-sea area (Fig. 8.6). The Iceland Sea relates neither to the continental margins nor to the deep-sea floor. This question will be examined below.

Results

Distribution of recent SR (Fig. 8.7) reveals a weak SR increasing toward the continental margins: the values, which exceed 2.5 or less often 5 cm/ky, were noted near Norway, Iceland and in some regions of the Greenland continental margin. There are vast areas without any recent or Holocene sedimentary cover on the Spitsbergen and East Greenland shelves. In these regions surface sediments were preserved either in the fjords or in the rare shelf depressions. A distinct increasing of SR is mapped in the Lofoten Basin, where it is caused by the presence of the Bear Island Trough Fan. Besides the mentioned areas, the studied region is covered with recent sediments, which are being accumulated with SR <2.5 cm/ky.

During MIS 1 the noted phenomenon of circumcontinental SR distribution was expressed much more definitely than in the recent epoch (Fig. 8.8). Regions with SR values >5 or even >10 cm/ky were located practically along the entire continental margins of Spitsbergen and Barents Sea, intermittently – along the continental margins of Norway, near Iceland and in some places in the continental margins of Greenland. In many cases the noted SR maxima are related to different fans, for example, of Is-fjord, Bear Island, the North Sea, and Scoresby Sund. Bear Island fan, outlined by the contour line of 5 cm/ky, was aligned across the entire Lofoten Basin.

Gradation of less than 2.5 cm/ky occupied large areas in the Iceland and Greenland seas, in the Norwegian Basin, in the southwest of the Lofoten Basin

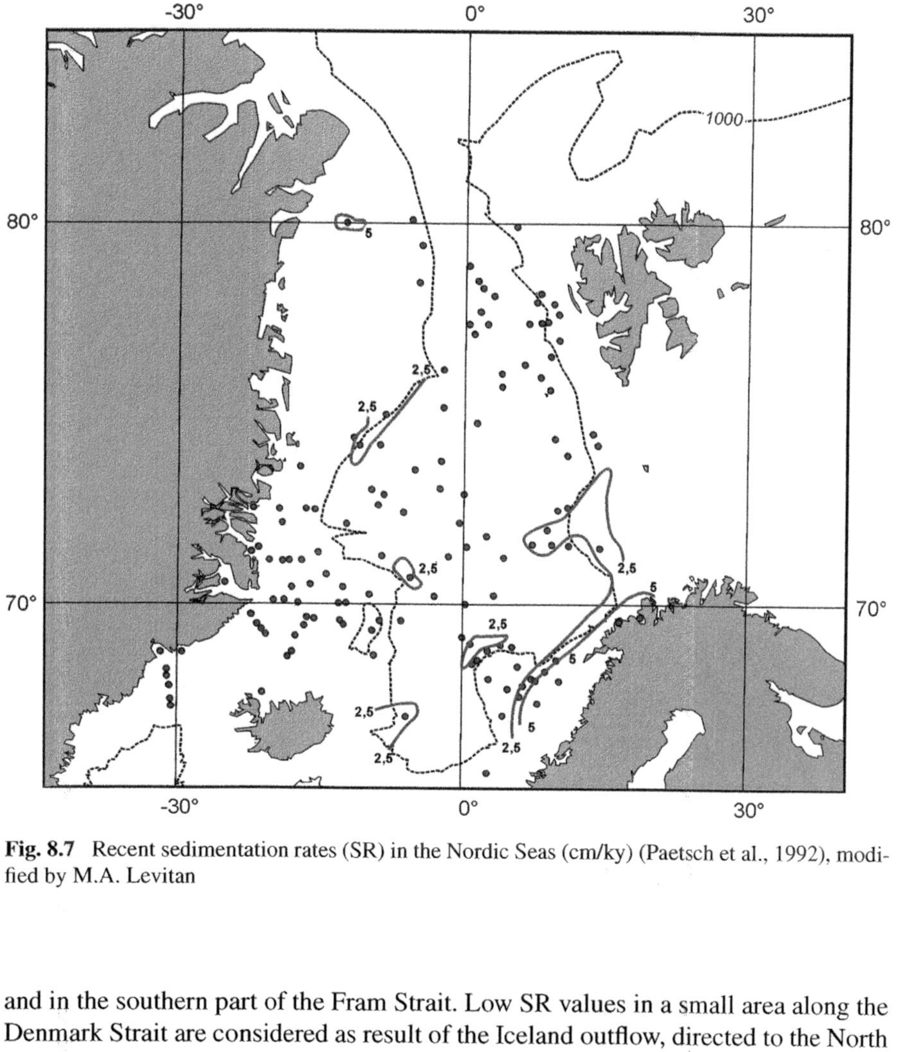

Fig. 8.7 Recent sedimentation rates (SR) in the Nordic Seas (cm/ky) (Paetsch et al., 1992), modified by M.A. Levitan

and in the southern part of the Fram Strait. Low SR values in a small area along the Denmark Strait are considered as result of the Iceland outflow, directed to the North Atlantic. It is obvious from comparison of Figs. 8.7 and 8.8 that SR during MIS 1 was higher on average than in the recent epoch.

Figure 8.9 demonstrates that SR has been, at least, twice more in MIS 2 than in MIS 1, in all areas of the studied region: both at the continental margins and in the pelagic zones. In some fjords of Spitsbergen, Norway and Greenland SR increased to some hundreds of cm/ky. Areas of SR <2.5 cm/ky distribution have been restricted to small areas in the Greenland and Iceland seas, along Jan-Mayen fault zone, and in the southern margin of the Vøring Plateau. The circumcontinental zonality of SR distribution was expressed very clearly, moreover, more sharply along the eastern continental margins than along the western ones.

In MIS 3 pattern of SR distribution was different from MIS 2, in particular: areas of SR >20 cm/ky distribution were absent, gradations of >10 cm/ky were devel-

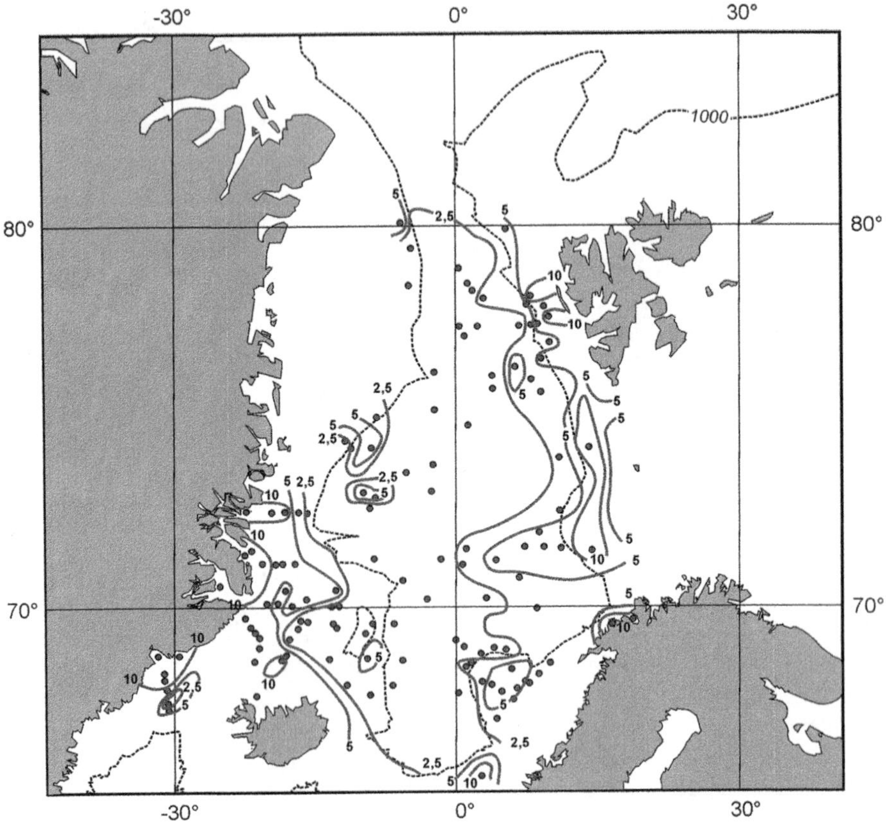

Fig. 8.8 MIS 1 sedimentation rates (SR) in the Nordic Seas (cm/ky) (for data source see text)

oped only seaward of central Norway, as well as between Greenland and Iceland (Fig. 8.10). As a whole, the eastern continental margins of the studied region were bordered by the contour line of 5 cm/ky, while along the western margins there were fluctuations from 5 to 2.5 cm/ky. In a narrow center area of the Greenland Sea, the Norwegian Basin, the western part of the Lofoten Basin, and in the north of the Iceland Sea SR values comprised <2.5 cm/ky.

Distribution map of SR in MIS 4 (Fig. 8.11) very clearly demonstrates SR decreasing comparatively to MIS 3: gradation >10 cm/ky is generally absent, values >5 cm/ky were noted only seaward of North Iceland, and, also, along the eastern continental margins. As a whole, the asymmetry of the continental margins was very well expressed from the point of view of SR distribution. The gradation of SR <2.5 cm/ky occupied the entire area of the Greenland Sea, north of Iceland Sea, Norwegian Basin and large part of the Lofoten Basin.

Comparison of SR distribution in MIS 4 and MIS 5 revealed a general weakly expressed tendency of SR decreasing in MIS 5 (Fig. 8.12). Actually, the contour line

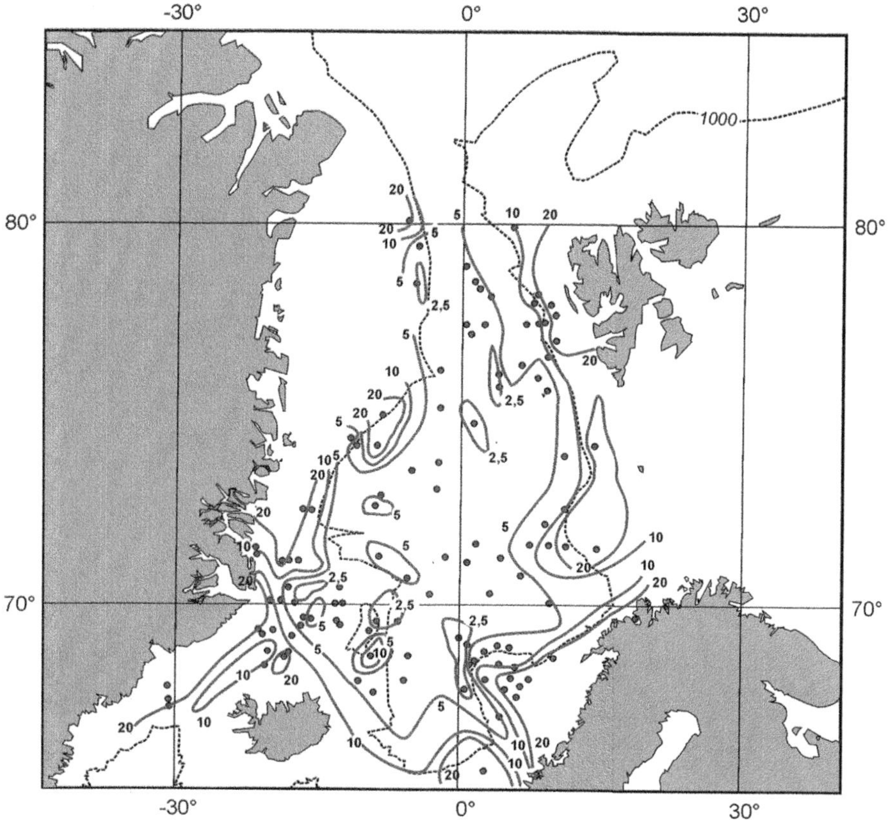

Fig. 8.9 MIS 2 sedimentation rates (SR) in the Nordic Seas (cm/ky) (for data source see text)

of 5 cm/ky, which bordered the eastern continental margins in MIS 4, was located further to the east in MIS 5, and SR values >5 cm/ky were absent seaward of North Iceland. Sea floor of almost entire studied region was covered by sediments, which were being accumulated with a rate <2.5 cm/ky. In this case the comparison of SR values for MIS 5 a-d (Fig. 8.13) and MIS 5e (Fig. 8.14) showed that they were slightly higher during MIS 5e than later. If for the MIS 5/MIS 6 boundary an older age (135 ky) would be used (Shackleton et al., 2002), SR values of MIS 5 would become somewhat lower than those based on our calculations, and average SR ratio MIS a-d and SR MIS 5e would be close to 1.0.

Discussion of Results

The general trend of SR changes appears as follows: increasing from MIS 5 to MIS 2 and sharp decreasing in MIS 1. This general trend is confirmed by the statistical data, based on calculation of arithmetical means, which not only quantitatively reflect the

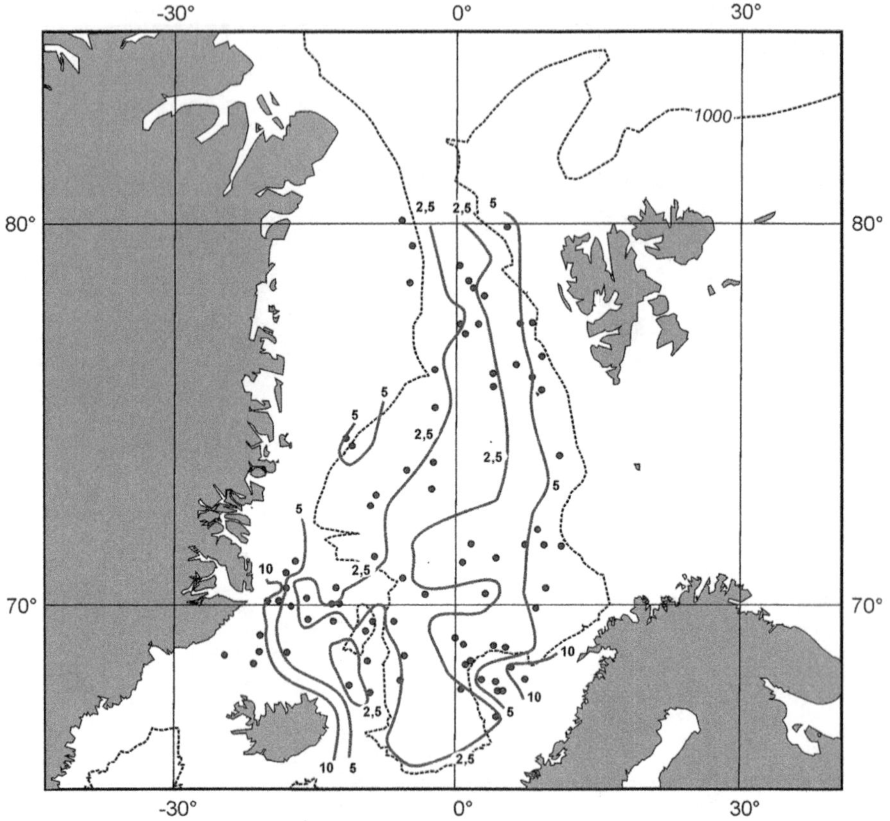

Fig. 8.10 MIS 3 sedimentation rates (SR) in the Nordic Seas (cm/ky) (for data source see text)

tendency mentioned above, but also make it possible to reveal numerous interesting details (Table 8.2).

Average SR ratio for MIS 5a–5d/MIS 5e (19 cores) is 0.72: 0.77 for pelagic zones (11 cores) and 1.63 for the Iceland Sea (5 cores). The generally higher SR values in the interglaciation substage 5e than in stadial-interstadial substages 5a–5d are related to Termination 2 at the MIS 5/MIS 6 boundary and increased advection of the Atlantic warm and salt water to pelagic zones at that time and the advection-related increase in carbonate productivity, which is reflected in sediments by higher $CaCO_3$ content (Henrich, 1998). Later, the advection was poorly expressed. The cause of a much higher SR ratio for the Iceland Sea is not clear yet. However, we may assume that Termination 2 was not so intensively manifested in the SR records off the territory of small Iceland as off zones characterized by well-developed Eurasian and Greenland ice sheets during the previous glacial MIS 6.

The general SR increasing at continental margins as well as in pelagic zones and in the Iceland Sea during the middle Weichselian as compared to the early Weichselian (see SR ratio for MIS 4/MIS 5 in Table 8.2) is certainly related to

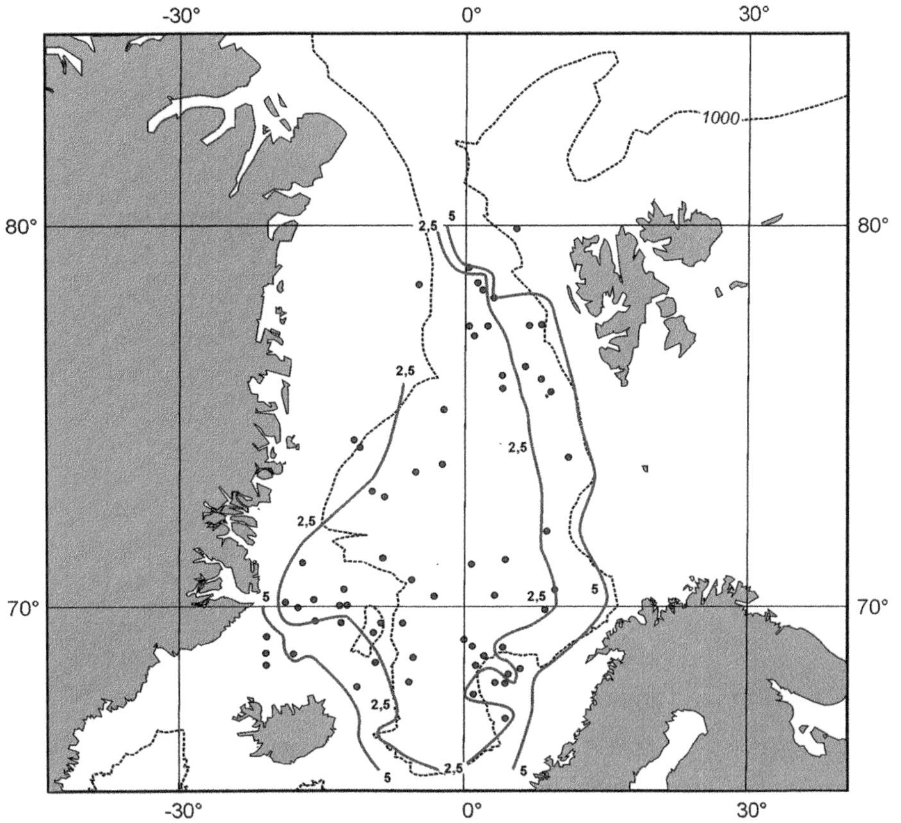

Fig. 8.11 MIS 4 sedimentation rates (SR) in the Nordic Seas (cm/ky) (for data source see text)

the glaciation in MIS 4. Very similar values of the SR ratio for different areas, such as continental margins (1.33), pelagic zones (1.32), and the Iceland Sea (1.29), point to a synchronism of ice-sheet growth for all continental masses surrounding the Norwegian-Greenland Basin. It is notable that the SR ratio is higher for the Greenland continental margin than for the Norwegian margin (1.63 and 1.02, respectively). A similar trend is observed for the Greenland and Norwegian seas (1.42 and 1.14, respectively). These data are likely to imply that dynamics of glaciation at that time was higher in Greenland than in Norway, because such dynamics governs the delivery of the siliciclastic material to the sedimentation basin. On the other hand, the comparison of SR ratios for MIS 4 and MIS 5 within pelagic zones revealed a northward increasing of this parameter: 1.14 for the Norwegian Sea, 1.42 for the Greenland Sea, and 1.46 for the Fram Strait. This phenomenon can be related to the northward decreasing of biogenic carbonates contribution to bottom sediments of MIS 5 (Henrich, 1998). We consider different dynamics of glaciation development as the major process and the influence of Atlantic waters during MIS 5 as a minor

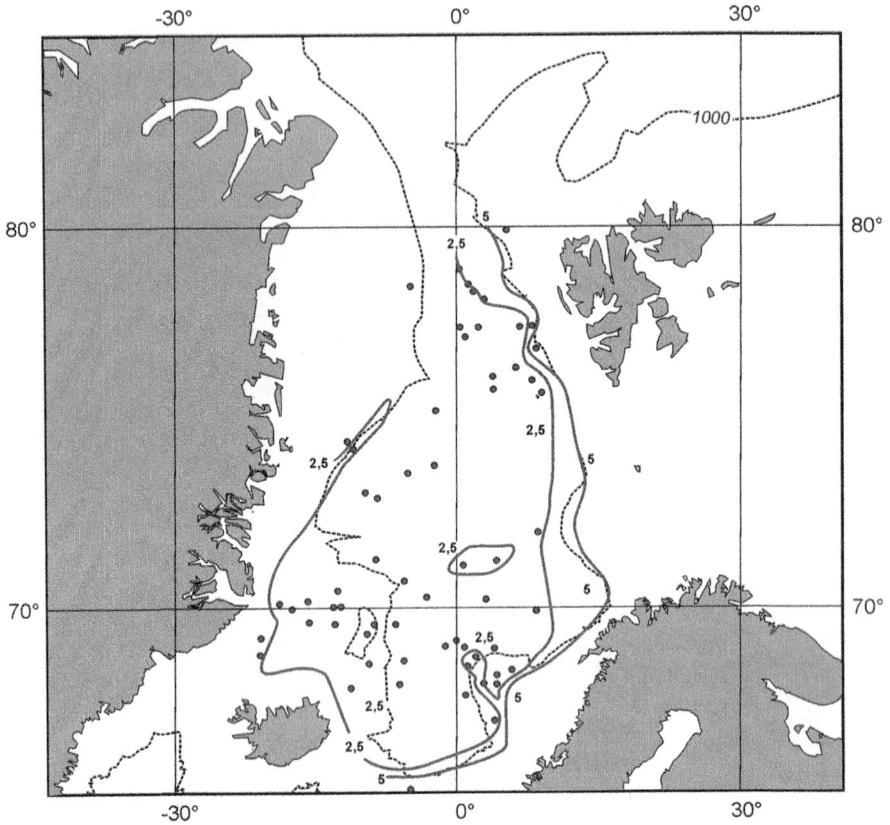

Fig. 8.12 MIS 5 sedimentation rates (SR) in the Nordic Seas (cm/ky) (for data source see text)

process. Moreover, calculations show that the carbonate material influx to sediments in pelagic zones was higher than at continental margins.

The Scandinavian ice sheet evolved more actively than the Greenland ice sheet in the middle Weichselian (SR ratios for MIS 3/MIS 4 is 2.27 for the Norwegian continental margin and 2.10 for the Greenland continental margin). In this epoch, like in the earlier one, the situation at continental margins was reflected in the adjacent pelagic zones (SR ratios are 1.98 and 1.68 for the Norwegian and Greenland seas, respectively). Carbonate content has been increased in pelagic sediments against terrigenous material. Comparison of average SR ratios for MIS 3/MIS 4 in pelagic zones revealed a northward decreasing in this parameter: 1.98 for the Norwegian Sea, 1.68 for the Greenland Sea, and 1.19 for the Fram Strait. Evidently, this trend is related to a decreasing in the flux of carbonates to sediments from the south to north in MIS 3 (Henrich, 1998). Since that time, the gradient of SR ratios between continental margins and pelagic zones became more prominent than in the previous period (average ratios are 2.18, 1.73, and 2.01 for continental margins, pelagic zones, and the Iceland Sea, respectively).

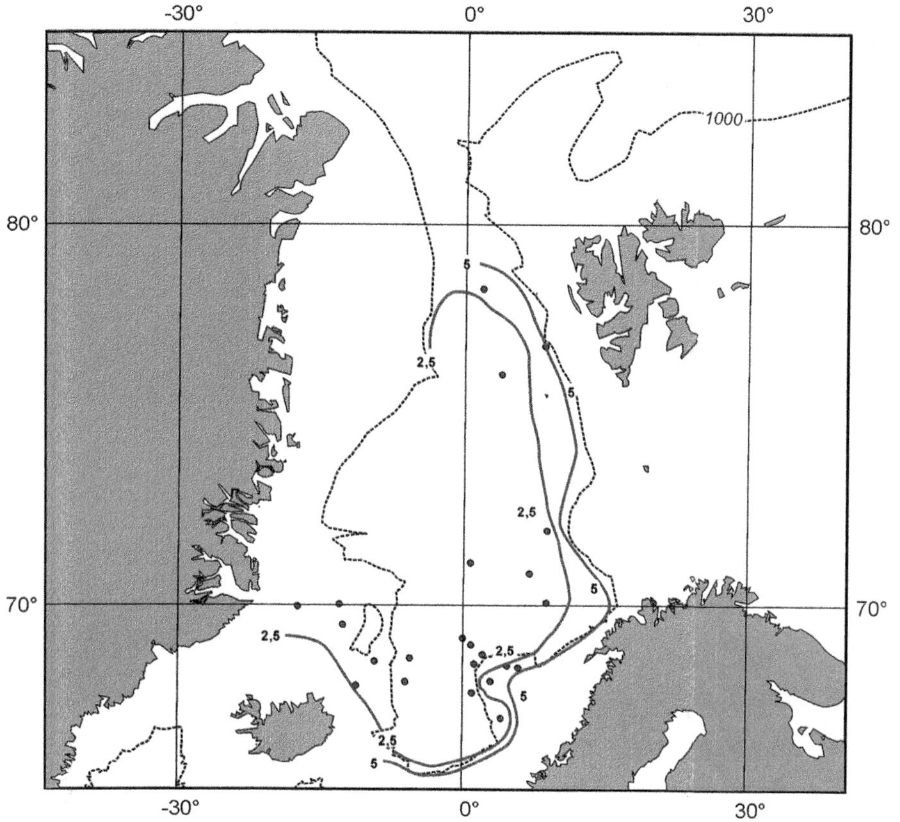

Fig. 8.13 MIS 5a-d sedimentation rates (SR) in the Nordic Seas (cm/ky) (for data source see text)

In MIS 2 highest SR values have were determined both at continental margins and in pelagic zones. First of all, it was related to the dynamics of the growth or breakdown of ice sheets in the late Weichselian, i.e., ice-sheet advances, retreats, and new advances; formation of icebergs; formation of ice streams in transverse troughs; development of debris- and mud flows on continental slopes and turbidity currents in channel systems. The "bulldozer effect" of the advances of the ice-sheet margin and the formation of basal moraines played a significant role on glacial shelves. The lateral motion of underwater landslides downward the continental slope and many other processes related to glaciation also played an important role (Taylor et al., 2002).

We would like to draw attention to one important sedimentation mechanism related to the melting of icebergs. Heinrich events (i.e., large-scale formation of icebergs in the process of surges) were commonly accompanied by the intense melting of icebergs and the appearance of substantial freshwater masses in the surface layer of water column (Bond et al., 1992; Sarnthein et al., 2003). We assume that

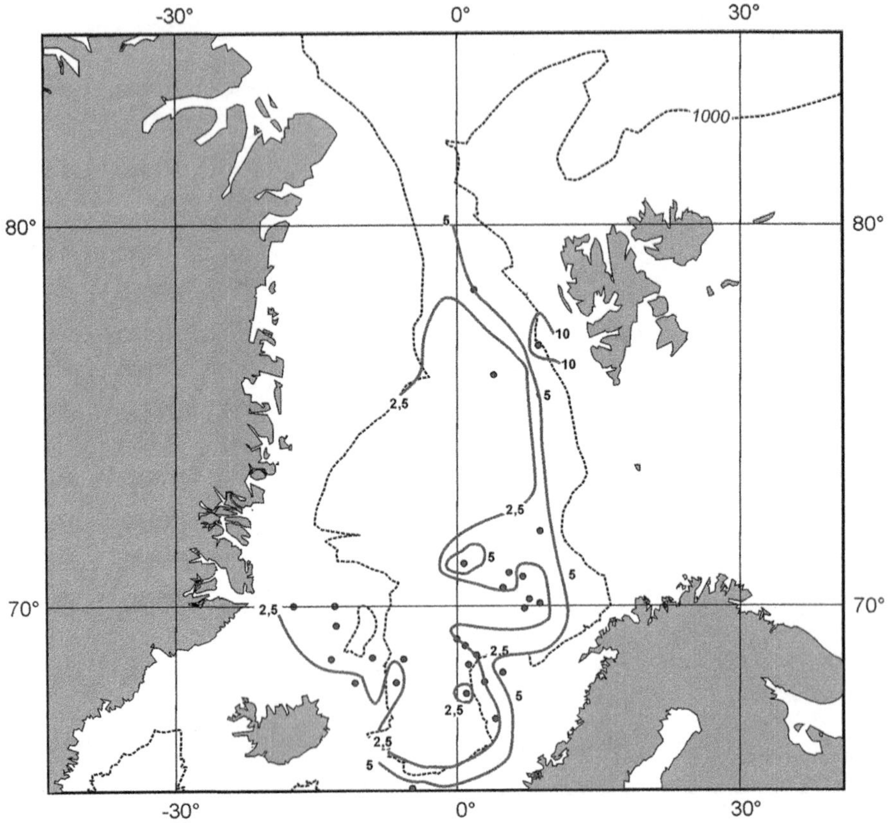

Fig. 8.14 MIS 5e sedimentation rates (SR) in the Nordic Seas (cm/ky) (for data source see text)

mixing of this freshwater with saline seawater fostered (and fosters) variation in discharge of clay minerals (as in the case of estuaries during the mixing of fresh and sea waters), their aggregation, fast sinking, and accumulation on the bottom. Clay fractions make up a considerable part of Heinrich layers, and sedimentation from a turbidity plume, which is formed during the larger-scale melting of icebergs, should substantially affect SR values.

We also would like to underline a great significance of the width of glacial shelves: shelves of Spitsbergen, Norway, and southeastern Greenland are relatively narrow; the ice sheet margin reached the shelf edge during the Last Glacial Maximum precisely in these zones (Elverhøi et al., 1998).

Consequently, the major part of the bottom areas with the highest SR values is confined to continental slopes and rises in these zones. However, a large part of the eastern Greenland shelf is very wide, and the margin of the Greenland ice sheet was probably located within the inner shelf in the Last Glacial Maximum (Stein et al., 1996). Along with other causes, this phenomenon could be responsible for

Table 8.2 Average values of SR ratio for MIS in the studied regions (by M.A. Levitan)

Region	MIS 1/ MIS 2	MIS 2/ MIS 3	MIS 3/ MIS 4	MIS 4/ MIS 5	MIS 5a–d/ MIS 5e	MIS 2/ MIS 4	MIS 1/ MIS 5e
Continental margin of Spitsbergen	0.48	–	–	–	0.33	–	–
Continental margin of the Barents Sea	0.42	0.99	–	–	–	–	–
Continental margin of Norway	0.38	2.39	2.27	1.02	1.18	2.83	–
Continental margin of Greenland	0.38	1.70	2.10	1.63	–	2.51	–
Continental margins	0.40	1.99	2.18	1.33	0.89	2.69	–
Norwegian Sea	0.79	1.44	1.98	1.14	0.63	2.53	1.22
Greenland Sea	0.69	1.46	1.68	1.42	2.42	1.94	–
Fram Strait area	0.52	1.94	1.19	1.46	0.39	1.78	0.88
Pelagic areas	0.70	1.54	1.73	1.32	0.77	2.15	1.10
Iceland Sea	1.15	1.79	2.01	1.29	1.63	2.61	1.69
Norwegian-Greenland Basin	0.67	1.72	1.91	1.32	0.72	–	1.46

Note: Blank means no data

the absence of deep-sea fans and large landslides in zones located off the shelf in the central part of eastern Greenland.

The Last Glacial Maximum ended 18 ky ago (Sarnthein et al., 2003). The subsequent deglaciation (Termination 1) (Fairbanks, 1989) lasted until the Early Holocene. Ratios of SR values for MIS2/MIS 3 (Table 8.2) suggest some details of sedimentation for that period. It should be noted that the spatial SR distribution demonstrates for MIS 2/MIS 3 more active evolution of the Scandinavian ice sheet (the average value is 2.39 for the Norwegian continental margin) as compared to the Greenland ice sheet (the average value 1.70 for the Greenland continental margin). This fact implies the following trends: asymmetry of the circum-continental zonation in the Norwegian-Greenland Basin; lower values for pelagic zones (average 1.54) than for continental margins (average 1.99) and the Iceland Sea (1.79); decreasing in the concentration of carbonate components in bottom sediments of MIS 2 in the northern pelagic zones (Henrich, 1998), and, consequently, increasing in SR ratios for MIS 2/MIS 3 in the northern pelagic zones (1.44 for the Norwegian Sea, 1.46 for the Greenland Sea, 1.94 for the Fram Strait).

SR values during MIS 1 considerably have been decreased as compared to MIS 2 (Table 8.2). In general, the SR ratio for MIS 1/MIS 2 is only 0.40 for continental margins and 0.70 for pelagic zones. Such a difference in SR values at continental margins and in pelagic zones is related to a higher concentration of biogenic $CaCO_3$ in pelagic sediments, especially for sediments of MIS 1, except for the Iceland Sea, where the SR ratio is higher (1.15) due to the following causes: (i) high volcanic activity in the region, particularly, in the southern Kolbeinsey Ridge area (Lackschewitz, 1991) and the consequent increase in the number of volcanic ash layers during MIS 1; (ii) higher concentrations (up to 55–60%) of biogenic $CaCO_3$ in Holocene sediments of the eastern and central parts of the Iceland Sea (Henrich, 1998); (iii) relatively small influx of the siliciclastic (terrigenous and volcano-terrigenous) material to distal northern pelagic zones of the Iceland Sea in MIS 2 as Iceland ice dome was too small and located too far from pelagic zones.

The glacioisostatic uplift, which provoked the drift of most sedimentary material to continental margins, was slightly more active at the Spitsbergen (average SR ratio is 0.48 for MIS 1/MIS 2) and Barents Sea continental margin (0.42) than at continental margins of Norway and Greenland (0.38 for both zones). A clear trend of northward decreasing in the SR ratios (0.79 for the Norwegian Sea, 0.69 for the Greenland Sea, and 0.52 for the Fram Strait) is obvious in pelagic zones. We interpret this phenomenon as a reflection of decreasing in amounts of carbonates in sediments of MIS 1 from south to north (Henrich, 1998; Taylor et al., 2002) nearly in the same way as in present-day sediments (Paetsch et al., 1992).

It was noted previously that comparison of stratigraphic units of different durations still remains a debatable issue. Therefore, we compared SR ratios for MIS stages (MIS 2/MIS 4) of equal (or nearly similar) duration. In addition, we also calculated data for interglaciations (MIS 1/MIS 5e).

Probably, Substage 5e is the closest climatic analog of MIS 1 (Kukla et al., 1997). The average SR ratio is equal to 1.46 for MIS 1/MIS 5e, 1.10 for pelagic zones, and 1.68 for the Iceland Sea. All these values were calculated without maximal and minimal values. Possible reasons of high SR ratios for the Iceland Sea have already been discussed above. The SR ratio value is most significant for pelagic zones and is very close to unity. This fact once more confirms the similarity of both interglacial epochs. Nevertheless, the SR ratio value was slightly higher during MIS 1 because of the following reasons. The first reason is evident: according to (Henrich, 1998), surface sediments of the zone between 65° and 73° N are slightly more enriched in carbonates than sediments of MIS 5e. The second reason is likely to be the influx of terrigenous material to the Norwegian–Greenland Basin from the Greenland ice sheet in MIS 1 and the absence of such influx in MIS 5e (Loutre and Berger, 2003).

Comparison of SR ratio values for the last two glaciations (MIS 2 and MIS 4) revealed several very clear tendencies. First, sedimentation rates were higher during MIS 2 relative to MIS 4 (the average SR ratio at MIS 2/MIS 4 is 2.69 for continental margins, 2.15 for pelagic zones, and 2.61 for the Iceland Sea). Hence, the late Weichselian glaciation in the study region was much more rigorous than the middle Weichselian one. This inference agrees well with (Svendsen et al., 2004). Distinctions between SR values at continental margins and in pelagic zones were appreciably lower in glacial epochs than in interglacials due to a greater advection of Atlantic waters (and, hence, a higher concentration of carbonates in sediments) during warmer periods. Second, the SR ratio at the Norwegian and Greenland continental margins, on the one hand, and in the Norwegian and Greenland seas, on the other hand, varied in a similar way. Hence, the influx of terrigenous material from glaciation zones mainly governed the SR trend and the influx from the territory of Norway was more intense than from Greenland during both glaciations. Third, the evident trend of the northward SR ratios reduction in pelagic zones (the average value is 2.53 for the Norwegian Sea, 1.94 for the Greenland Sea, and 1.78 for the Fram Strait) might be accounted, at least, for two reasons. As it was mentioned above, the first reason is related to decreasing from south to north influx of terrigenous material from glacial zones. The second reason is probably related to the following phenomenon: the so-called "Nordway events" or "High-Productivity events" (occasional intensification of the advection of Atlantic water to the Norwegian–Greenland Basin during glacial periods) were developed more intensely in MIS 2 than in MIS 4 (Hald, 2001). These events are represented in sediments by enhanced amount of biogenic carbonates. Naturally, the degree of manifestation of these events decreased from south to north, because Atlantic water ran into the Norwegian-Greenland Basin from the North Atlantic.

It would be appropriate to compare general trends of the SR history described for the Norwegian-Greenland Basin with more general regularities. For instance, M. Loutre and A. Berger (2003) elaborated the model of variations in the volume of continental ice of the Northern Hemisphere for the last 570 ka. Let us consider the results for the last 130 ka (i.e., for the last five oxygen isotope stages). The volume of glaciers was close to nul during MIS 5e, MIS 5c, and MIS 5a. However, the volume reached nearly 20 mln km^3 during Substage 5d and 3–5 mln km^3 during Substage 5b. During MIS 4, the volume of glaciers rapidly grew from 0 to 27–28 mln km^3. The short period of slight reduction of the volume of glaciers at the beginning of MIS 3 was followed by its intense growth during the remaining episodes of MIS 3 and MIS 2 right up to the end of the Last Glacial Maximum, when the volume reached maximal values $\sim$ 48 mln km^3.

Thereafter, the volume of continental glaciers of the Northern Hemisphere rapidly reduced to 3 mln km^3 in the current epoch. It is clear that these data are given for the whole hemisphere. However, their comparison with our results revealed a very interesting regularity: general trends of the SR history in the Norwegian-Greenland Basin and variations in the volume of continental glaciers of the Northern Hemisphere were practically similar (Fig. 8.15). Hence, the SR history was mainly governed by the history of variations in the volume of continental glaciers, and the influx of terrigenous material from glaciers to sea basins played the leading role in the SR history. In this process, sea level fluctuations were of great importance: increase in the volume of continental glaciers promoted fall of the sea level and increase in the SR, while reduction of volume fostered rise of the sea level and decrease in the SR. Contribution of biogenic carbonates had an insignificant influence on the SR and was most prominent during interglacials and, probably, interstadials in pelagic zones of the Norwegian-Greenland Basin. Figure 8.15 demonstrates that the SR/glacier volume ratio was higher during MIS 1 and MIS 5 but lower during MIS 2–MIS 4. This phenomenon is likely to reflect the greater role of biogenic sedimentation in warm epochs.

The asymmetric distribution of both SR values and their ratios at the eastern and western continental margins of the basin is a significant feature of sedimentation in the study region. This phenomenon is related to difference in the glaciation history of Greenland relative to Spitsbergen, Barents Sea, and Scandinavia. This issue

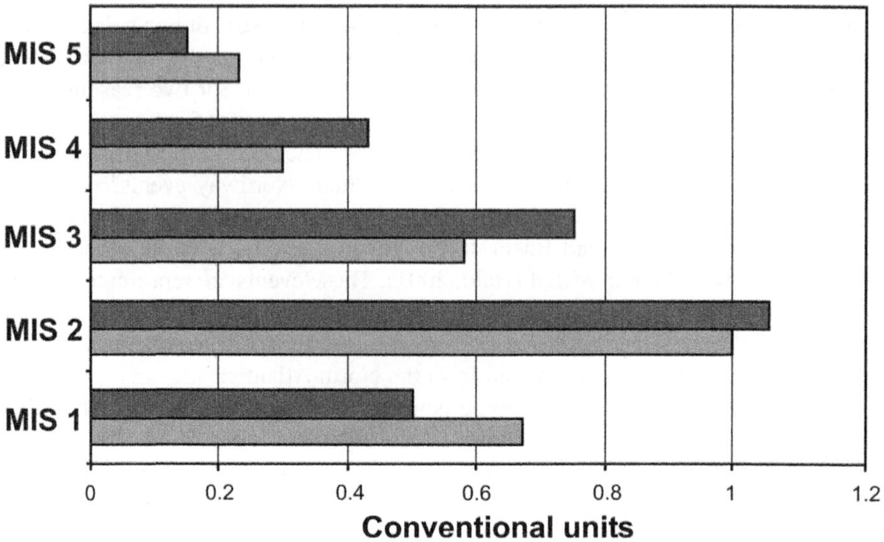

Fig. 8.15 Ratio of ice volume in the Northern Hemisphere (dark) (Loutre, Berger, 2003) and average sedimentation rates in the Nordic Seas (light) for the last five MIS (conventional units) (Levitan et al., 2007a)

is discussed in a series of publications (Elverhøi et al., 1998; Funder et al., 1998; Bischof, 2000). According to literature data, the onset of glaciation and the duration of glacier growth in eastern Greenland and Spitsbergen during MIS 5d and, probably, 5b are rather close. However, the glaciation history became quite different after 65 ka. We pointed out this episode of the SR history, but it was expressed in other terms.

The Greenland ice sheet remained stable since ~65 ka to 10 ky ago. At that time, the ice sheet slightly decreased due to the melting of ice or breakup of icebergs. In contrast, variation of eastern ice sheets was much more active. The Greenland ice sheet lost about 2 mln km^3 of ice, whereas the Spitsbergen-Barents Sea and Scandinavian ice sheets lost about 8 mln km^3 of ice. Hence, the above-mentioned asymmetry of the circum-continental zonation expressed in the SR distribution and their ratios in the Norwegian-Greenland Basin over the last 65 ka is governed by different glaciation histories in Greenland (especially, in the central and northern parts of eastern Greenland (Elverhøi et al., 1998) and in eastern continental and marine zones. In general, these distinctions are related to different oceanographic and climatic conditions: Greenland is subjected at present (and was subjected previously) to the influence of East Greenland Current, which carries cold Polar water, whereas eastern margins are exposed to the highly dynamic Norwegian and West Spitsbergen currents, which carry warm and saline Atlantic water (Elverhøi et al., 1998).

It was repeatedly mentioned above that the Iceland Sea exhibits specific features that do not allow us to consider this zone either as continental margin or pelagic zone. The reason is rather evident: Iceland is a relatively small island; therefore, its influence on the SR is confined mainly to inner zones of the Iceland Plateau. The northern zones represent an arena of "pelagic" sedimentation. It should be noted that the term "pelagic zones" in this work is used conventionally because of a high ratio of the total water- (ice-) shed area to the sedimentation basin area. Such way we understand that sedimentation in this area corresponds to the hemipelagic type.

Lastly, we would like to consider the influence of currents on SR values and their distribution. In our opinion, the Norwegian and West Spitsbergen currents affect the SR in the eastern, southeastern, and central pelagic zones of the Norwegian–Greenland Basin via an increase in the $CaCO_3$ content in bottom sediments during interglaciations and interstadials, on the one hand, and during Nordway events at the time of glacials and glaciations, on the other hand. The East Greenland Current influenced vast areas of the Greenland shelf and eroded bottom sediments and/or washed out their fine fractions, which can accumulate only in different shelf depressions. Present-day contour currents, the velocity of which may reach 100 cm/s in the bottom water layer, are recorded along the continental slope in southeastern Greenland (Rasmussen et al., 2003). They affect the bottom landforms and transport the sedimentary material to south. The powerful Iceland deep-water current flowing from the Norwegian–Greenland Basin to North Atlantic affected bottom sediments of the Denmark Strait and decreased the SR value in the Holocene (Völker, 1999; Kuijpers et al., 2003). We failed to find any serious evidence in favor for a strong influence of cryozoles carried by ice of the Transpolar Drift and the East Greenland Current on the SR in the Fram Strait area. Some sediment accumulations at this zone are confined to either pockets on the slope of the Yermak Plateau (Levitan et al., 2002) or local depressions on plain areas of the seafloor.

The results obtained allow to make the following conclusions:

(1) SR variations in the Norwegian–Greenland Basin during the last glacial/interglacial cycle were mainly due to changes in the volume of continental glaciers of the Northern Hemisphere (Fig. 8.15). Hence, they were controlled by climatic conditions.

(2) Advection of the Atlantic water, which promoted the delivery of biogenic carbonates to bottom sediments, had an insignificant influence on the SR. This process was mainly reflected in pelagic zones in the course of interglaciations and interstadials, and "Nordway events" or "High-Productivity events" during glaciations and stadials.

(3) Circumcontinental zonality of SR distribution has existed during an entire studied period. MIS 5 and MIS 4 periods were characterized by weak and symmetrical zonality, while MIS 3–1 were characterized by strongly expressed asymmetric zonality. The reason for the noted phenomenon serves different histories of the Greenland ice sheet formation, from one side, and of eastern ice sheets (Scandinavian, Svalbard, Barents-Kara), from other side.

After showing that during the last climatic cycle terrigenous sedimentary material has entered into the Norwegian-Greenland Basin mainly from the eastern and western continental margins, it is necessary to examine the processes of sedimentation at these margins in some more detail.

History of Sedimentation on the Vøring Plateau During the Last 25 ka

The works by R. Henrich et al. (1989) played an important role in reconstruction the history of sedimentation in the region, in particular, focusing on sedimentary facies of glacial-interglacial cycles in the Norwegian Sea during the last 350 ka, that were subsequently refined for the Vøring Plateau (Dahlgren and Vorren, 2003). It is worth noting that sedimentary facies in the above mentioned publications correspond to core cross-sections of bottom sediments with different structural features and composition caused by various sedimentary factors. For example, they include turbidites and iceberg-rafted sediments. We believe that the facies identified in these works correspond to different genetic types of marine sediments (Frolov, 1984).

The Vøring Plateau is a key area for solving a paleocirculation issue during the last glacial maximum (LGM). The model proposed by J. Bischof (2000) based on studies of coarse IRD points out widely developed surface currents in the eastern part of the Norwegian–Greenland basin directed southward. The other reconstruction that combines data on IRD, from the one hand, and oxygen isotopes in foraminifers, from the other hand, asserts the existence of circulation which is very similar to the present situation (Hebbeln et al., 1998). The latest reconstruction by M. Sarnthein and his co-authors (Sarnthein et al., 2003) based on a comprehensive database of oxygen isotopes determined in planktonic foraminifers introduces into the present circulation pattern the powerful current which carries the Atlantic waters from the south into the region of the Denmark Strait and abeam of North Iceland steeply turning to the east, where approximately in the area of the Vøring Plateau merges with the Norwegian Current and directs northward. Therefore, it is of vital importance to study the connection of special features of composition and structure of bottom sediments with the synchronous climatic and oceanological changes and development of proxies of such changes for the contemporary stage of lithological investigations of the eastern Norwegian Sea.

Furthermore, it is useful to study sediments of the Vøring Plateau because this plateau occupies a unique position in the system of the continental margin of the Norwegian Sea. Indeed, the detailed geophysical and geochronological investigations in the recent years have revealed that sediments of this margin, which have been accumulating over the last 30 ka (except for sediments of the Vøring Plateau), belong to numerous submarine slumps, deep-sea fans, and channels. Consequently, the "background" hemipelagic sediments, which have been accumulating particle-by-particle, account for only 7% of sediments of this period (Vorren et al., 1998; Ó Cofaigh et al., 2002; Taylor et al., 2002a). Therefore, it is difficult to overestimate the significance of the Vøring Plateau for the investigation of background sedimentation and its relation with climate and oceanographic environment during the last 25–30 ka.

Materials and Methods

The obtained results are based on the study of five bottom sediments cores retrieved from the Vøring Plateau area during cruises 12 and 16 of the R/V *Academik Sergey Vavilov* (Shirshov Institute of Oceanology, Russian Academy of Sciences) in 1997 and 2003, respectively (Fig. 8.16). Figure 8.16 shows the locations of sediment cores retrieved by the Norwegian researchers during cruises of the R/V *Marion Dufresne* (Dokken and Jansen, 1999) and *Jan Mayen* (Dahlgren and Vorren, 2003). Radiocarbon datings are available for these cores. We scrutinized the reference section of sediment core ASV 1372. This core was correlated with other cores using analytical data on the chemical composition, physical properties, and lithological descriptions. Specific attention was focused on marker intervals (iceberg-rafted sediments, turbidites, carbonate-rich sediments, and others).

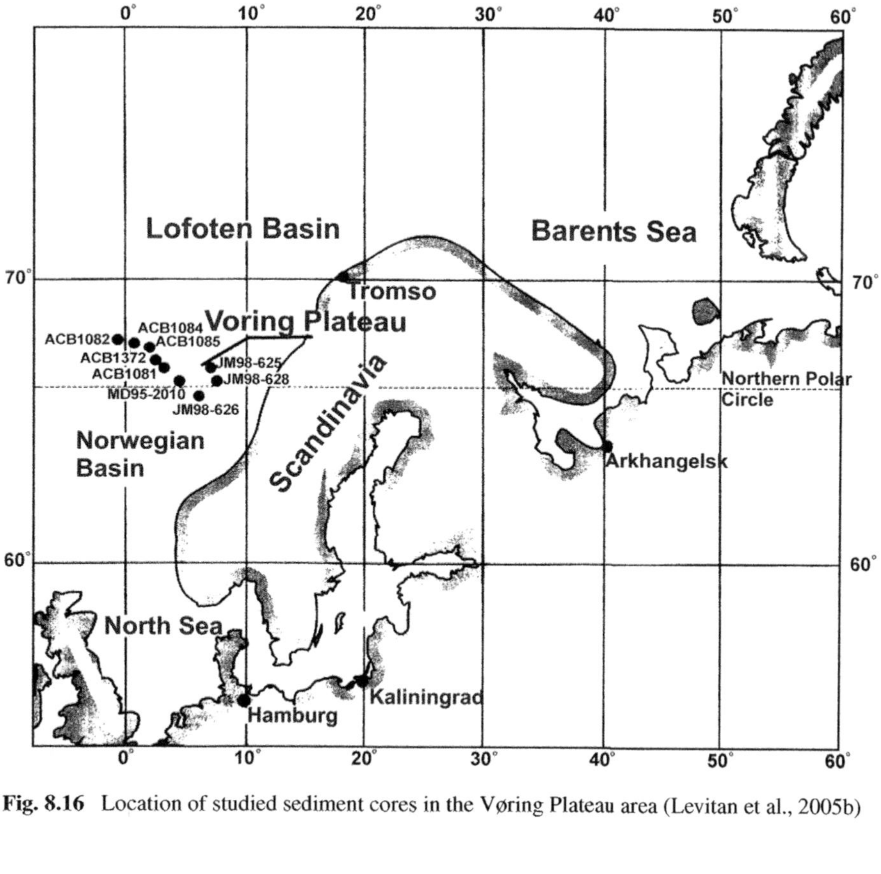

Fig. 8.16 Location of studied sediment cores in the Vøring Plateau area (Levitan et al., 2005b)

Modern Sedimentation Environment

The Vøring Plateau is a sub-horizontal block of the continental slope off central Norway (35000 km^2 in area) located at a depth of 1000–1500 m in the Norwegian Sea. The shelf edge is located at a depth of 250 m. The upper continental slope is gentler than the lower slope (below the Vøring Plateau). The plateau is bounded by the Lofoten Basin in the north and is truncated by the Jan Mayen Fault Zone in the southwest and west. The fracture zone separates the plateau from the Norwegian Basin.

The surface water circulation in the study area includes two major northward currents. The Norwegian Coastal Current (salinity less than 35‰, up to 32‰ in summer) (Vogt, 1986) passes over the shelf, while the Norwegian Current (salinity more than 35‰) transports Atlantic waters into the Norwegian-Greenland Basin. The major branch of the current flows over the Vøring Plateau and the Norwegian Basin. The water column of the studied region can be divided into the following water masses (from top to bottom): the North Atlantic Surface Water with salinity equal to 35‰ or more and temperatures of 9–10°C in summer and 4.6–6.2°C in winter (Sarnthein et al., 2003); the Norwegian Sea Deep Water (salinity 34.92‰; temperature < 0°C); the Norwegian Sea Bottom Water (salinity ~35‰, temperatures from –1.4 to –0.9°C). At present, the Norwegian Sea Deep Water washes the Vøring Plateau summit (Fig. 8.17).

Sea ice is absent in the study area the whole year round. Even during the Little Ice Age in 1800–1818, the eastern boundary of the sea ice was located far west of the Vøring Plateau (Sarnthein et al., 2003).

The plateau is covered with a fine layer of calcareous clay with the $CaCO_3$ content of 10–30% (Hebbeln et al., 1998). Carbonates are mainly composed of planktonic foraminifers dwelling in the North Atlantic water column. Benthic foraminifers and calcareous nannoplankton are subordinate. Diatoms are also present. IRD is absent. Clay minerals are mainly represented by the illite-chlorite assemblage (Berner, 1991). It is believed that the terrigenous sedimentary material was mainly delivered from Norway and its continental margin (Hebbeln et al., 1998). Thus, lateral (relative to the continental margin) transport of sedimentary material has played a major role.

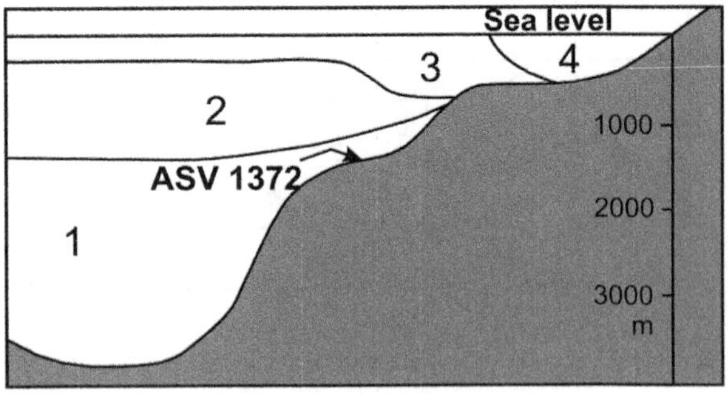

Fig. 8.17 Water-mass stratification in the Vøring Plateau area (Levitan et al., 2005b). 1 – Bottom Water; 2 – Deep Water; 3 – Atlantic Water; 4 – Norwegian Coastal Water

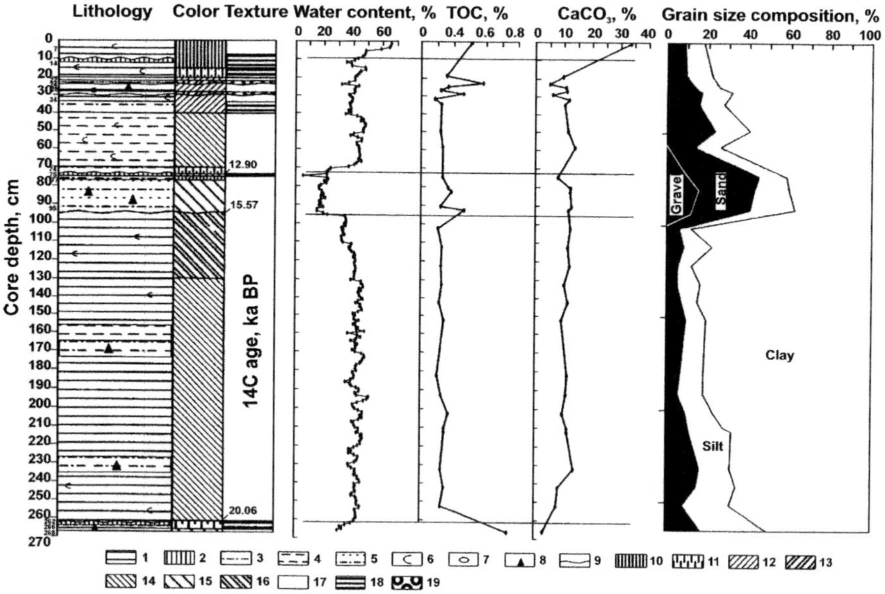

Fig. 8.18 Lithology, TOC and $CaCO_3$ amount, and grain-size distribution in sediment core ASV 1372 (Levitan et al., 2005b). 1–8 – lithology: 1 – clayey mud, 2 – hardground, 3 – sandy-silty clay, 4 – silty clay, 5 – sandy-silty-clayey mictite, 6 – marly mud (10–30% $CaCO_3$), 7 – fragments of dense clay, 8 – IRD; 9 – erosion boundary; 10–16 – color: 10 – brown, 11 – brownish-gray, 12 – olive-gray, 13 – greenish-gray, 14 – light gray, 15 – dark gray, 16 – light yellowish-gray; 17–19 – texture: 17 – homogenous, 18 – laminated, 19 – bioturbated. ^{14}C dates concern iceberg sediments from «Marion Dufresne» and «Jan Mayen» sediment cores (their location is shown at Fig. 8.16)

Results of the study of sediment core ASV 1372 are presented in Figs. 8.18, 8.19, 8.20, 8.21 and 8.22 and Tables 8.3 and 8.4.

The obtained results allow us to identify the following lithostratigraphic units in the cross-section (from top to bottom):

(1) Unit 1 (0–9 cm) is brown marly (33.7% of $CaCO_3$) clayey silt (Fig. 8.19a) with high content of TOC (0.62%) and water (63.5%). The wet bulk density is low (1.60 g/cm^3). This unit contains approximately 40% biogenic micro fossils represented by the following varieties (in decreasing order): calcareous nannoplankton, planktonic foraminifers, diatoms, and sponge spicules). The contents of planktonic and benthic foraminifers are 2100 and 600 tests/g, respectively. The content of clastic mineral grains in >0.1 mm fraction is minor (approximately 1000 grains/g). The benthic foraminifers are mainly represented by *Cassidulina teretis*(up to 40%) and *C. reniforme.*

The abiogenic portion is mainly composed of the following clay minerals: kaolinite (16%), smectite (18%), illite (29%), and chlorite (37%). Reconstructions based on planktonic foraminifer assemblages (Barash and Os'kina, 1978)

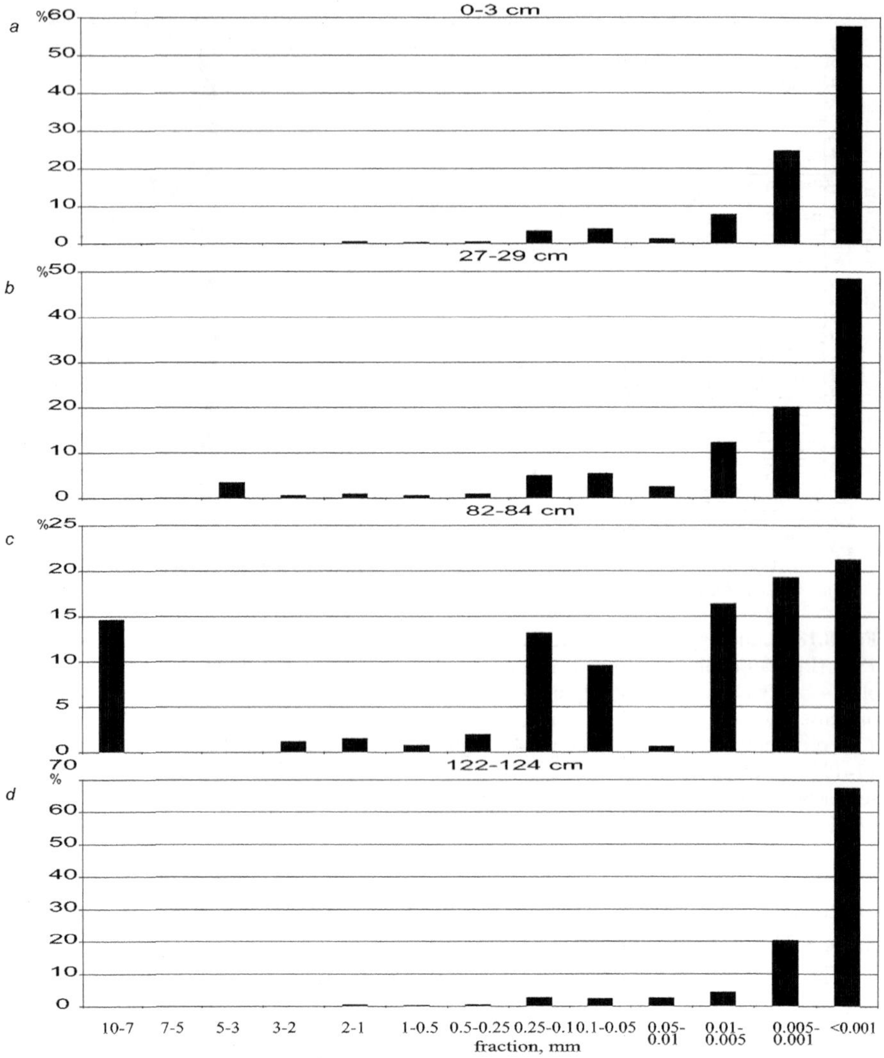

Fig. 8.19 Examples of grain-size distribution (histograms) from sediment core ASV 1372 (Levitan et al., 2005b). *a–d* – core depths of samples, cm: *a* – 0 – 3, *b* – 27 – 29, *c* – 82 – 84, *d* – 122 – 124

suggest that the mean annual sea surface temperature (SST) was 6°C. These temperatures are insufficiently high relative to the present-day measured SST. In the Holocene, they probably corresponded to the pre-Boreal, according to the chronostratigraphic scale for the last 14 ka (Merkt, 1999). Thus, Unit 1, probably, corresponds to the earliest Holocene.

(2) Unit 2 (9–30 cm). This horizon is of complex structure and consists of the following layers (from top to bottom):

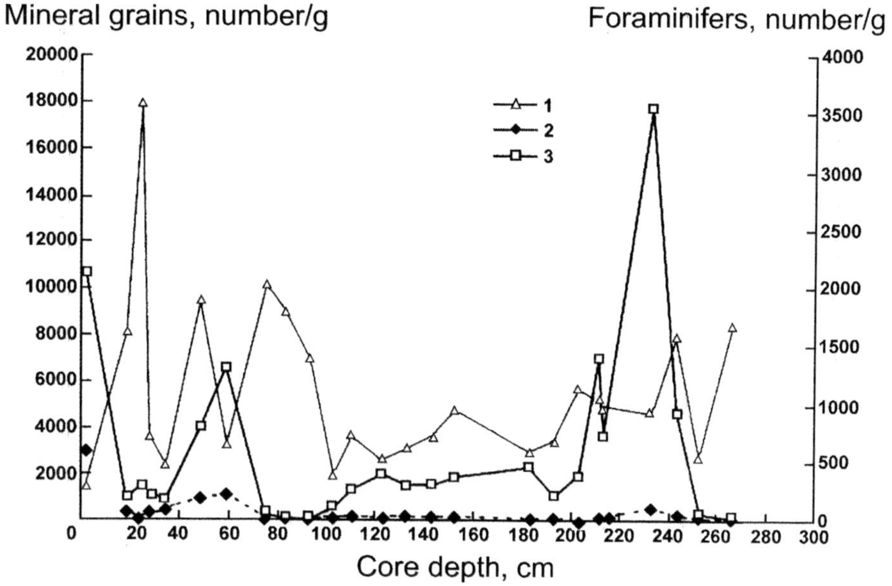

Fig. 8.20 Concentration of mineral grains and foraminifers in the >0.1 mm fraction (Levitan et al., 2005b). 1 – mineral grains, 2 – benthic foraminifers, 3 – planktonic foraminifers

9–11 cm. Dark grayish brown hardground (water content 35%)

11–22 cm. Pale grayish brown terrigenous clayey silt (9.73% $CaCO_3$) with a small amount of TOC (0. 31%) that becomes yellowish at the interval of 16–19 cm. Water content is 48%. The content of biogenic micro-fossils does not exceed 5–6%, and siliceous microfossils disappear at the base. The unit includes both rounded and angular quartz and feldspar grains, sandstone fragments, and calcite crystallites (dissolution residue of planktonic foraminifer tests). Clay minerals are mainly represented by a smectite–illite assemblage.

22–24 cm. Greenish gray intensely bioturbated terrigenous (4.91% $CaCO_3$) clayey silt with silt and sand and high TOC content (0.76%). The sediment is slightly consolidated (water content ~30%). This interval resembles the interval 11–22 cm in terms of composition.

24–29 cm. Olive-gray (10.8–11.4% $CaCO_3$) bioturbated sandy-silty clay (Fig. 8.19b) with a chalk fragment approximately 2 cm in size. The TOC content is low (0.24–0.33%). The water content ranges from 38 to 42%. The number of planktonic foraminifers is 250 tests/g. The benthic foraminifers are virtually absent. The clastic component is abundant (up to 18000 grains/g). The benthic assemblage is strongly depleted in *Cassidulina* and enriched in agglutinated forms indicating an intense dissolution in the bottom layer. The content of clay minerals is strongly variable. For example, the smectite content locally drops to 11% due to increase in contents of kaolinite and chlorite up to 23 and 18%, respectively.

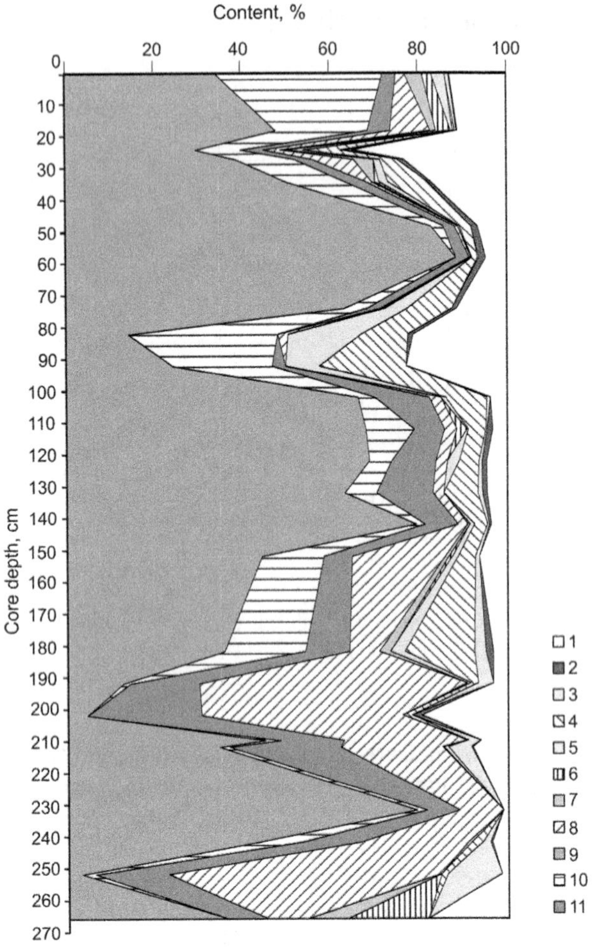

Fig. 8.21 Composition of benthic foraminifer assemblages in sediment core ASV 1372 (Levitan et al., 2005b). 1 – *Islandiella spp.;* 2 – *Fissurina/Parafissurina;* 3 – *Elphidium clavatum;* 4 – *Cibicides wuellerstorfi;* 5 – *Epistominella exigua;* 6 – *Epistominella pusilla;* 7 – *Melonis barleeanus;* 8 – *Oridorsalis umbonatus;* 9 – *Cassidulina reniforme;* 10 – *Cassidulina teretis*; 11 – other species

29–30 cm is characterized by intensively bioturbated sediments resembling those at interval 22–24 cm. Contents of $CaCO_3$ and TOC are equal to 6.07% and 0.52%, respectively.

In a whole, sediments of Unit 2 are characterized by high content of reworked pre-Quaternary benthic foraminifers. Based on the available data, we suggest that these sediments have been accumulated during the Younger Dryas cooling (Merkt, 1999).

(3) Unit 3 (30–70 cm) is olive or light gray, layered, very soft and stickysilty clay (10.39–14.14% $CaCO_3$, 0.17–0.27% TOC) with abundant clay lumps and sand in the middle part (49–52 cm). The water content varies from 39 to 45%; wet bulk density – from 1.57 to 1.59 g/cm^3; content of planktonic foraminifers increases down cross-section to as much as 1250 tests/g dry sediment at the base. In the same direction, the number of benthic foraminifers also slightly

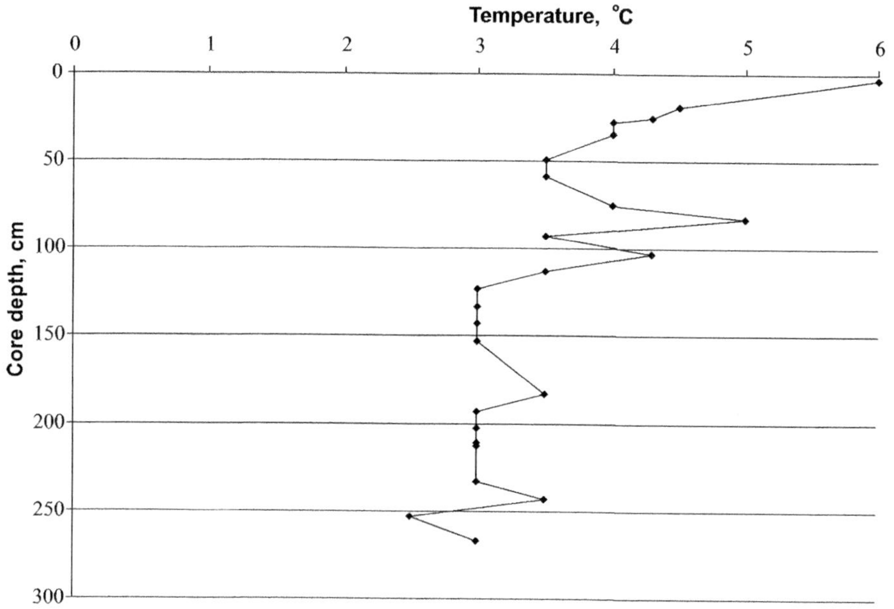

Fig. 8.22 Reconstruction of annual SST based on planktonic foraminifers (Levitan et al., 2005b)

increases and reaches 200 tests/g at the base. The content of clastic component is approximately 2000–3000 grains/g dry sediment and increases to nearly 10000 grains/g in the middle (coarser-grained) part. This unit includes numerous *C. teretis* (up to 87% of the total assemblage), indicating the maximum of Atlantic waters advection. The composition of clay minerals is variable. Contents of smectite and illite are very low, while the chlorite content increases in the middle part. Sediments of this unit probably formed during the Bølling-Allerød warming episode, whereas sediments of the middle part accumulated during the short-term Old Dryas cooling episode (Merkt, 1999).

Table 8.3 Physical properties in sediment core ASV 1372 [*Levitan et al.*, 2005]

Lithostratigraphic unit, cm	Water content, wt. %	Bulk density, g/cm^3
1–5	63.5	1.6
31–35	39.48	1.59
58–62	44.57	1.57
86–90	18.62	2.13
121–125	42.56	1.59
165–169	43.31	1.57
233–237	43.7	1.55
251–255	44.86	1.52
268	24.4	–

Table 8.4 Clay mineral composition in sediment core ASV 1372 [*Levitan et al.*, 2005]

Core depth, cm	Smectite	Illite	Kaolinite	Chlorite
0–3	18.4	28.6	16.1	36.9
18–20	19.3	50.1	13.8	16.8
24–26	11.0	48.3	23.2	17.5
27–29	19.3	53.8	13.0	13.9
34–36	16.4	57.9	9.7	15.9
45–47	12.7	58.5	16.4	12.4
58–60	5.5	48.0	17.2	29.3
74–76	5.8	51.0	9.8	33.4
82–84	4.1	50.7	13.4	31.8
92–94	13.4	57.3	11.1	18.2
102–104	7.9	67.0	11.0	14.2
112–114	7.5	65.9	15.7	10.9
122–124	9.6	67.1	9.4	13.9
132–134	9.6	66.9	12.5	11.0
142–144	5.8	63.9	9.8	20.5
182–184	6.9	55.4	11.3	26.5
192–194	12.4	52.8	12.8	22.0
200–202	7.3	56.2	12.5	24.0
202–204	15.1	68.0	6.1	10.9
212–214	14.3	53.7	10.5	21.4
222–224	18.0	56.3	7.3	18.4
232–234	18.2	61.1	7.1	13.7
242–244	12.9	53.9	13.0	20.2
252–254	15.1	37.1	25.6	22.3

(4) Unit 4 (70–77 cm) is brownish gray terrigenous silty clay with silty–sandy admixture and a rather compact dark brown hardground represented by gravelly mixtite (8.07% $CaCO_3$ and 0.27% TOC) at the interval of 74–75 cm. In terms of the composition of clay minerals, the hard ground resembles sediments from the middle part of Unit 3. However, the kaolinite content is significantly decreased.

(5) Unit 5 (77–95 cm) is dark gray gravelly mixtite or diamicton (Fig. 8.19c) with abundant carbonates (fragments of shelf mollusk shells, 12–13%) and TOC varying from 0.54% at the base of the unit to 0.25–0.38% at the top. The unit is characterized by low water content (18.62%), high wet bulk density (2.13 g/cm^3), abundant clastic component (up to 10000 grains/g), and virtual absence of planktonic foraminifers. The benthic foraminifer assemblage is characterized by a low content of *C. teretis* (up to 16%) and high contents of *C. reniforme, Epistominella pusilla, E. exigua,* and pre-Quaternary and agglutinated forms. This implies both intense dissolution and input of several species from the shelf. The clay mineral assemblage reflects a change in their provenance in the course of the accumulation of sediments. Layer 82–84 cm includes a smectite–kaolinite–illite assemblage, whereas layer 92–94 cm contains a kaolinite–smectite–chlorite–illite assemblage. The latter layer is depleted in chlorite and represents iceberg-rafted sediment that probably

corresponds to the HA1 layer in the adjacent cores (Dahlgren and Vorren, 2003) and the Heinrich 1 (H1) layer in the North Atlantic (Hald, 2001).

(6) Unit 6 (95–129 cm) is light gray, slightly calcareous, soft and sticky clayey silt (Fig. 8.19d). The carbonate content varies from 11.80 to 12.64%, while the TOC content varies from 0.22 to 0.27%. The carbonates are represented by abundant planktonic foraminifers (up to 500 tests/g at the base) and remains of calcareous nannoplankton. The number of benthic foraminifers does not exceed several dozen tests/g, and the content of *C. teretis* is approximately 60%. The content of clastic component varies from 2000 to 4000 grains/g. The water content slightly increases downward the section (42.56% at 121–125 cm), and the wet bulk density is 1.59 g/cm^3. Clay minerals are mainly represented by a smectite-kaolinite-chlorite-illite assemblage with a very high content of illite and low content of smectite. It should be emphasized that the zone below the level of 129 cm is characterized by a significant warming of the surface water column. Thus, this level possibly marks the LGM termination. The average water content is 43.31%. The wet bulk density is 1.57 g/cm^3. The clay mineral assemblage is very variable. The upper and lower parts are characterized by abundance of smectite and low contents of chlorite and kaolinite, while the middle part (142–202 cm) is depleted in chlorite (and illite) and enriched in chlorite. Temperature of the surface water column did not exceed 3°C. The sediments, presumably, correspond to the LGM episode.

(7) Unit 7 (129–204 cm) is gray clayey silt with some admixture of sand and gravel. The $CaCO_3$ content varies from 9.73 to 11.98%; the TOC, from 0.22 to 0.36%. Content of clastic component (including dolomite "rhombs") is increased (up to 6000 grains/g at the base). The number of planktonic foraminifers does not exceed 420 tests/g. Content of benthic foraminifers is very low. The content of *C. teretis* rapidly decreases downward the section, while the content of *O. umbonati* s and *M. barleeanus* increases.

(8) Unit 8 (204–266) cm is characterized by gray clayey silt (with silty-sandy admixture), sandy-silty, and silty-sandy clays with local gravel dissemination. The average content of clastic material is approximately 6000 grains/g. Glauconite grains and rare fragments of shelf mollusk shells are also found. At the base of the unit, the color of sediments changes to dark brownish gray. The $CaCO_3$ content varies from 8.23 to 14.30%, and the highest carbonate content at interval 230–235 cm is related to a very high number of planktonic foraminifers (as much as 3600 tests/g). Such interlayers are known in the Norwegian–Greenland Basin as "Nordway Events" (Vorren et al., 1998) or "High Productivity Events (HP)" (Hald, 2001). The possibility of correlation of the HP1 layer (Hald, 2001), which is slightly enriched in benthic foraminifers, with the Dansgaard–Oeschger 2 event will be considered below. The variation of benthic foraminifer assemblages is very regular. The content of *C. teretis* increases from the base of the unit to as much as 75% at interval 230–235 cm and then decreases to 3% at the top of the interval. This interval is virtually devoid of *E. clavatum* that is abundant in the overlying sediments. The TOC content varies from 0.27 to 0.31%. The water content and wet bulk density

are 43–45% and 1.52–1.55 g/cm^3, respectively. The clay mineral assemblage is generally dominated by smectite and illite, with the illite content decreasing toward the base. The content of kaolinite and chlorite is variable. Paleotemperatures of surface waters range from 2.5 to 3.5°C, suggesting the accumulation of sediments during the LGM.

(9) Unit 9 (266–268 cm). The base of this unit includes iceberg-rafted dark gray gravelly mixtite (diamicton) with low water content (24.4%). According to (Dahlgren and Vorren, 2003), the wet bulk density is as high as 1.92 g/cm^3 in the HA2 layer. These sediments contain numerous sand-sized grains of quartz, feldspars, glauconite, dark-colored minerals, rock clasts, and mollusk shell fragments (as much as 8000 grains/g). The clayey fraction is enriched in very fine clastic material and depleted in clay minerals. The $CaCO_3$ content is 3.57%, while the TOC content is high (1.09%). Planktonic foraminifers are virtually absent. The benthic foraminifers are dominated by pre-Quaternary and agglutinated forms. The clay mineral assemblage is characterized by a very low content of smectite, high contents of chlorite and illite, and a moderate content of kaolinite. Based on literature data, these sediments, possibly, correspond to the HA2 layer (Dahlgren and Vorren, 2003) that does not coincide in age with the Heinrich 2 (H2) layer (Hald, 2001) and predates the LGM event.

Age Model, Sedimentation Rates, and Mass Accumulation Rates

If we assume that Unit 1 (0–9 cm) corresponds to the pre-Boreal, its age should range between 9.0 and 11.5 cal. kyrs BP (Klitgaard-Kristensen et al., 2001). Correspondingly, sedimentation rate is equal to 3.6 cm/ky (or more if the unit includes only the pre-Boreal section). In this case, the mass accumulation rate of terrigenous material (without $CaCO_3$ and TOC) is equal to 1.38 g/cm^2/ky. Mass accumulation rates of $CaCO_3$ and TOC are 0.71 and 0.013 g/cm^2/ky, respectively. One should take into consideration that the biogenic opal makes up 3–5% and is included in the terrigenous material composition. If the biogenic opal is omitted, mass accumulation rate of the terrigenous material is equal to 1.30 g/cm^2/ky (Table 8.5).

Unit 2 (9–30 cm), probably, corresponds to the Younger Dryas. This is suggested by the presence of regional hardground at the top of this unit in the eastern Norwegian Sea and on the Yermak Plateau (Levitan et al., 2000). Moreover, the two-member structure of this unit noted above is typical for the Younger Dryas on the Vøring Plateau (Dahlgren and Vorren, 2003). If we accept that the studied core includes the complete stratigraphic volume of the Younger Dryas, the age of its top and bottom corresponds to 11.5 and 12.7 cal. kyrs BP, respectively (Klitgaard-Kristensen et al., 2001). Then, the sedimentation rate reaches 17.5 cm/ky. Mass accumulation rates of terrigenous material, carbonates, and TOC are 16.44, 1.56, and 0.078 g/cm^2/ky, respectively (Table 8.5). As it was mentioned above, biogenic opal is absent in both the studied and older sediments.

Unit 3 (30–70 cm) and, probably, Unit 4 (70–77 cm) correspond to the Bølling/Allerød episode in our model, i.e., the period of 14.4–12.7 cal. kyrs BP

Table 8.5 Sedimentation rates and mass accumulation rates of major sediment components in sediment core ASV 1372 [*Levitan et al.*, 2005]

Lithostratigraphic unit	Interval, cm	Sedimentation rate, cm/ky	Mass accumulation rate, g/cm^2/ky		
			Terrigenous	$CaCO_3$	TOC
1	0–9	3.6	1.3	0.71	0.013
2	9–30	17.5	16.44	1.56	0.078
3	30–77	27.6	22.4	2.84	0.061
4	77–95	4.9	7.43	1.07	0.032
5	95–129	30.9	24.62	3.47	0.071
6	129–204	34.7	27.46	3.41	0.087
7	204–266	34.7	26.46	3.3	0.086
8	266–268	6.9	9.55	0.36	0.011

(Klitgaard-Kristensen et al., 2001). This conclusion is supported not only by results of the foraminifer analysis presented above, but also by the age of the underlying unit in adjacent cores from the Vøring Plateau (Dahlgren and Vorren, 2003). Moreover, the middle part of this unit contains an interlayer of coarser-grained sediments with the characteristic clay mineral composition that can logically be related to the Older Dryas cooling 13.9–13.6 cal. kyrs BP (Klitgaard-Kristensen et al., 2001). The presence of hardground (74–75 cm) suggests a hiatus of uncertain duration. It is impossible to confidently judge about this hiatus without datings. Therefore, calculations of sedimentation rate and mass accumulation rates were based on the assumption of a virtual lack of hiatus. Thus, we obtained the following values (Table 8.5): sedimentation rate is 27.6 cm/ky; mass accumulation rate of terrigenous material, $CaCO_3$, and TOC are 22.40, 2.84, and 0.061 g/cm^2/ky, respectively. If the hiatus is present, all calculated values will naturally increase.

The corrected age of Unit 5 (77–95 cm) composed of iceberg-rafted sediments was dated in several cores of the study region, is 12.90–15.57 ^{14}C ky (Dahlgren and Vorren, 2003) or 14.34–18.00 cal. kyrs BP (Stuiver et al., 1998). Consequently, sedimentation rate is 4.92 cm/ky, while mass accumulation rates of terrigenous material, $CaCO_3$, and TOC are equal to 7.43, 1.07, and 0.032 g/cm^2/ky, respectively. A similar unit is recorded at approximately the same level in two adjacent cores (Dahlgren and Vorren, 2003).

Unit 6 (95–129 cm) is composed of sediments attributed to the early deglaciation phase formed 18.0–19.1 cal. kyrs BP (Dahlgren and Vorren, 2003; Weinelt et al., 2003). These values yield a sedimentation rate of 30.9 cm/ky. Mass accumulation rates of terrigenous material, $CaCO_3$, and TOC are equal to 24.62, 3.47, and 0.071 g/cm^2/ky (Table 8.5).

According to our reconstructions, Unit 7 (129–204 cm) belongs to sediments of the later half of the LGM period. The age of its base (21.26 cal. kyrs BP) has been tentatively determined based on sedimentation rate during the LGM, on the whole (Dahlgren and Vorren, 2003), and thickness of the unit. Results of the calculation indicate very high values of all estimated parameters: sedimentation rate is

34.7 cm/ky; mass accumulation rates of terrigenous material, $CaCO_3$, and TOC are 27.46, 3.41, and 0.087 g/cm^2/ky, respectively.

Sediments of Unit 8 (204–266 cm) formed during the earlier half of the LGM. The base of this unit is 23.05 cal. kyrs BP old. The sedimentation rate is similar to that of Unit 7. Mass accumulation rates of terrigenous material, $CaCO_3$, and TOC are also similar (26.46, 3.30. and 0.086 g/cm^2/ky, respectively). Calendar age of sediments with a high content of planktonic foraminifers (230–235 cm) turns out to be 1.40 ky younger than the Dansgaard-Oeschger 2 event in the North Atlantic (Weinelt et al., 2003). Therefore, these sediments cannot be correlated with the latter event.

Unit 9 (iceberg-rafted sediment) has only been retrieved in the upper part (266–268 cm). Corrected ages reported in literature for the top and bottom of analogous units in the adjacent cores are 20.06 and 21.2 ^{14}C ka, respectively (Dahlgren and Vorren, 2003). Comparison of unit calendar age with an age of layer H2 in the North Atlantic (Thiede and Tiedemann, 2001) shows that iceberg-rafted sediments in the Norwegian Sea are nearly 2 ky younger than their counterparts in the North Atlantic (Hald, 2001). Calculations based on these parameters indicate that mass accumulation rates of terrigenous material, $CaCO_3$, and TOC are equal to 6.9 cm/ky, 9.55, 0.36, and 0.011 g/cm^2/ky, respectively.

The character of mass accumulation rates of all components of bottom sediments is governed by the sedimentation rate. The highest sedimentation rates are recorded for the LGM. In general, the sedimentation rate decreases upward the section and is notably lower in the iceberg-rafted sediment.

Since the sediments are mainly composed of terrigenous material that, in turn, is primarily delivered from central Norway and its shelf (see discussion below), the sedimentation rate history is primarily a function of the evolution of delivery of terrigenous material in the course of glaciation in the study region. This inference is consistent with results of the analysis of sedimentation rate history in the Norwegian–Greenland Basin during the last climatic cycle (Levitan and Stein, 2005).

Differences in the $CaCO_3$ distribution along the section were described above. They suggest that carbonates are represented by planktonic foraminifer remains with a minor contribution of benthic foraminifers. However, the role of calcareous nannoplankton significantly increased in the Holocene. IRD was dominated by fragments of shelf mollusk shells, whereas foraminifers were dissolved. The LGM sediments locally include dolomite "rhombs" repeatedly described as a clastic material from the eastern Norwegian Sea (Vogt, 1997).

Distribution of TOC in the core section (Fig. 8.18) shows that sediments enriched in IRD (Younger Dryas iceberg-rafted sediment) are also enriched in organic carbon, suggesting the primarily terrigenous nature of organic matter in the studied sediments. At the same time, the TOC distribution indicates that other intervals of the section are dominated by marine organic matter. This fact is known way back for the eastern Norwegian Sea (Vogt, 1997). These specific features of the distribution of carbonate and organic carbon should be taken into consideration in the qualitative or quantitative reconstructions of paleoproductivity.

History of Sedimentation

At first, we consider some methodical aspects related to clay mineral data interpretation. Correlation analysis revealed the absence of positive correlation between any pair of clay minerals; i.e., each major clay mineral has its own source province. Significant negative correlations are noted for the following pairs of clay minerals: smectite/chlorite (–0.62), illite/chlorite (–0.72), and illite/kaolinite (–0.61).

Study of clay mineral distribution in surface sediments of the Norwegian Sea (Berner, 1991) and analysis of our core data (Table 8.4) demonstrated that smectite is primarily derived from Iceland and sediments of the Iceland shelf. Kaolinite is probably delivered from the North Sea, as suggested by kaolinite content in sediments of the Younger Dryas that contains a chalk fragment. J. Bischof (2000) demonstrated that the chalk might be delivered as IRD to the Norwegian-Greenland Basin only from the North Sea due to erosion and abrasion of the chalk formation in Denmark, southern Norway, and Great Britain. Chlorite is probably derived from Paleozoic rock of central Norway and is concentrated in the iceberg-rafted sediment.

Finally, illite is presumably derived from Mesozoic rock developed on the central Norwegian shelf (Bischof, 2000). This is indicated, for example, by the considerable content of illite in sediments accumulated during the late LGM and early deglaciation. Of course, one should take into consideration that each of the provinces mentioned above delivers not only the major clay mineral, but also a certain amount of other minerals. Our data on the Vøring Plateau revealed differences in the distribution of clay minerals, relative to the central Norwegian-Greenland Basin that received kaolinite mainly from the Barents Sea (Kuhlmann et al., 1993). In both areas, smectite is concentrated in sediments formed under relatively warm conditions. Provenance of clay minerals has not changed during the geological history, but their transport agents and, sometimes, transport pathways changed. At present, for example, the studied region is devoid of sea ice or icebergs that played a significant role during glaciation period.

Let us consider sedimentation history of the study region in the chronological succession (Levitan and Stein, 2005), taking into account that ideas on advection of Atlantic waters are based on mass accumulation rates of $CaCO_3$ combined with the data on carbonate development and results of foraminifer analysis (Fig. 8.23).

Iceberg-rafted sediment (HA2) accumulated at the end of the Late Weichselian Fennoscandian Ice Sheet growth (Hald, 2001). The western margin of the ice sheet was located at that time in the central Norwegian shelf area (Dahlgren and Vorren, 2003). These facts convincingly indicate that central Norway and its shelf served as the major source province. Sedimentation rates and mass accumulation rates of major sedimentary components were rather low. Transport agents were primarily represented by icebergs that were widespread during surges, i.e., large-scale breakup of marine margins of ice sheets. Appearance of huge freshwater masses owing to the melting of icebergs promoted a strong stratification of the water column, resulting in a nearly complete breakdown of thermohaline circulation (Sarnthein et al., 2001), and rapid precipitation of clay minerals due to changes in their charges during the interaction of melted water with sea water. The sea surface over the

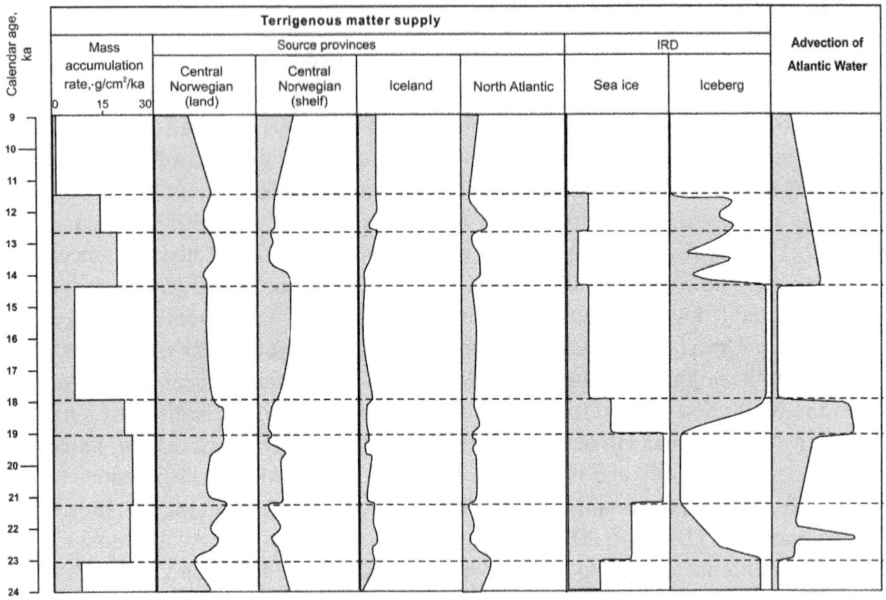

Fig. 8.23 Compilation of data on mass accumulation rates and source provinces of terrigenous matter as well as advection of Atlantic Water in the area of sediment core ASV 1372 for the time interval 9–14 cal. kyrs. BP (Levitan et al., 2005b)

Vøring Plateau was covered with icebergs and sea ice. Also, one cannot rule out the existence of small areas of open sea that are essential for the formation and drift of icebergs. Benthic foraminifers and relatively high kaolinite content suggest that input of Atlantic surface waters from the south was insignificant at that time, because these waters were located in the subsurface layer (Bischof, 2000). The extremely low smectite content implies that sedimentary material was not delivered to the Vøring Plateau area from Iceland. High TOC contents in sediments promoted intense input of CO_2 into the bottom water during diagenesis and dissolution of carbonate foraminifer tests.

The formation of iceberg-rafted sediment was postdated by more severe climatic conditions and onset of the LGM. The western boundary of the Fennoscandinavian ice sheet shifted to the shelf edge (Dahlgren and Vorren, 2003). The area occupied by open sea and number of icebergs have been abruptly decreased. Sedimentary material was mainly transported by sea ice and sub-ice flows that delivered sedimentary material to the continental slope owing to the "bulldozer" effect during oscillations of the glacier edge. Consequently, sedimentation rates and sedimentary material fluxes have been sharply increased. This is also indicated by the abundance of *M. barleeanus*. Terrigenous material was actively delivered by icebergs from central Norway at the initial LGM stage. This was followed by an abrupt enhancement of delivery from the shelf province. High smectite content supports the early LGM scenario proposed by (Sarnthein et al., 2000), according to which the Atlantic waters flowed to the Norwegian–Greenland Basin mainly via the passage between Iceland and Greenland, abruptly turned to east in the Iceland Plateau area, and reached the Vøring Plateau. This is indicated by very high content of *C. teretis* that decreased only during the onset and end of the period corresponding to the first half of the LGM. Judging from maximal contents of *O. umbonatis*, the bottom water was very cold (probably, less than 0°C). The extremely irregular distribution of kaolinite in sediments of the lower LGM implies that the Atlantic waters could occasionally penetrate from the North Sea. In particular, sediments of HP1 with abundant planktonic foraminifers are also enriched in kaolinite. It is possible that these sediments were deposited according to the following scenario suggested by (Bischof, 2000): the influence of catabatic winds promoted the formation of polynyas and development of zones with high primary production therein.

In the second half of the LGM, icebergs completely disappeared from the Vøring Plateau and gave way to an almost permanent ice cover. The participation of sub-ice material promoted the fast accumulation of fine sediments mainly dominated by materials from the central Norwegian shelf. In particular, variations in contents of allochthonous benthic foraminifers *E. clavatum* and *C. reniforme* record oscillations in the supply of sedimentary material. This period was marked by a lower contribution of the Iceland branch of Atlantic waters and a higher contribution of the North Atlantic branch. The distribution of *O. umbonatis* suggests a gradual and irregular increase in temperature of the bottom water. High contents of *C. teretis* indicate a simultaneous intensification of the advection of Atlantic intermediate waters.

During the early deglaciation, the bottom water temperature virtually did not differ from the late LGM temperature, while the surface water temperature gradually increased. At the same time, IRD were absent in sediments. The sea surface was completely covered with sea ice that was probably thinner than in the LGM. Sedimentary material was mainly delivered by sub-ice flows formed during the melting and eastward retreat of the ice sheet margin toward the shelf. They supplied the nepheloid layer with very fine suspended matter including downslope density flows. This mechanism was less powerful than the sedimentary material transport during the LGM. Therefore, sedimentation rates and mass accumulation rates were lower. Actually, the subsurface circulation at this time did not differ from the late LGM pattern.

The process of deglaciation was interrupted by deposition of iceberg-rafted sediments over more than 3.5 ky. The scenario of this process was very similar to that of HA2 described above. Both layers are similar with respect to grain-size distribution, physical properties, TOC content, carbonate composition, low sedimentation rates, and sedimentary fluxes. Differences are mainly related to a significantly greater sediment thickness of HA1, which makes it possible to judge about several variations during HA1 development. For example, the middle interval was marked by a notable short-term warming of subsurface waters, intense input of allochthonous shallow-water benthic foraminifers and iceberg-rafted material from central Norway, and an almost complete termination of the Atlantic water input from the Iceland branch. These data confirm the interpretation of paleoceanographic evolution and change of source provinces during accumulation of Heinrich layers (Scourse et al., 2000).

The Bølling-Allerød warming period was accompanied by an increase in SST. Judging from the high content of *C. teretis*, the Vøring Plateau area was characterized by the most intense advection of Atlantic warm saline waters in the studied geological time interval. Allochthonous benthic foraminifers virtually disappeared. By the end of this period, the ice sheet margin retreated to land and iceberg calving ceased. Consequently, the input of Iceland sedimentary material increased due to the contribution of surface currents. During the middle Bølling-Allerød warming period, a short-term (300 yr) cooling episode of Old Dryas intensified the formation of sea ice and the input of sedimentary material by the North Atlantic branch of Atlantic waters. Bottom sediments of this period are characterized by a coarser grain-size distribution relative to the early deglaciation period, suggesting an abrupt reduction of the sea ice cover (up to the point of its disappearance by the end of this period), and probable acceleration of bottom currents. One cannot rule out that intensification of hydrodynamic activity of the bottom layer was responsible for the hardground formation at the Bølling-Allerød base and the hiatus. However, we do not know the duration of this episode. In general, sedimentation rates and sedimentary material fluxes during this period were considerably lower relative to those during the early deglaciation and higher relative to HA1 values.

The two-stage Younger Dryas cooling promoted a short-term formation of sea ice. However, data on IRD and clay minerals suggest the apparent absence of icebergs during this period. The advection of Atlantic waters notably decreased, and the lateral input of allochthonous benthic foraminifers from the shelf slightly increased. Dissolution of foraminifer tests could also slightly increase, for example, due to increase of the TOC content of sediments. The episode of hardground formation at the end of the Younger Dryas may indicate a notable reduction of sedimentation rates and acceleration of bottom currents. As was mentioned above, sedimentation rates and sedimentary material fluxes during the Younger Dryas were lower than those during the Bølling-Allerød warm period.

Thus, the major conditions of sedimentation in the early Holocene were already similar to the present-day ones. Rapid reduction in terrigenous material fluxes against the background of decrease in sedimentation rates and sedimentary material fluxes provoked an abrupt increase in the content of biogenic sedimentary components (HPE-0, according to (Hald, 2001)). The warming could be accompanied by increase of primary production of siliceous algae. The Arctic Front migrated to the west and reached the Vøring Plateau longitude (judging from the sufficiently high smectite content). The formation of sea ice and icebergs ceased. The Atlantic water dominated in the surface water. The lateral transport of sedimentary material is also indicated by the high content of *C. reniforme* in the benthic foraminifer assemblage. Later on, the Arctic Front migrated further to the west and blocked the intervention of smectite into the Vøring Plateau area.

Lithostratigraphic Correlation

Analytical data used for the lithostratigraphic correlation are given in publications cited in the section "Materials and Methods" of the present study. Among the studied cores, three cores (ASV 1081, ASV 1085, and ASV 1372) are located on the Vøring Plateau, while two cores (ASV1082 and ASV 1084) are retrieved from the lower part of the continental slope. Not all marker horizons are traceable in these cores: hardground layers are primarily confined to the plateau itself; turbidite interlayers are only detected on continental slopes; and nearly black hydrotroilite-rich layers are recorded only in cores ASV 1081 and 1085. Finally, a 18-cm-thick volcanic ash layer was only found in sediment core ASV 1082 beneath the Holocene sequence. Judging from its stratigraphic position, this layer may be related to the regional Vedde volcanic ash with the age estimated at 10600 ^{14}C yr (Dahlgren and Vorren, 2003).

Figure 8.24 shows schematic correlation of cores retrieved during cruises of the R/V *Academik Sergey Vavilov*. IRD are present in all cross-sections. We could not detect sediments older than HA2 on the Vøring Plateau, whereas iceberg-rafted sediments with an age up to HA4 have been found on its slope (Core ASV 1084). The same trend is typical of carbonate-rich sediments: their age is restricted by HP1 and HP3 on the Vøring Plateau and its slope, respectively.

The thickness of the Holocene section ranges from 5 to 20 cm. Thickness of iceberg-rafted sediments varies from several centimeters to 28 cm. Turbidite cyclites are also characterized by similar thicknesses. In general, thickness of synchronous sediments along the profile is notably variable (for example, between HA1 and HA2), reflecting the intricate history of sedimentary material delivery from various sources and redistribution of sediments on the seafloor under the influence of diverse currents and flows. In addition, cross-sections from the continental slope show that turbidites often overlie iceberg-rafted sediments. On the Vøring Plateau summit, the iceberg diamicton is overlapped by sediments enriched in coarse-grained clastic material. The termination of surges could be accompanied by dilution of the associated suspension to concentrations sufficient for the formation of relatively coarse-grained sediments on the plateau and turbidites on the slope.

Thus, one can affirm that the oldest sediments in cores retrieved from the Vøring Plateau are only slightly older than HA2 sediments. As it was already noted, bottom sediments from the lower part of the continental slope contain much older sediments that can be referred to iceberg-rafted sediments of the HA3 layer with a corrected age of 25.12–27.00 ^{14}C ky and the HA4 layer with a corrected age of 33.30–35.08 ^{14}C ky (Dahlgren and Vorren, 2003). Thus, if we take into consideration the core length, average sedimentation rates on the Vøring Plateau during the last 25 cal. kyrs BP were higher relative to those on the lower continental slope, probably, because of the steepness of the slope. This circumstance was also responsible for the development of turbidites on the continental slope.

The sedimentary sequence in the upper part of the continental slope between diamictons HA1 and HA2 includes debris-flow deposits (poorly sorted clastic material) that correspond to the LGM moraine on the shelf (Dahlgren and Vorren, 2003).

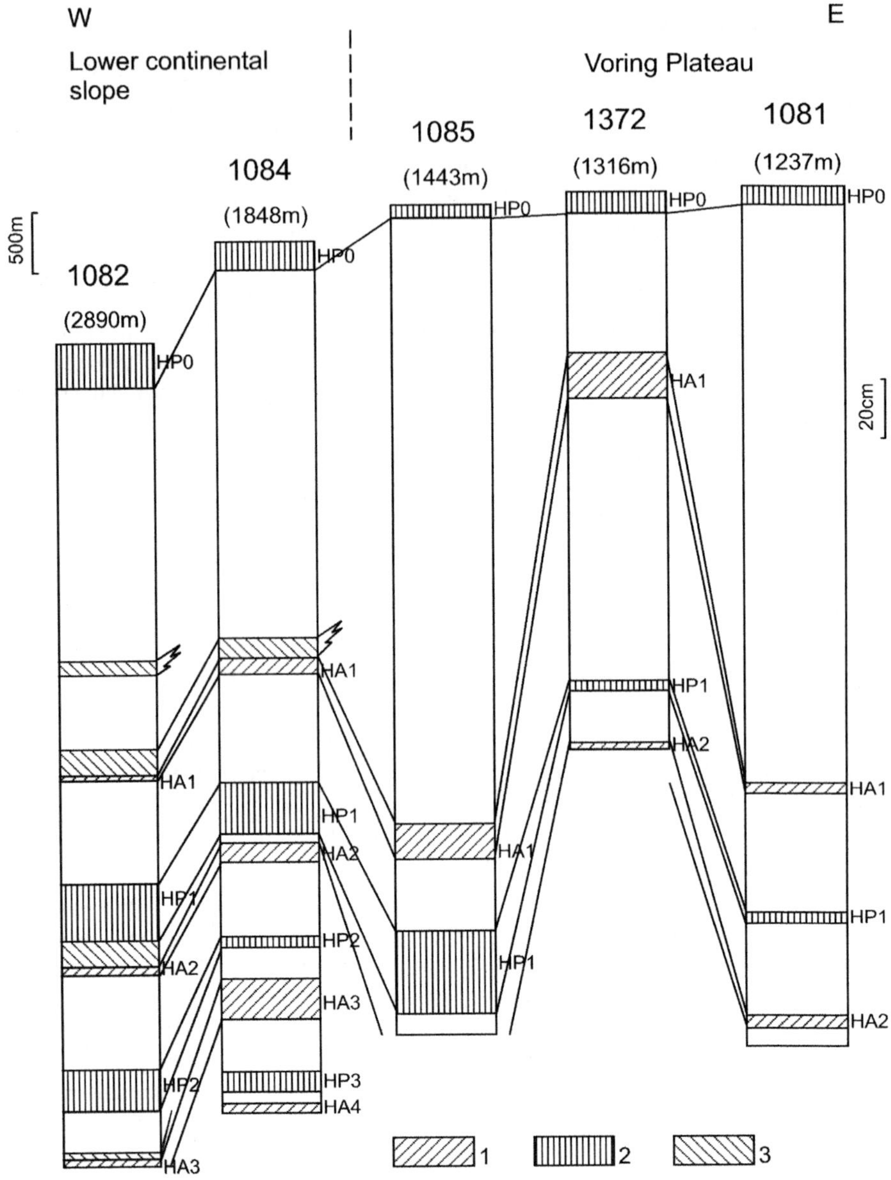

Fig. 8.24 Lithostratigraphic correlation of sediment cores along the Vøring Plateau latitudinal transect (see Fig. 8.16 for location of sediment cores) (Levitan et al., 2005b). 1 – iceberg deposits, 2 – "high productivity" sediments enriched in $CaCO_3$, 3 – turbidites. For letter designations see text. In parentheses – water depths, m

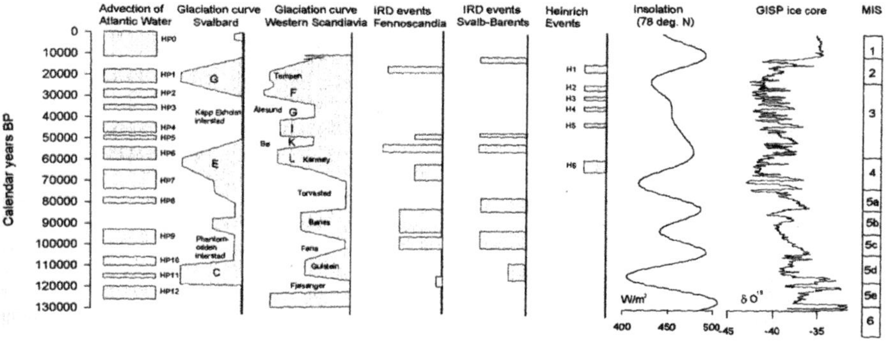

Fig. 8.25 Comparison of main paleoceanographic events in the Nordic Seas, glaciation history of Svalbard and western Scandinavia, insolation change at 78° N, and oxygen-isotope record of the Greenland Ice Core during the last climatic cycle (Hald, 2001)

When the glacier reached the continental shelf edge, the moraine saturated with water was transported downslope as debris flow. The shelf sediments also include some sandy interlayers deposited in the proximal periglacial zone.

Thus, specific features of climate changes in Fennoscandia during the last glacial cycle (Fig. 8.25) are reflected in structural properties and compositions of sediments accumulated on the Vøring Plateau and adjacent areas of the continental margin off central Norway. Iceberg-rafted diamictons HA4-HA2 accumulated during surges of the Fennoscandian Ice Sheet edge, whereas sediments of HA1 were deposited during the onset of the ice-sheet breakup. The timing of LGM is recorded by moraine complexes on the shelf, debris-flow deposits on the upper continental slope, and sand-rich sediments on the Vøring Plateau. Turbidites make up paragenetic complexes with iceberg-rafted sediments on the lower continental slope. The study of foraminifers and clay minerals indicated that the intricate (in terms of both qualitative and spatial structure) advection of Atlantic waters in the Vøring Plateau area during the last 25 ka was governed by climate changes. At the growth stage, the Fennoscandian Ice Sheet was relatively independent of the Laurentide Ice Sheet, whereas the breakup of both sheets was nearly synchronous. Our results confirmed the previous inference concerning a crucial role of the Northern Hemisphere glaciation in sedimentation processes within the Norwegian–Greenland Basin during the last glacial cycle (Levitan and Stein, 2005).

History of Sedimentation at the Continental Margin of Eastern and South-Eastern Greenland During the Last 130 ka

History of sedimentation in this particular region is examined based on sediment cores located along a number of cross-sections at the Greenland continental margin between 68° and 72° N (Fig. 8.26). These cores were retrieved during RV *Polarstern* cruises in 1988, 1990 and 1994 and were described in detail in a number

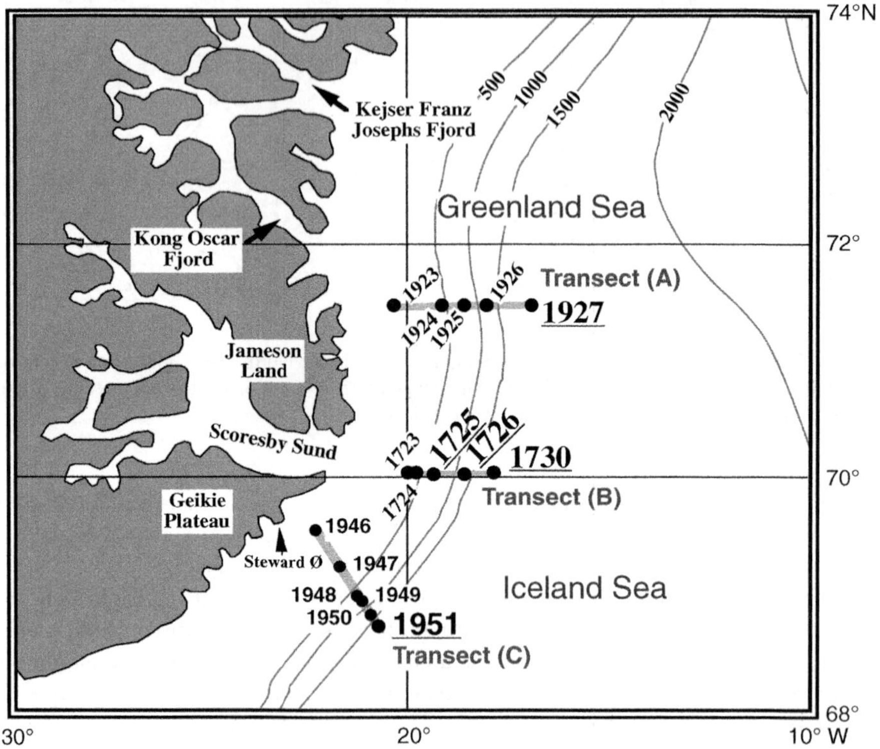

Fig. 8.26 Location of studied sediment cores at the East Greenland continental margin off Scoresby Sund (Nam et al., 1995)

of publications (Evans et al., 2002; Funder et al., 1998; Nam et al., 1995; Nam and Stein, 1999; Stein et al., 1996).

As already mentioned above, the East Greenland Current directed toward the south prevails in the surface water mass, bringing cold (lower than 0°C) and slightly freshened (less than 34.4‰) waters from the Arctic Ocean. These waters at a depth of approximately 200 m are underlain by the weak (to 150–160 m thick) layer of the Arctic Intermediate Waters, warmer (0–2°C) and salty (34.5–34.8‰) (Andrews et al., 1996). The colder deep waters of the Greenland Sea and finally bottom waters are located below. Ice cover exists nine months a year (from October to the end of June), moreover minimal size of sea-ice area occurs in September and maximum in March (Nam and Stein, 1999). The summer edge of continuous ice is located to the west of the Polar Front and the winter boundary – to the west of the Arctic Front (Fig. 8.2). Icebergs, in essence, are supplied from the Greenland ice sheet along the fjord systems, which emerge in the shelf, and further on are caught by the East Greenland Current.

Terrigenous bottom sediments that have been accumulated during the last glacial-interglacial cycle are mainly layers of silty-clayey mud and sandy clays, in which gravel-sized material is found in various proportions (Nam et al., 1995). From the point of view of genetic type sediment studies, a development of different moraine varieties of the last glaciation in the fjords and in the internal shelf is obvious. Fine-layered sediments of deglaciation lie on top of the moraines, and in some areas there are bioturbated pelitic silts in shelf depressions. In the upper part of the continental slope there are numerous bodies, formed by sedimentary mass flows; at the lower continental slope there are layers of hemipelagic fine sediments, which contain a small amount of coarse-grained IRD, with sporadically developed interlayers of debris flows and turbidites (Evans et al., 2002). At the lowermost continental slope and at the continental rise contourites were discovered by means of seismoacoustic profiling (Rasmussen et al., 2003). Complete absence of carbonate or marl sediments should be noted.

Sediment age of the upper part of the cross-sections is established by numerous AMS^{14}C-radiocarbon datings and for underlying sediments by oxygen-and carbon-isotope stratigraphy (Andrews et al., 1996; Evans et al., 2002; Funder et al., 1998; Nam and Stein, 1999; Stein et al., 1996). According to (Funder et al., 1998), the Langelandselv Interglaciation period (MIS 5e), when summer temperatures were higher than Holocene optimum by 2°–3°S, sea level has exceeded the modern level by 20 m, and sediments on the continental slope contained noticeable quantities of the secretion benthic foraminifers *Cibicides wuellerstorfii* (just as in MIS 1). The Holocene sediments, furthermore, have an increased content of dinoflagellates and coccoliths (Matthiessen and Baumann, 1997).

Meaningful results were obtained by IRD studies, which were calculated on radiographs (quantity of particles >2 mm in size for 10 cm^3) (Nam and Stein, 1999; Stein et al., 1996). IRD enrichments in interlayers are concentrated in the sediments of the Weichselian interval (MIS 5d–2). A more detailed examination of sediments of the last 60 ka (Fig. 8.27) showed that such interlayers were concentrated in the sediments, which have been accumulated during an active advance and intensive retreat of the Greenland Ice Sheet in the Late Weichselian, i.e., there are the same regularities, which were noted for the eastern continental margins of the studied basin in the previous chapter. Furthermore, there was a correspondence between accumulation of the majority of interlayers with periods of atmosphere temperature drop, recorded in the ice core of Greenland (Fig. 8.28). Some interlayers are synchronous to the Heinrich layers in the North Atlantic, some are not, but in essence they have been accumulating approximately in 1 thousand years, in contrast to the Heinrich layers, which are divided approximately by 10 thousand years (Stein et al., 1996).

Sedimentation rates and mass accumulation rates of sedimentary material as a whole and separately of sandy-gravel fractions (>63 μm) in the studied region are distributed in accordance with the descriptions given above, i.e., all these parameters have been growing from MIS 5 to MIS 2 and sharply fall in MIS 1 (Figs. 8.29 and 8.30).

Thus, the following very similar regularities are characteristic for the sedimentation history of the continental margins of the Norwegian-Greenland Basin during the last climatic cycle:

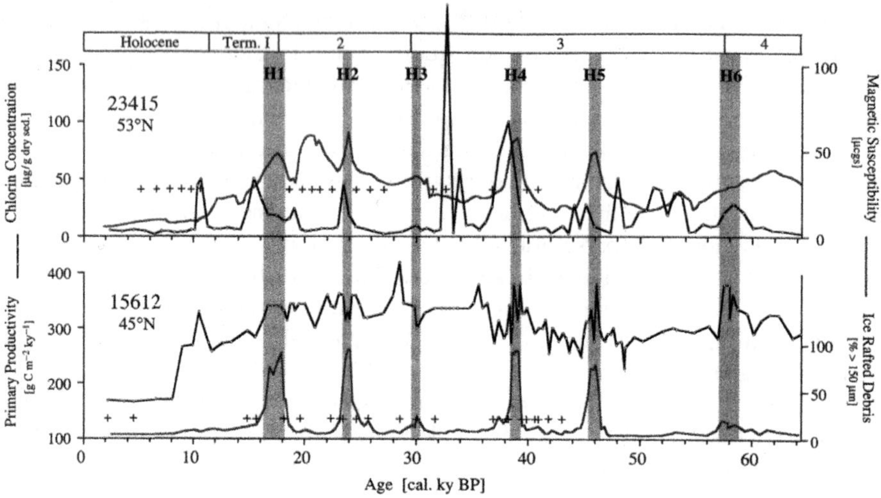

Fig. 8.27 Comparison of paleoproductivity and IRD in the North Atlantic during the last 60 ka (Sarnthein et al., 2001)

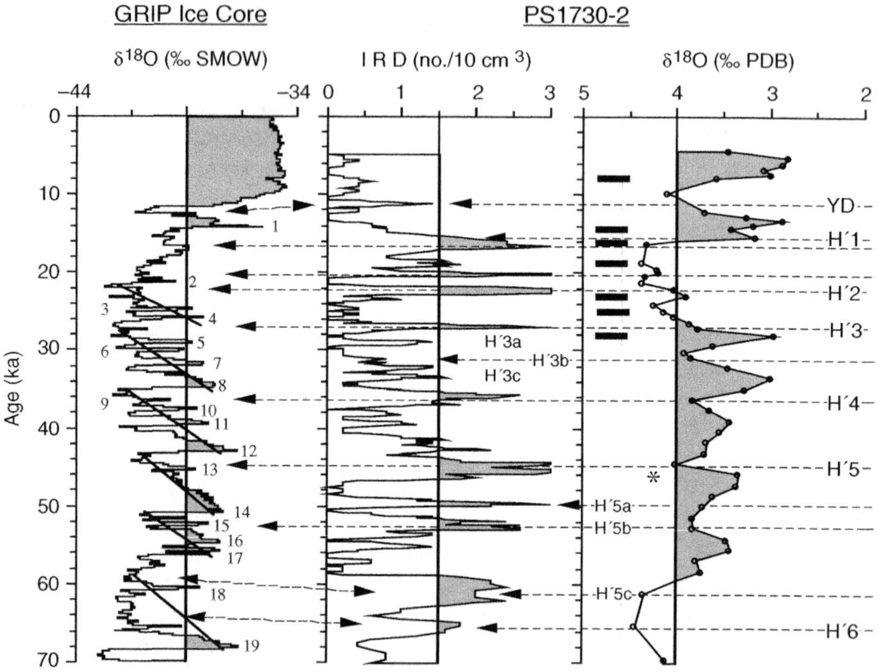

Fig. 8.28 Oxygen-isotope record of the GRIP Ice Core (Dansgaard et al., 1993), IRD content and magnetic susceptibility in sediment core PS1726 (from (Stein et al., 1996)). Numbers in column for ice core: number of Dansgaard-Oeschger interstadials (warming periods)

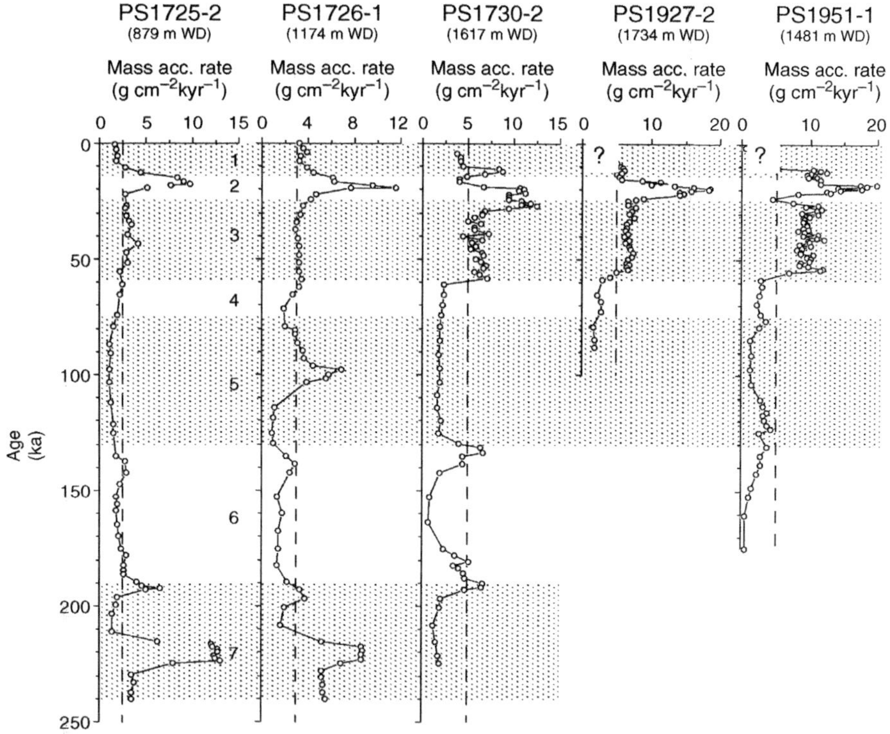

Fig. 8.29 Variations of sedimentation rates (cm/ky) in selected sediment cores from the East Greenland continental margin (Nam and Stein, 1999). In parentheses – water depths, m

(1) dominance of terrigenous material delivery in the lateral direction from the adjacent continental margins;
(2) connection of episodes of intensive delivery of iceberg material with periods of advance and retreat of the ice sheet;
(3) actual synchronization of these events for both margins, in spite of the much greater dynamics of glaciation at the eastern continental margin;
(4) increase of biogenic component role in the sediment composition, which have been accumulated during interglacial period in comparison with the sediments of glacial epochs.

The most essential difference is obvious by the absence of carbonate (marlaceous) layers in the west of the basin in contrast to its eastern part. A special study of interglacial sediments based on the example of the sediments, which have been developed in the Norwegian-Greenland basin during MIS 11, MIS 5 and Holocene, allowed H. Bauch (oral communication, 2006) to make a conclusion about a clearly expressed trend of mean temperature and paleoproductivity increase upward the cross-section.

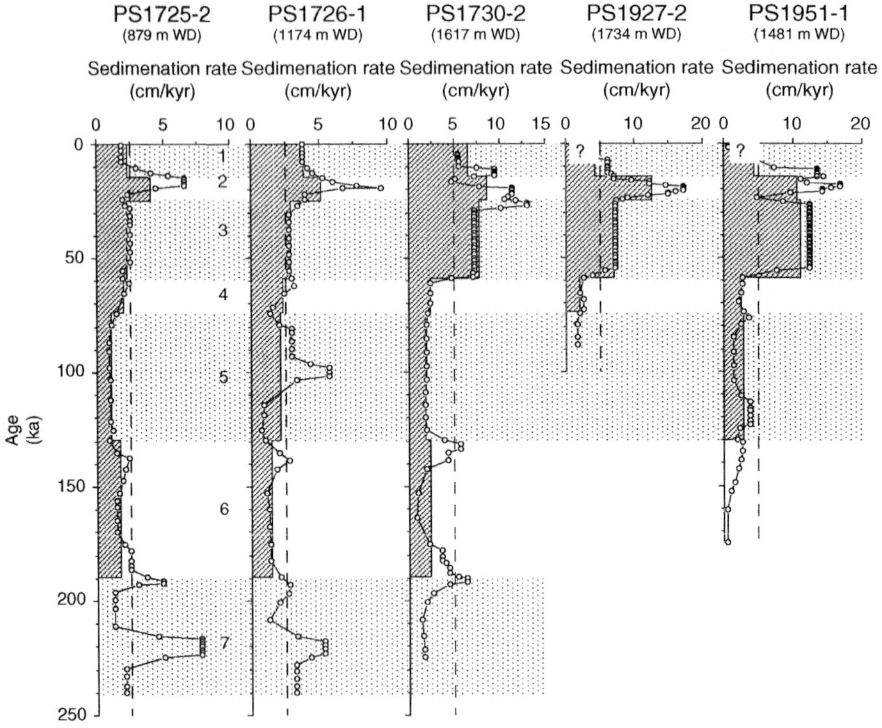

Fig. 8.30 Variability of mass accumulation rates (g/cm^2/ky) in selected sediment cores from the East Greenland continental margin (Nam and Stein, 1999)

Consequently, for the duration of the entire studied period in the Norwegian-Greenland Basin the character of sedimentation processes practically has not changed, that was caused by the preferred delivery of terrigenous material by glaciers from the western and eastern continental margins, by dominance of the cold East Greenland Current directed southward in the west and the Norwegian warm waters directed northward – in the east (Hebbeln et al., 1998). The mentioned phenomena, in particular, the "Nordway events" and growth of the role of "Irminger" branch of the Norwegian Current during maximum of the last glacial cycle, in essence, did not change the regularities of sedimentation particularities.

Chapter 9
The Arctic Ocean

The Arctic Ocean consists of extensive shelf seas, continental slopes, continental rises (in the sum of the components composing the continental margins) and deep-sea floor, which is made of ridges of different genesis and deep-water basins. In this chapter we not will examine the history of sedimentation in the continental shelves, since their more detailed description is given in the subsequent chapters. Based on (Jacobson et al., 2004) the total area of the Arctic Ocean is 9534 thousand km^2 which can be divided into the shelves (5025 thousand km^2 or 52.7%), the continental slopes (541 thousand km^2 or 5.7%), the continental rise (1095 thousand km^2 or 11.5%), the ridges (1506 thousand km^2 or 15.8%), and the deep-water basins (1367 thousand km^2 or 14.3%). The Arctic Ocean in the west is limited by the Fram Strait (connection to the Atlantic Ocean) and in the east by the Bering Strait (connection to the Pacific Ocean), and is bordered by the continents of Eurasia, North America and North Greenland.

Recent Environment

Morphostructure, Oceanographic and Sea-Ice Setting, Recent Sediments and Their Mineral Composition

The deep-sea floor consists of the following basic morphostructural elements consecutively from the Fram Strait to the Bering Strait: the Nansen Basin, the Gakkel Ridge, the Amundsen Basin, and the Lomonosov Ridge (the components the Eurasian Basin having a Cenozoic age), and the Makarov Basin, the Alpha-Mendeleev Ridge, and the Canada Basin (the components of the Amerasian Basin having a the Late Jurassic – Early Cretaceous age) (Fig. 9.1). It is well known that the Lomonosov ridge is a microcontinent. The remaining above-mentioned structures, most likely but still under discussion, have an oceanic crust of different structure. On average the sea floor of the deep basins has a water depth of about 3.5–4.2 km whereas the depth of the ridges varies between about 1.0 km (Lomonosov Ridge) and about 3.4 km (Gakkel Ridge).

M.A. Levitan, Y.A. Lavrushin, *Sedimentation History in the Arctic Ocean and Subarctic Seas for the Last 130 kyr*, Lecture Notes in Earth Sciences 118,
DOI 10.1007/978-3-642-00288-5_9, © Springer-Verlag Berlin Heidelberg 2009

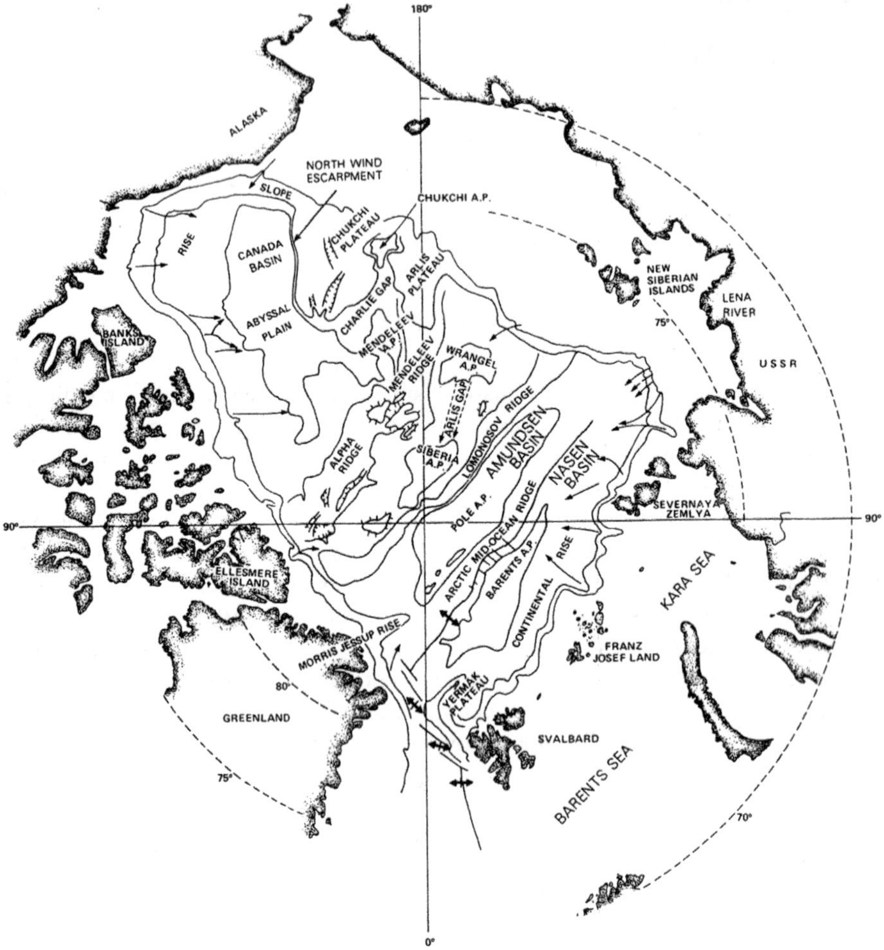

Fig. 9.1 Elements of the Arctic Ocean morphostructure (Johnson, 1990)

In winter times, the ocean surface is covered with ice (in the most severe winters the total area of sea-ice cover of the Northern Hemisphere has reached 15.1 million km^2), while in summer times this area is substantially less (about 8.4 million km^2 in the warmest summer seasons) (Fig. 9.2) (Macdonald et al., 2004). The sea-ice cover has steadily been decreasing during 20th century (based on average annual observations as well as for summer and winter seasons separately) (Macdonald et al., 2004). Its average thickness is 3 m, however, thickness is smallest in the center of the ocean (in summer to 1.0–0.5 m or ice is even absent in openings, which exist even in the region of the North Pole), and in winter nearer to the surrounding continental masses the thickness of ice reaches 4–6 m. The basic sources of sea ice are located in the Laptev Sea (about 50% production) and in the Kara Sea (about 25% production) (Eicken, 2004).

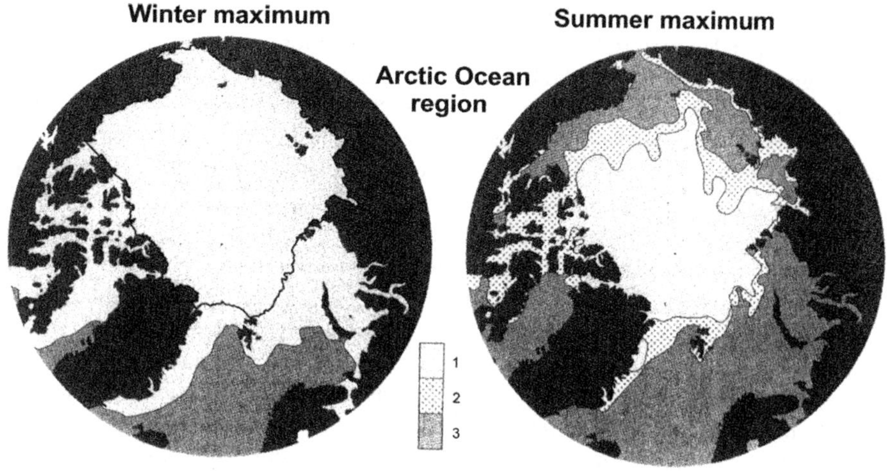

Fig. 9.2 Distribution of winter and summer sea ice in the Arctic Ocean (Macdonald et al., 2003). 1 – compact ice; 2 – non-compact ice (>15 %); 3 – open sea surface

The age of the sea-ice cover has been increasing from the Eurasian margin to the North America and Greenland margins approximately from 1 year to 5 years (Untersteiner, 1990). As a whole, the ice drifts and, after all, either partially melts in the ocean, or it is carried through the Fram Strait into the Norwegian-Greenland Basin, where it finally melts. In contrast to glacial periods, today icebergs are of secondary importance. At present their most active sources are the glaciers of Severnaya Zemlya, Northern Island of Novaya Zemlya, Franz-Joseph Land, Spitsbergen, the Canadian Arctic and Greenland (Bischof, 2000).

Researchers have repeatedly noted areas of so-called "dirty ice" in the Arctic Ocean, i.e., sea ice containing significant amount of sedimentary material (cryozoles) of different grain size (from the pelite particles to the chunks). Mechanisms of sediment entrainment into the sea ice and icebergs are very diverse: anchored shore ice material of surface sediments, freezing of sea water containing suspension during fall ice formation (especially in polynyas), river ice discharge, gravitational transfer from the rocky slopes of fjords during winter times, and eolian dust storms, and – for icebergs – inclusion of rock substratum in the composition of continental glaciers with their motion and subsequent shaping of icebergs.

A common interest of sedimentologists is related to the role of cryozoles in the contemporary Arctic Ocean sedimentation. Based on literature data (Levitan et al., 1995a, also see references therein), we accepted a maximum area of sea-ice development of 15.1 million km^2; an area of "dirty ice" development of 10% of the total area; a 10 cm thick upper layer of ice enriched by cryozoles; and a mean concentration of cryozoles of 400 mg/l. Simple calculation shows that with the maximum ice formation the total mass of cryozoles is 60.4 million tons. Considering the ice age in different regions of the Arctic sea-ice cover for calculating annual sea ice production forces to decrease this value 3–4 times, thus there is 15–20 million t of cryozoles per year.

According to (Gordeev and Rachold, 2004), annual river discharge is 227.3 million t of suspended material to the Arctic Ocean, i.e., more than one order of magnitude higher in comparison to the IRD input. Annual delivery of terrigenous material by processes of coastal abrasion is about 430.8 million t (Grigoriev et al., 2004), i.e., it is approximately 20 times higher than cryozoles. Finally, eolian delivery constitutes to about 5.7 million t per year (Shevchenko and Lisitzin, 2004). These data have to be taken into account when discussing the genesis sources of sedimentary material to contemporary sedimentation for different regions of the Arctic Ocean. However, on the basis of sedimentation rate data on different morphostructures, it is obvious that the relative role of cryozoles is increasing toward the pelagic environment (see later discussion for more details).

A well stratified water layer is located under the sea-ice cover in the Arctic Ocean (Fig. 9.3). Water exchange with the other oceans (with the Pacific through the shallow Bering Strait and with the Atlantic Ocean through the deep Fram Strait) is well balanced: approximately 4 Sv (millions of cubic m/s) enter the Arctic Basin and about the same amount of water is flown out of the Arctic Ocean (Coachman and Aagaard, 1974). Moreover, basic water exchange (approximately three fourth from the total) is achieved with the Atlantic (Norwegian-Greenland Basin).

It is possible to isolate three basic water masses in the water column from top to bottom: the surface polar (Arctic) water reaching a depth of approximately 200 m (on the surface the water temperature composes usually 1–2°C, more deeply it becomes colder, up to –1.0°C; salinity depends on proximity to the sources of fresh water and varies from 27 to 34.5‰); the intermediate Atlantic water reaching approximately a depth of 900 m (temperature is positive, and salinity varies in the different areas from 34.5 to 35.0‰), and a deep-water mass with approximately the same salinity as Atlantic waters, but with negative temperatures (from –0.45 to –0.9°C) (Coachman and Aagaard, 1974).

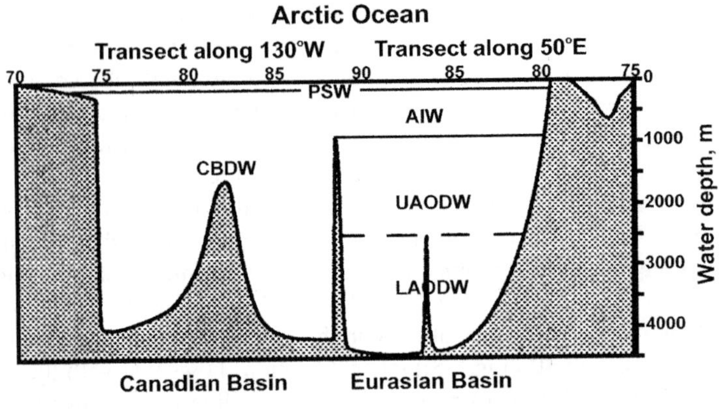

Fig. 9.3 Stratification of the Arctic Ocean water column (Aagaard and Carmack, 1989). PSW – Polar Surface Water; AIW – Arctic Intermediate Water; CBDW – Canada Basin Deep Water; UAODW – Upper Arctic Ocean Deep Water; LAODW – Lower Arctic Ocean Deep Water

There are small regional special features of the described stratification: for example, a distinct halocline between the Polar and Atlantic water masses has been recorded in the Amerasian Basin, caused by admixture of Pacific Ocean waters. Characteristics of deep and bottom waters in the Eurasian and Amerasian basins display minor differences (Macdonald et al., 2004). The character of surface-water circulation in the pelagic ocean part, and the total of intermediate and deep-water masses are radically various. In the surface layer there are two main systems of circulation: the Beaufort Gyre and the Transpolar Drift. The latter consists of two branches, the Polar Branch and the Siberian Branch. The boundary between these branches on the shelf is passing approximately in the central part of the Laptev Sea (Aagaard and Carmack, 1989). The Polar Front is located between both systems (Fig. 9.4).

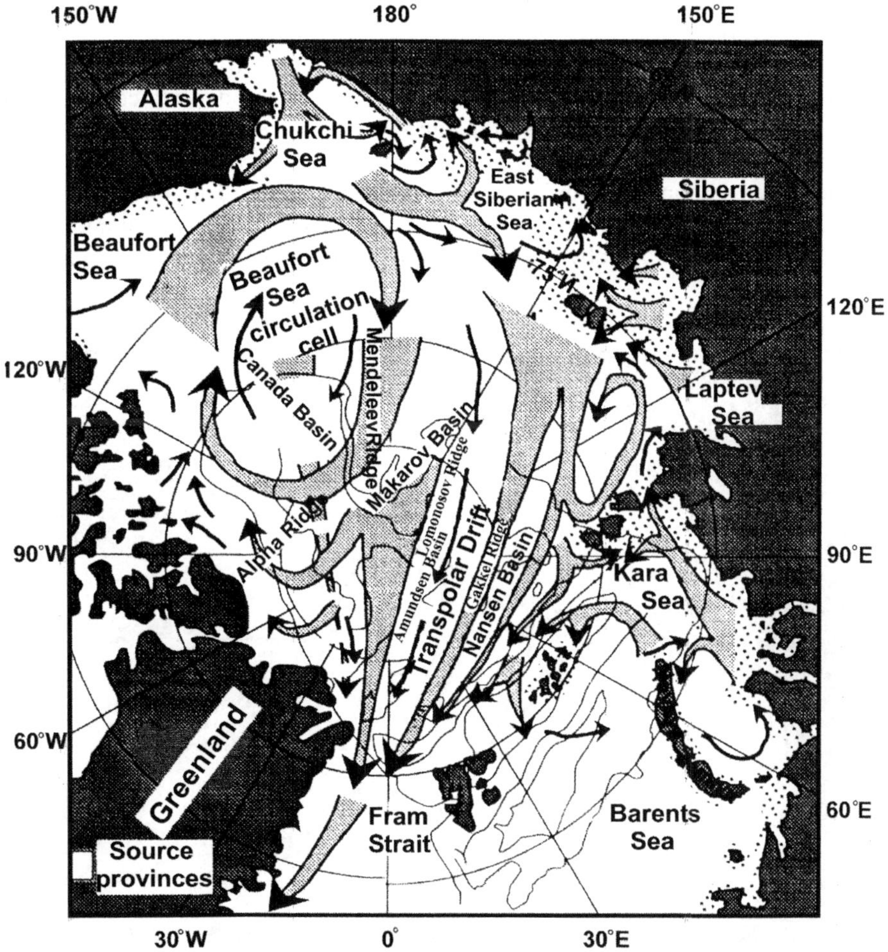

Fig. 9.4 Arctic Ocean surface-water circulation (Aagaard and Carmack, 1989)

The atmospheric Arctic Oscillation (AO) exerts strong influence almost on all components of the climate: hydrological, ice and oceanographic situations. A positive AO contributes to an increase of river discharge into the basin, a freshening of surface water, a reduction of ice-cover area, an expansion of circulation of the Beaufort Gyre and a displacement of the Polar Front toward Eurasia (Macdonald et al., 2004). A negative AO leads to the opposite results.

Upon transfer of the Norwegian-Greenland Basin to the Arctic Ocean in the region of the Fram Strait the Atlantic waters of the West Spitsbergen Current are immersed under the Polar waters and turn to the east, following along the upper part of the continental slope of Eurasia, and then they turn to the west in the Alaskan region and return to the Norwegian-Greenland Basin along the upper part of the continental slope of North America. During their flow in the interlayer these waters are being transformed (in particular, become colder) and form the sub-nucleus of recurrent circulation in the deep-water basins of the Arctic Ocean (Rudels et al., 1994) (Fig. 9.5).

In the described trajectory of flow, the deeper waters and cold brines of shelf regions are being involved as well. Cold brines are formed under fall ice formation and fallen by the mechanism of cascading into the deep water regions, flowing along the slope downward.

The lowered salinity of surface water was noted above. This phenomenon is caused by river discharge and melting of sea ice and icebergs in summer, and, according to some data (Bobylev et al., 2003), the latter may not be just compared

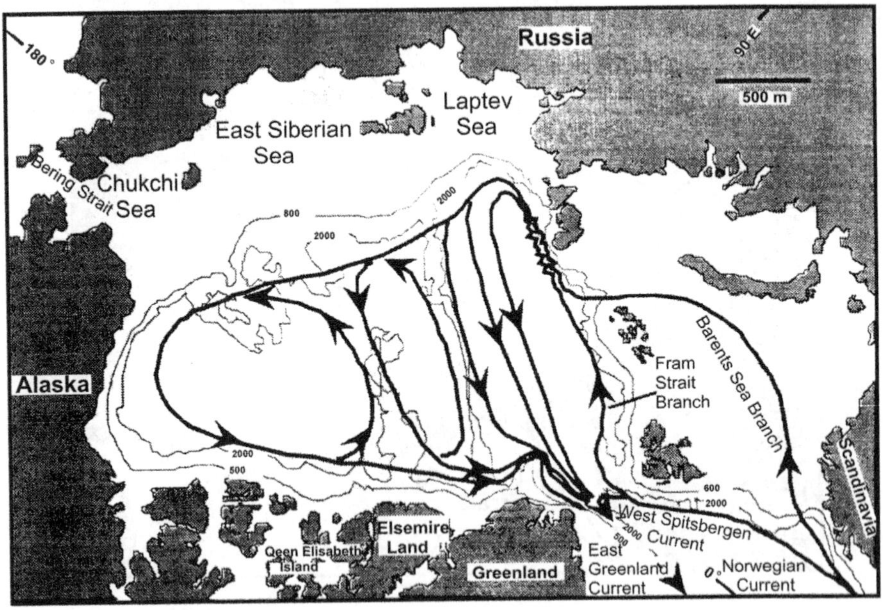

Fig. 9.5 Arctic Ocean Intermediate-Water circulation (Rudels et al., 1994)

with the first one, but may even exceed it. Discharge of the Arctic rivers has been extensively studied recently. In the work (Gordeev and Rachold, 2004) it was indicated, that total annual freshwater discharge into the Arctic Ocean is about 3300 km^3, from which the portion of the Pechora River is 131 km^3, the Severnaya Dvina River is 110 km^3, the Ob River is 404 km^3, the Yenisei River is 620 km^3, the Lena River is 523 km^3, the Kolyma River is 122 km^3 and the Mackenzie River is 330 km^3. The main part of the discharge is achieved in 2–3 summer months, especially during June and July.

Arctic primary production is low corresponding to the oligotrophic regions: about 11 gC/m^2/year in the central Arctic (Sakshaug, 2004; Subba Rao and Platt, 1984) and in average about 32 gC/m^2/year for the Arctic shelves (Sakshaug, 2004). Shelf seas characterized by relatively increased productivity values are the Chukchi and Barents Sea, where nutrient-rich Pacific and Atlantic waters, respectively, penetrate into the Arctic Ocean. Furthermore, a narrow strip of increased primary production characterizes the marginal ice zone due to the increased delivery of nutrients by sea-ice melting in summer.

The sea floor of the deep-water part of the Arctic Ocean is mainly covered by terrigenous hemipelagic silty-clayey mud and sandy silty-clayey mud, which contain significant amount of coarse-grained IRD (Belov and Lapina, 1961). Concentration of IRD decreases from the ridges to the abyssal plains. Massive and laminar diamictons are described in the ridge zones (Svindland and Vorren, 2002).

Sand content is usually higher on the ridges; however, in the interlayers of turbidites, which are very typical for all deep-water basins (their occurrence may reach 40% and more of the sedimentary sections in the recovered sediment cores; Darby et al., 1989) its concentration is also increased. The total amount of biogenic components (planktonic and benthic foraminifers, ostracod, small mollusks, dinoflagellates, radiolarians, sponge spicules, etc.) usually do not exceed 1–2% (Darby et al., 1989), with some exceptions of up to 10%.

The composition of IRD reflects the Beaufort Gyre and Transpolar Drift circulation systems particularly in the sandy and gravel-pebble fractions of the surface sediments. Fine-grained dolomites and dense limestones from the Paleozoic rock of Alaska and Canadian Arctic and also gray sandstones and siltstones predominate in the coarse-grained material under the circulation of the Beaufort Gyre (Bischof, 2000; Darby et al., 1997).

A dominance of quartz (Bischof, 2000), red-colored detrital rock from the Eurasian sources and fragments of the iron-manganese crusts has been recorded in surface sediments under the flow of the Transpolar Drift (Darby et al., 1997). Based on heavy-mineral associations in surface sediments (Fig. 9.6) (Belov and Lapina, 1961) it is clearly obvious that the projection of the Beaufort Gyre circulation is reflected in the abundance of garnet and dolomite, while under the Transpolar Drift sediments are enriched by associations of epidote, amphiboles, black ore minerals and clinopyroxenes. Moreover, minerals of the epidote-amphibole association dominate in the sediments accumulated under the Polar Branch, whereas under the Siberian Branch an association of black ore minerals, epidote and clinopyroxenes is dominant in the surface sediments.

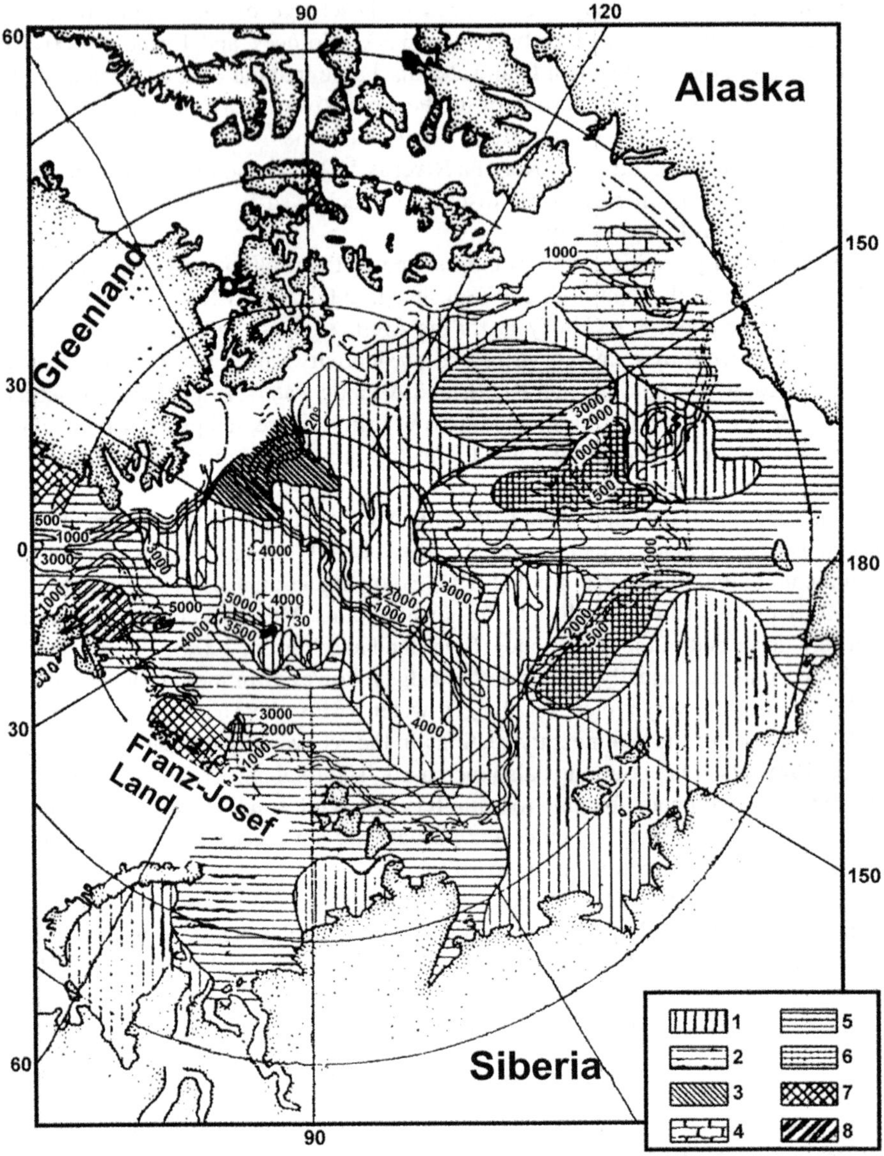

Fig. 9.6 Heavy-mineral distribution in Arctic Ocean surface sediments (Belov and Lapina, 1961). 1– 8– heavy-mineral associations: 1 – epidot-pyroxene, 2 – black ore-epidot-pyroxene, 3 – chloritoid, 4 – dolomite, 5 – garnet, 6 – black ore minerals, 7 – pyroxene, 8 – pyroxene-amphibole

The distribution of clay minerals in surface sediments, due to transport not only by sea ice and icebergs, but also by different oceanic currents, is much more complex. Our study on the distribution of clay minerals in surface sediments from the continental margin and deep-sea basin areas of the Arctic Ocean is based on own and literature data (Levitan et al., 1995b, see references therein). Clay minerals were mainly analyzed in the fraction <2 μm (following the method of (Biscaye, 1965)). To obtain comparable values in the fraction <1 μm correlation coefficients of different clay minerals in both fractions from the same samples were used. For statistical data processing 443 analyses from the continental margins and 61 analyses from deep-sea areas were carried out. Results are presented in Tables 9.1 and 9.2.

The results of Table 9.1 allow the following general statements. First, there is a very variable character of clay-mineral associations, which reflects, first of all, differences in the composition of the source provinces and the low degree of differentiation of clay material. Secondly, there is a common dominance of chlorite and illite paragenesis as result of the climatic conditions of sedimentation. Thirdly, noticeable differences in the composition of clay-mineral associations in different parts of a single basin of sedimentation (for example, in the Barents, Kara and Laptev seas) are caused by special features of regional circulation systems and by local sources of clay minerals.

Data of Table 9.2 show an obviously much clearer homogeneity of clay mineral associations in the deep-sea areas: they are barely distinguished in the Eurasian and Amerasian basins, which indicate advanced mineralogical differentiation and mature processes of mineral mixing from different sources in comparison with the continental margins. Paragenesis of chlorite and illite clearly prevails. There is a certain tendency of smectite enrichment in sediments from deep-sea basins, and by chlorite and kaolinite – in the sediments of the ridges of different genesis. The noted phenomenon is probably caused by differences in the hydraulic coarseness of these

Table 9.1 Average composition (%) of clay mineral associations in fraction <2 mkm of surface layer sediments at the continental margins of the Arctic Ocean [*Levitan et al.*, 1995b], with changes

No n/n	Number of samples	Smectite (Sm)	Illite (Il)	Chlorite (Ch)	Kaolinite (K)	Sm/Il	Ch/K	Sm/Ch	Association
1	90	11	59	19	11	0.19	1.7	0.6	(K,Sm)–Ch–IL
2	11	13	55	21	10	0.24	2.1	0.6	K–Sm–Ch–IL
3	31	6	60	27	6	0.10	4.5	0.2	(K,Sm)-Ch-IL
4	36	4	68	22	6	0.06	3.7	0.2	(Sm,K)-Ch-IL
5	8	2	68	20	9	0.03	2.2	0.1	Sm–K–Ch–IL
6	16	26	45	20	9	0.58	2.2	1.3	K–Ch–Sm–IL
7	43	48	31	15	6	1.55	2.8	3.2	K–Ch–IL–Sm
8	43	34	40	14	10	0.85	1.4	2.4	K–Ch–Sm–IL
9	85	16	53	20	11	0.30	1.8	0.8	K–Sm–Ch–IL
10	39	27	32	20	21	0.84	1.0	1.4	(Ch,K)-Sm-IL
11	38	5	61	19	14	0.09	1.4	0.3	Sm–K–Ch–IL
12	3	4	63	22	11	0.06	2.0	0.2	Sm–K–Ch–IL
13	443	18	51	19	11	0.35	1.7	0.9	K-(Sm,Ch)-IL

Note: 1 – rivers of Alaska; 2 – shelf of North Alaska; 3 – Chukchi Sea; 4 – East Siberian Sea; 5, 6 – southeastern (5) and central (6) parts of the Laptev Sea; 7, 8 – northern (7) and western (8) parts of the Kara Sea; 9, 10 – eastern (9) and northwestern (10) parts of the Barents Sea; 11 – northern continental margin of Spitsbergen; 12 – northeastern continental margin of Greenland; 13 – continental margins of the Arctic Ocean

Table 9.2 Average composition (%) of clay mineral associations in the fraction <2 mkm of surface layer sediments in the deep-water areas of the Arctic Ocean [*Levitan et al.*, 1995b], with changes

n/n	Number of samples	Smectite (Sm)	Illite (Il)	Chlorite (Ch)	Kaolinite (K)	Sm/Il	Ch/K	Sm/Ch	Association
1	11	7	59	19	15	0.2	1.3	0.4	Sm–K–Ch–IL
2	6	5	59	20	17	0.08	1.2	0.2	Sm–K–Ch–IL
3	18	7	55	24	13	0.13	1.8	0.3	Sm–K–Ch–IL
4	8	11	52	24	14	0.21	1.7	0.4	Sm–K–Ch–IL
5	43	7	56	22	14	0.12	1.6	0.3	Sm–K–Ch–IL
6	3	7	59	23	11	0.12	2.1	0.3	Sm–K–Ch–IL
7	1	1	51	24	24	0.02	1.0	0.04	Sm–(Ch-K) –IL
8	14	10	58	19	15	0.17	1.3	0.5	Sm–K–Ch–IL
9	18	9	58	20	15	0.16	1.3	0.4	Sm–K–Ch–IL
10	61	7	56	21	14	0.12	1.5	0.3	Sm–K–Ch-IL

Note: 1 – Nansen Basin; 2 – Gakkel Ridge; 3 – Amundsen Basin; 4 – Lomonosov Ridge; 5 – Eurasian Basin – average data 1–4; 6 – Makarov Basin; 7 – Alpha and Mendeleev Ridges; 8 – Canadian Basin; 9 – Amerasian Basin – average data 6–8; 10 – deep-water part of the Arctic Ocean – average data from 5 and 9.

minerals and by a different reaction to the hydrodynamic regime of bottom waters which is more active on rises.

As a whole the sediments of the Arctic continental margins are different from the deep-sea sediments by higher content of smectite, while the minerals of the other three basic groups, in opposite, are enriched in the deep-sea areas. A comparison of average clay-mineral data in the Arctic Ocean with other ocean basins (Table 9.3) indicates that the chlorite-illite paragenesis only prevails in this basin. According to average contents of smectite and illite, the clay-mineral composition in Arctic Ocean sediments is most similar to that determined in sediments from the North Atlantic, which is not surprising because of the active water exchange between these basins.

The described tendencies in the distribution of clay minerals in the subcolloidal fraction of the surface sediments for different regions of the Arctic Ocean are more clearly represented in schematic maps (Wahsner et al., 1999) (Figs. 9.7, 9.8, 9.9 and 9.10). From the map of smectite distribution (Fig. 9.7) it is obvious that the basic regions of its accumulation are located in the Kara Sea (especially its southern and

Table 9.3 Average composition (%) of clay mineral associations in surface layer sediments of the main oceans [*Levitan et al.*, 1995b], with changes

Oceanic basins	Smectite (Sm)	Illite (Il)	Chlorite (Ch)	Kaolinite (K)	Associations
North Atlantic	16	55	10	20	Ch–Sm–K–Il
South Atlantic	26	47	11	17	Ch–K–Sm–IL
Northern Pacific	35	40	18	8	K–Ch–Sm–IL
Southern Pacific	53	26	13	8	K–Ch–IL–Sm
Indian Ocean	41	33	12	17	Ch–K –IL–Sm
Arctic Ocean	17	52	19	11	K–Sm–Ch–Il

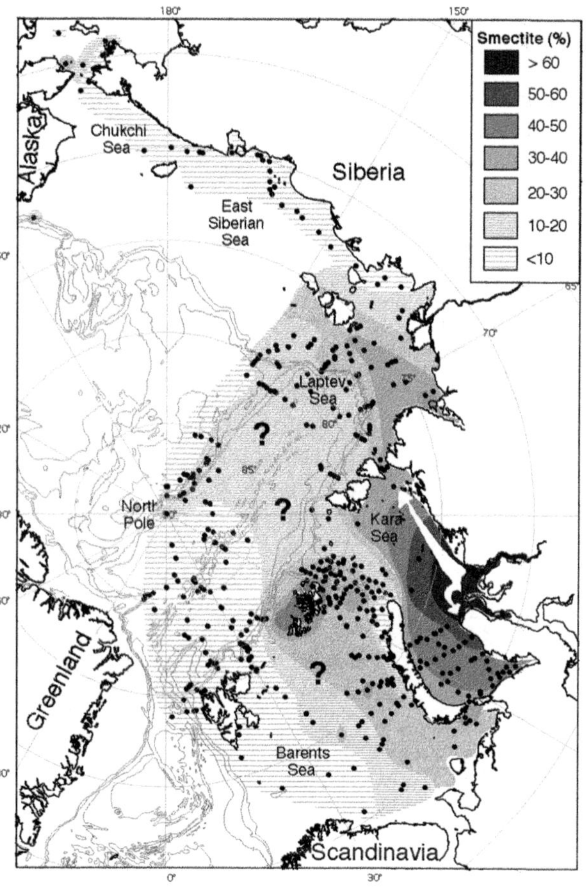

Fig. 9.7 Smectite distribution in eastern and central Arctic Ocean surface sediments (Wahsner et al., 1999)

southeastern areas), the western part of the Laptev Sea and the area around Franz-Joseph Land. The concentrations of smectite are minimal in the deep-sea areas. The basic sources of illite are related to the coast of the East Siberian Sea and the Chukchi Sea; local sources are the Novaya Zemlya Islands, Scandinavia, and Spitsbergen (Fig. 9.8). Deep-sea sediments are characterized by high content of this mineral.

The relatively high concentrations of chlorite (Fig. 9.9) gravitate towards practically exclusively to the sediments of the ocean floor and the transverse troughs (St. Anna Trough and Voronin Trough). The basic source of kaolinite is the archipelago of Franz-Joseph Land (Fig. 9.10). Its content is increased in the eastern part of the Kara Sea and in the western part of the Laptev Sea as well as in a larger part of the Barents Sea.

According to (Stein et al., 1994c) and (Levitan et al., 1995a) the composition of clay minerals in the cryozoles of the pelagic zone of the Arctic Ocean strongly

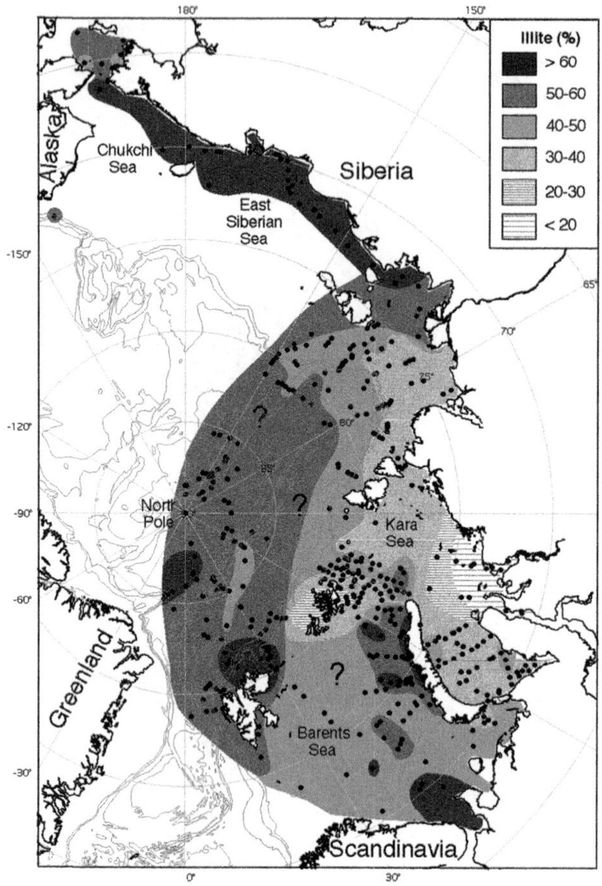

Fig. 9.8 Illite distribution in eastern and central Arctic Ocean surface sediments (Wahsner et al., 1999)

differs from the associations of clay minerals in surface sediments of the corresponding areas. Clay minerals enter bottom sediments not only due to the melting of sea ice and icebergs (in contrast to the silty- sandy and coarser material), but also through transport by different oceanic currents.

In the following, the facies structure of recent sedimentation of the Arctic Ocean will be shown in an example in sediments from the Yermak Plateau.

Facies Variations of Holocene Sediments on the Yermak Plateau (According to Study Data of >63 mkm Fraction)

The Yermak Plateau is a marginal rise located above the Nansen Basin in the north and east, the Spitsbergen continental rise in the south, and the Lena Deep in the west (Fig. 9.11). The almost flat ridge is located at a depth of approximately 800

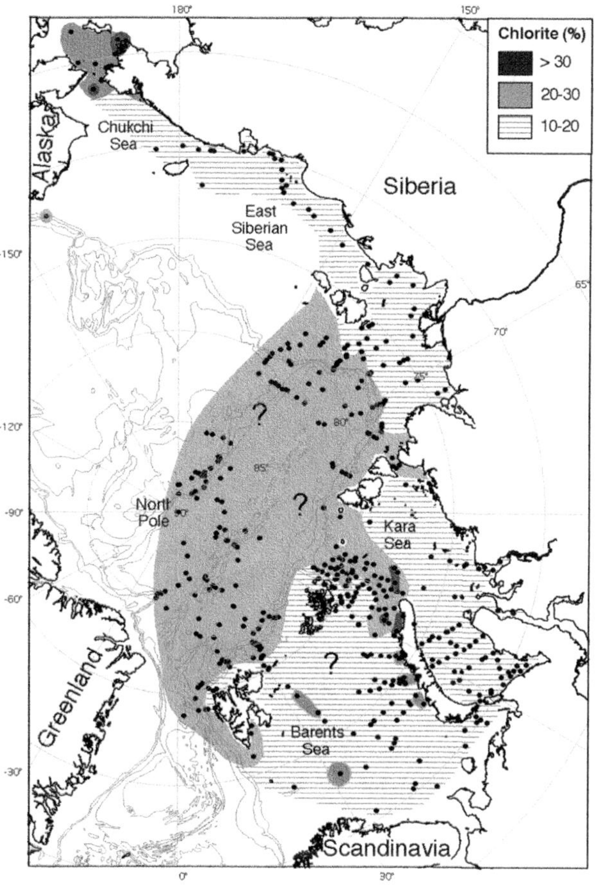

Fig. 9.9 Chlorite distribution in eastern and central Arctic Ocean surface sediments (Wahsner et al., 1999)

m, while the maximal depth of adjacent heterogeneous basins ranges from 2000 to 2500 m (Fig. 9.12). According to (Hevrøy et al., 1996), this type of geomorphologic setting practically rules out the delivery of sand and gravel to the plateau surface by oceanic currents.

The Yermak Plateau is located not only in a transition zone between Atlantic and Arctic Oceans, but also between two systems of glacial sedimentation: with seasonal sea-ice and permanent cover. The seasonal system (Norwegian-Greenland Basin) is characterized by ephemeral (seasonal) existence of sea ice, delivery of icebergs during warm seasons by glaciers from Spitsbergen and Greenland, and transport of lithogenic sedimentary material mainly by currents. Gravitational processes are developed on slopes of continents, islands, submarine ridges and rises. Sedimentation of planktonic carbonates plays an essential role, particularly in zones with warm surface currents (Henrich et al., 1989). These features are primarily typical

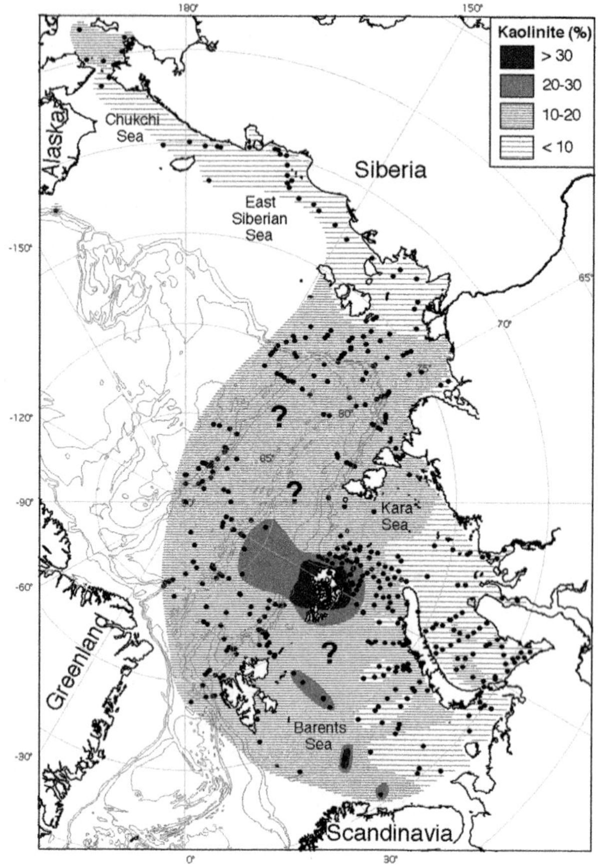

Fig. 9.10 Kaolinite distribution in eastern and central Arctic Ocean surface sediments (Wahsner et al., 1999)

of sediments accumulated during interglaciations which are similar to the modern environment. Sediments of colder (glacial) epochs differ in color, grain size, structure, and IRD composition. They generally contain a lesser amount of carbonate material (Henrich et al., 1989; Spielhagen, 1991).

The sedimentation system of the Central Arctic, on the other hand, is characterized by the presence of perennial sea-ice cover with relatively small and transit open-water areas, the prevalence of IRD on submarine ridges and rises, the dominance of material delivered by bottom currents and turbid flows in deep-water basins, and strongly reduced biogenic sedimentation. Sediments of cold and warm stages differ from each other primarily in color but not so sharply as in the Norwegian-Greenland Basin (Clark, 1971; Nørgaard-Pedersen, 1997).

The location of the Yermak Plateau slightly northward from the summer boundary of the sea-ice domain in the marginal zone of the Arctic ice cover (Fig. 9.11) makes us to conclude that lithogenic grains and clasts might be delivered to the ridge surface only during the thawing of sea ice and icebergs. Thus, the Yermak Plateau, as

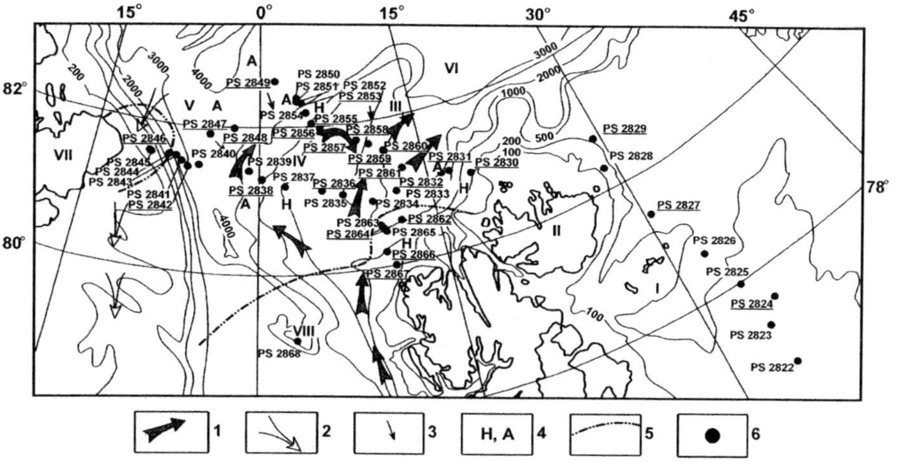

Fig. 9.11 Location of studied sediment cores at Yermak Plateau and elements of surface circulation (Levitan et al., 2000b). Isobaths are given in meters. 1 – West Spitsbergen Current; 2 – East Greenland Current; 3 – Arctic Ocean Deep Water currents; 4 – Bottom Water of the Norwegian Sea (N) and the Arctic Ocean (A); 5 – sea-ice margin in summer of 1997; 6 – RV *Polarstern* sediment cores. I – Barents Sea; II – Spitsbergen; III – Sophia Deep; IV – Yermak Plateau; V – Fram Strait; VI – Nansen Basin; VII – Greenland; VIII – Lena Deep

well as Lomonosov, Gakkel, and Mendeleev submarine ridges of the Arctic Ocean, may be considered as a natural laboratory useful for the study of particularities of glacial and iceberg sedimentation controlled by the history of ice-cover formation in the Arctic Ocean and the evolution of ice sheets in northern Eurasia, North America, and Greenland. These issues are obviously essential for understanding the nature of glacial sedimentation and studying paleoceanography and paleoclimatology of the Arctic.

It is well known that the Arctic climate is strongly affected by sea-ice cover that increases albedo and hampers the heat-gas exchange between ocean and atmosphere. The appearance or expansion of open-water areas within the ice cover leads to changes in climatic characteristics. Data on supply of Atlantic warm and salt waters to the Arctic Ocean, as well as materials concerning the history of river discharge, are of similar importance for the study of the Arctic paleoclimate.

Icebergs is considered to deliver relatively coarse lithogenic material, therefore, fractions coarser than 150, 500, 2000, or 4000 μm were thoroughly investigated (Bischof, 1990; Spielhagen, 1991; Nørgaard-Pedersen, 1997; Phillips and Grantz, 2001). The study of heavy minerals in the fraction 63–125 μm in reliably dated cores of Arctic bottom sediments has been initiated by (Behrends, 1999). The determination of sea ice and iceberg relationship by heavy-mineral composition in the Arctic geological history along with elucidation of source provenance and transport pathways and their evolution during the alternation of glacial and interglacial epochs are crucial issues. In particular, the relative role of material delivered from the Canadian High Arctic, history of IRD delivery, significance of different currents of the Transpolar Drift, sources of sediment-forming minerals, and peculiarities of sedimentation during warm and cold epochs are among the issues that have been outlined but insufficiently studied for the Yermak Plateau.

Materials and Methods

Materials and research methods applied in the present study, are described in the preliminary cruise report of R/V *Polarstern* Cruise ARK-XIII/2 carried out in the area of the Fram Strait and the Yermak Plateau (Stein and Fahl, 1997). The locations of geological stations are shown in Fig. 9.11. We have mainly used the following data for this study: (1) bathymetric data (Niessen and Kleber, 1997) obtained by the Parasound high-frequency parametric seismoacoustic profiler; (2) temperature and salinity measurements in the water column by the Niel Brown CTD device (Rudels et al., 1997); (3) data on the composition of sediments recovered by box corer (sections up to 80 cm long) and by multicorer (sections up to 26 cm long); and (4) the sand and gravel fractions (>63 μm) washed out from multicorer sediment samples. The coarse fraction samples taken every 2 cm throughout the section, were studied using a binocular microscope. Overall, 248 coarse-grained-fraction samples were studied from 24 multicorer sections (Fig. 9.11). The relative proportions of the >2 mm and 2.0–0.063 mm fractions were estimated. Content and composition of rock fragments, biogenic remains, and mineral grains were determined in both fractions.

Recent Environment

The study area is characterized by a highly variable sedimentation environment. The sources of the terrigenous material can be subdivided into three groups: regional, local, and remote. The regional sources include Spitsbergen and Greenland. The local sources are considered to be the exposures of sea bottom supplying edaphogenous material as a result of erosion processes. The remote sources are located southward and northward of the study area, and sediment is transported by sea ice, icebergs and currents. This classification (regional, local and remote sources) might be applied as well for cryozoles of the study area. In this case, the local sources are the sea waters of the ice-margin area and the remote sources are located in Siberia where the sediments are being transported not only by ice-rafting mechanism but also by eolian processes.

The northern part of the Fram Strait studied during the cruise is covered by year-round ice. Only two areas are ice-free in summer: the first area is located near the northeastern coast of Greenland known as the Greenland Polynya, while the second area stretched along the Spitsbergen coast (Fig. 9.11). According to observations during the cruise, an area of dirty ice (i.e., sea ice containing sedimentary material) apparently decreases with distance from Spitsbergen. The Arctic Ocean, as well as aquatory off Spitsbergen and Greenland, is the major sea-ice source in the region. The bulk ice delivery from the north is 0.15 mln. t/s (Rudels, 1987).

In general, the water column can be subdivided into three water masses (Rudels, 1987): (1) surface water (negative temperature, salinity <34.7‰); (2) Atlantic water (temperature between 0 and +3°C, salinity >34.9‰); and (3) deep water (negative temperature, salinity >34.7‰). The main picnocline occurs at depths of 500–1000 m. The background hydrodynamic structure in the Fram Strait includes two large-scale drifts: a warm, high-salinity, northward West Spitsbergen Current and a cold southward East Greenland Current, contouring the continental slope of Spitsbergen and Greenland, respectively. According to oceanographic data (Rudels, 1987), the West Spitsbergen Current discharge is 2.8 Sv, the East-Greenland Current discharge is 2.9 Sv.

The northward-flowing stream of the West Spitsbergen Current meets with the southward flow of cold and freshened waters from the Arctic Ocean at about 80°N. As a result of mixing the core of Atlantic waters submerges, cools down and freshens up. Continuing its northward flow in the subsurface layer, the West Spitsbergen Current is pushed against the returning Atlantic water flow from the Arctic Ocean. This results in a subdivision of the West Spitsbergen Current into two branches. The main eastern branch continues along the Spitsbergen continental slope and further on penetrates into the Nansen Basin through the Sofia Deep. The second branch envelopes the Yermak Plateau from the west, intensely mixing with returning Atlantic waters from the Nansen Basin (Rudels et al., 2000). Therefore, an anticyclone gyre is formed above the Yermak Plateau. Further westward, up to the Greenland continental slope, an outflow from the Arctic Ocean predominates (Levitan et al., 2000b).

Deep waters of Arctic origin predominate here at depths >2000 m. A deep-water flow from the Norwegian Sea, which is observed in the east along the western slope of the Yermak Plateau, penetrates northward to the Nansen Basin and flows into the Sofia Deep (Fig. 9.11). Depths and temperature-salinity characteristics of bottom waters are given in Table 9.4.

The above-mentioned and well-known regularities of intensive primary production relation to the waters washing off the ice edge from one hand and Atlantic waters from the other hand (Stein et al., 1994) allow us to suggest a high plankton productivity in the eastern part of the study region (along the anticyclone gyre periphery) and the Greenland polynia area.

Based on bottom topography, recent sedimentation environments, lithology, chemical composition, and clay mineral assemblages of sediments, the study region can be divided into the following major zones of Holocene sedimentation (Hebbeln and Berner, 1993): (1) northwestern part of the Barents Sea; (2) northern continental slope of Spitsbergen; (3) Sofia Deep; (4) slopes of the Yermak Plateau; (5) crest area of the Yermak Plateau; and (6) Greenland continental margin. Brief characteristics of sedimentation conditions in these zones are given in Table 9.4.

Table 9.4 Temperature and salinity of bottom waters [*Levitan et al.*, 2000b]

Geological station	Water depth, m	T, °C	S, ‰
PS 2824	194	0.371	34.783
PS 2829	464	0.911	34.859
PS 2830	506	1.308	34.946
PS 2831	972	−0.496	34.908
PS 2832	2067	−0.829	34.925
PS 2836	636	0.007	34.910
PS 2838	2426	−0.825	34.905
PS 2846	275	0.641	34.832
PS 2847	4165	−0.647	34.937
PS 2848	2574	−0.799	34.929
PS 2849	3254	−0.744	34.942
PS 2853	2117	−0.864	34.916
PS 2855	1380	−0.909	34.911
PS 2856	913	−0.798	34.909
PS 2857	774	−0.334	34.908
PS 2858	819	−0.584	34.909
PS 2859	1398	−0.806	34.917
PS 2860	1967	−0.874	34.919
PS 2861	2277	−0.818	34.927
PS 2862	1025	−0.675	34.908
PS 2864	767	−0.465	34.907
PS 2867	512	1.038	34.944

Facies Variability of Holocene Sediments

The structure of the upper part of sedimentary cover is outlined in Fig. 9.13, showing lithostratigraphic cross-sections recovered by box corer along the profile II – II. The figure demonstrates that the coarsest sediments (sand with gravel, and gravel) are related to the upper continental slopes of Greenland and Spitsbergen. Finer grained terrigenous hemipelagic sediments (silt and clayey silt) have been accumulating in the lower parts of the same slopes, as well as on the eastern slope of the Yermak Plateau and on a local submarine hill on the plateau crest (station PS 2836). Relatively coarse-grained layers and lenses are found as well within some core sections from local sedimentary pockets on the lower continental slope (station PS 2832) and from the Lena Deep (station PS 2840). At the surface the rest of the profile is represented by hemipelagic (terrigenous) clayey mud (in the Sofia Deep, Lena Deep, and the western part of the Yermak Plateau crest). Hardground layers composed of stiff ferruginized clay, 1–7 cm thick, are rather widespread in core sections here.

The proportion of the >0.063 mm fraction does not exceed a few percent in the studied multicorer samples. The fraction >2 mm mainly consists of biogenic remains (first of all, *POLYCHAETA* tubes and large benthic foraminifer tests) and

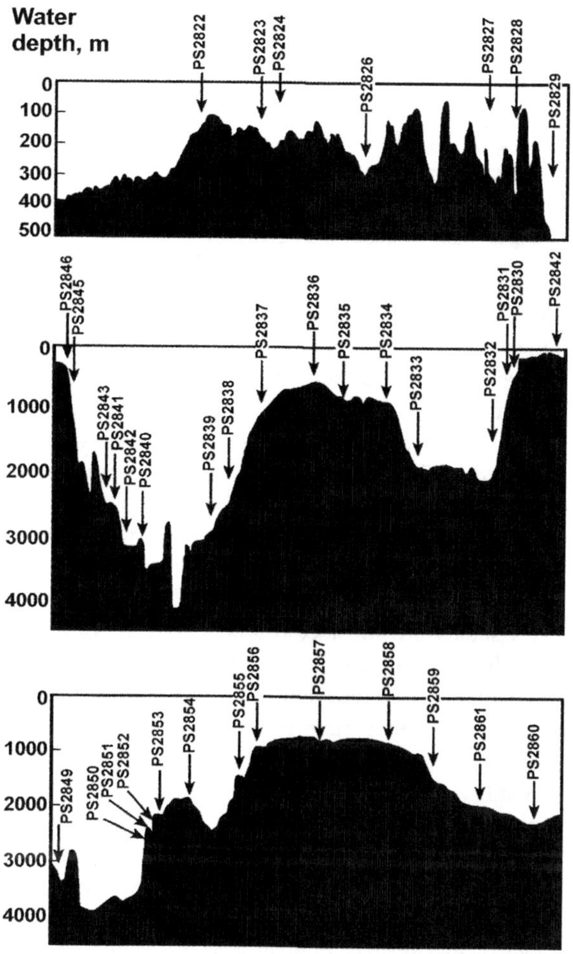

Fig. 9.12 Bathymetric profiles across the Yermak Plateau (Stein and Fahl, 1997)

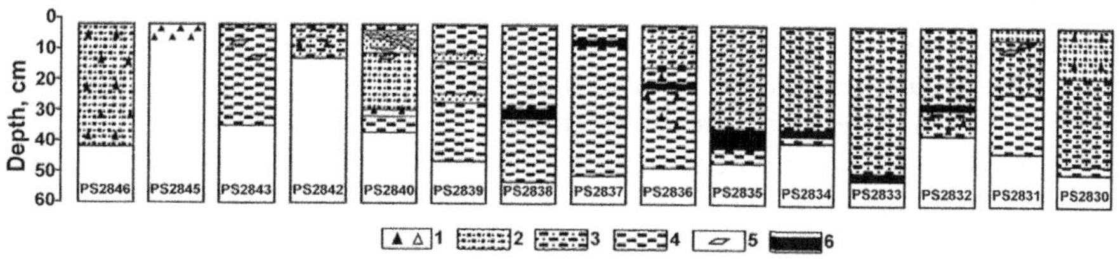

Fig. 9.13 Lithological profiles across the Yermak Plateau from box core records; for location of cores see Fig. 9.12 (Levitan et al., 2000b). 1 – unrounded coarse rock fragments, gravel, and pebble; 2 – sand; 3 – silts and silty clay; 4 – hemipelagic mud; 5 – lenses; 6 – hardground

rock fragments. Quartz grains also occur in rare amounts. Planktonic and benthic foraminifer tests, as well as quartz and feldspar grains, predominate in the 2.0–0.063 mm fraction. Rock fragments, accessories, and black ore minerals constitute a minor proportion.

Composition of Rock Fragments

In the fraction >2 mm, rock fragments 2–12 mm in size, largely subangular with minor well-rounded ones occur. Glacial striation is observed on several fragments collected near source zones (stations PS 2827 and PS 2846). Rock fragments with a fine oxidized film at their surface (like desert varnish) formed in subaerial environment, are rather common in the region.

Among 25 studied stations, both metamorphic and sedimentary rocks were described at 15 stations (metamorphic rocks predominated at 8 stations, equal amounts of metamorphic and sedimentary rocks were found at 5 stations, and sedimentary rocks prevailed at 2 stations); only metamorphic rocks occurred at 7 stations; and only sedimentary rocks were described at 3 stations. Overall, metamorphic rock fragments predominate. Among them, fragments of various shales and micaceous shales are most common, whereas argillaceous shales, micaceous schists, crystalline schists, and schistose sandstones and siltstones occur in minor amounts. Fragments of quartzite and marble are rare. Most of the marble fragments were found on the Spitsbergen continental slope and on the eastern slope of the Yermak Plateau.

Sandstones, bituminous shales, siltstones, silty clay, and coal predominate among fragments of sedimentary rocks. Limestone fragments are rare. Fragments of hematitized sandstone and siltstone attract one's attention by their "crimson" color. Limestone fragments occur only on the Spitsbergen continental slope. Hematitized sandstone fragments also occur mainly here. Coal fragments are only found on slopes of the Yermak Plateau. Specific compact pink feldspar-quartzose sandstone fragments associated with greenish gray quartzite were described only from the Greenland continental slope (station PS 2846).

Spitsbergen and its basement serve as a major source of gravel- and sand-size grains in the studied region. Bituminous shale fragments widely distributed throughout the region are likely of either Late Jurassic or Middle Triassic age; these rocks are well known in the northwestern Barents Sea (Geological structure of the USSR and regularities of mineral deposits distribution, 1984). Coal fragments are likely derived from well-known coal deposits in Spitsbergen. Most of the above-listed petrographic rock varieties belong to Paleozoic (regionally metamorphosed) sedimentary series of Spitsbergen.

The Spitsbergen-derived coarse material in sediments from the Yermak Plateau crest possibly underwent a passive enrichment due to bottom erosion and winnowing of more fine-grained fractions. The material might also be reworked by submarine slumping and various density flows on the plateau slopes. The influence of Greenland is likely restricted to its continental margin. It is presently impossible to evaluate the role of remote material sources, e.g., that of Scandinavia, using the data of rock fragments.

Therefore, we could reveal the following facies indicators of sedimentary material sources based on rock associations: (1) the north Spitsbergen continental slope – limestone, marble, and hematitized sandstone; (2) the Yermak Plateau slopes – coal and marble; and (3) the Greenland continental margin – association of compact pink sandstone and greenish gray quartzite. Rock associations serving as facies indicators have not yet been revealed for other zones of Holocene sedimentation. Subaerial glaciers of Spitsbergen and Greenland apparently played an important role in the coarse material supply, whereas icebergs and sea ice actively participated in the material transport, and their melting resulted in supply of IRD onto the seafloor.

Composition of Biogenic Components

The following groups of organisms predominate among the studied biogenic components in a >0.063-mm fraction: polychaets, planktonic foraminifers, calcareous benthic foraminifers, agglutinating benthic foraminifers, and siliceous sponges (Fig. 9.14). Besides, mollusk shells (bivalves and gastropods), sea urchin spines, ostracod valves, deep-water coral fragments, and fish teeth or jaw fragments also occur in the coarse fraction of samples.

The obtained data show a distinct facies pertinence of the biogenic components. Benthic foraminifers and sponge spicules predominate at shallow depths in the northwestern part of the Barents Sea. Polychaets and agglutinating benthic foraminifers mainly occur at the northern continental slope of Spitsbergen. This

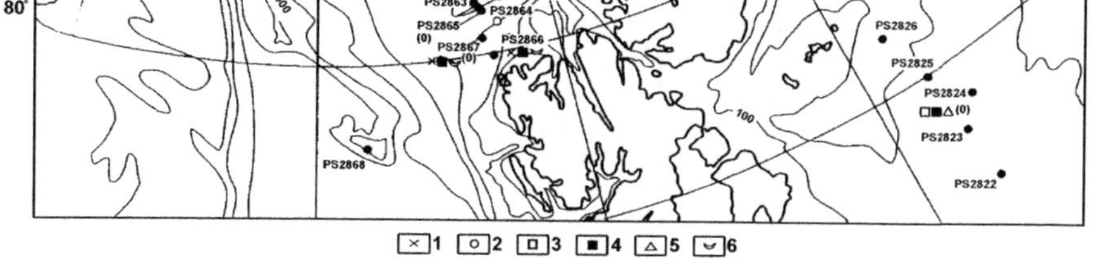

Fig. 9.14 Distribution of biogenic remains in the >0.063 mm fraction of surface sediments (Levitan et al., 2000b). 1 – polychaets; 2 – planktonic foraminifers (number in parentheses – amount in fraction 2.0–0.063 mm, %); 3 – benthic foraminifers including *Biloculina* and *Triloculina*; 4 – agglutinated benthic foraminifers; 5 – silica sponges; 6 – bivalves

assemblage is associated with sponge spicules in the northeast; and with mollusk shells in the northwest. Large (2–4 mm) calcareous benthic foraminifers (mainly *Biloculina* and *Triloculina*) prevail in the Sofia and Lena deeps at depths more than 2000 m. An assemblage of planktonic and agglutinating benthic foraminifers is developed on the eastern slope of the Yermak Plateau. The same assemblage is found within a major part of the plateau crest, but to the south and on the lower part of the western slope is replaced by an assemblage of planktonic and calcareous benthic foraminifers. It is important to note that planktonic foraminifers predominate in the sand fraction from the upper continental slope of Greenland and that siliceous sponge spicules are observed all over the slope area.

It is of interest that polychaets, sponges, and mollusk shells mainly are found on continental slopes and the Yermak Plateau slope. Abundant sponges and polychaets are related to high hydrodynamic activity of bottom waters leading to nepheloid layer formation (Stein and Fahl, 1997). The widespread occurrence of mollusk shells is possibly related to their redeposition from the adjacent shelves. Agglutinating benthic foraminifers predominate among the biogenic components of sediments accumulated at depths shallower than sediments with *Biloculina* and *Triloculina*. It is quite meaningful to observe an occurrence of burrowing agglutinating benthic foraminifers *Rheophax sp.* and *Atlantica atlantiella,* these large (2–4 mm) tests are found in subsurface layers (from 2–4 up to 6–16 cm) of bottom sediments at stations PS 2856–2864. The stations are located in different facies zones at various depths, but are distinguished by similar characteristics of bottom waters – saline ($S > 34.9‰$) and cold ($T < 0^{o}C$) (Table 9.4). On the whole, the overall distribution pattern of benthic organisms is mainly controlled by their feeding mode (Stein and Fahl, 1997).

Table 9.5 Major zones of Holocene sedimentation [*Levitan et al.*, 2000b], with changes

Zone of Holocene sedimentation	Terrigenous matter sources	Transport agents of terrigenous matter	Primary production
Northwestern part of the Barents Sea	Spitsbergen, sea bottom	sea ice, icebergs, bottom currents	low
Northern continental slope of Spitsbergen	Spitsbergen, sea bottom	sea ice, icebergs, slope processes	low
Sophia Deep	Spitsbergen, Yermak Plateau, Siberia (?)	sea ice, icebergs, currents	high
Yermak Plateau slopes	Spitsbergen, Greenland, Yermak Plateau, Siberia, sea bottom, Scandinavia	sea ice, slope processes, currents	high
Crest of the Yermak Plateau	Spitsbergen, Greenland, Siberia, sea bottom, Scandinavia	sea ice, surface and bottom currents	average and high
Continental slope of Greenland	Greenland, sea bottom	sea ice, surface and bottom currents, slope processes	low and average

The distribution of planktonic foraminifers mainly represented by *N. pachyderma* (sin.) in the fraction 2.0–0.063 mm of the surface sediments (Fig. 9.14), shows that their highest concentrations exactly mark the position of major Atlantic water currents. Indeed, planktonic foraminifer tests are practically absent in sediments from the northwestern part of the Barents Sea and the northern continental slope of Spitsbergen. They appear in considerable amounts in the central part of the Sofia Deep (station PS 2861) and are abundant in the northeastern part of the Yermak Plateau. Further westward, the content of planktonic foraminifers decreases to 1% at station PS 2856 and to 20% at station PS 2855, and increases again up to 65% and even to 80% (station PS 2847) on the western slope of the Yermak Plateau and in the Lena Deep. Low concentrations (4–10%) are observed in the southern part of the Yermak Plateau, as well as on the lower continental slope of Greenland (station PS 2842). We believe that a high content of planktonic foraminifer tests (50%) on the upper continental slope of Greenland (station PS 2846) is related to the location of this station within the Greenland polynia area, beneath high-productivity waters near the edge of a huge sea-ice massif.

Therefore, composition of biogenic remains in sand and gravel fractions and content of skeletal elements of some organisms are important facies indicators for the studied region reflecting seafloor environment (benthic organisms), structure and productivity of surface waters (planktonic organisms).

Relationship Between Clastic and Biogenic Components in Surface Sediments

The clastic to biogenic components (C/B) ratio varies from 90/10 (stations PS 2822 and PS 2824) to 40/60 (station PS 2829) in the northwestern part of the Barents Sea eastward of Spitsbergen (for location of stations see Fig. 9.14). The ratio is 75/25 on the northeastern Spitsbergen slope. Values of about 20/80 are characteristic for the northwestern Spitsbergen continental slope. Biogenic remains predominate (10/90) on the eastern slope of the Yermak Plateau. They still prevail on the plateau crest, but the proportion of rock fragments somewhat increases here; the average C/B value equals 15/85. On the western Yermak Plateau slope, the ratio is almost the same, as that on the eastern slope. The proportion of terrigenous clastic material in the >0.063-mm fraction is slightly higher (up to 25%) in the Lena Deep. The average C/B value equals 75/25 in sediments from the Greenland continental slope.

Therefore, terrigenous clastic components apparently predominate on continental slopes near major terrigenous material sources (Spitsbergen and Greenland). Their proportion considerably decreases in those parts of continental slopes which are likely located beneath the relatively elevated primary production zone, and are characterized by an increased zoobenthos biomass. Biogenic components distinctly predominate far off the regional terrigenous material sources. Hence, the C/B ratio in sand and gravel fractions of the surface layer of sediments from the study region serves as a good facies indicator reflecting the distance from terrigenous material sources and a relationship between fluxes of terrigenous and biogenic components.

History of Sedimentation

History of Sedimentation Rates During the Last 130 ka

As it is well known, the calculation of sedimentation rates mainly depends on accepted age scales, which are based on the applied stratigraphic scales. At the beginning of the seventies of the last century D. Clark (1971) proposed the lithostratigraphic scale for sediment cores from the Arctic Ocean, in which the fundamental units were designated by the letters A (end of the Pliocene) to M (Upper Pleistocene-Holocene). Subsequently, additional lithostratigraphic subdivisions were added for the Miocene-Pliocene (AE, A2b, A2a, and A1) (Bischof et al., 1996 and further references therein).

Clark's lithostratigraphy was related to magnetostratigraphic data (Clark et al., 1980; Nowaczyk and Baumann, 1992; Nowaczyk et al., 1994; Schneider et al., 1996). A principal role in stratigraphic reconstructions was played by isotope-oxygen studies (Nørgaard-Pedersen, 1997, Nørgaard-Pedersen et al., 2003; Spielhagen et al., 2004). For the upper part of the sedimentary sections AMS ^{14}C datings are the most important dating tool. Furthermore, biostratigraphy (remains of planktonic and benthic foraminifers, calcareous nannoplankton, ostracods, diatoms) (Backman et al., 2004; Cronin et al., 1995; Polyakova, 1997) as well as chemostratigraphy (for example, Mn cycles; Jakobsson et al., 2000), Be isotope stratigraphy (Aldahan et al., 1997), and Sr and Nd stratigraphy (Tütken et al., 2002) were applied.

As a result of numerous stratigraphic studies there has been a long-lasting and still ongoing controversial discussion whether the central Arctic Ocean sediments were deposited under conditions characterized by very low sedimentation rates or conditions characterized by significantly higher sedimentation rates The low-sedimentation-rate scenario is mainly derived from cores obtained from ridges in the Amerasia Basin and implies that the majority of cores presently available extend well into Plio–Pleistocene (Clark, 1970; Clark et al., 1980; Scott et al., 1989). On the other hand, higher sedimentation rates of about one to a few cm/ky were mainly determined in cores from ridges and basins in the Eurasian part of the central Arctic Ocean (Gard, 1993; Stein et al., 1994a, 1994b; Jakobsson et al., 2000, 2003). M. Jakobsson et al. (2000) and J. Backman et al. (2004) have shown that earlier determinations of sedimentations rates mainly based on magnetostratigraphy and the identification of negative polarity changes representing complete reversals, have to be revised due to the fact that many of the negative polarities recorded in the upper part of Arctic Ocean sedimentary sections, were caused by excursions of the magnetic field within the Brunhes Epoch. These results indicate that higher sedimentation rates of cm/ky are much more typical than previously thought. More details about the stratigraphy of Arctic oceanic sediments and related problems are discussed in synthesis papers of J. Backman et al. (2004) and R. Stein (2008).

Detailed regional lithostratigraphic studies of deep-sea sediment cores showed that distinct lithostratigraphic division is based on temporal changes of color, structural features, sediment composition, physical properties and other parameters, for example, oxygen-isotope composition of planktonic foraminifer tests. In the majority of sediment cores an alternation of brownish and yellowish/olive-grayish sediments has been noted. The color changes are probably caused by variations in the content of MnO_2. Brownish sediments, as a rule, were accumulated more during interglacial periods, when riverine delivery of dissolved manganese into the Arctic Ocean was increased and more intensive ventilation of bottom waters contributed to the accumulation of this element in the oxidized layer of the surface sediments (Jakobsson et al., 2000).

Lithological descriptions and grain-size data show that sediments of glacial intervals are usually coarser-grained than those deposited during interglacial (interstadial) intervals. The observed layers of iceberg sediments are related to history of ice sheets in North Eurasia and, as well, to the Arctic part of the Laurentide Ice Sheet and Innuit Ice Sheet of Greenland. Based on data of IRD composition and interpretation in terms of sediment sources, D. Darby et al. (2002) proposed that usually intensive iceberg supply from the Innuit Ice Sheet occurred somewhat earlier than that from the Laurentide Ice Sheet. Furthermore, these authors observed substantial age differences between the Heinrich events in the North Atlantic and episodes of intensive iceberg discharge into the Arctic Ocean from the Laurentide Ice Sheet (Table 9.6).

All iceberg events related to discharge from the Arctic part of the Laurentide Ice Sheet, occurred earlier than the corresponding Heinrich events related to the history of the more southern part of the Laurentide Ice Sheet, by several thousand years (AL1 – H0 = 2180 years; AL2 – H1 = 2770 years; AL3 – H2 = 1897 years; AL4 – H3 = 3285 years). This means that they relate to the previous Dansgaard-Oeschger cycles, and that the waves of ice-sheet instability may have been propagated from north to south.

According to a summary paper by R. Spielhagen et al. (2004), IRD maxima in deep-sea cores from the central Arctic Ocean were concentrated around 190–130 ka (i.e., during MIS 6), 90–80 ka (i.e., substage 5b), 75 ka (i.e., at the boundary of MIS 5 and MIS 4), and 65–50 ka (i.e., at the end of MIS 4 – beginning of MIS 3). In sediments of the LGM, IRD is noticeably less than in the previous glacial periods, although – in comparison with the Holocene – this value is still substantially higher (Bischof, 2000; Spielhagen et al., 2004). These data coincide very well with the

Table 9.6 Age difference between Heinrich events in the North Atlantic and iceberg sediments related to the Arctic part of the Laurentian Ice Sheet [*Darby et al.*, 2002], simplified

Average age, years	H0	H1	H2	H3	AL1	AL2	AL3	AL4
Radiocarbon scale	10977	14922	20631	27955	12873	15550	21960	29770
Calendar scale	12200	15500	23280	31600	14380	18270	25177	34885

Note: H0–H3 – Heinrich events; AL1–AL4 – iceberg events, related to history of the Arctic part of the Laurentian Ice Sheet.

reconstructions of the extent of major ice sheets in Eurasia during late Saalian to Weichselian times based on terrestrial field work (Svendsen et al., 2004).

Furthermore, biogenic components also contribute to a change in the sediment composition within the sedimentary sections: as a rule, the contents of planktonic and benthic foraminifers, calcareous nannoplankton and deep-water ostracods are increased in sediments of MIS 1, middle MIS 3 and MIS 5e (Backman et al., 2004; Cronin et al., 1995; Nørgaard-Pedersen et al., 2003; Polyak et al., 2004), which is interpreted by increased productivity due to Atlantic-water inflow and more seasonally open sea-ice areas. These results are confirmed by oxygen-isotope data (Spielhagen et al., 2004). In Holocene sediments radiolarians were found, which are completely absent in MIS 2 sediments. The increased content of biogenic components in the sediments of interglacial stages is confirmed by $CaCO_3$ and TOC data (Stein et al., 1994b). Furthermore, oxygen-isotope studies showed that inflow of fresh waters into the Arctic Ocean, connected with the breakthrough of the glacial lakes, was highest at 130, 80–75, and 52 ka and substantially lower at 18 ka (Spielhagen et al., 2004).

All available data on sedimentation rates from Arctic cores dated by oxygen isotopes and/or AMS ^{14}C methods, are presented in Table 9.7. Based on these data, a number of hiatuses in the sedimentary records of many cores is obvious. The most accurate values of sedimentation rates (SR) are obtained by AMS ^{14}C method, i.e., for MIS 1, MIS 2 or parts of MIS 3 sediments. When using AMS ^{14}C ages, one should have in mind that the reservoir effect is likely to be higher for glacial periods than for the interglacial periods (Backman et al., 2004). For the continental margins we considered SR only for the continental slopes and the continental rises (without the shelves) (for locations of sediment cores see Fig. 9.15). In other respect, the methodology was similar to that described in the chapter about the sedimentation rates in the Norwegian-Greenland Basin.

For the first time, schematic maps of the distribution of sedimentation rates have been compiled for each of the five last MIS as a result of the accomplished work (Levitan and Stein, 2007). Thus, history of SR for the last 130 ka has been illuminated for the Arctic Ocean.

According to the SR map for the MIS 5 (Fig. 9.16) is evident that the gradation of more than 5 cm/ky is located only in the area of the Fram Strait, and also, in the southern end of the Lomonosov Ridge. The gradation 1–5 cm/ky is located at the continental margin of the Northern Greenland and the Barents-Kara basin, and the extensive areas of the Nansen Basin and Lomonosov Ridge. SR distribution of less than 1 cm/ky is observed at the Gakkel Ridge, in the Amundsen and Makarov Basins.

The MIS 4 is characterized by somewhat different nature of SR distribution (Fig. 9.17). The gradation of more than 5 cm/ky in the Fram Strait was inherited from the previous stage, and it is also possible to assume its development along the continental margin of the Barents, Kara and Laptev Seas. The minimum (less than 1 cm/ky) SR distribution areas are contoured around the Gakkel Ridge and in the Makarov Basin. Distribution areas of SR from 1 to 5 cm/ky cover the extensive areas of the Nansen Basin and Amundsen Basin sea floor, and are found at the Lomonosov Ridge.

Table 9.7 Sedimentation rates (cm/ky) in the Arctic Ocean during the last five marine oxygen isotope stages (MIS) (compiled by R. Stein and M.A. Levitan)

Sediment core	Northern latitude	Longitude	Water depth, m	MIS 1	MIS 2	MIS 3	MIS 4	MIS 5a-5d	MIS 5e	MIS 5	Reference
Yermak Plateau											
PS2837	81°13'	02°22'?E	1042	23	4	–	–	–	–	–	Birgel and Stein, 2004
PS2122	80°23,4'?	07°33,0'?E	705	9	7.8	–	–	–	–	–	Birgel and Stein, 2004
PS2123	80°10,4'?	09°51,4'?E	571	14.17	8.75	6.43	2.86	–	–	–	Birgel and Stein, 2004
PS 2213	80°27,6'	08°02,6'E	853	2.5	5.0	2.7	?	–	–	2.4	Aldahan et al., 1997
ODP 910	80°16'	06°35'E	552	3.75	2.60	2.60	2.75	–	–	1.84	McManus et al., 1996
ODP 912	79°57,557'	05°27,390'E	1036	5.00	9.58	6.00	7.50	–	–	6.34	Hevrøy et al., 1996
PS 2212	82°04,2'	15°43,0'E	2550	3.33	5	4	3.57	–	–	>1.79	Vogt, 1997
PS 1533	82°01,9'	15°10,7'E	2030	4.17	5.42	3.14	3.21	2.62	1.55	2.59	Vogt, 1997
Morris-Jessup Rise											
PS2200	85°19,6'?	14°00,0'?W	1074	0.2	0.2	?	2.94	1.15	1.82	1.28	Spielhagen et al., 2004
PS2202	85°06'	14°25'W	1081	–	–	0.59?	–	–	–	–	Stein et al., 1996
Continental slope of Northern Greenland											
PS 2876	81°45,7'	09°24,0'W	1991	4.68	18.83	15.4	–	–	–	–	Nørgaard-Pedersen et al., 2003
Continental slope of Northern Spitsbergen											
PS1533	82°01,9'?	15°10,7'?E	2030	4.17	5.42	3.14	3.21	2.22	3.18	2.31	Spielhagen et al., 2004
PS2212	82°04,2'?	15°51,2'?E	2550	3.33	5.00	4.00	3.57	–	–	>1.79	Vogt et al., 2001
Northern continental slope of the Barents Sea											
PS2138	81°32,1'?	30°35,6'?E	995	5.4	21.2	4.4	2	–	–	1.7	Knies et al., 2000
PS2446	82°23,8'?	40°54,5'?E	2022	10	>30	–	–	–	–	–	Knies et al., 2000
PS 1519	81°47,4'	31°30,1'E	2998	6.67	5.83	6.71	3.93	–	–	–	Kubisch, 1992
PS 1521	82°56,5'	32°05,2'E	3752	2.08	7.00	3.31	3.57	–	–	2.32	Kubisch, 1992
ACB 880	79°50'	47°E	388	27.65	43.18?	–	–	–	–	–	Duplessy et al., 2001
Continental slope of the Kara Sea											
PL 94-08	81°71,1'	67°45,6'E	617	3	>27.27	–	–	–	–	–	Polyak et al., 1997
PL 94-60	76°99,4"	70°06,2'E	572	36	–	–	–	–	–	–	Hald et al., 1999

Table 9.7 (continued)

Continental slope of the Laptev Sea											
PS51/154-11	77°15,6'?	120°36,6'?E	270	18	–	–	–	–	–	–	Bauch et al., 2001
PS2778–2	77°58,7'?	113°03,9'?E	341	38	–	–	–	–	–	–	Stein and Fahl, 2000
PS2476–4	77°23,4'?	118°11,6'?E	521	50	–	–					Stein and Fahl, 2000
PS2458–4	78°10,0'?	133°23,9'?E	983	34	66.8?	–	–	–	–	–	Bauch et al., 2001
PS2474–3	77°40,2'?	118°34,5'?E	1494	18.9	8.5	10.9	–	–	–	–	Stein et al., 2001
PS2742–5	80°47,3'?	103°48,9'?E	1890	22	–	–	–	–	–	–	Stein and Fahl, 2000
Continental rise of the Laptev Sea											
PS2741	81°06,0'?	105°22,0'?E	2400	11.9	2	2.4	2.7	–	–	2.7	Stein et al., 2001
PS2471	79°09,0'?	119,77,4'?E	3047	2	5.8	1.86	4.29	–	–	0.54	Stein and Fahl, 2000
Continental slope of the East Siberian Sea											
PS2763	80°16,9'?	150°26,1'?E	1591	12.6	10.2	13.6	–	–	–	–	Stein et al., 2001
Continental slope of the Chukchi Sea											
92BC17	76°05,2'?	164°50,2'?W	402	2	–	–	–	–	–	–	Grantz et al., 1999
Northwind Rise											
88BC6	74°35,9'?	158°02,2'?W	954	3.2	–	–	–	–	–	–	Grantz et al., 1999
88BC22	71°03'	161°60.5'	?	3.45	–	–	–	–	–	–	Darby et al., 1997
PI–88–AR–5	74°37,35'	157°53,04'W	1089	0.63	–	–	–	–	–	–	Poore et al., 1999
Mendeleev ridge											
94BC16	80°20,3'?	188°42,7'W	1533	1.1	0.5 (hiatus)	0.74	–	–	–	–	Grantz et al., 1999
94BC17	81°15,91'	178°58,1'?E	2217	1.5	0.3 (hiatus)	1.17	–	–	–	–	Grantz et al., 1999
94BC19	82°26,8'	175°45,5'E	2400	0.41	Hiatus	0.75	–	–	–	–	Poore et al., 1999
NP26-32	79°19,4'	178°04'W	1610	0.85	Hiatus	0.71	–	–	–	–	Polyak et al., 2004
Lomonosov ridge											
PS2177	88°02,2'?	134°55,1'?E	1388	0.9	0.3	–	–	–	–	–	Nørgaard-Pedersen et al., 2003
PS2185	87°31,9'?	144°08,4'?E	1073	0.79	0.3	0.9	2.94	1.42	2.88	1.50	Spielhagen et al., 2004
94BC28	88°52,3'?	140°22,8'?E	2015	1.35	0.10	–	–	–	–	–	Darby et al., 1997
PS2757	81°09,6'?	140°12,1'?E	1230	5.2	5.8	3.4	4	–	–	5.2	Stein et al., 2001
96/12–1	87°05,9'	144°46,4'E	1003	2.80	2.80	2.80	2.80	–	–	0.80	Jakobsson et al., 2000
PS2186	88°30,9'	140°29,4'E	2036	1.00	0.50	0.49	–	–	–	–	Nørgaard-Pedersen, 1997
PS2179	87°44,8'	138°01,7'E	1230	1.00	0.50	0.31	–	–	–	–	Nørgaard-Pedersen, 1997
PS2184	87°36,7'	148°08,4'E	1640	0.33	0.25	0.34	0.21	–	–	–	Nørgaard-Pedersen, 1997

Table 9.7 (continued)

Gakkel ridge											
PS2163	86°14,5'?	59°12,9'?E	3040	0.7	<0.7	–	–	–	–	–	Stein et al., 1994a
PS2166	86°51,6'?	59°45,9'?E	3636	0.9	0.4	0.67	–	–	–	–	Nørgaard-Pedersen et al., 2003
PS2206	84°16,7'?	02°30,3'?W	2993	0.7	0.5	–	–	–	–	–	Stein et al., 1994a
PS 1524	85°21,8'	26°12,9'E	3646	1.20	0.35	0.35	0.35	–	–	0.60	Eisenhauer et al., 1994
PS 1525	85°30,8'	25°17,8'E	3180	0.75	1.25	–	–	–	–	–	Kubisch, 1992
PS 1528	86°07,8'	23°09,5'E	3972	0.75	0.42	0.34	0.86?	–	–	>1.25	Kubisch, 1992
PS 1529	85°22,9'	21°42,5'E	2896	0.67	0.42	0.17	0.64	–	–	–	Kubisch, 1992
PS 2164	86°20,1'	59°16,0'E	2030	0.17	0.33	0.23	–	–	–	–	Kubisch, 1992
Nansen Basin											
OD-041-04	84°01,8'?	11°14,3'?E	3344	0.84	0.57	–	–	–	–	–	Nørgaard-Pedersen et al., 2003
PS1527	86°05,8'	22°01,0'E	3704	0.50?	0.42	0.56?	–	–	–	–	Nørgaard-Pedersen et al., 2003
PS2208	83°38,6'	04°39,5'E	3682	1.25	3.10	2.4	?	–	–	1.4	Aldaham et al., 1997
Amundsen Basin											
PS2170	87°35,8'?	60°53,7'?E	4226	0.9	0.5	–	–	–	–	–	Stein et al., 1994a
PS2195	86°13,7'?	09°35,6'E	3873	0.6	0.3	0.57	–	–	–	–	Nørgaard-Pedersen et al., 2003
FRAM1/4	84°29,9'	08°58,7'W	3820	0.67	1.46	–	–	–	–	–	Nørgaard-Pedersen et al., 2003
PS2192	88°15,7'	09°52,7'E	4375	4.00	–	–	–	–	–	–	Nørgaard-Pedersen et al., 2003
PS2172	87°15,4'	68°22,7'E	4391	1.00	0.67	0.46	–	–	–	–	Nørgaard-Pedersen, 1997
PS2196	85°57,1'	00°06,9'E	3958	0.50	0.42	0.23	0.14	–	–	–	Nørgaard-Pedersen, 1997
PS2193	87°31,1'	11°15,5'E	4337	0.67	0.58	–	–	–	–	–	Nørgaard-Pedersen, 1997

Table 9.7 (continued)

Makarov Basin											
PS2761	81°11,5'?	150°29,8'?E	2640	2	3.4	4.2	3.3	–	–	3.8	Stein et al., 2001
PS2178	88°02,2'	159°14,0'E	4009	–	–	6.17	3.91	–	–	3.16	Spielhagen et al., 2004
PS 51/038	85°08,1'	171°26,4'E	1473	0.5	0.17	1.11	0.5	–	–	0.80	Spielhagen et al., 2004
PS2180	87°37,6'	156°40,5'E	4005	0.80	0.33	0.88	–	–	–	–	Nørgaard-Pedersen, 1997
Wrangle abyssal plain											
94BC20	83°10,6'?	174°5,9'?E	3145	1	0.2	–	–	–	–	–	Darby et al., 1997
Canadian abyssal plain											
88PC10	74°43,5'?	156°08,5'?W	3899	145	–	–	–	–	–	–	Grantz et al., 1999
93BC13	75°48,3'?	155°07,6'?W	3831	1	–	–	–	–	–	–	Grantz et al., 1999
Continental slope of the Beaufort Sea											
P189AR–P45	70°33,03'	141°52,08'W	405	135	–	–	–	–	–	–	Andrews and Dunhil, 2004

Note: Blank – no data

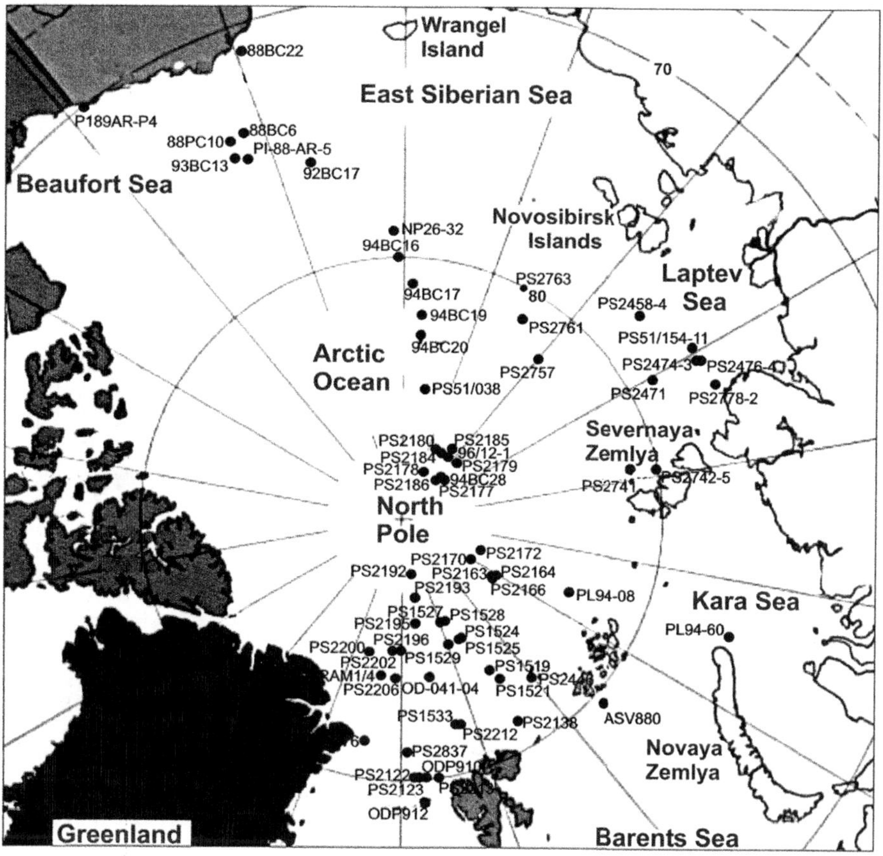

Fig. 9.15 Location of studied sediment cores in the Arctic Ocean

During MIS 3 the areas of SR distribution with more than 10 cm/ky (Fig. 9.18) have appeared. They are related to the continental margin of northeastern Greenland and the Laptev Sea. The gradation 5–10 cm/ky borders the previously contoured gradation for MIS 4 from the north and towards the central area of the Fram Strait. There is a reason to consider that this gradation has been also extended along the continental margin of the Barents, Kara and Laptev Seas. Furthermore, a separate spots of SR from 5 to 10 cm/ky were observed in the northern part of the Lomonosov Ridge. The distribution pattern of gradation 1–5 cm/ky is located to the north from the areas of development SR with 5–10 cm/ky, occupying the Yermak Plateau, the southern part of the Nansen Basin and the large part of the Lomonosov Ridge. The very extensive areas of sea floor in the Nansen, Amundsen and Makarov basins, and also on the Gakkel Ridge and Mendeleev Ridge have been the arena of the low SR, in particular, with less than 1 cm/ky. Uneven distribution of sedimentation processes along the extension of the Lomonosov Ridge is noteworthy.

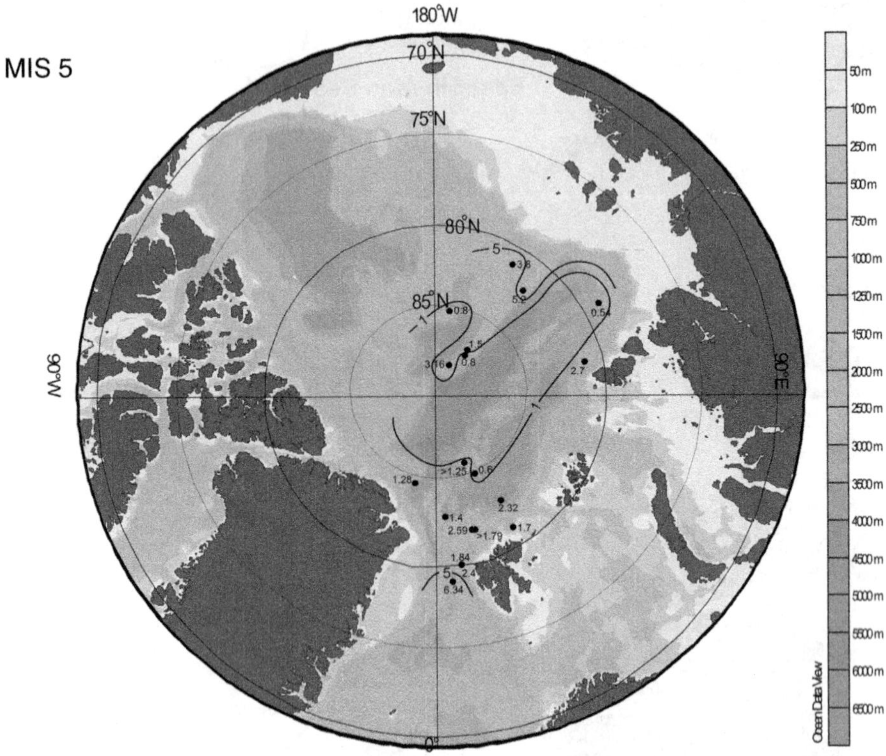

Fig. 9.16 MIS 5 sedimentation rates (SR) (cm/ky) in the central Arctic Ocean (Levitan and Stein, 2007)

During MIS 2 the areas with SR distribution of more than 20 cm/ky have been located, for example, in the continental margin of the Barents Sea and, to a lesser degree, in the Laptev Sea (Fig. 9.19). Gradations 10–20, 5–10 and 1–5 cm/ky have been consecutively bordering the above mentioned areas northwards. Furthermore, as in the MIS 3, the gradation of 10–20 cm/ky was observed in the continental margin of northeastern Greenland. Almost entire central Arctic (Gakkel Ridge, Amundsen Basin, northern half of the Lomonosov Ridge, Makarov Basin, and Mendeleev Ridge) was occupied by SR distribution of less than 1 cm/ky.

During MIS 1 the SR distribution of more than 20 cm/ky are reliably charted in the continental margin of the Laptev Sea and Barents Sea (Fig. 9.20). In the north of the Kara Sea such high values are known only directly to the north from Novaya Zemlya (in the southern part of the St. Anna Trough) (Hald et al., 1999). SR has reached 135 cm/ky in the fan of the Mackenzie River. As an exception, on this map we showed the results of SR calculations for the sediment core on the continental rise of the marginal Northwind Plateau located in the Chukchi Sea, composed by turbidites, which sedimentary matter was brought from the area of the underwater delta of the Mackenzie River (Grantz et al., 1999). The noted strip of

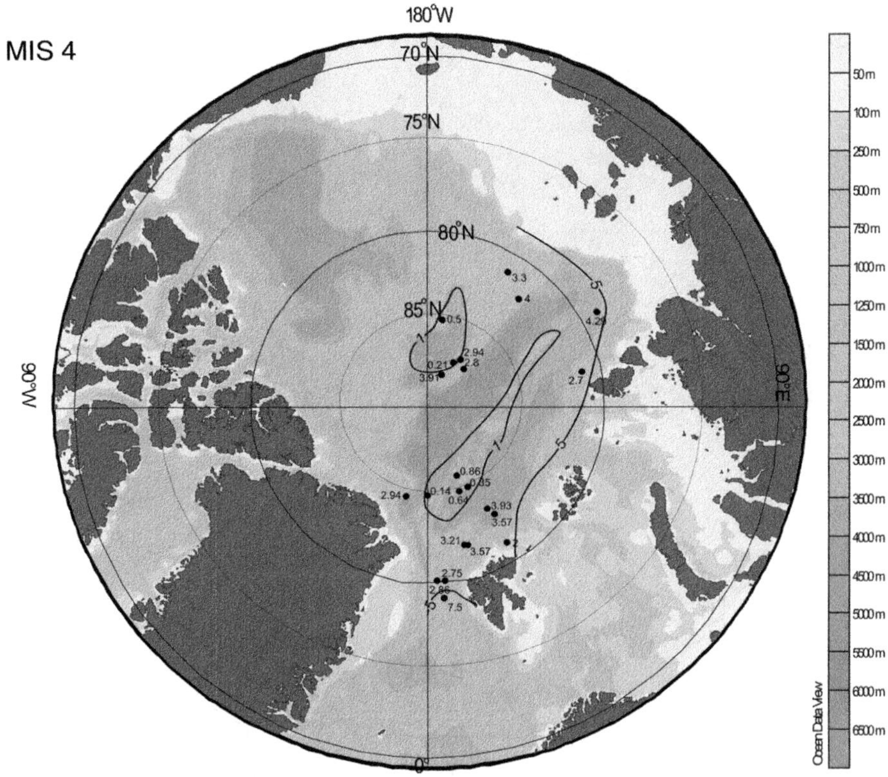

Fig. 9.17 MIS 4 sedimentation rates (SR) (cm/ky) in the central Arctic Ocean (Levitan and Stein, 2007)

gradation distribution of more than 20 cm/ky along the continental margin of Eurasia is contoured northwards by areas of development of SR distribution with 10–20 and 5–10 cm/ky. At this time the northeastern part of Greenland has no longer been a sufficiently powerful source of terrigenous matter. The small field of values of more than 10 cm/ka in the western slope of the Yermak Plateau and in the Fram Strait is rather connected with the unloading of sedimentary matter from the melted sea ice and icebergs, which were carried by the Transpolar Drift from the Central Arctic. The distribution pattern of SR gradation 1–5 cm/ky clearly reflects the circumcontinental sedimentation zonality; however, the sedimentation rates of this order are quite widely developed at the Lomonosov Ridge as well. Actually, with this exception, during MIS 1 the entire Central Arctic was the arena of slow sedimentation rates of less than 1 cm/ky.

The results show unambiguously that the SR value in both types of continental margins was higher in each of five MIS than in pelagic zones; i. e. circumcontinental zonation was prominent during the last climatic cycle. A basically different SR distribution by age is observed at glacial and nonglacial margins (Fig. 9.21). For instance, average SR values increase at glacial margins from MIS 5 to MIS 2 and

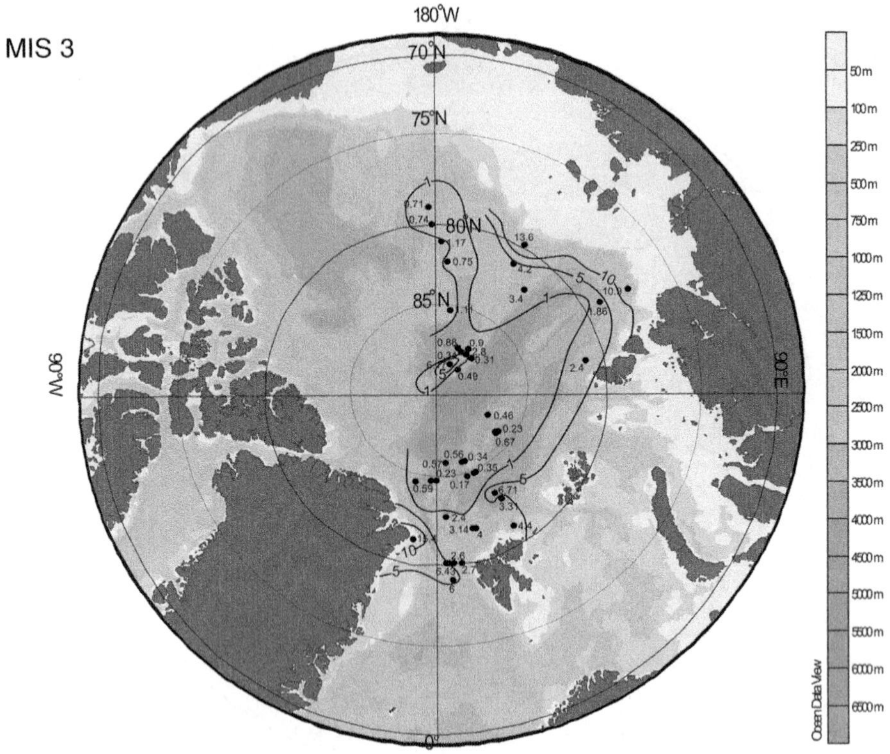

Fig. 9.18 MIS 3 sedimentation rates (SR) (cm/ky) in the central Arctic Ocean (Levitan and Stein, 2007)

then decrease during MIS 1. This regularity completely coincides with the pattern described for the Nordic Seas (their continental margins and pelagic zones) (Levitan and Stein, 2007, 2008; Levitan et al., 2007). Hence, like in the Nordic Seas, the SR history within glacial continental margins of the Arctic Ocean was governed, first of all, by the history of glaciations of the Northern Hemisphere, i.e., the history of variations in the volume of ice sheets. The SR value of deglaciated margins also increased from MIS 5 to MIS 3, slightly decreased during MIS 2, and sharply increased during MIS 1. Hence, the glaciation history of the Northern Hemisphere was no longer crucial for margins of that type. Along with the ice factor, other mechanisms were also in force and they need further investigations. If we perform smoothing of data on all continental margins of the Arctic Ocean, the time distribution of SR values will resemble the regularity obtained for glacial margins (Table 9.8).

Temporal variation trends of SR in pelagic zones coincide for both ridges and deep-sea basins: SR values has been steadily decreasing from MIS 5 to MIS 2 and then increasing during MIS 1 (Fig. 9.22). At the same time, average SR values are usually higher in deep-sea basins than on ridges, except for MIS 1. Thus, an inverse

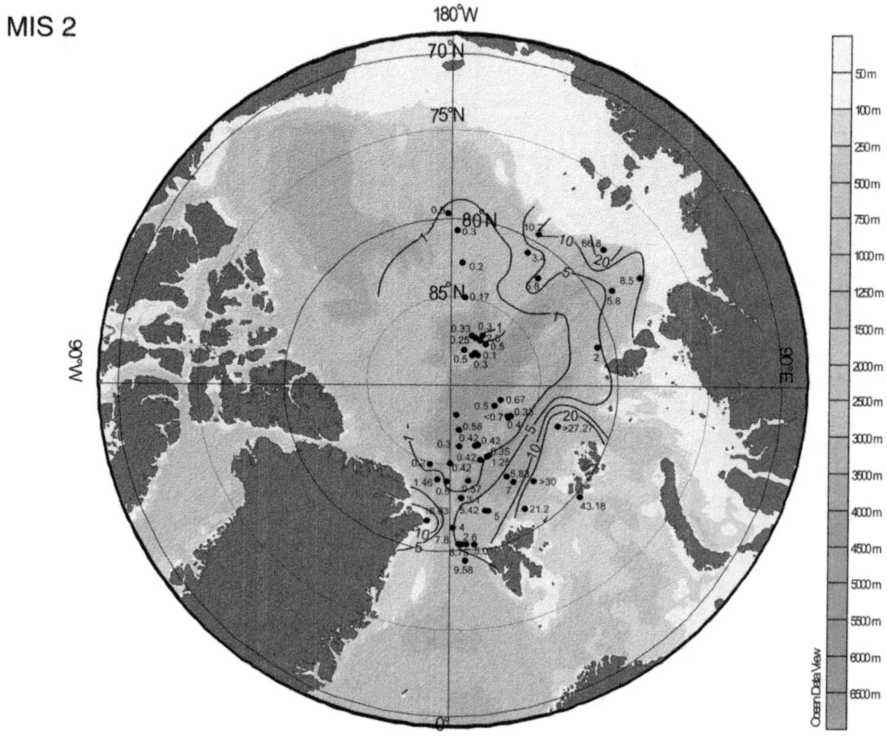

Fig. 9.19 MIS 2 sedimentation rates (SR) (cm/ky) in the central Arctic Ocean (Levitan and Stein, 2007)

correlation is clearly defined in pelagic zones between average SR values and the ice volume in glaciers of the Northern Hemisphere.

In our opinion, the opposite temporal variation trends of SR for continental glacial margins and pelagic zones are accounted for the climate-related glaciation history of the Northern Hemisphere, but the mechanisms of SR variations were different. For instance, as was discussed above, the lateral supply of terrigenous matter by glaciers and subglacial flows ("the bulldozer effect" included) from the surrounding land plays the leading role for continental slopes and rises of glacial margins. In pelagic zones, the thickness and compactness of ice, which are directly proportional to severity of the climate (the extent of glaciation development), correlate negatively with the SR due to a lesser intensity of melting during cold epochs. Hence, the ocean floor should receive less sedimentary matter in glaciation periods than in warm epochs. This conclusion is also confirmed by numerous hiatuses in sediment sequences of ridges and basins that are often confined to MIS 2 (Poore et al., 1999), i.e., the epoch of the most severe glaciation in the last climatic cycle.

Turning back to the discussion about the SR typical of Arctic deep-water sediments and based on our generalization of data (Table 9.8), we can conclude that the

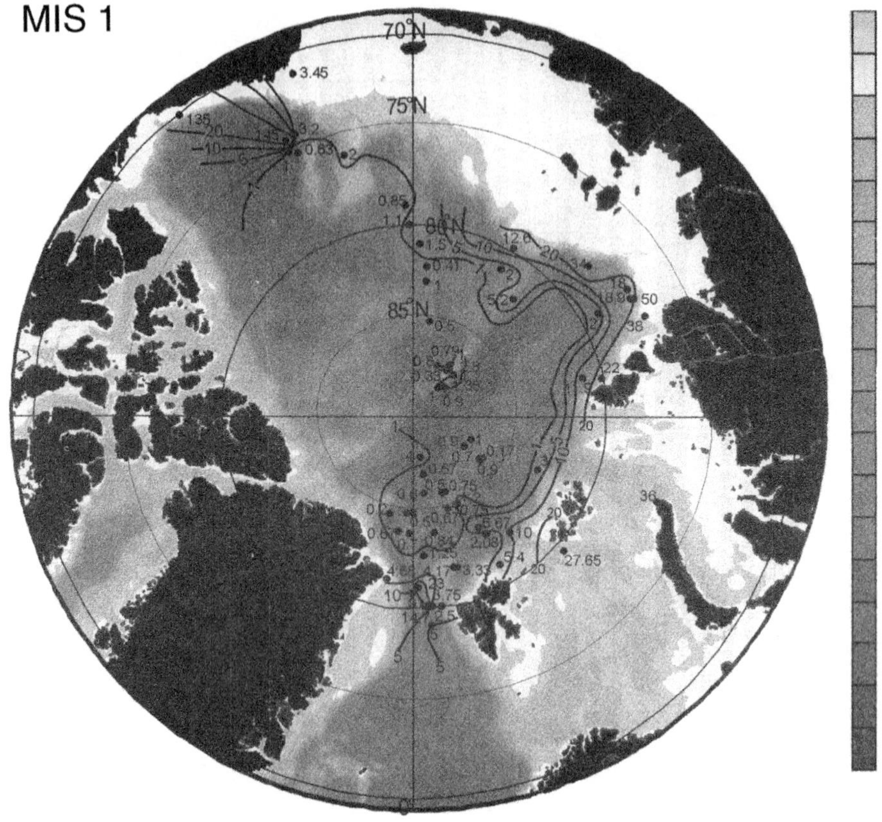

Fig. 9.20 MIS 1 sedimentation rates (SR) (cm/ky) in the central Arctic Ocean (Levitan and Stein, 2007)

SR values have comprised cm/ky in most zones during the last climatic cycle, and the SR has commonly decreased to mm/ky during MIS 2.

In the following, we consider examples of sedimentation history in the Arctic Ocean at the Yermak Plateau and Central Arctic.

History of Sedimentation on the Yermak Plateau During the Last 190 ka

Lithology and Mineral Composition of Middle Pleistocene-Holocene Sediments

During *Polarstern* Cruise ARK-XIII/2, sediment cores were retrieved with gravity corers and Kastenlot corers (elongated box corers) on the Yermak Plateau (Stein and Fahl, 1997) and its slopes and comprehensively (layer-by-layer) described

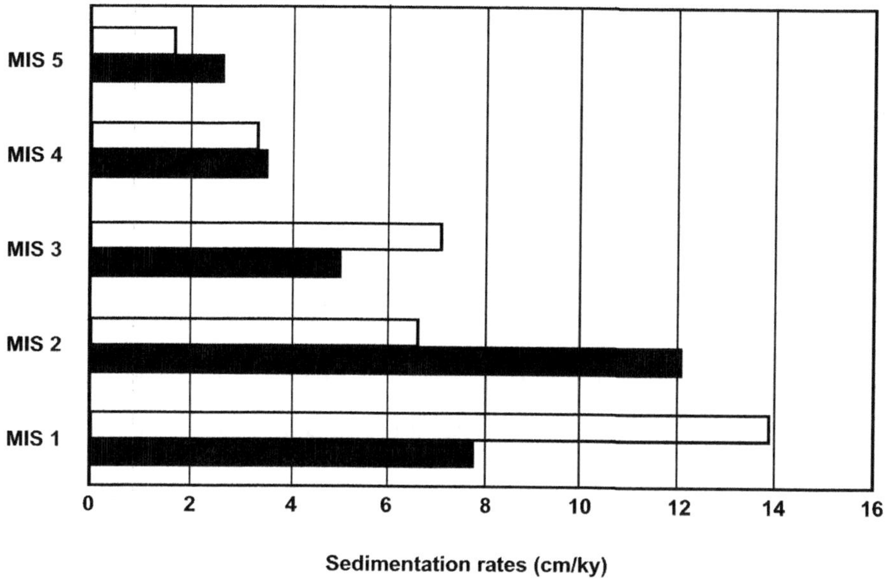

Fig. 9.21 Histograms of average SR (cm/ky) distribution for glacial (black) and non-glacial (light) continental margins of the Arctic Ocean for the last five marine isotope stages (MIS) (Levitan and Stein, 2007)

(Fig. 9.11). Cores PS2837, PS2834, and PS2859 have been comprehensively studied. The key core for stratigraphic age model is Core PS2837 (see below).

Sediment Core PS2837 (Water Depth 1023 m)

This core is unique for the Yermak Plateau. The Holocene sedimentation rate typically does not exceed a few centimeters per thousand years in this region (Levitan et al., 2000b). However, in sediment core PS2837 the Holocene sequence is about 2.5 m thick (Nørgaard-Pedersen et al., 2003); i.e., the sedimentation rate is an order of magnitude higher than the normal value. Sediments were probably accumulated in a very limited area in the pocket in the upper western slope of the Yermak Plateau.

Stratigraphy

The age model of sediment core PS2837 is based on 15 AMS ^{14}C datings and oxygen-isotope stratigraphy (Hass, 2002; Nørgaard-Pedersen et al., 2003). Primary values of radiocarbon ages were corrected for a reservoir effect of 400 yr (Stuiver and Braziunas, 1993). Age boundaries of marine isotope stages (MIS) are according to (Martinson et al., 1987). The oldest date (more than 49 ka) corresponds to sensitivity limit of the applied method and, therefore, was omitted in the age model. Based on the available oxygen isotope and IRD distribution data (see below), we assume that sediments of the core bottom did not reach Substage MIS 5e (i.e.,

Table 9.8 Spatial and temporal distribution of sedimentation rates (cm/ky) and its ratio in the Arctic Ocean during the last five MIS (by M.A. Levitan)

Regions		Sedimentation rates (cm/ky)					Ratios of sedimentation rates for pairs of MIS			
		MIS 1	MIS 2	MIS 3	MIS 4	MIS 5	MIS 1/ MIS 2	MIS 2/ MIS 3	MIS 3/ MIS 4	MIS 4/ MIS 5
I	max	27.65	43.18	15.40	7.50	6.34	–	–	–	–
	min	0.00	0.00	0.59	2.00	1.28	–	–	–	–
	average	7.71	12.06	4.89	3.59	2.50	0.64	2.47	1.36	1.44
	n	16	16	10	9	8	16	10	9	8
II	max	38.00	10.20	13.60	4.29	2.70	–	–	–	–
	min	0.63	2.00	1.86	2.70	0.54	–	–	–	–
	average	13.89	6.63	7.19	3.50	1.62	2.09	0.92	2.05	2.16
	n	12	4	4	2	2	4	4	2	2
III	average	10.36	10.97	5.55	3.57	2.32	0.94	1.98	1.55	1.54
	n	28	20	14	11	10	20	14	11	10
IV	max	5.20	5.80	3.40	4.00	5.20	–	–	–	–
	min	0.17	0.00	0.17	0.21	0.60	–	–	–	–
	average	1.15	0.79	0.89	1.69	1.87	1.46	0.89	0.53	0.90
	n	20	20	15	7	5	20	15	7	5
V	max	4.00	3.40	6.17	3.91	3.80	–	–	–	–
	min	0.00	0.00	0.23	0.14	0.80	–	–	–	–
	average	1.01	0.85	1.37	1.96	2.29	1.19	0.62	0.70	0.86
	n	16	14	9	4	4	14	9	4	4
VI	average	1.09	0.81	1.07	1.79	2.06	1.35	0.76	0.60	0.87
	n	36	34	24	11	9	34	24	11	9
VII	average	5.15	4.57	2.72	2.68	2.20	1.13	1.68	1.01	1.22
	n	64	54	38	22	19	54	38	22	19

Note: I – continental margins, which were subjected to glaciation during LGM (Greenland, Spitsbergen, Barents and Kara Seas); II – continental margins, which were not subjected to glaciation during LGM (Laptev Sea, East Siberian and Chukchi Seas); III – continental margins of the Arctic Ocean; IV – underwater ridges (Mendeleev, Lomonosov, and Gakkel); V – deep-water basins (Nansen, Amundsen, Makarov, Wrangel, and Canadian); VI – pelagic areas; VII – Arctic Ocean.

Eemian Interglaciation). The calculated sedimentation rates indicate increasing sedimentation from MIS 4 to MIS 1 (Table 9.9). The sedimentation rate was less precisely determined for MIS 5 relative to the younger sediments. Nevertheless, one may suggest that it was higher relative to MIS 4.

As it follows from Table 9.9, sedimentation rates are typical for the Yermak Plateau up to the end of MIS 3 in the area of Core PS2837 (Nowaczyk et al., 1994; Flower, 1996; Hevrøy et al., 1996). A change of the local relief in the upper part of the western Yermak Plateau slope, the formation of the above-mentioned "pocket", probably occurred sometimes at the beginning of MIS 2.

Lithology

Sediments of MIS 1 (0–248 cm) consist of homogeneous, dark olive-gray, terrigenous clayey mud with rare interlayers of silty-clayey mud (Fig. 9.23). The sedimen-

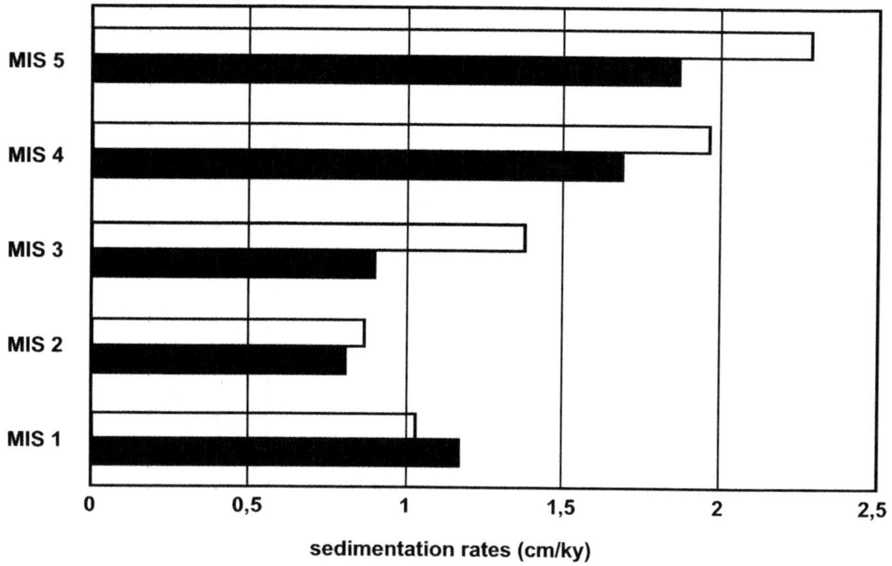

Fig. 9.22 Histograms of average SR (cm/ky) distribution for deep-sea basins (black) and ridges (light) of the Arctic Ocean for the last five marine isotope stages (MIS) (Levitan and Stein, 2007)

tary section below 207 cm contains abundant black coating of hydrotroilite. The microscopic study revealed that sediments are mainly composed of clay minerals, ferruginous-clayey aggregates, and quartz and feldspar grains. Accessory minerals include transparent heavy minerals, black ore minerals, rock fragments, volcanic glass, planktonic foraminifer tests, coalified plant remains, sponge spicules, and coccoliths. Plant detritus has not been found below 182 cm.

The MIS 2 sequence (248–402 cm) consists of two units. The upper unit (290–386 cm), practically indiscernible from the overlying unit in composition, color, and structure, is referred to MIS 2 exclusively based on radiocarbon data. The lower unit (386–416 cm) is composed of coarser sediments mainly observed as olive-gray sandy-silty clay mixtites (black-colored at the base). The planktonic foraminifer content is slightly increased in some interlayers. Rare fragments of IRD

Table 9.9 Sedimentation rates (cm/ky) on the Yermak Plateau

No. of cores and holes	MIS 1	MIS 2	MIS 3	MIS 4	MIS 5	MIS 6	MIS 7
PS 2837	20.7	12.8	5.5	2.8	about 5	–	–
PS 2834	3.4	5.9	3.5	5.1	3.1	>3.2	–
PS 2859	3.9	3.7	4	4.3	3.4	2	–
PS 2212	2.5	5.8	4	3.3	1.8	–	–
ODP 912A	5	8.3	8.4	7.3	6.4	5.2	2.2
ODP 910C	2.1	6.2	2	5	1.8	1	1.2

Note: Blank – lack of sediments.

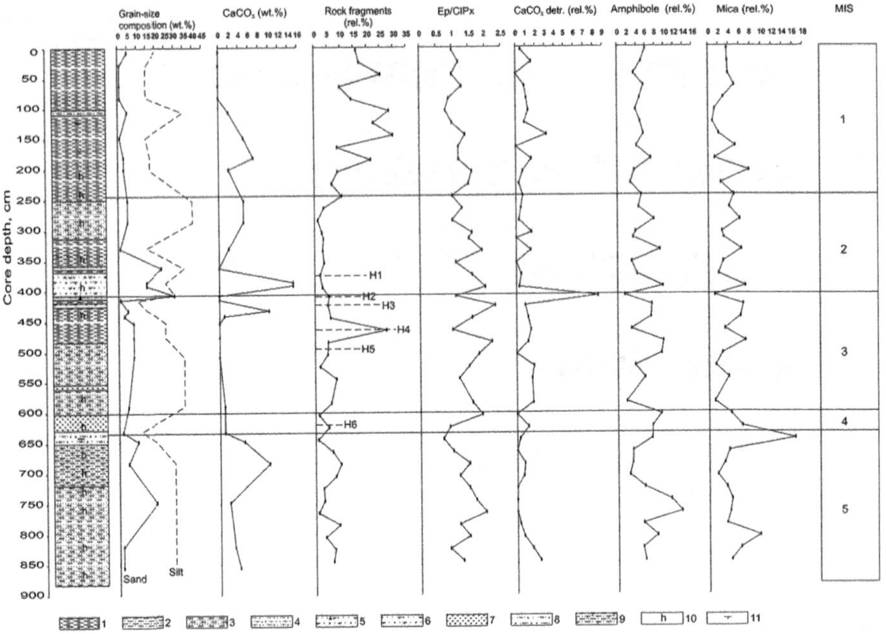

Fig. 9.23 Lithology of sediment core PS2837 (Levitan et al., 2002a). 1 – clayey mud; 2 – silty clay; 3 – fine silt; 4 – coarse silt; 5 – mictite (sandy-silty clay); 6 – mictites with IRD; 7 – silty clay with sand admixture; 8 – clay with silt admixture; 9 – silty clay with carbonate admixture; 10 – hydrotroilite; 11 – planktonic foraminifers. H1-H6 – Heinrich events (Thiede and Tiedemann, 2001)

(black sandstone) are also present. Interlayers with very coarse sediments reveal a "cottage-cheese structure" related to aggregation of clay minerals (Hass et al., 2000). All layers are characterized by sharp boundaries. Sediments of both MIS 1 and MIS 2 are actually composed of similar components. However, their proportions are different. Sediments of MIS 2 are considerably depleted in clay minerals and ferruginous-clayey aggregates. Consequently, they are enriched in rock fragments and quartz-feldspar clastic material.

Sediments of MIS 3 (402–598 cm) consist of dark olive-gray and sufficiently homogeneous clayey mud. It is replaced upward the sedimentary section by laminated clayey mud with alternation of olive-gray, dark yellowish brown and brown layers. The sediments are enriched in planktonic foraminifers only at the level of 433 cm.

Sediments of MIS 4 (598–643 cm) are composed of olive-gray and dark yellowish brown clayey and silty-clayey mud with increased contents of sandy and silty fractions. The upper half of this unit is characterized by a laminated structure, whereas the lower half is slightly enriched in planktonic foraminifers.

Sediments of MIS 5 (643–880 cm) include two units of olive-gray clayey mud (653–782 and 828–880 cm) containing black lenses of sandy-silty clays (0.5–3.0 cm thick) with pyrite (up to 30%) separated by coarser-grained silty-clayey mud without pyrite lenses. The uppermost part of MIS 5 is slightly enriched in planktonic foraminifers and coccoliths.

Mineral Composition of Heavy Fraction

Many researchers, including authors of this work (Levitan et al., 1999a, 1999b) traditionally recalculated heavy-fraction data for determination clastogenic material. We used primary analytical results in this work. The validity of this approach is based on the following phenomenon observed in surface sediments from the Saint Anna Trough: the domains of increased rock-fragment concentration in the light fraction of fine sand fit the zone of maximum iceberg development (Kukina et al., 1999). We also revealed that the highest content of rock fragments in the light fraction is confined to iceberg-derived sediment interlayers rather than turbidites in sediment cores from this trough (Levitan and Kukina, 2002). Since one of the main objectives of our work was to distinguish indicators of IRD, it was essential to take into account the amount of rock fragments. Generally, the total content of unidentifiable grains and biogenic remains is approximately 2%. Therefore, we did not recalculate the primary data.

Figure 9.23 shows the distribution of different heavy-fraction components in sediment core PS2837. Maximal concentration of rock fragments is confined to the Middle Holocene (according to data of (Nørgaard-Pedersen et al., 2003)), immediately above the Thermal Optimum that took place in this region 7–8 cal. kyrs. BP (Duplessy et al., 2001) and later. Downward the MIS 1 section, the concentration drastically decreases and reaches minimal values at the MIS 1/MIS 2 boundary. In MIS 2, the concentration is generally insignificant and comparable with that in MIS 4. The concentration of rock fragments appreciably increases in MIS 3. It is slightly higher in MIS 5 relative to MIS 4. Some maxima fit approximately the stratigraphic position of respective Heinrich events (Fig. 9.23) that represent the large-scale thawing of icebergs located south of Iceland in the North Atlantic (Thiede and Tiedemann, 2001). The core bottom is not characterized by increase in rock-fragment content, which is typical for Substage MIS 5e sediments in the Ocean Drilling Program (ODP) drill core from the Yermak Plateau (Flower, 1996; Hevrøy et al., 1996) and in Core PS 2185–3/6 from the Lomonosov Ridge in the Central Arctic (Nørgaard-Pedersen, 1997). The impressive similarity of our data with the distribution of IRD in the studied sections may allow the following conclusions. First, rock-fragment content in the heavy subfraction of the 63–125 mkm fraction is an indicator of the supply of IRD. Second, variations in the supply of IRD on the Yermak Plateau are similar to those on the Lomonosov Ridge. Hence, sediments from Core PS 2837 did not attain the Eemian Interglaciation (MIS 5e).

Sediment core PS2837 basically reveals a similar distribution of iron oxides. According to (Phillips and Grantz, 2001), sediments of the Central Arctic include IRD from Eurasian provinces consisting of sandstone and siltstone with a red color owing to the ferruginous matrix. In contrast, IRD from source provinces in North America and Greenland contain abundant gray clastic rocks in addition to numerous fragments of carbonates (limestones and dolomites). Our data on iron oxide distribution possibly suggests that IRD in the Yermak Plateau region is mainly delivered via the Transpolar Drift system. This assumption is also supported by the abundance of iron oxides (up to 10%) in sediment core PS 2892 recovered from the Voronin Trough located west of Severnaya Zemlya (Levitan et al., 1999b). Therefore, one cannot rule out that precisely the Severnaya Zemlya Archipelago served as the main source of icebergs transported by the Siberian Branch of the Transpolar Drift to the Yermak Plateau. The low content of carbonate rock fragments (Fig. 9.23), which become abundant only near the MIS 3/MIS 2 boundary, indicates a subordinate role of American provinces and the Beaufort Gyre in the transportation of sea ice to the Yermak Plateau.

The distribution pattern of rutile and sphene is also similar to that of rock fragments in Core PS 2837. This relationship is less typical for staurolite, sillimanite, garnet, and apatite. Thus, these minerals are also likely to be transported by icebergs. The sharp garnet increase in the heavy fraction of MIS 3 sediments suggests a contribution of icebergs moving from the North Kara Rise (Levitan et al., 1999a).

Several components of the heavy fraction, such as grains with iron oxide coating, epidote, amphiboles, chlorite, and biotite, reveal a weak increasing trend with depth (Fig. 9.23). Black ore components, epidote/clinopyroxene (Ep/Cpx) ratio, and other parameters tend to increase in sediments of glacial periods (MIS 2 and MIS 4) relative to interglacial periods (Fig. 9.23). Concentration of clinopyroxene remains sufficiently stable throughout the sedimentary section.

Thus, clastogenic minerals generally consist of a clinopyroxene-epidote assemblage with the content of individual minerals exceeding 10%. Grains with iron oxide coating are significant in sediments of MIS 2-MIS 5; iron oxides prevail in sediments of MIS 1 and MIS 2; and rock fragments are abundant in the upper part of MIS 1 section.

Sediment Core PS2834 (Water Depth 1024 m)

Stratigraphy

This core was obtained from the upper eastern slope of the Yermak Plateau. Sediments from this core were not examined for radiocarbon dates or oxygen-isotopic composition in planktonic foraminifers. Therefore, our age model is based on lithostratigraphy and the correlation of reference horizons. For example, the MIS 1/MIS 2 boundary is drawn at the base of the regional hardground layer (Levitan et al., 2000) estimated to have an age of about 12 cal. kyrs. BP (Kassens, 1990). The MIS 2/MIS 3 boundary is based on the so-called "black layer," i.e., dark gray organic-rich sandy-silty clay, which is widespread in the studied region and has a radiocarbon age of 21–23 ka (Stein and Fahl, 1997). Cottage-cheese layers were also used as reference horizons. In addition, we took into account data on the distribution of wet bulk density (Niessen and Kleiber, 1997); IRD in Hole 912A (Hevrøy et al., 1996); and $CaCO_3$, grain size, epidote, rock fragments, Ep/Cpx ratio, garnet, apatite, total sillimanite and staurolite, and biotite in sediment core PS2837. The obtained

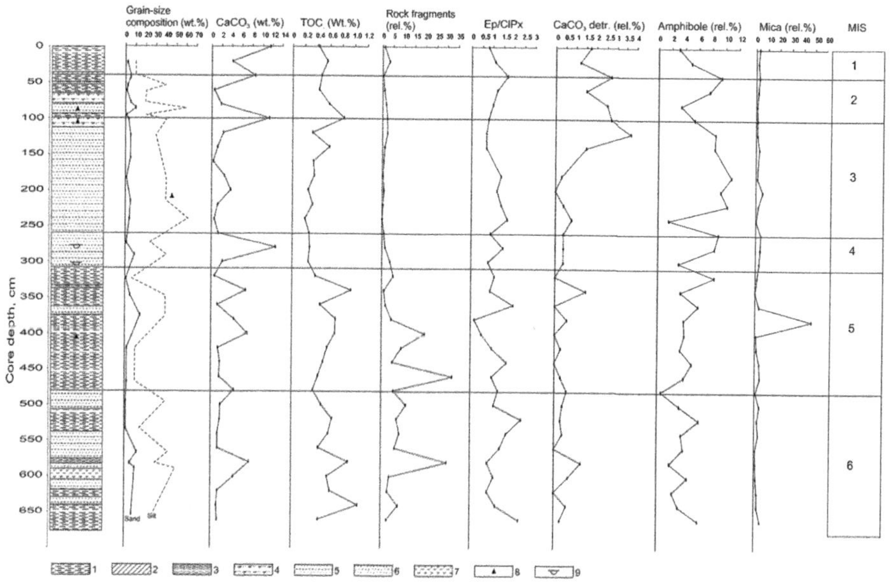

Fig. 9.24 Lithology of sediment core PS2834 (Levitan et al., 2002a). 1 – clayey mud; 2 – hardground; 3 – silty clay; 4 – silty clay with carbonate admixture; 5 – clayey silt; 6 – mictite; 7 – clayey mud with carbonate admixture; 8 – IRD; 9 – bivalves shells

results suggest that sediments of MIS 5e and MIS 6 are also present in sediment core PS2834. The sedimentation rate, typical of the Yermak Plateau, was higher during glacial intervals relative to interglacial ones.

Lithology

Sediments of MIS 1 (0–41.5 cm) include dark grayish brown terrigenous clayey mud that is bioturbated in the upper part (Fig. 9.24). They are composed of clay minerals (75%), iron oxides (10%), and ferruginous-clayey aggregates (10%). Clastic material (quartz and feldspar) along with colored nonmetallic and black ore minerals are subordinate. The upper half of the section is slightly enriched in foraminifer tests and marked by rare dolomite grains. This soft mud is underlain by a thin layer of very massive, dark reddish brown, clayey silt (hardground) at a depth of 40–41.5 cm.

The underlying sediments of MIS 2 (41.5–112 cm) consist of dark brown and brownish gray sediments (dominating in the upper 10 cm) with 7- to 15-cm-thick gray and olive-gray sediment layers. In terms of grain-size distribution, these sediments correspond to clayey mud, silty-clayey mud, and silt. The sediment components are similar to those in MIS 1, but their relationships depend on the sediment types. Coarser-grained varieties are enriched in rock fragments, quartz, feldspars, and other clastic material. The clay-mineral content is minimal. In silty-clayey and clayey mud at some levels (78 and 102 cm), foraminifer tests and their fragments are abundant; clay minerals prevail in the lithogenic part; furthermore, rare grains of volcanic glass were observed. Increased iron-oxide content was only found at the top of MIS 2 (43 cm). Horizon 94–99 cm is composed of almost black organic-rich silty clay. IRD (massive clay, gravelite, and clay) is present at 91–94 and 104 cm.

Sediments of MIS 3 (112–234 cm) are composed of olive-brown, grayish brown and gray layers of silty-clayey mud. The thickness of layers varies from 6–8 to 30 cm. Iron-oxide content is increased in brown varieties. The ratio of sandy-silty and clay fractions in different layers ranges from 3:1 to 1:1.

Sediments of MIS 4 (234–310 cm) are divided into two members. The upper unit (234–258.5 cm) includes dark brown, olive-brown, and dark grayish brown banded (cottage-cheese) clayey silt layers with distinct tops and bottoms. The mineral composition of the sediments includes feldspar-quartz aggregates (35%), rock fragments (25%), clay minerals (25%), ferruginous-clayey aggregates (5–10%), black ore minerals (7–8%), and colored minerals (3–5%). The lower unit (258.5–310 cm) is composed of an alternation of gray, olive-brown, brown, and reddish brown clayey and silty-clayey mud (sometimes silt) with planktonic foraminifers and rare amounts of small shells of bivalve mollusks. Black coating of diagenetic sulfides is typical. Coarser-grained sediments are enriched in rock fragments, quartz, feldspars, and black ore minerals. The clay-mineral content does not exceed 25%. In contrast, finer mud is mainly composed of clay minerals of up to 45–50%.

Sediments of MIS 5 (310–484 cm) are divided into three units. The upper unit (310–364 cm) is distinguished from the lower unit of MIS 4 only by an appreciably lower content of biogenic carbonates. The second unit (364–398 cm) is composed of brown silt and sandy silt with a "cottage-cheese" structure (364–373 cm) and dark gray mixtite with sandstone fragments (373–398 cm). The composition is as follows (in decreasing order): quartz-feldspar clastics, clay minerals, rock fragments, colored minerals, and ore minerals. We correlate this unit with Substage 5d. The third unit (398–484 cm) consists of gray and olive-gray, fine clayey bioturbated mud with pockets of black sulfides. The mineral composition includes clay minerals (up to 65%), quartz and feldspars (20%), rock fragments (5%), black ore minerals (5%), and colored minerals (5%). Volcanic glass fragments are rare. These sediments were probably deposited during MIS 5e.

Sediments of MIS 6 (484–675 cm) represent the lowermost part of sediment core PS2834. They are composed of dark grayish brown, olive-brown, dark gray and gray layers of silty-clayey and clayey mud, clayey silt, and mixtites. A weak bioturbation is observed at some intervals. Patches and pockets of iron oxides and sulfides are present. The brown variety is appreciably enriched in iron oxides. Feldspar and quartz fragments predominate in coarser-grained sediments (up to 50%). Finer varieties are mainly composed of clay minerals. The sediments occasionally contain a minor amount of planktonic foraminifer remains.

Results of the $CaCO_3$ analysis (Fig. 9.24) somewhat refine the data presented above. Against the background average value of 1–4%, some higher carbonate contents were determined in subrecent sediments of MIS 1, lower parts of MIS 2 and MIS 4 (as much as 12–13%), MIS 5, and middle MIS 6 (6–7%). In general, sediments of interglacial stages and terminations I (MIS 2/MIS 1 transition) and II (MIS 6/MIS 5 transition) are enriched in biogenic carbonates.

Figure 9.24 shows that TOC content is generally less than 0.8% in the studied sediments. TOC peaks are observed in the black layer at the base of MIS 2, upper unit of MIS 5, and within MIS 6. Rock-Eval pyrolysis data obtained from sediments of the adjacent sediment core PS2212 (Vogt, 1997) suggest that the organic matter has an obviously terrigenous origin in the black layer (MIS 2) and in the sediments at 640 cm level (MIS 6), but a marine origin in sediments from 580 cm core depth (MIS 6). In Core PS2212, peak values did not occur in MIS 5, probably due to a hiatus determined at this level.

Mineral Composition of the Heavy Fraction

As it was mentioned above, rock fragments (rf) in the 63–125 mkm fraction of sediments from Core PS2837 and ice-rafted fragments (rf/g) in the 500–1000 mkm fraction of sediments from ODP Hole 912A reveal a similar distribution pattern. For example, the concentration of ice-rafted fragments is maximal (30–35%) in sediments from the lower half of MIS 5 and MIS 6. The distribution of rock fragments, iron oxides, grains with ferruginous coating, and total sillimanite and staurolite along the section of Core PS2834 matches the above-described trend (Fig. 9.24). In general, iron-oxide content is very high in this section.

Epidote and clinopyroxene, which are the two most widespread mineral groups, demonstrate a high correlation. Figure 9.24 shows that the Ep/Cpx ratio, however, obviously increases in sediments of glacial periods and decreases in sediments of interglacial periods.

Concentrations of apatite, total rutile and sphene, biotite, and chlorite are also almost similarly distributed in the section (Fig. 9.24). These minerals are characterized by a single distinct peak, e.g., in the lower part of MIS 2 for apatite, in the upper part of MIS 4 for rutile and sphene, and in the middle of MIS 5 for biotite (up to 45%).

The garnet distribution is similar to that of the Ep/Cpx ratio with maxima during glacial periods. Additional peaks corresponding to terminations I and II are obvious. The garnet content is particularly high in sediments of MIS 4 and 6.

Amphiboles, black ore minerals, and carbonate fragments show a general downward trend in cross-section (Fig. 9.24). Content of black ore minerals is maximal in sediments of MIS 1 and 2 (particularly in the upper part of MIS 2). Maximal concentration of carbonate fragments is observed at the top of MIS 3.

The analysis of stratigraphic distribution of heavy minerals revealed a predominance of the following assemblages: black ore minerals + iron oxides + Ep + Cpx (MIS 1); black ore minerals + Cpx + Ep (MIS 2); iron oxides + Cpx + Ep (MIS 3); iron oxides + garnet (Gr) + Cpx + Ep (MIS 4); Cpx + Ep (MIS 5); and Gr + Cpx + iron oxides + Ep (MIS 6).

Sediment Core PS2859 (Water Depth 1120 m)

Stratigraphy

The stratigraphic subdivision of sediment core PS2859 located on the eastern slope of the Yermak Plateau, is based on the lithostratigraphic correlation with sediment core PS2834. The correlation was carried out based on the following units: hardground layer at the base of MIS 1; succession of coarse-grained sediments (with a minor content of biogenic $CaCO_3$) and black layer at the base of MIS 2; layer with a high content of planktonic foraminifers at the base of MIS 4 and the top of MIS 5; coarse-grained sediments and coal fragments in MIS 6; and a layer with an increased $CaCO_3$ concentration at the top of MIS 6. We also compared the grain-size distribution and content of $CaCO_3$, as well as heavy-mineral assemblages in sediments of both cores. The presence of turbidites in sediment core PS2859 complicated the lithostratigraphic correlation with sediment core PS2834. Nevertheless, we could elaborate a hypothetical model of sediment core PS2859 suggesting the presence of the oldest sediments among those described so far (the base of MIS 6 or even the top of MIS 7) at the bottom of sediment core PS2859.

Based on the stratigraphic model, sedimentation rates in sediment core PS2859 increased from MIS 6 (2 cm/ky) to MIS 4 (4.3 cm/ky) and then remain relatively stable during MIS 3 to MIS 1 (around 3.7–4 cm/ky; Table 9.9).

Lithological Characteristics

Sediments of MIS 1 (0–47 cm) consist of bioturbated brown clayey and silty-clayey mud underlain by a hardground layer (46–47 cm) composed of massive, dark reddish brown, sandy-silty clay (Fig. 9.25).

Sediments of MIS 2 (47–91 cm) include gray homogeneous soft clayey mud (47–70 cm) underlain by a unit of alternating sandy-silty-clayey sediments with 3- to 4-cm-thick dark olive-gray and grayish brown layers characterized by the presence (or absence) of "cottage-cheese" structure. This unit is underlain by the black layer (82–86 cm) with "cottage-cheese" structure. At its base, the MIS 2 interval incorporates a layer of homogeneous gray silty-clayey mud enriched in planktonic foraminifers (86–91 cm). Sediments of MIS 2 occasionally contain dolomite, glauconite, and coal grains.

Sediments of MIS 3 (91–231 cm) consist of strongly bioturbated brown clayey mud that includes a small amount of silty material and rare fragments of gray stiff silty clay (2–3 cm in diameter) in the upper part of the unit (125–140 cm).

Sediments of MIS 4 (231–295.5 cm) consist of four members (from top to bottom): dark grayish brown sandy-silty clay (231–234 cm); dark grayish brown, olive-brown, and olive-gray sandy silt with graded structure (234–266 cm); dark gray silty clay with planktonic foraminifers and sandstone fragments 1 cm in diameter (266–275 cm); and olive-brown sandy silt with graded structure (275–295.5 cm). A 12 to 34 cm thick turbidite interval (layers 2 and 4) contains rock fragments, coal grains, plant remains, and volcanic glass particles with a ferruginous coating.

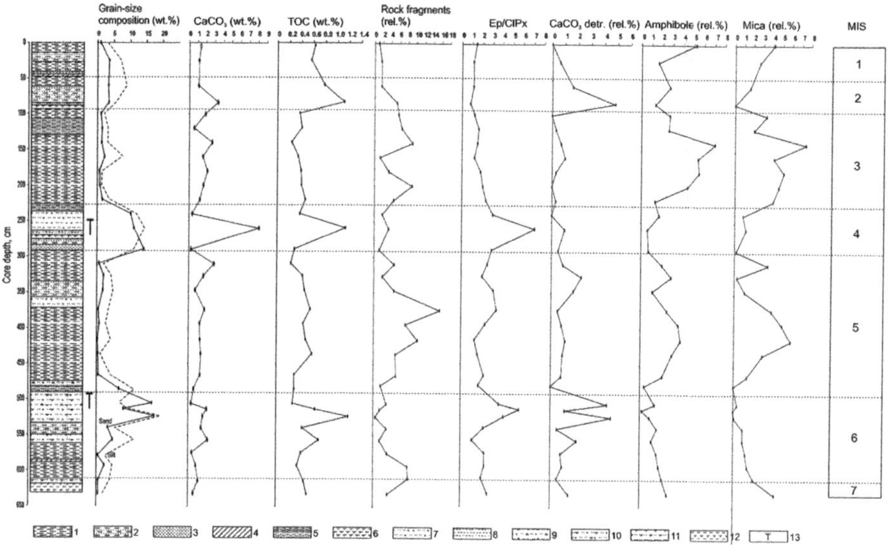

Fig. 9.25 Lithology of sediment core PS2859 (Levitan et al., 2002a). 1 – clayey mud; 2 –silty clay with sand admixture; 3 – sandy silt; 4 – hardground; 5 – silty clay; 6 – clayey mud with carbonate admixture; 7 – silt with sand admixture; 8 – silty clay; 9 – sandy-silty clay; 10 – sandy silt with carbonate admixture; 11 – silty clay with carbonate admixture; 12 – clayey mud with carbonate admixture; 13 – turbidites

Their grain size changes from sandy silt in the basal layer of cyclite to silty clay at the top.

Sediments of MIS 5 (295.5–488 cm) are composed of olive-brown, olive, and olive-gray clayey mud (dark gray at the base) with interlayers of olive-colored and slightly bioturbated sandy silt in the middle part (356–370 cm). In general, this unit is characterized by the development of differently colored centimeter-scale bands. Clayey mud at the base (422–488 cm) includes numerous patches of black sulfides. Sandy silt layers with abundant rock fragments are underlain and overlain by fine mud layers with planktonic foraminifer remains. These silt layers were probably deposited during MIS 5d.

Sediments of MIS 6 (488–611 cm) contain two units. The upper unit (488–532 cm) includes gray, olive-gray, and grayish brown turbidites with coal fragments up to 1.2 cm in diameter. An interlayer of yellowish brown sandy silt without graded structure is observed at the interval of 500–512 cm. The coarser-grained variety of turbidite is enriched in rock fragments, coal grains, glauconite, and pyrite. The turbidite of the lower part is slightly enriched in biogenic $CaCO_3$. Turbidite members, 12–20 cm thick, are composed of cyclites (thickness 6–7 cm) with 1–2.5 cm thick layers. The turbidite includes sediments ranging from sandy silt at the cyclite base to silty clay at the top.

The lower unit (532–611 cm) consists of an alternation of olive-brown mixtites (mainly down to a level of 562 cm) with gray and olive-gray clayey mud. The mixtites contain abundant biogenic calcite and rock fragments with ferruginous coating. Dolomite and glauconite grains are subordinate. An ice-rafted sandstone fragment was registered at the level of 576 cm. The hardground layer composed of dark yellowish brown and massive silty clay is located at 593–595 cm.

Presumably sediments of MIS 7 are exposed at the base of sediment core PS2859 (611–636 cm). They consist of homogeneous dark gray clayey silt and silty-clayey mud with numerous black sulfide patches. These sediments are mainly composed of rock fragments and mineral grains (quartz, feldspar, calcite, and iron oxides). Some grains have a ferruginous coating. The clay-mineral content does not exceed 25–30% even in the finest fraction.

Results of the granulometric and chemical analyses (Fig. 9.18) supplement the lithological characteristics of the studied core. The gravel-sized grains generally account for <0.5%. They increase up to 2.3–2.4% in the black layer (MIS 2 base) and turbidites of MIS 6 and account for 4.3% in MIS 4 turbidites. In general, the sand concentration correlates with the gravel content variation. Against the background value of 2%, sediments of MIS 2 display a slight enrichment up to 3%. The maximal content (14–18%) is observed in turbidites of MIS 4 and MIS 6. It is worth to mention that sediments of MIS 6 are enriched in sand beyond the turbidite zone as well.

Diagrams of $CaCO_3$ distribution show maxima in the sedimentary section of sediment core PS2859 at the base of MIS 2 (3.3%) and in sediments of MIS 4 (8%) and MIS 6 (2%). A high $CaCO_3$ concentration (up to 3%) is also noted at the top of MIS 5. The maximum carbonate content is registered in silty clay layers between turbidites in MIS 4 (266–275 cm).

The maximum content of organic carbon is confined to $CaCO_3$-rich interlayers with a centimeter-scale precision. In sediments of MIS 5, the TOC behavior is more stable relative to $CaCO_3$. Based on sediment core PS2212 data (Vogt, 1997), all of the measured TOC peaks are related to a terrigenous organic-matter source.

Mineral Composition of the Heavy Fraction

The content of rock fragments is increased in sediments of MIS 3 and MIS 5 and drastically decreased in turbidites of MIS 4 and MIS 6 (Fig. 9.25). The sandy fraction of turbidites in the Saint Anna Trough is also depleted in rock fragments (Levitan and Kukina, 2002). Variation of the total content of sillimanite and staurolite along the section is similar to that of rock fragments. The content of iron oxides and carbonate fragments is maximal in sediments of the black layer in the lower part of MIS 2, the upper part of MIS 5, and turbidites of MIS 6. Contents of epidote, clinopyroxenes, and amphiboles slightly increase downcore. Their concentration is distinctly increased during interglacial stages (Fig. 9.25).

The Ep/Cpx ratio and the content of particles with ferruginous coating slightly increase downward and demonstrate distinct peaks in the turbidites of MIS 4 and MIS 6 (Fig. 9.25). An even more distinct correlation with sediments of glacial periods is revealed in the distribution of garnet and black ore minerals. The content of particles with ferruginous coating is also generally increased in the sediments of sediment core PS2859.

Interglacial stages distinctly correlate with maximum concentrations of micas (Fig. 9.25) and apatite (except for MIS 1). The total content of rutile and sphene in sediments of sediment core PS2859 does not demonstrate such a distinct correlation with climate. Their maximal contents are registered in sediments of MIS 2, MIS 4, and MIS 5.

The turbidites of MIS 4 and MIS 6 have different mineral composition. Heavy-mineral assemblage, with maximum concentration observed in the turbidites of MIS 4, includes rutile, sphene, black ore minerals, garnet, and grains with ferruginous coating. The turbidites of MIS 6 are characterized by an association of carbonate fragments, iron oxides, garnet, and grains with ferruginous coating. Thus, each turbidite unit is probably related to a specific source of material.

The distribution of main heavy minerals suggests the following typical assemblages in sediments of different oxygen-isotope stages: Cpx + particles with ferruginous coating + Ep (MIS 1); Gr + Cpx + Ep + particles with ferruginous coating (MIS 2); Cpx + particles with ferruginous coating + Ep (MIS 3); Gr + Ep + particles with ferruginous coating (MIS 4); Ep+ particles with ferruginous coating (MIS 5); and Gr + Ep + particles with ferruginous coating (MIS 6).

Paleoceanographic Interpretation

Stratigraphy

The described sediment cores strongly differ from each other by many parameters, particularly by the distribution of mass accumulation rates in the Late Quaternary period. In sediment core PS2837, accumulation rates began to increase since MIS 2, probably reflecting local topographical changes in the upper western slope of the Yermak Plateau that in turn resulted in "pocket" development. The turbidites were found in Core PS2859, in layers correlated to MIS 4 and MIS 6 that significantly complicated the general scenario of hemipelagic sedimentation. Just based on sedimentation rates, sediment core PS2834 can be considered to some extent as a typical core of glacial sedimentation on the Yermak Plateau.

The lithostratigraphic correlation of the three described sediment cores as well as other cores recovered during *Polarstern* Cruise ARK-XIII/2 (Stein and Fahl, 1997) suggest that the age of the oldest sediments has not exceed MIS 7 neither in the northern profile (Fig. 9.26) nor in the southern one (Fig. 9.27). Note that the abundance of turbidites on the slopes of the Yermak Plateau hampers the chronological interpretation of the sedimentary records.

The low sedimentation rate during MIS 5 of sediment core PS2212 (see Fig. 9.11 for location) is related to a hiatus detected at this level (Vogt 1997). In general, sedimentation rates calculated on the basis of hypothetical age models of sediment cores PS2834 and PS2859 correspond to standard values for reliably dated sections in the Yermak Plateau area. A slightly increase of sedimentation rate during MIS 4 of sediment core PS2859 is attributed to the presence of turbidites.

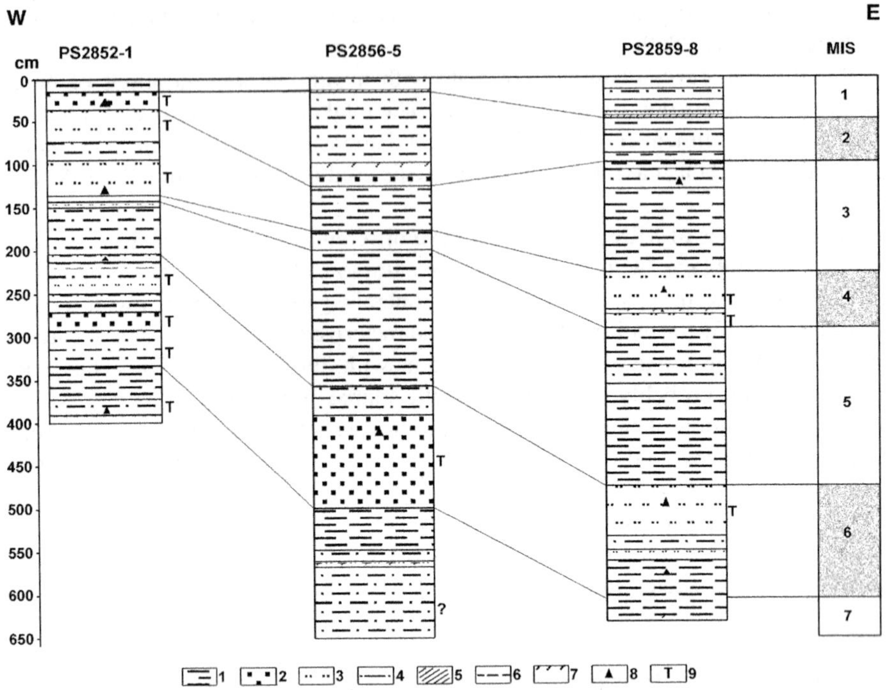

Fig. 9.26 Lithostratigraphic correlation of sedimentary sections along to the northern transect at Yermak Plateau (Levitan et al., 2002b). Location of profiles is shown at Fig. 9.11. 1 – clayey mud; 2 – sand; 3 – silt; 4 – silty clay; 5 – hardground; 6 – clayey mud with carbonate admixture; 7 – silt with carbonate admixture; 8 – IRD; 9 – turbidites

The comparison of data from Holes 912A (Hevrøy et al., 1996) and 910C (Flower, 1996) (see Fig. 9.11 for locations of holes) clearly shows that sedimentation rate was lower during all oxygen isotope stages for Hole 910C located on the plateau crest relative to Hole 912C drilled on the western slope. However, an actual distribution of sedimentation rate is specific for each section. Based on an overview of available data, one can assume that the sedimentation rate tends to increase during glacial periods (at least during the last climatic cycle, i.e., from MIS 5 to MIS 1).

Lithology

Based on the data presented above, general facies differences of marine sediments accumulated during glacial and interglacial periods on the Yermak Plateau can be indentified. Sediments of glacial intervals are obviously coarser-grained than interglacial and interstadial sediments. The presence of mixtites in respective parts of the sedimentary sections indicates a poorer sorting of glacial sediments. This is also confirmed by the occurrence of the most amount of IRD during glacial periods.

The interpretation of color variations in sediments is much less obvious. On the one hand, gray-colored sediments are slightly increased during glacial periods

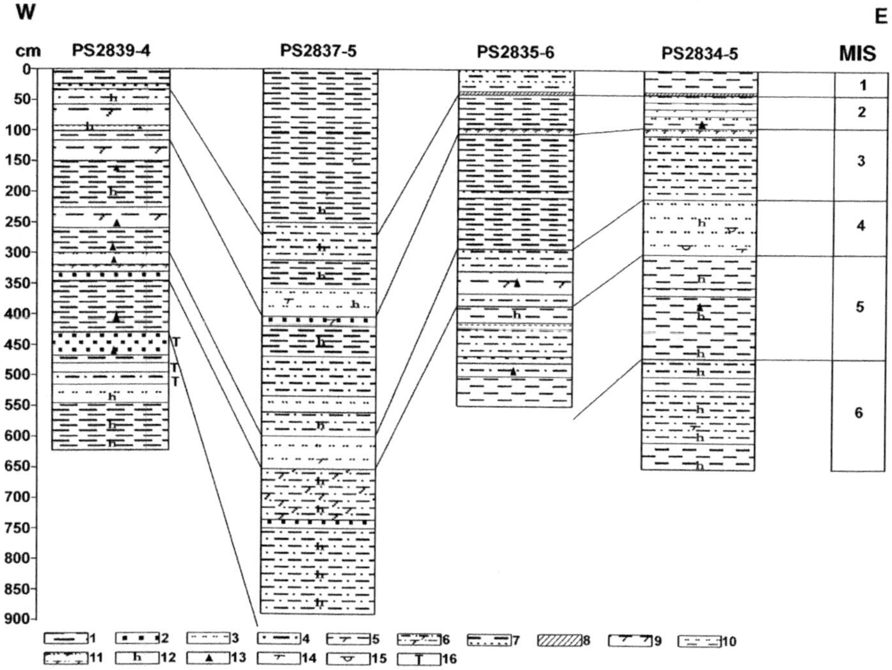

Fig. 9.27 Lithostratigraphic correlation of sedimentary sections along to the southern transect at Yermak Plateau (Levitan et al., 2002b). Location of profiles is shown at Fig. 9.11. 1 – clayey mud; 2 – sand; 3 – silt; 4 – silty clay; 5 – clayey mud with foraminifers admixture; 6 – silty clay with foraminifers admixture; 7 – interlayering of clayey mud and sandy silt; 8 – hardground; 9 – clayey mud with carbonate admixture; 10 – interlayering of silt and clayey mud; 11 – silt with carbonate admixture; 12 – hydrotroilite; 13 – IRD; 14 – admixture of foraminifers; 15 – bivalves shells; 16 – turbidites

relative to interglacial ones. On the other hand, the available data on sedimentation rates, IRD, and lithological composition indicate that the color variations are governed not only by primary sedimentary factors but also by diagenetic processes. The iron oxides probably supplied by icebergs play a significant role. It appears that, in general, sedimentary processes were presumably more essential than the diagenetic ones, because reddish to brown and gray sediments occur in sedimentary sequences of both glacial and interglacial periods.

Furthermore, structural differences are very distinctive. Bioturbation is more often and intensely developed in sediments of interglacial stages as well as diagenetic structures caused by the presence of iron sulfide patches and lenses. The sediments of glacial periods are characterized by fine layering, abundance of "cottage-cheese" structures, and graded bedding.

Concerning the abundance of turbidites at the slopes of the Yermak Plateau (Figs. 9.26 and 9.27), we suppose that the presence of turbidites in MIS 4 and MIS 6 of sediment core PS2859 is not accidental. Turbidites (glacio-turbidites) were mostly accumulating on continental slopes of the Arctic Seas especially during glacial stages of the oceanic periglacial environment (Matishov, 1984). Their development is associated with glacioeustatic sea-level falls. The exposed looze sediments were eroded, transported, and accumulated on the continental slopes and rises, in particular by turbid currents.

However, we suggest that a different mechanism of turbidite sedimentation probably occurred on the slopes of the Yermak Plateau during glacial periods. The shelf glaciers and sea ice that existed at that time in the outer marginal shelf and the upper continental slope directly overlapped bottom sediments of older epochs. The sea ice included much IRD that was precipitating on the Yermak Plateau during drift and melting of sea ice under relatively warm conditions in the Fram Strait area. Precipitation of large amounts of sedimentary material on slopes caused a development of suspended flows including turbid currents. The accumulated sediments were complemented by locally derived sediments due to the bottom erosion upon flows moving.

Thus, the origin of coarse-grained material encountered in basal layers of turbidite units cannot be related just to IRD as it is used to suggest. Glacio-turbidites of continental slopes of the Arctic Seas could not be accumulating on the slopes of the Yermak Plateau during the glacial periods due to the described above geomorphologic reasons. However, during the interglacial periods, prevalence of different mechanism of sediment incorporation into the sea-ice (freezing of sea-water containing suspended matter, particularly in polynyas) contributed to the supply of finer-grained sediments in much lesser amount than it would be sufficient for turbidite development.

Discussing the given issue, one cannot ignore the presence of deep plough marks on the crest surface of the Yermak Plateau. The plough marks were left during the movement of either the Barents Sea Ice Sheet or huge deep-draft icebergs during past glacial periods (Vogt et al., 1994). The eroded material could be transported onto the slopes of the plateau to form local turbidites. The plough marks were also registered in Parasound records during *Polarstern* Cruise ARK-XIII/2 (Stein and Fahl, 1997, and further references therein). To determine during which glacial period (i.e., the age) the plough marks were formed, is an important issue. In ODP Hole 9I0C Quaternary sediments with normal physical properties overlay extremely stiff consolidated sediments. Consolidation was attributed to the impact of a thick ice cover or deep-draft icebergs (Flower, 1996; Kristoffersen et al., 2004). Simultaneously, strong erosion occurred on top of the plateau surface. The age of this event was dated to be 660 ka or older (MIS 16 or MIS 19/20?) (Flower, 1996; Knies et al., 2007). As it was mentioned above, the studied sediment cores do not contain sediments older than MIS 7. Hence, the mechanism proposed by (Flower, 1996) could not have developed the turbidites found in these sediment cores.

Distribution of $CaCO_3$ mainly composed of planktonic foraminifer tests revealed the confinement of maximum $CaCO_3$ concentrations in the studied cores to the following episodes: middle MIS 6 (Substage 6c); lower, middle, and upper MIS 5 (substages 5e, 5c, and 5a); uppermost MIS 3 (3a); lower MIS 2 (termination IA); and MIS 1 (termination IB and upper part of Holocene).

High concentrations of planktonic foraminifers are locally found in sediments of MIS 4 in sediment core PS2837. Seasonal periods of active Atlantic-water inflow are marked by maximal concentrations of planktonic foraminifers. These periods occasionally induced the development of ice-free water and promoted an increase in primary production. These events took place mainly during the interglacial periods and transition phases between interglacial and glacial stages (particularly during terminations). It should be emphasized that even glacial maxima (e.g., MIS 2) included short-period episodes of intense delivery of Atlantic waters into the Yermak Plateau area.

Based on the study of planktonic foraminifers, a number of researchers have come to similar conclusion (Spielhagen, 1991; Flower, 1996; Hald and Aspeli, 1997; Hebbeln and Wefer, 1997). According to (Wollenburg et al., 2001) the reconstructed paleoproductivity maxima based on benthic foraminifers on the Yermak Plateau do not correspond to the episodes of intense planktonic foraminifer concentration in bottom sediments, although the maxima appear to be confined to the above-indicated time intervals in less detailed chronological correlations.

Mineralogy

We have determined approximately 40 components (mainly minerals) in the heavy subfraction of the 63–125 mkm fraction. All of the studied cores contain the following major components (10% or more): iron oxides, grains with ferruginous coating, rock fragments, epidote, clinopyroxene, and garnet. Hypotheses on sources of these components and pathways of their transportation are based on the distribution pattern of heavy minerals in surface sediments in the Arctic Ocean and its marginal seas (Belov and Lapina, 1961; Gurevich, 1995; Kulikov, 1961; Levitan et al., 1996, 1999a; Behrends et al., 1999; see Stein, 2008 for most recent synthesis). Areas of the Laptev Sea and, to a lesser degree, Kara Sea serve as the main source of sea ice today (Gordienko and Laktionov, 1969). Spitsbergen, Franz-Josef Land, Northern Island of Novaya Zemlya, and Severnaya Zemlya are the main producers of icebergs at the Eurasian margin of the Arctic Ocean (John and Sugden, 1975).

The assemblage of black ore minerals, epidote, and clinopyroxene, developed in the surface sediments along the Siberian Branch of the Transpolar Drift, occurs on the Yermak Plateau as well (Belov and Lapina, 1961). Sources of these minerals are located in the eastern Kara Sea and western Laptev Sea. As mentioned above, this mineral assemblage also dominates in the studied core sections. Therefore, one can suppose the prevalence of Eurasian sources of lithogenic sandy material for sediments on the Yermak Plateau and the major role of sea ice of the Siberian Branch for transportation of sandy material. The dominance of sea ice of the Siberian Branch for the Yermak Plateau region is also supported by independent clay-mineral data in the sedimentary material of the Arctic sea ice (Dethleff et al., 2000).

The present northern slope of Spitsbergen is a zone of accumulation of an amphibole-pyroxene assemblage (Belov and Lapina, 1961). Therefore, Spitsbergen can hardly be considered as major province of heavy minerals for the Yermak Plateau. However, the invasion of icebergs via the Barents Sea is quite probable.

Based on heavy-mineral data (Belov and Lapina, 1961; Behrends, 1999) it is possible to assess the contribution of the Polar Branch of the Transpolar Drift in the delivery of heavy minerals to the Yermak Plateau. The heavy-mineral assemblages in the surface sediments beneath this current system indicate that sea ice in the Polar Branch mainly transports minerals of the epidote-amphibole assemblage. As noted above, this assemblage was not detected just by itself in the studied sediment cores. However, the heavy-mineral distribution in sediment core PS2837 located on the western slope of the Yermak Plateau slightly differs from the distribution pattern in sediment cores PS2834 and PS2859 recovered from the eastern slope. For example, sediments of MIS 3 and particularly MIS 5 from sediment core PS2837 reveal a synchronous increase in the concentration of amphiboles and epidote against background values of rather stable clinopyroxene content.

These results suggest a low contribution of IRD from the Polar Branch of the Transpolar Drift. Such phenomenon is missing in the cores PS2834 and PS2859. The concentration of micas is increased in sediments of interglacial stages due to their delivery from the eastern Laptev Sea (Behrends, 1999; Peregovich, 1999); i.e., this phenomenon is also related to the Polar Branch ice. The detailed study of heavy minerals in sediment core PS2185, which was recovered from the Lomonosov Ridge located beneath the Polar Branch, and sediment core PS2757, which was also recovered from the same ridge but beneath the Siberian Branch, revealed significant differences (Behrends, 1999). In general, the studied sediment cores from the Yermak Plateau are more similar to sediments of sediment core PS2757 in terms of heavy-mineral assemblages.

Detrital $CaCO_3$ concentration is an indicator of the American and Greenland provinces in the studied fraction (Darby et al., 2002). In general, $CaCO_3$ concentration does not exceed 0.5–1.5%, suggesting a negligible contribution of the above-mentioned provinces and a minor role of the Beaufort Gyre in the delivery of sedimentary material to the Yermak Plateau. The role of the gyre slightly increased only at the MIS 2/MIS 3 boundary, which is registered in all of the studied sediment cores. This phenomenon is also recorded in sediment core PS2185 from the Lomonosov Ridge (Behrends, 1999). Smaller peaks were also found at 7 ka BP, Termination IB, top of MIS 5, and Termination II.

We propose the total content of rock fragments and iron oxides in the heavy fraction to be a proxy of IRD. Maximum values of this parameter in sediment core PS2837 correspond to stratigraphic positions of a number of the Heinrich Layers in the North Atlantic (Bond et al., 1992), i.e., H0 (Younger Dryas), H2, H4 H5, and H6 events (Fig. 9.28). Hence, the proposed proxy can be actually applied for the study of IRD distribution. Additionally, we found that some Heinrich events coincide with periods of iceberg production maxima in remote northern areas of the Arctic Europe. According to data by (Grousset et al., 2000), these events took place during the first few hundred years of short-term cooling episodes and marked the unstable state of continental glaciers.

Thus, maximum IRD concentrations on the Yermak Plateau (Fig. 9.28) are related to sediments of MIS 6 (without turbidites), lower MIS 5, MIS 3 (particularly, H4 and H2 levels), and the upper half of MIS 1. The comparison of actual

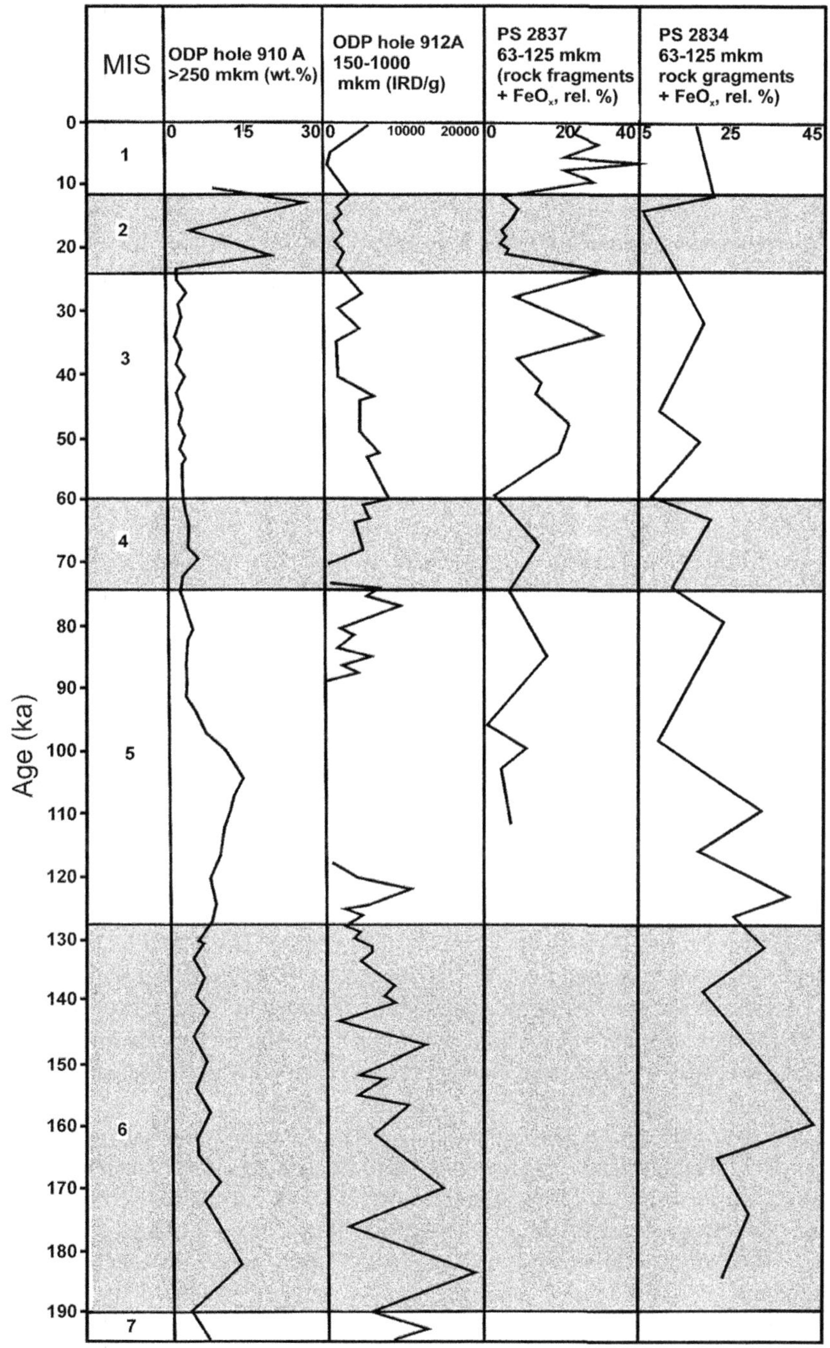

Fig. 9.28 Distribution of IRD in sedimentary sections from Yermak Plateau (Levitan et al., 2002b). For core locations see Fig. 9.11

data shows that maximum concentrations of $CaCO_3$ and IRD have never been found in the same layer though they are often alternated in the sediment sequence. This fact is optional evidence in favor of our interpretation; i.e., increase in carbonate content is related to warming (interglacials) while IRD concentration is connected to cooling (glacial) periods. This conclusion is consistent with data on the Central Arctic presented in (Nørgaard-Pedersen, 1997; Polyak et al., 2004; Spielhagen et al., 2004).

At present, main source of clinopyroxenes found in sediments from the Eurasian continental margin and adjacent deep-sea areas, is the Putoran Plateau in Siberia, washed out by tributaries of the Yenisei, Yana, and Khatanga rivers. The Putoran Plateau is known to be composed of Permian-Triassic trap rock. Trap rock of Taimyr and Jurassic-Cretaceous plateau basalts of Franz-Josef Land play a subordinate role in the supply of clinopyroxenes. Highest concentrations of the clinopyroxene group and lowest Ep/Cpx values are characteristic for present sedimentation environment in the eastern Kara Sea and western Laptev Sea. The comparison of main heavy-mineral assemblages in sediments of the Yermak Plateau with those in these areas of the Eurasian margin as well as the consideration of the major role of this area in the production of present sea ice, allow the following conclusion.

The western Laptev Sea and the eastern Kara Sea served as the main source provinces for sea ice that obviously dominated over icebergs during the studied time interval. At the same time, it is evident that the sediments on the Yermak Plateau mainly derived from Eurasian sources. Therefore, sedimentary material in the studied region has a mixed composition. This is also indicated, in particular, by the Ep/Cpx value which is significantly higher relative to that in discharge areas of rivers that drained the Putoran Plateau (Levitan et al., 1996; Behrends et al., 1999).

Distribution of heavy minerals in sediments of glacial and interglacial periods is a complicated issue. The same minerals (e.g., epidote group) can be concentrated in sediments of glacial stages in one sediment core or in sediments of interglacial stages in another core. In the third core, they can behave indifferently relative to climatic fluctuations. Nevertheless, based on the generalization of all data on heavy minerals, one can conclude that the Ep/Cpx ratio can be used as an indicator for glacial periods, whereas the garnet content is less applicable. Micas, rock fragments, clinopyroxenes, and amphiboles are typical indicators for interglacial intervals. The Ep/Cpx ratio, which is increased in cold epochs, is probably one of the most distinct climate proxy (Table 9.10).

Suggesting that discharges of the Ob, Yenisei, Yana, and Khatanga rivers were obviously higher during interglacials relative to glacials, this should also be reflected in the Ep/Cpx ratio, i.e., the ratio should be decreased in interglacial sediments. This assumption is supported by data in Table 9.10. At the same time, it is reasonable to suppose a strengthening of discharges of the Lena, Indigirka, and other rivers, which drain into the marginal seas influenced by the Polar Branch of the Transpolar Drift, during interglacial intervals. This phenomenon – as suggested – is reflected in interglacial sediments by increased concentrations of amphiboles and micas as also shown in our data (Table 9.11) based on recalculation of the total content of these minerals into IRD-free material (i.e., without consideration of rock fragments and

Table 9.10 Epidote/pyroxene ratio in sediments of different marine isotope stages [*Levitan et al.*, 2002b]

MIS	PS 2837	PS 2834	PS 2859
1	1,2	0,9	1,2
2	1,5	1,2	1
3	1,7	1	1,7
4	0,8	1,2	4,3
2–4	1,5	1,1	2,2
5	1,4	1,1	2,2
6	–	1,4	2,9
7	–	–	2,6

Note: Blank means lack of sediments.

Table 9.11 Total content of mica and amphiboles (rel. %) in sediments of the Yermak Plateau calculated for IRD-free matter [*Levitan et al.*, 2002b]

MIS	PS 2837	PS 2834	PS 2859
1	12.8	9.9	8.3
2	11.4	7.9	4.4
3	13.5	15.0	11.0
4	16.6	10.9	2.3
2–4	13.0	11.9	7.7
5	15.1	14.3	7.3
6	–	9.5	3.0
7	–	–	9.6

Note: Blank means lack of sediments

iron oxides). This approach naturally increased the relative importance of the Polar Branch in sea ice transportation of sedimentary material to the Yermak Plateau; i.e., the drift trajectory shifted to the south.

Thus, our mineralogical data indicate an increase of discharge of Siberian rivers during interstadial and interglacial periods (Table 9.11). The increased delivery of fresh and warm waters during such periods inevitably altered density stratification of the Arctic water column and its thermohaline circulation. The distribution of amphiboles and micas in the studied cores suggests a sporadic nature of these processes.

The present (MIS 1) Putoran Plateau is known to be free of glaciers and easily eroded by rivers. The plateau was only partly covered by glacier during MIS 4, but was completely covered by glaciers during the Glacial Maximum of MIS 6 (Late Saalian) (Svendsen et al., 2004). However, the entire plateau was free of ice during interstadial and interglacial periods. Variations in Ep/Cp ratio throughout the sediment core PS2834 (Table 9.10) are quite well explained by the Middle-Late Quaternary history of the Putoran Plateau. Thus, Ep/Cpx value serves as an additional argument in favor of the Siberian origin of the major part of heavy minerals in fine sands of the Yermak Plateau.

We assume that iron oxides in sediments of the Yermak Plateau were derived from both the North Kara Rise and the Severnaya Zemlya Archipelago, while garnets were mainly supplied from the North Kara Rise (Levitan et al., 1999b). Additional data are necessary to specify sources of other heavy minerals.

Finally, concluding the discussion of sources of heavy minerals, we should definitely admit the major role of the Eurasian provinces. Icebergs and sea ice transported lithogenic sand via the Siberian Branch of the Transpolar Drift to the Yermak Plateau area from the sources mentioned above.

An overview of reference data regarding the mineralogical composition of silty-clay fractions in sediment core PS2212 is given in (Vogt, 1997) and summarized in the following. The silt is composed of the following minerals (in decreasing order): quartz, phyllosilicates (clay minerals and micas), and feldspars (plagioclases and potassic feldspars). Quartz prevails in sediments of the glacial intervals whereas phyllosilicates dominate in sediments of interglacial (interstadial) stages. There are observed irregular variations of quartz/feldspar and K-feldspar/plagioclase ratios throughout the sedimentary sections. As in the 63–125 mkm fraction, the silty fractions of interglacial stages are characterized by high mica contents. The similarity is emphasized by the presence of a strong biotite peak at the middle part of the MIS 5 period. Therefore, we suggest that lithogenic silt was mainly transported by sea ice from Eurasian sources. This inference is supported by the similarity of distribution pattern of sandy and silty fractions throughout the sediment core PS2834 (Levitan et al., 2002a) and the abundance of reworked Upper Cretaceous palynomorphs derived from the Eurasian shelves in the Upper Quaternary sediments of the Yermak Plateau (Willard, 1996).

The clay fraction is mainly composed of clay minerals (illite, chlorite, kaolinite, and smectite), feldspars, and quartz. Smectite and quartz prevail, and the kaolinite/chlorite ratio is high in sediments of glacial stages. Sediments of interglacial stages are characterized by high illite contents and low kaolinite/chlorite, smectite/kaolinite, and smectite/(illite + chlorite) ratios (Vogt, 1997). In contrast, bottom sediments recovered from the continental slope of the Laptev Sea clearly show high clay fraction content in sediments of warm stages and high illite content in sediments of cold stages (Müller, 1999). We believe that this discrepancy might be explained by the existence of different provinces on the Yermak Plateau for sandy and silty fractions of sediment on the one hand, and for the clay fraction, on the other hand. As mentioned above, the coarser-grained fractions were probably derived from the Eurasian provinces and subsequently transported by sea ice and icebergs. Provinces of the clay fraction of sediments on the Yermak Plateau were presumably located in the North Atlantic (including the Norwegian-Greenland Basin) and its continental margins. The minerals were mainly transported by oceanic current systems. Previously, H. Berner came to a similar conclusion based on clay mineral research in the surface sediments from the Fram Strait (Berner, 1991).

Finally, the geographic location of the submarine Yermak Plateau in the transitional zone between glacial sedimentation systems with seasonal and permanent ice covers played a crucial role in the evolution of sedimentary processes on the plateau throughout the last 190 ka. As today, periods of seasonal ice cover were characterized by intense accumulation of fine-grained sediments and remains of planktonic and benthic foraminifers, whereas periods of permanent ice cover were marked by

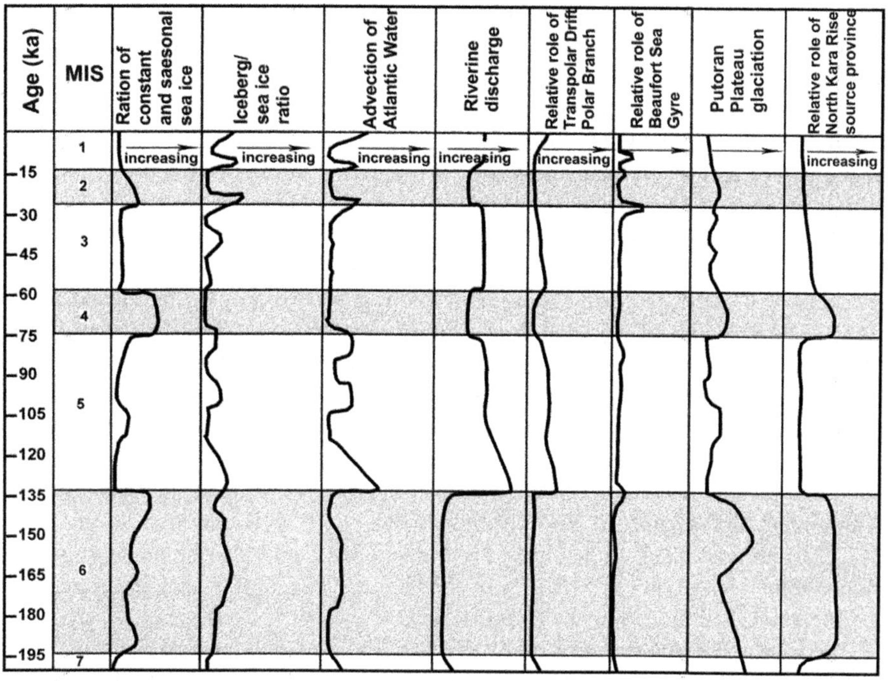

Fig. 9.29 Evolution scheme of some paleoceanographic and paleogeographic parameters in the Arctic (Levitan et al., 2000b)

the deposition of gravelly, sandy, and probably silty fractions. Evidently, the firstly mentioned processes prevailed and the latter sedimentation processes diminished during the interglacial stages (Fig. 9.29), whereas glacial stages were marked by an opposite trend. This conclusion is also supported by differences in the variation of sedimentation rates over the last 80 ka on the Yermak Plateau (Levitan et al., 2002b) and in the Norwegian-Greenland Basin regions (McManus et al., 1996) as well as in the Central Arctic region (Nørgaard-Pedersen, 1997). In general, sedimentation rates are higher during the glacial stages relative to the interglacial ones.

Influx of warm Atlantic waters into the Yermak Plateau area reconstructed by carbonate content in bottom sediments, was most intensive during the following stages or phases: middle MIS 6; lower, middle, and upper MIS 5; lower MIS 4; MIS 3/MIS 2 transition; termination I; and the upper part of Holocene. The simultaneous increase in primary production is demonstrated by the development of diagenetic sulfide lenses and bioturbation structures. Based on the study of calcareous nannoplanktons in bottom sediments from the Arctic Ocean, G. Gard and J. Backman (1990) obtained similar results related to the advection of Atlantic waters into the Central Arctic during the period from MIS 3 to MIS 1. Our data regarding the history of the North Atlantic water influx into the Arctic Ocean also well coincide with oxygen-isotope data on foraminifer tests in sediments from the Central Arctic (Nørgaard-Pedersen, 1997). Based on an overview of comprehensive material from the Norwegian-Greenland Basin, D. Hebbeln and G. Wefer (1997) demonstrated that the warmest Atlantic water were widely developed during the MIS 5e period (Eemian Interglaciation).

As mentioned earlier, maximum IRD concentrations on the Yermak Plateau were found in sediments of MIS 6, lower MIS 5, MIS 4 (maximum at H6), MIS 3 (maximum at H4), and second part of MIS 1. These events were controlled by the instability of continental ice sheets around the Arctic Ocean both during the progressive advance to the ocean and during the breakup of ice sheets and domes followed by Retreat of their margins. The correlation of instability of ice sheets with intense production of icebergs during short-term cooling is corroborated by (Bond et al., 1992). The presence of ice-free areas was essential for the formation of icebergs. The large-scale thawing of icebergs was accompanied by changes in oxygen-isotope composition of the surface waters, reflected in the oxygen-isotope composition of planktonic foraminifer tests (Nørgaard-Pedersen, 1997; Poore et al., 1999).

The most probable sources of icebergs were located in Severnaya Zemlya, including those related to the breakup of ice sheets during MIS 4 and MIS 6 stages (Knies et al., 2000), and the North Kara Sea (Fig. 9.29). IRD enriched in fragments of Upper Jurassic black shales from the Barents Sea could also be present in sediments of the second part of MIS 6 (150–170 ka), the black layer of MIS 2 base, and sediments of the Late Holocene (Vogt, 1997).

The sandy material transported by sea ice might also be derived from northern Eurasia. As today, at that time sea ice was mainly concentrated in the Laptev and Kara seas. The paleogeographic environment was crucial for heavy-mineral assemblages and grain-size distribution of bottom sediments on the Yermak Plateau. Paleogeography of glacial and stadial periods was governed by sea-level falls, expansion of shelf and continental glacier areas, the expansion of glaciation on the Putoran Plateau, and the low river discharge. At the same time, the relative share of sedimentary material in the ice cover increased as a result of bottom sediments freezing in the "anchored" ice. Interglacial periods promoted the rise of sea-level, maximum transgression during MIS 5e (Arkhipov et al., 1986) and other transgressions, deglaciation of the Putoran Plateau (Fig. 9.22), and increase in river discharge. They also boosted the mechanism of sedimentary material incorporation into sea ice during the formation of ice crystals in polynyas (Nürnberg et al., 1994).

Areas of free-ice surface occasionally appeared or expanded during interglacial periods in the Central Arctic (Nørgaard-Pedersen et al., 1998; Poore et al., 1999). The Polar Branch of the Transpolar Drift has been episodically shifting to the south (Fig. 9.22). The Beaufort Gyre domain might have been expanded more rarely at Termination IB (Nørgaard-Pedersen et al., 1998) and during mid-Holocene, and significantly at the MIS 2/MIS 3 boundary (maximum expansion), upper MIS 5, and Termination II. These expansions could provoke the invasion of deep waters from the Canadian Basin to the Eurasian Basin. This is reflected, for example, in variations of deep-water ostracod assemblages between 8.5–7.0 and 5.0–4.0 ka (Cronin et al., 1995). In contrast, the Beaufort Gyre has diminished during the glacial periods (Bischof et al., 1996).

However, in terms of major parameters, circulation systems of the Central Arctic (Phillips and Grantz, 2001) and the Fram Strait area (Hebbeln and Wefer, 1997) retained their structure during the last climatic cycle.

Organic-Geochemical Sediment Studies of the Eastern Part of the Central Arctic

During ARCTIC 91 Expedition, an extensive sampling program was carried out onboard R/V *Polarstern*, using piston, gravity, box, and multicorer coring gears (Fütterer, 1992). Material was recovered from the eastern Arctic Ocean basins and ridges/plateaus, i.e., the Nansen, Amundsen, and Makarov Basins, the Gakkel and Lomonosov Ridges, and the Morris-Jesup-Rise and the Yermak Plateau (see Fig. 9.15 and Table 9.12 for location of geological stations). Here, we present some results of a detailed study on the amount and composition of organic carbon in the uppermost 40 cm of the sedimentary sequences recovered by multicorer device (Stein et al., 1994b; Schubert et al., 2001). In each multicorer sedimentary section, samples were taken every 2 cm and analyzed for $CaCO_3$, TOC (total organic carbon), C/N ratios, and hydrogen indices as well as oxygen and carbon isotopes in the tests of planktonic and benthic foraminifers. Radiocarbon (AMS ^{14}C) age were determined on selected samples to get an age control (Fig. 9.30) (Stein et al., 2004b).

In general, TOC contents are higher in the basin cores (Cores PS2159, PS2161, PS2170, PS2176: 0.3–1%) and lower in the cores from the oceanic ridges (Cores PS2163, PS2165, PS2185: 0.2–0.6%) (Fig. 9.30), with one exception. In the Makarov Basin Core PS2178, TOC values are significantly lower in comparison to the other basin cores, ranging between 0.05 and 0.45%. Although differences in absolute values, however, the TOC curves look similar in most of the cores. From low values occurring in the lower part, TOC contents distinctly increase in the upper interval. The uppermost increase partly coincides with the (biogenic) carbonate increase and correlates with Termination Ib (Stein et al., 1994b). Based on low hydrogen index values (Fig. 9.30) (Stein et al., 1994b, 2004b), as well as biomarker data (Schubert, 1995; Schubert and Stein, 1996), the organic matter deposited in the basins as well as on the ridges is mainly of terrigenous origin. In the lower part of some of the records probably representing MIS 3, higher amounts of marine organic matter may have been preserved as suggested from elevated hydrogen index values. In general, C/N ratios vary between 4 and 15 (Fig. 9.30). When using C/N ratios for sediments from the central Arctic Ocean sediments, however, the potential contribution from inorganic nitrogen needs to be weighed (Stein and Macdonald, 2004 and further references therein). With significant amounts of inorganic nitrogen, the "C_{org}/N_{org}" ratios will be higher than the C/N ratio and, accordingly, will support a more mixed terrigenous/marine origin of the organic matter in general.

In the uppermost (Holocene) interval, also detectable amounts of marine organic carbon are preserved as suggested from Rock Eval data (Stein et al., 1994b) and

Table 9.12 Location of the studied stations [*Stein et al.*, 1994a]

Station number	Latitude, N	Longitude	Water depth, m
PS 2159	83º9,60'	30º3,70'E	4055
PS 2161	85º4,50'	44º4,22'E	4005
PS 2163	86º2,42'	59º2,15'E	3040
PS 2165	86º4',47'	59º9,60'E	2011
PS 2170	87º5,90'	60º7,66'E	4226
PS 2175	87º5,68'	103º5,66'E	4378
PS 2177	88º0,36'	134º9,26'E	1388
PS 2178	88º0,05'	159º1,68'E	4009
PS 2179	87º7,46'	138º0,29'E	1230
PS 2184	87º6,11'	148º1,40'E	1640
PS 2185	87º5,29'	144º1,66'E	1073
PS 2190	90º0,00'	90º0,00'E	4240
PS 2192	88º2,60'	9º8,57'E	4375
PS 2193	87º5,12'	11º4,75'E	4399
PS 2196	85º9,62'	0º1,65'E	3958
PS 2198	85º5,65'	9º0,58'W	3820
PS 2200	85º3,28'	14º0,22'W	1073
PS 2202	85º1,09'	14º3,69'W	1081
PS 2204	85º0,58'	13º0,35'W	3899
PS 2205	84º6,44'	6º7,67'W	4283
PS 2206	84º2,78'	2º5,05'W	2993
PS 2208	83º6,40'	4º6,03'E	3681
PS 2209	83º2,25'	8º5,73'E	4046
PS 2210	83º0,45'	10º1,25'E	3949
PS 2212	82º0,24'	15º6,72'E	2531

results of biomarker determinations (Schubert, 1995; Schubert and Stein, 1996). The contemporaneous increases in biogenic carbonate and organic carbon during the Holocene are interpreted as increased surface-water productivity and reduced sea-ice cover at that time (Stein et al., 1994b). This interpretation is supported by carbon and nitrogen isotope values also increasing steadily to modern values (Schubert et al., 2001; Stein and Macdonald, 2004). In general, the increase in $\delta^{15}N$ values during the late Holocene shows the higher nutrient utilization in surface waters due to more open ocean water conditions enabling the phytoplankton to receive enough light for reproduction.

TOC records representing older time intervals (MIS 4 and older) are still very rare for the central Arctic Ocean and adjacent continental slope (Stein and Macdonald, 2004 and references therein). On the Alpha and Lomonosov ridges (cores PS2185 and PS051/38; see Fig. 9.15 for location), the TOC contents vary between 0.05 and 0.3% whereas in the Amundsen Basin Core PS2174 (see Fig. 9.15 for location) TOC contents are significantly higher ranging between 0.4 and 1.5% (Fig. 9.31) (Schubert, 1995; Stein et al., 2004b). In the latter core, maximum TOC contents coincide with low hydrogen index values of < 100 mgHC/gOC and high C/N ratios of 10 to > 20 (Fig. 9.31), indicating high input of terrigenous organic

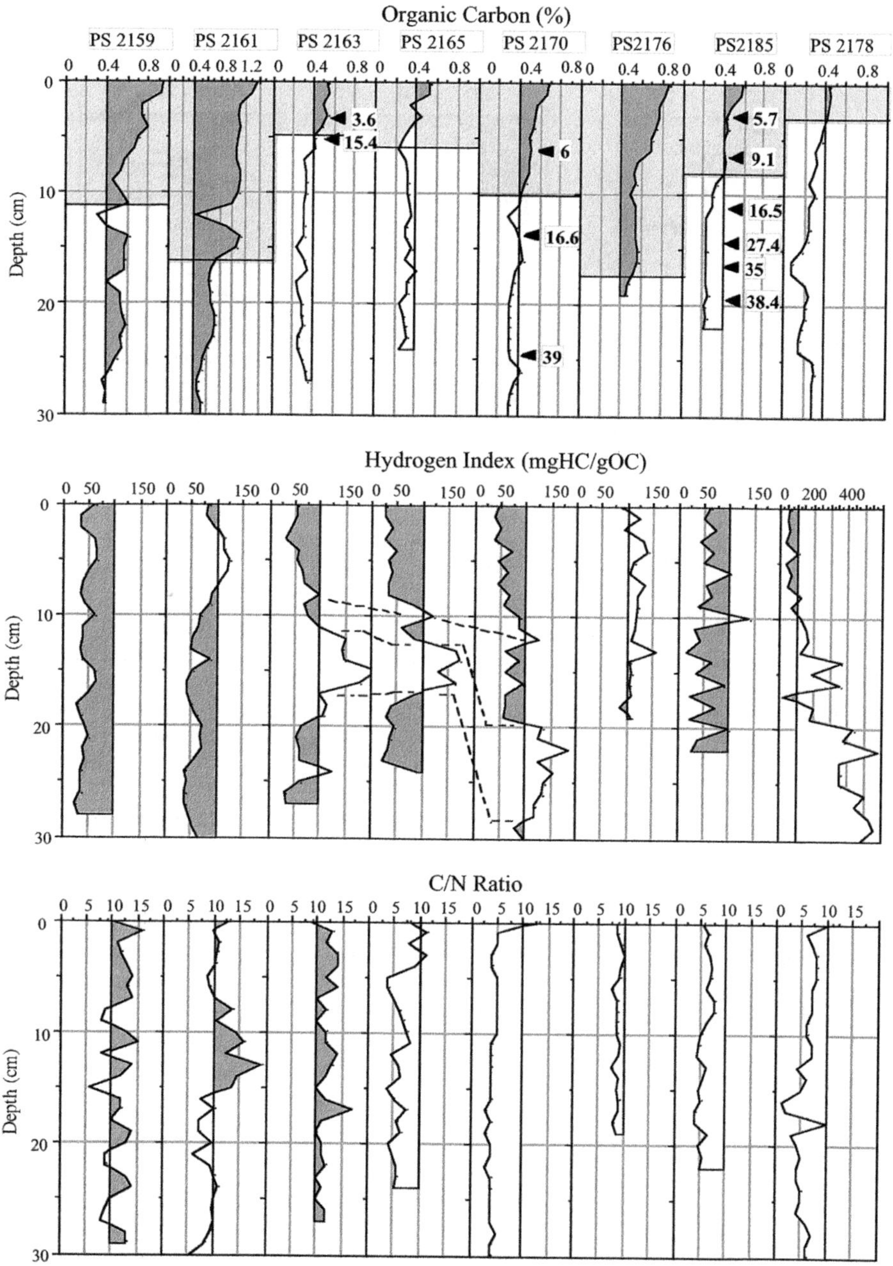

Fig. 9.30 TOC, hydrogen index (HI) values, and C/N ratios in organic matter of Arctic Ocean sediments representing the last about 30 ka (Stein et al., 2004b). Arrows show ^{14}C dates; light gray color matches MIS 1

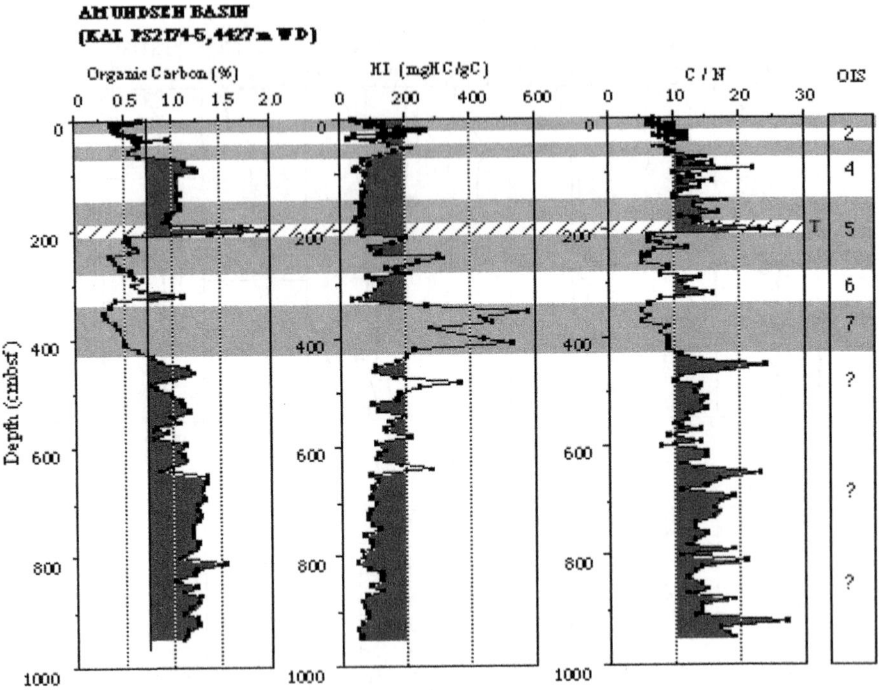

Fig. 9.31 TOC, hydrogen index (HI) values, and C/N ratios in organic matter of the Amundsen Basin core PS2174-5 possibly representing the last 7–8 MIS (Stein et al., 2004b)

matter. The clear dominance of long-chain *n*-alkanes over the short-chain *n*-alkanes also supports a terrigenous origin of the organic matter (Schubert and Stein, 1996). Intervals with low OC contents most typical for the interglacials MIS 5 and MIS 7 show higher hydrogen index values and lower C/N ratios (Fig. 9.31), probably due to higher surface-water production. Higher surface-water production for these intervals is also suggested from higher concentrations of short-chain *n*-alkanes and biogenic opal (Schubert and Stein, 1996).

Chapter 10
The Western Arctic Seas

Recent Sedimentation Environment

The Barents Sea

The Barents and Kara seas are traditionally referred to as the western arctic seas. The Barents Sea is limited by the Norwegian Sea in the west, by Spitsbergen and Franz-Joseph Land archipelagoes and by the Nansen Basin in the north, by the Novaya Zemlya Archipelago in the east, and by the northern coast of Europe in the south. Two transverse troughs pass from the shelf to the continental slope: the Bear Island Trough between Norway and Spitsbergen in the west and the Franz Victoria Trough between Spitsbergen and Franz-Joseph Land in the north. The Kara Sea and the Barents Sea are connected through the Kara Gates Strait in the southeast and through the extensive transverse St. Anna Trough in the northeast. The southeastern part of the Barents Sea, located between the southern island of Novaya Zemlya, the Kara Gates, the coast of the Bolshezemelskaya Tundra and the Kolguyev Island is called the Pechora Sea.

The Barents Sea has a size of 1512 thousand km^2, a volume is 302 thousand km^3, and an average water depth of 200 m (Jakobsson et al., 2004). Freshwater discharge by rivers (together with the White Sea) is equal to 463 km^3/yr, and the riverine suspended matter supply reaches 17.9 million t/yr (Gordeev and Rachold, 2004). 119 million t/yr of solid sedimentary material is supplied into both seas by coastal abrasion processes (Grigorjev et al., 2004). The delivery of sea ice from the Barents Sea to the Arctic Ocean composes 35 km^3/yr (Eicken, 2004). This value should be added to the delivery by icebergs from the Northern Island of Novaya Zemlya, Franz-Joseph Land, Spitsbergen and North Norway. Large part of the terrigenous material, however, is already accumulating in the natural sedimentation traps of fjords. In addition, 0.904 t/yr of solid sedimentary matter is delivered to the Barents Sea due to the eolian supply (Shevchenko and Lisitsin, 2004). Finally, one should not exclude the edafogenic mechanism of the sedimentary delivery, to which a special attention was given by V.I. Gurevich (1995).

M.A. Levitan, Y.A. Lavrushin, *Sedimentation History in the Arctic Ocean and Subarctic Seas for the Last 130 kyr*, Lecture Notes in Earth Sciences 118,
DOI 10.1007/978-3-642-00288-5_10,

In general, the central part of the Barents Sea could be considered as "starving" from the point of view of modern sediment accumulation because a substantial part of the Barents Sea seafloor does not contain a Holocene sedimentary cover, but older Quaternary formations or bedrock are exposed at the surface. This phenomenon is especially characteristic for the relatively shallow arch rises. Intensive accumulation of recent sediments is restricted to the fjords, Pechora Bay and the central part of the Pechora Sea.

The bottom of the Barents Sea is characterized by distinct roughness of the relief: there are a number of isometric and, less often, elongated underwater rises (Persey, Central, Admiralty Swell, Murmansk Swell, Goose Shoal), dividing troughs and depressions (Southern and Northern Barents depressions, Persey Trough, Aldanov Trough, Bear Trough, Franz Victoria Trough, West and Southern Novozemelsky Troughs). Relative elevation in this case means several hundred meters. In general, the Pechora Sea is characterized by shallow depth (a few tens of meters), flatness, slight slope to the north and the presence of a number of bars along the shore.

Heat advection and saline Atlantic-water inflow play a most important role for the climate of the Barents Sea (Matishov et al., 1994). The Atlantic waters known as Norwegian Current move along Scandinavia, and continue as Cape Current and further on as Murmansk Current to the east. They almost reach the Novaya Zemlya Archipelago, but sharply turn to the north and actually dissipate in the north of the Barents Sea. Another branch of Atlantic waters is split off at traverse of Nordcape Cape and moves to the northeast, forming a jet stream, isolated by the Arctic Front from the fresher and colder Barents Sea waters to the south. In this case, Atlantic waters are partially transformed into the Barents Sea waters. Furthermore, the Atlantic waters of the West Spitsbergen Current in the area of the Fram Strait are included in the composition of intermediate water mass and follow to the east along the continental slope of North Eurasia, recirculating in the Franz Victoria Trough and St. Anna Trough. The described system of Atlantic-water penetration into the Barents Sea ensures not only the most favorable ice conditions among all Arctic seas (including the year-round ice-free water area of the Kola Bay), but also increased primary production (about of 90 $gC/m^2/yr$) (Sakshaug, 2004).

The Kara Sea

The Kara Sea (which, in particular, includes the St. Anna Trough) is limited by the archipelago of Novaya Zemlya in the west, by the Nansen Basin in the north, by the archipelago of Severnaya Zemlya in the east, and by the coast of West Siberia in the south. The St. Anna Trough is located between the Northern Island of Novaya Zemlya, the archipelago of Franz-Joseph Land and the North Kara Rise. One additional transverse trough of the Kara Sea (Voronin Trough) is located between the North Kara Rise and the archipelago of Severnaya Zemlya.

The area of the Kara Sea is equal to 926 thousand km^2, its volume is 121 thousand km^3, and the average water depth is 131 m (Jakobsson et al., 2004). Annual river discharge is equal to 1480 km^3, and the riverine suspended-matter supply is 30.9 million t/yr (Gordeev and Rachold, 2004). 109 million t/yr of sedimentary material enter the Kara Sea due to the coastal abrasion (Grigorjev et al., 2004). In all Arctic shelf seas the value of annual underground discharge composes approximately one tenth part of the riverine freshwater discharge (Gordeev and Rachold, 2004). 240 km^3/yr of sea ice is carried out from the Kara Sea (Eicken, 2004). Delivery of terrigenous material by icebergs from Severnaya Zemlya plays a leading role in the northern part of the sea. Eolian input is estimated to be about 0.553 million t/yr (Shevchenko and Lisitzin, 2004).

The Kara Sea consists of two very different parts. The western part is characterized by rough relief (for example, the sufficiently deep East Novozemelsky Trough and relatively shallow Central Kara Rise), and the basic source of sedimentary material is thermo-abrasion of the southern coast. The eastern part is the shallow (a few tens of meters) Ob-Yenisei Shoal that is weakly inclined northward and is split by the numerous channels of valleys of the Pre-Ob and Pre-Yenisei. Here, sedimentary material is mainly delivered by the great Siberian rivers Ob and Yenisei. Accordingly, the average salinity of surface water in this part of the sea is considerably lower than in the west. In the western part, an anticyclonic circulation prevails, whereas the eastern part is subjected to a general northeastern transfer of waters. The Atlantic waters of the St. Anna Trough are not delivered to the south of the Kara Sea. Hence, there are more severe ice conditions and lower primary production of about 40 gC/m^2/yr (Sakshaug, 2004).

Surface Sediments of the Pechora Sea

This paragraph is mainly based on data by (Khasankaev, 1978; Gurevich, 1995; Tarasov, 1996; Levitan et al., 2003b). Figure 10.1 shows the distribution of lithotypes of surface sediments in the Pechora Sea (Tarasov, 1996). Practically all sediments are noncarbonate and do not contain a biogenic opal. Boulder-pebble material in large quantities is encountered only in the Czesh Inlet (Khasankaev, 1978). According to our observations, its concentration usually sharply decreases at a distance of the first kilometers from the eroded coast. Sand and silty sand are widely developed, mainly, in the central part of the sea.

Middle- and fine-grained sand prevails among the sandy fractions, and fine silt dominates among the silty fraction. Closer to the periphery of the basin these sediments are changed to sandy silt and mictites of sandy-silty-pelite composition (Gurevich, 1995). Fine clayey silt or silty- clayey muds are rarely encountered. They are related either to noticeable depression in the relief (to the northeast of the Kolguyev Island, in the Southern Novozemelsky Trough, etc.), or to outcrops of older Holocene sediments (Levitan et al., 2000). Results of our sediment studies, obtained in 1998 during the 13th cruise of RV *Academik Sergey Vavilov* and cruise *G-230* (Figs. 10.2 and 10.20; Tables 10.1 and 10.2), confirm the described scheme of the lithotype distribution.

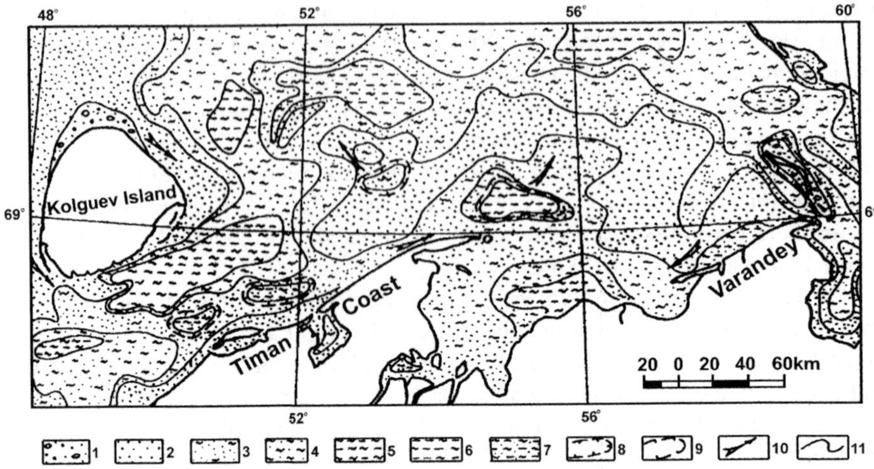

Fig. 10.1 Scheme of the Pechora Sea lithodynamics (Tarasov, 1996). 1 – sandy-gravely-pebbly sediment; 2 – sand; 3 – silty sand; 4 – sandy silt; 5 – clayey silt; 6 – silty clay; 7 – sandy-silty-clayey sediment; 8 – areas of bottom erosion of Pleistocene deposits; 9 – areas of bottom erosion of Paleozoic rock; 10 – transportation directions of sand and silt; 11 – lithological boundaries

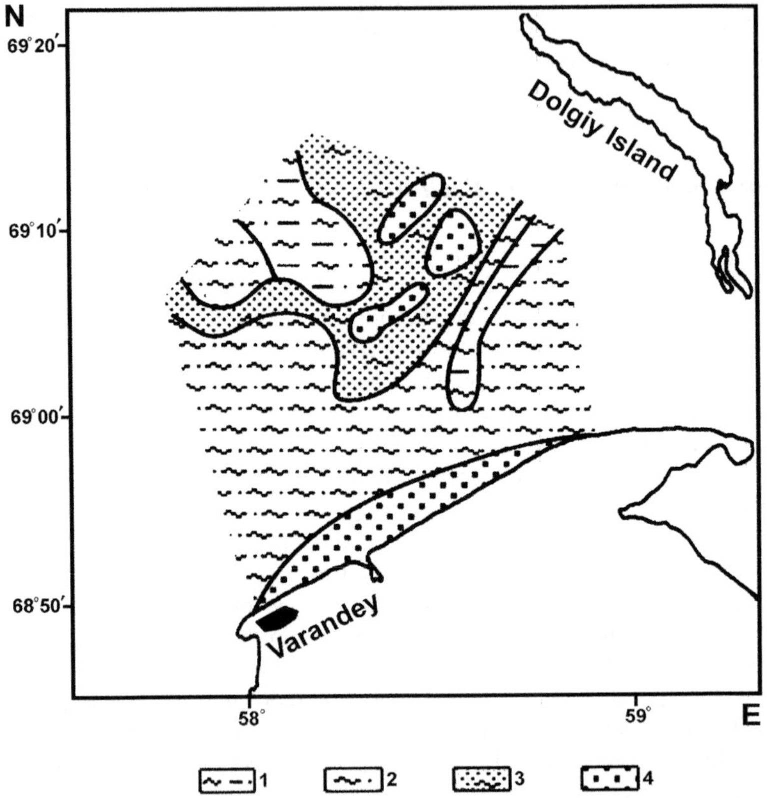

Fig. 10.2 Lithology of surface sediments of Varandey Site (Levitan et al., 2003b). 1 – clayey-silty sand; 2 – small-fine sand; 3 – fine-small sand; 4 – small-grained sand

Table 10.1 Content of grain-size fractions (wt. %) in the surface layer sediments of the Pechora Sea [*Levitan et al.*, 2003b]

Station ASV	2–1 mm	1–0.5 mm	0.5–0.25 mm	0.25–0.1 mm	0.1–0.05 mm	0.05–0.01 mm	0.01–0.005 mm	0.005–0.001 mm	<0.001 mm	Lithotype
1088	1.43	1.30	7.05	45.63	28.70	4.70	1.95	3.94	5.90	Fine-medium sand
1092	1.33	0.48	4.74	86.78	6.57	0.10	–	–	–	Medium sand
1093	0.28	0.33	7.69	73.05	8.51	0.10	2.11	3.55	4.28	Medium sand
1094	–	0.31	0.75	19.27	62.22	1.56	3.70	3.76	8.43	Medium-fine sand
1095	–	–	0.56	24.78	54.90	0.51	5.67	5.98	7.60	Medium-fine sand
1096	0.05	0.21	1.03	45.71	39.30	0.46	2.74	3.86	6.64	Fine-medium sand
1097	0.31	0.37	1.82	27.80	40.92	1.12	7.53	8.83	11.30	Clayey-silty sand
1098	0.57	0.62	7.27	45.81	35.29	0.28	2.54	3.45	4.17	Fine-medium sand
1099	0.58	0.48	4.02	57.50	25.67	0.58	2.03	3.61	5.53	Fine-medium sand
1100	0.40	0.05	–	40.80	48.85	0.80	2.54	3.66	2.90	Medium-fine sand
1101	0.11	–	0.61	34.76	54.51	0.71	2.13	3.06	4.11	Medium-fine sand
1102	0.21	0.27	2.98	27.71	41.97	2.24	5.05	7.15	12.42	Clayey-silty sand
1103	–	0.33	0.95	37.32	45.90	1.11	3.17	5.04	6.18	Medium-fine sand
1104	–	0.10	0.53	72.22	15.96	0.69	2.19	3.35	4.96	Medium sand
1105	–	0.22	0.66	17.09	56.63	3.15	5.07	7.14	10.04	Clayey-silty sand
1106	–	0.17	–	26.23	52.45	0.11	6.40	8.14	6.50	Medium-fine sand
1112	–	0.25	7.53	79.31	7.27	0.06	1.56	1.75	2.27	Medium sand
1117	–	–	0.78	90.85	3.20	0.20	1.31	1.96	1.70	Medium sand
1118	–	–	1.64	88.93	4.27	0.18	1.46	1.76	1.76	Medium sand
1119	–	1.02	4.42	77.32	9.07	0.54	1.19	2.98	3.46	Medium sand
1123	–	0.52	0.97	87.26	5.07	0.07	0.82	1.34	3.95	Medium sand
1124	–	–	2.28	85.78	4.14	0.14	1.45	2.20	4.01	Medium sand
1125	–	–	1.14	88.29	6.60	1.84	0.71	0.64	0.78	Medium sand
1126	–	–	–	40.94	39.61	5.07	2.15	4.00	8.23	Fine-medium sand
1127	–	–	1.09	42.29	38.67	1.98	2.47	5.14	8.36	Fine-medium sand
1128	–	–	0.53	77.39	8.20	0.71	1.41	4.29	7.47	Medium sand
1129	–	0.68	2.59	87.23	3.15	0.12	1.54	1.85	2.84	Medium sand
1130	–	–	3.13	86.94	2.78	0.12	1.92	1.74	3.37	Medium sand
1137	–	–	3.99	63.96	8.41	2.99	2.35	7.38	10.92	Medium sand
1150	–	1.08	3.96	81.86	5.40	0.11	1.62	2.17	3.82	Medium sand
1151	–	1.14	1.02	74.87	14.66	0.06	1.71	1.97	4.57	Fine-medium sand

Note: Blank means lack of fraction.

Thus, if we exclude from examination the older sediments, it turns out that the finer-grained recent sediments are being developed on the periphery of the Pechora Sea, and this is partially caused by bathymetric control. The coarser-grained sediments are concentrated in the middle part of the basin. If we compare a similar distribution of lithotype with the distribution of bottom current rate values (Gurevich, 1995), one could make a paradoxical conclusion about the reverse correlation of the sediment coarseness and rate of bottom currents. This dependence might be explained by a larger value of mass accumulation rate/rate of bottom current ratio in the peripheral parts of the basin than in its central part.

In more detail the distribution of lithotype was studied at the site *Prirazlomnoe* (see data at stations 1115–1127) and the *Varandey* site (see data at stations 1095–1112 and station G-230) (Table 10.1) (Levitan et al., 2003b). It was

Table 10.2 Calculated grain-size parameters [*Levitan et al.*, 2003b]

Station number ASV	Sorting coefficient, So	Md, mm	Mode (Mo), mm	Asymmetry coefficient, Sk	Excess, Ek
1088	1.70	0.11	0.15	0.89	0.25
1092	1.30	0.15	0.13	1.17	0.30
1093	1.38	0.15	0.13	1.03	0.21
1094	1.37	0.07	0.06	1.08	0.13
1095	1.68	0.09	0.10	0.98	0.25
1096	1.73	0.09	0.07	0.93	0.24
1097	3.32	0.08	0.08	0.17	0.26
1098	1.60	0.10	0.09	1.26	0.23
1099	1.65	0.13	0.18	0.79	0.26
1100	1.53	0.09	0.08	1.20	0.22
1101	1.42	0.08	0.08	1.21	0.32
1102	3.23	0.08	0.06	0.27	0.31
1103	1.53	0.08	0.07	1.31	0.21
1104	1.42	0.14	0.15	0.79	0.22
1105	4.20	0.07	0.06	9.00	2.84
1106	1.40	0.08	0.07	0.96	0.15
1112	1.26	0.16	0.18	0.94	0.22
1117	1.28	0.17	0.15	0.89	0.35
1118	1.18	0.18	0.18	0.75	0.24
1119	1.30	0.18	0.21	0.84	0.27
1123	1.17	0.16	0.15	0.99	0.22
1124	1.30	0.18	0.16	0.84	0.30
1125	1.24	0.16	0.13	1.40	0.29
1126	1.60	0.09	0.06	1.10	0.22
1127	1.50	0.09	0.06	1.00	0.18
1128	1.30	0.14	0.13	0.98	0.18
1130	1.20	0.15	0.14	1.12	0.24
1137	1.76	0.15	0.15	0.55	0.27
1150	1.20	0.15	0.13	1.08	0.22
1151	1.40	0.15	0.16	0.91	0.25

demonstrated at the Varandey site that the coarsest-grained sands are accumulated in the contemporary bars or at the apexes of structural rises. Large part of sea floor is covered with finer (clayey-silty) sands. The finest sediments are revealed in the relief depressions, where they are connected either with the outcrops of Middle Holocene black clays (station 1105), or with the recent accumulation of fine-grained material (station 1102).

Useful sedimentological information can be obtained from Table 10.2. First of all, there is a close direct correlation (r^2=0.81) between the median diameter and the mode. The absence of any correlation between the median diameter and the coefficient of sorting (So) is interesting. The finest sediments are characterized by the worst sorting; however, the interval of So values from 1.1 to almost 1.8 is characteristic for all types of sediments (with mean diameters from 0.06 to 0.18 mm). At the same time a study of distribution of sorting for the sediments at the Varandey Site (Levitan et al., 2003b) showed that So values of 3.0 are related to the fine sediments, and So values less than 1.4 correspond to the coarse-grained sediments.

In the sands at the site Prirazlomnoe the coefficient of sorting is often less (1.17–1.18; Table 10.2). Thus, the best sorting is observed among sands, than in finer-grained sediments. This makes it possible to assume a delivery of minor amount of silt and clay from the sediments by bottom currents and redeposition of these fractions in the zones of recent accumulation of fine sediments.

At the same time, different genesis and age of fine mud development in the surface sediments leads to the conclusion that only contemporary fields can serve as accumulation zones of fine silt and clay. As known, absorption processes of natural and technogenic contaminants are most frequently related to fine sediments. Thus, the areas of fine sediment distribution and absorption zone of pollutants might correspond to each other in the case of recent silty-clayey mud, but do not coincide in the case of outcrops of older clay sediments). This conclusion is of definite interest for geo-ecology of the Pechora Sea. The other important conclusion is that finer fractions of the suspended matter will be carried away from the Prirazlomnoe oil field area beyond its limits, since in recent time there have been accumulated only well sorted fine-grained sands in this area.

Mineral composition of recent sediments (especially heavy minerals) is one of the most important proxies for determination of source provinces and transport pathways of sedimentary material. The content of the light subfraction of the 0.1–0.005 mm fraction in the studied sediments of the Pechora Sea varies from 80 to almost 100% (Levitan et al., 2003b). Most important components are quartz (49.5–60.6%) and potassium feldspar with $n<1.54$ (12.0–21.9%), followed by acid plagioclases (usually less than 3–4%). Basic plagioclases, microcline and grains of carbonates only occur in very rare amounts. Over the area of the Pechora Sea the composition of the light subfraction is remarkably uniform.

In general, there is no significant correlation between the total content of the heavy subfraction and the content of concrete grain-size fractions. A close positive correlation, however, exists between the total content of the heavy subfraction up to 5% and the content of the 0.25–0.1 mm fraction up to 58%. There is no relationship between an increase in the abundance of this fraction (especially to 80% and above) and the total content of the heavy subfraction. Thus, grain-size and gravitational differentiations coexist only to a definite limit (increase of fine sand abundance to 58%).

The abundance of indeterminable grains in the heavy subfraction varies from 2.8 to 16.5%. Among the numerous heavy minerals determined in the surface sediments of the Pechora Sea, representatives of the four basic groups prevail: black ore minerals, garnet, hornblende and epidote. Besides the enumerated minerals zircon plays a notable role, whereas monocline pyroxene, siderite, apatite, and sphene only occur in rare amounts.

The increased concentrations of black ore minerals (>25%; Table 10.3) are related to the peripheral parts of the basin, increasing towards to the source provinces. This phenomenon can be explained by gravitational differentiation and high hydraulic coarseness of the studied minerals. The sediments of the western part of the sea are enriched by garnet, which is related to the delivery of material from the granite-metamorphic complexes of the Baltic Shield (Tarasov, 1996) and a general transfer of waters of the Pechora Sea to the east. The zone of increased

Table 10.3 Content of main heavy minerals (rel. %) in the surface sediment layer of the Pechora Sea [*Levitan et al.*, 2003b]

Station number ASV	Output of heavy subfraction, wt.%	Black ore minerals	Garnet	Hornblende	Epidote	Zircon
1088	2.32	29.5	21.0	14.2	17.5	15.2
1092	16.26	23.6	16.8	8.6	20.9	8.0
1093	6.77	26.3	16.2	4.2	20.8	5.3
1094	1.04	25.8	11.8	8.3	32.6	2.5
1095	0.62	20.8	11.8	20.0	34.3	2.5
1096	3.82	24.3	12.4	6.1	33.4	10.0
1097	2.45	27.7	14.6	7.3	31.0	8.2
1098	1.08	19.2	17.4	5.2	24.2	6.5
1099	2.11	35.8	19.1	2.9	16.7	17.8
1100	3.02	30.1	20.5	3.3	14.5	11.2
1101	2.65	28.0	16.1	6.9	15.3	12.0
1102	1.82	21.1	9.9	10.5	25.5	7.1
1103	1.70	20.7	16.2	9.3	26.1	8.3
1104	7.00	24.9	16.5	3.9	20.6	11.6
1105	0.30	19.0	19.0	9.5	31.7	7.8
1106	2.06	21.1	12.5	11.9	21.1	7.8
1112	4.55	19.2	23.1	4.9	23.6	5.8
1117	5.88	20.4	20.8	6.3	25.0	3.6
1118	15.29	11.0	15.9	16.2	31.2	1.7
1119	22.28	28.6	15.6	4.4	19.6	11.9
1123	16.45	15.4	16.4	12.1	27.2	3.9
1124	13.85	14.9	17.5	8.9	26.7	5.3
1125	11.76	19.1	14.5	14.2	23.1	4.3
1126	3.48	19.3	13.6	10.1	25.5	5.3
1127	3.30	17.4	15.4	14.0	25.2	6.4
1128	9.27	18.6	15.8	11.6	24.7	4.6
1129	19.05	18.2	22.5	9.3	20.2	4.9
1130	20.33	28.4	19.9	4.7	14.5	7.1
1137	26.71	27.4	21.1	5.6	13.5	6.3
1150	14.04	30.5	20.3	7.3	16.8	7.0
1151	8.84	22.8	20.8	4.2	14.7	7.6
1157	1.75	15.0	19.5	5.0	13.0	0.5
DM 4376	2.02	15.2	11.1	11.3	25.2	0.7

Note: In the sediments of sediment core ASV 1151 – 12.6% of siderite, sediments of sediment core ASV 1157 contain 17.0% of monocline pyroxenes.

content of amphiboles (hornblende) is related to the central part of the basin, probably, in connection with the increased buoyancy of hornblende grains and their capability for distant transport.

The increased concentrations of epidote (>20%) clearly increase towards the eastern part of the sea. This distribution confirms significant contribution of sedimentary material from the greenstone-changed sedimentary Paleozoic complexes, developed at the Polar Urals, Pai-Khoy, Vaygach and Novaya Zemlya (Tarasov, 1996). Zircon is enriched in an even narrower strip of sediments in the eastern periphery of the basin, increasing towards the Vaygach Island and sands of the

Table 10.4 Location of geological stations of RV *Polarstern* in the eastern Kara Sea

Station	Northern Latitude	Eastern Longitude	Water depth, m
PS 2718	77°31′05″	97°04′04″	153
PS 2719	77°36′00″	97°32′00″	135
PS 2792	79°33′05″	87°33′01″	245

central part of the Varandey site. The increased content of clinopyroxenes of paleotypic kind in the sediments south of Novaya Zemlya, most likely, is caused by erosion processes of Lower Paleozoic basaltoids developed at the Southern Novaya Zemlya Island (Tarasov, 1996). The enrichment of siderite in sediments of the coastal zone of Kolguyev Island, is related to erosion of Upper Pleistocene moraines (Tarasov, 1996).

Based on the combination of distribution patterns of the basic heavy mineral groups mentioned above, specific terrigenous-mineralogical provinces can be distinguished in the surface sediments of the Pechora Sea (Fig. 10.3) as follows (in

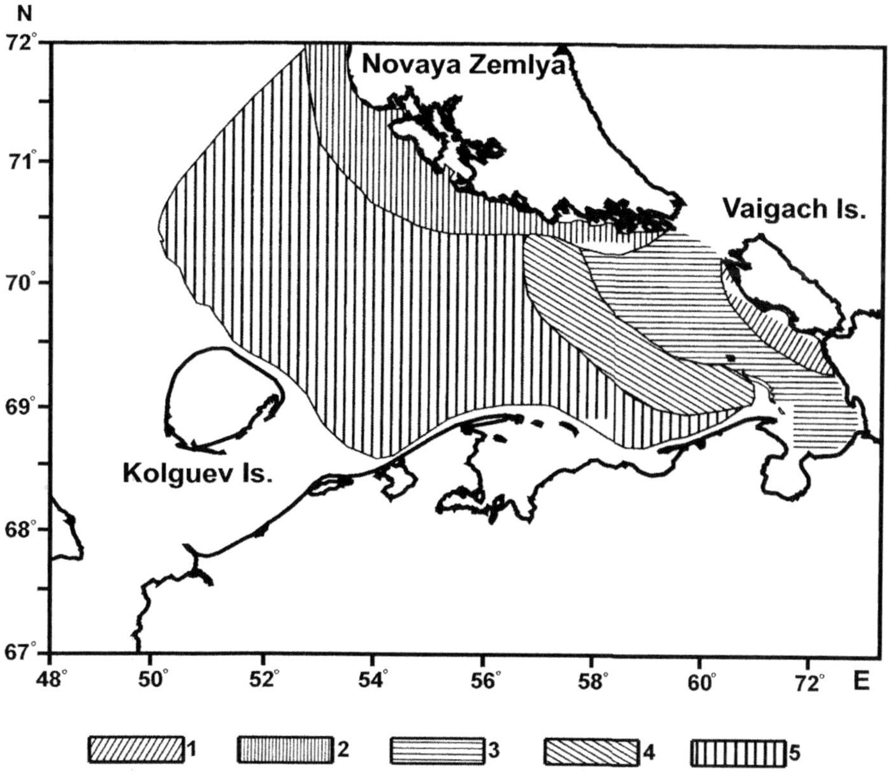

Fig. 10.3 Terrigenous-mineralogical provinces of the Pechora Sea sediments (Levitan et al., 2003b). 1 – Vaigach; 2 – South Novaya Zemlya; 3 – East Pechora; 4 – Central Pechora; 5 – West Pechora

the names of associations content of heavy minerals increases from left to right): Vaygach (zircon-garnet-epidote-ore association); Southern Novozemelsky (epidote-ore-clinopyroxene-garnet association); Western-Pechora (epidote-garnet-ore association); Central-Pechora (ore-garnet-epidote association); and Eastern-Pechora (hornblende-garnet-ore-epidote association). Within the limits of the detailed ranges it is possible to isolate additional associations.

In general, the dominance of an acid/intermediate plagioclases-potasssium feldspar-quartz association in the light subfraction and a hornblende-garnet-black ore association (to a lesser degree also – zircon, apatite and sphene) in the heavy fraction indicate substantially a granitoid composition of source provinces of the Pechora Sea sediments. A second-order province area is represented by the altered greenstone Hercynide formations of Pai-Khoy, Vaygach, and Novaya Zemlya (Tarasov, 1996).

Clay minerals in the surface sediments of the Pechora Sea are represented by a monotonous association of kaolinite, magnesium-ferrous chlorite and illite (Levitan et al., 2003b). Smectite was only found in a minor quantity, and mixed-layered clay minerals are absent.

Surface Sediments of St. Anna Trough

Transverse troughs of the Barents and Kara Sea (Bear Island Trough, Franz Victoria Trough, St. Anna Trough, and Voronin Trough) serve as basic transit ways for sedimentary material and cold bottom brines from the shelf area into the surrounding deep water basins (Levitan et al., 1999). In this paragraph we discuss the mineral composition of the surface sediments of the St. Anna Trough and the influence of sediment supply from the avalanche accumulation zones located in the estuaries of the Ob and Yenisei rivers, on the recent sedimentary environment in the St. Anna Trough.

Looking at the bathymetric map compiled by G.A. Tarasov (Fig. 10.5) (Levitan et al., 1999) and following the 400 m isobath, the St. Anna Trough is stretched to the north from 77° to 82° N, and its central part is located between 66° and 73° E. In the upper part the trough is connected to several troughs of higher order from the southwest and southeast. In general, the slopes of the trough are parallel and sufficiently steep (on average, the eastern slope is somewhat steeper). The bottom is wide, flat, and a bit uneven at the northern end, and inclination to the north is well expressed. Maximum water depths so far observed in northern part of the trough, exceed 550 m.

Materials and Methods

The studied surface sediments (0–2 cm) were recovered during the MMBI expeditions with RV *Dalnye Zelentsy* (1992) and RV *Academik Golitsyn* (1994), and St. Petersburg Polar Sea Expedition, with RV *Professor Logachev* (1994). Locations of the geological stations are shown at Fig. 10.4.

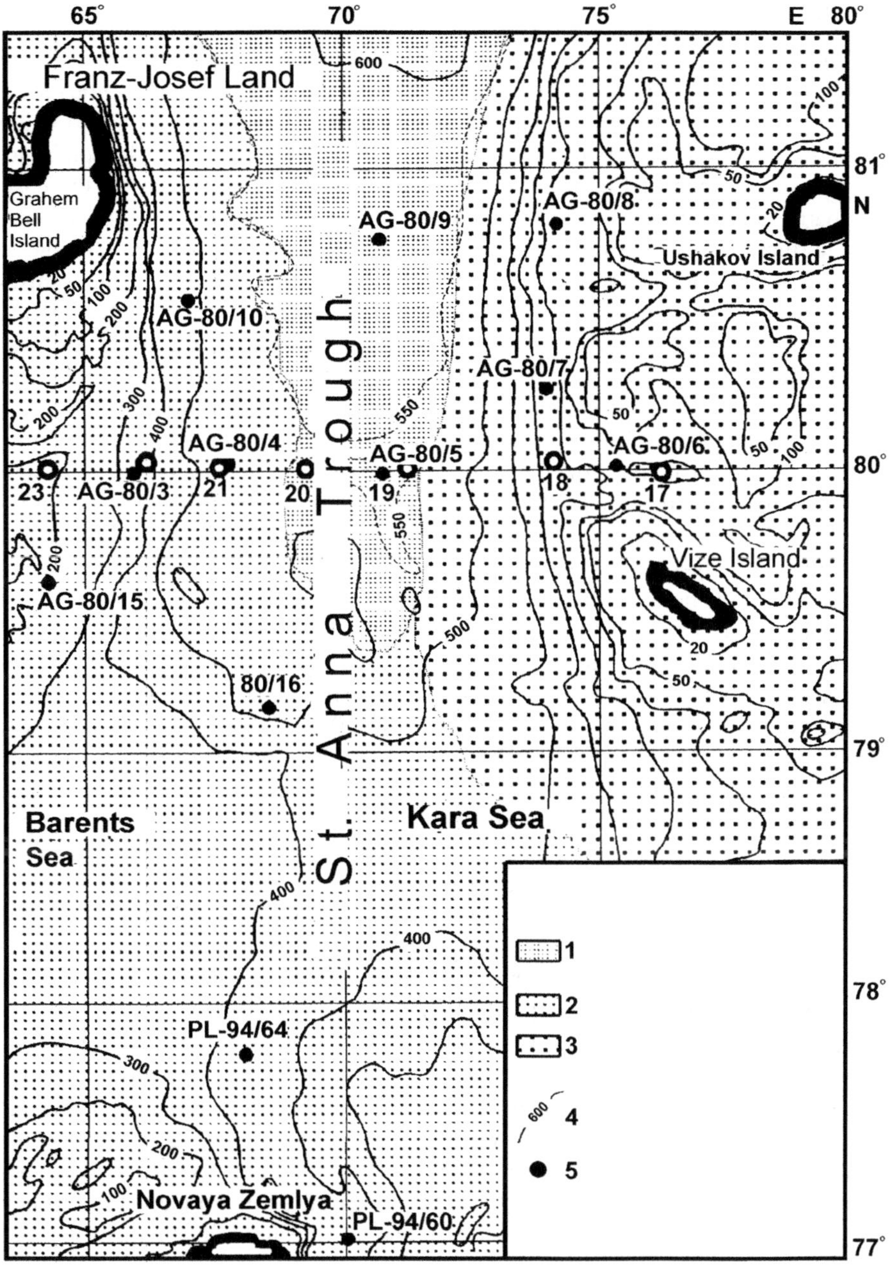

Fig. 10.4 Light-mineral assemblages in surface sediments from the St. Anna Trough (Kukina et al., 1999). 1 – quartz, feldspars, rock fragments; 2 – quartz and feldspars; 3 – feldspars, rock fragments, quartz; 4 – isobaths, m; 5 – geological stations

Recent Sedimentation Environment

The adjacent archipelagoes of Franz-Joseph Land, Novaya Zemlya and North Kara Rise are certainly dominant source areas for the supply of terrigenous sedimentary material into the St. Anna Trough. The Lower Cretaceous trapp formation prevail in the composition of the central and western regions of Franz-Joseph Land, overlain by a cover of Cenozoic and Quaternary sediments, whereas in the eastern part mainly Triassic and, partly, Jurassic terrigenous complexes are exposed. The northern end of Novaya Zemlya is mainly composed of Upper Proterozoic to Lower Paleozoic terrigenous formations, which went through weak metamorphism of greenstone phase. Islands of Vize and Ushakov, situated on top of the North Kara Rise, consist of a Lower Cretaceous terrigenous clastic formation (Geology of Quaternary deposits of the north-western European part of the USSR, 1967; Geological structure of the USSR and regularities of mineral deposits distribution, 1984).

Based on bathymetric, seismoacoustic and geological data, the corresponding outcrops of bedrock are frequently found on the sea foor around the islands mentioned above (Matishov et al., 1994). It is very probable that the more remote sources of material (such as, for example, the Ob River) play a less important role in the delivery of sandy-silt fractions; however, these source may become much more important for the delivery of clay fractions, especially clay minerals.

An important factor that mainly determines the sedimentary environment of the St. Anna Trough region is the ice regime. The studied region is seasonally covered with pack ice; only two months a year the area is free of ice (Levitan et al., 1999). There were some years (for example, 1980, 1982, 1983, 1986), when the trough was covered with all year-round ice (Levitan et al., 1999 and references therein). The role of iceberg supply is sufficient in the west-south-western part of the studied area due to the delivery of material from Franz-Joseph Land (average annual iceberg supply is 2.26 km^3) and from the Northern Island of Novaya Zemlya (2.0 km^3) (Levitan et al., 1999 and references therein). The sedimentary environment is not just influenced by the modern ice regime, but also by the older regimes. Geo-chronometrically dated moraine formed during the Last Glaciation on the western slope of the St. Anna Trough (Polyak and Mikhailov, 1996), are exposed at the sea floor and can be considered as an additional source of sedimentary material.

Based on data obtained in April 1984 (Levitan et al., 1999 and references therein), stratification of water column in the trough appears as follows: surface water mass is presented by Arctic waters with low temperature and salinity (temperature of -1.75 to -1.90°C and salinity of <34.25‰); warmer (to +2.8°C) and saline (up to 34.90‰) Atlantic waters at a depth of 150–500 m; bottom waters, which are colder (to -0.6°C) and less saline (to 34.60‰). In the southern part of the trough this stratification is expressed less distinctly, and the transformed Atlantic waters are located at shallower depth.

In general, the stratification of water column and the circulation system display a clear seasonality and strongly depend on pressure field in the atmosphere (Levitan et al., 1999 and references therein). When anticyclone rotation in the atmosphere of the Arctic Basin is developed very weakly, waters of the Transpolar Drift, directed to the Fram Strait, occupy utmost northern position, and the northward St. Anna Current in the surface water occupies the water area of the St. Anna Trough. In this case the inflow of warm Atlantic waters is strengthened along the trough to the south. During periods of intensive development of anticyclone circulation, the area of the St. Anna Trough is occupied with the Arctic waters, the St. Anna Current is connected with them between Novaya Zemlya and Franz-Joseph Land, and Atlantic waters are only observed in the deepest part of the trough. The partially transformed Atlantic waters enter the St. Anna Trough from the southwest as well as from the Barents Sea in the intermediate layer.

During winter, shelf Arctic waters are formed to a considerable degree around polynyas in the shoal due to cooling and intensive convective mixing (Levitan et al., 1999 and references therein). They have a high salinity (about 35.0‰) and a low temperature (–1.9°C and below), resulting in a high density. Thus, water masses sink down and reach the basic inflows of the St. Anna Trough deep water regions. These waters probably serve as one of the main transportation mechanisms of sedimentary material from the shelf to the pelagic regions.

Furthermore, coastal erosion, under-ice currents, waves and surface flow, bottom tidal currents, slope processes, eolian supply, and vital activity of plankton should be mentioned as additional factors influencing the sedimentary material transport to the St. Anna Trough. To estimate quantitatively the role of these different processes is more or less impossible yet.

Results

The bed of trough is covered by terrigenous clayey and silty-clayey mud of brown color with insignificant admixture of sandy particles. Along the slopes the fine sediments are altered by coarser-grained variations, fine-grained silts dominate, and at some areas there are coarser silts, which contain substantially more sandy particles. The lithological composition of surface sediments on the "shoulders" of the trough differs in terms of considerably larger complexity and, first of all, is governed by hydrodynamics of sedimentation environment.

According to the results of grain-size analysis, there is a clear transverse zonality of the composition of recent sediments of the St. Anna Trough. On the slopes the content of gravel can reach 5%, the concentration of sandy particles varies from 10 to 40%, silt fraction composes 25–45% of the sediment, and clay fraction content varies from 20 to 40%. In the sediments of trough, the gravel, as a rule, is absent; the content of sand is usually lower than 10%, silt is about 20–30%, and clay is 50–70%.

Quartz, potassium feldspar, plagioclases, rock fragments prevail in the light subfraction of the fine sand. At the northern stations, numerous grains of biogenic carbonate occur. Sporadically spindle-shaped aggregates of clay minerals were found (see discussion below). The grains of quartz are characterized by the best roundness in the sediments from the central part of the trough compared to its slopes.

Based on mineral composition data of the fine sand fraction and quartz/feldspar (Q/FS) ratios in each sample, three main mineral associations were identified

(Fig. 10.4) (Levitan et al., 1999). The first is composed (in the ascending order) of feldspar, quartz and biogenic carbonates and an average Q/FS ratio of 1.6. This association occupies the central and deepest (>500 m water depth) part of the northern area of the trough. The second association is dominated by quartz and feldspar; the Q/FS ratio is less or equal 1.0. This association is clearly related to the eastern slope of the trough. Finally, the third association is mainly composed quartz and rock fragments in variable proportions, and there is almost no feldspar. The Q/FS ratio is equal or exceeds 5.0, moreover this ratio decreases from west to east. The examined association occupies not only the western slope, but also almost the entire area of the central part of the trough, with exception of its northern end.

Among the heavy minerals of the 63–125 mkm fraction, epidote, monocline pyroxene, black ore minerals (in essence, ilmenite), hornblende and garnet predominate. The iron hydro-oxides are sporadically developed. It is possible to isolate six mineral associations (Fig. 10.5). The distribution area of the first association (ore-epidote-clinopyroxene) occupies the middle part of the center of the trough and is elongated in the submeridional direction (Levitan et al., 1999). The ratios of the group of epidote (Ep) to the group of monocline pyroxene (ClPx) vary from 0.51 to 0.86. The distribution area of the second association, epidote-clinopyroxene-ore, borders the zone of the first association along its periphery; Ep/ClPx ratios vary between 0.51 and 0.89.

The distribution area of the third association (epidote-ore-clinopyroxene) is related to the western border and the "shoulder" of the trough, which adjoins Franz-Joseph Land. The Ep/ClPx ratio is about 0.5. The distribution area of the fourth association (ore-amphibole-epidote-clinopyroxene) is merged with the zone of the third association from the south; its continuation to the east is not clear due to the lack of data. The Ep/ClPx ratio is 0.73. The zone of the fifth association (clinopyroxene-epidote-ore) borders the northern end of the northern island of the Novaya Zemlya archipelago. The Ep/ClPx ratio is 1.18. Finally, the distribution area of the sixth association, garnet-epidote-clinopyroxene-ore, at some areas with a significant amount of iron oxides and iron hydroxides, is mainly located on the North-Kara Rise. The Ep/ClPx ratio varies between 0.61 in the north and 1.06 in the south.

In Fig. 10.6 a schematic distribution map of the clay mineral associations in surface sediments is shown from the St. Anna Trough area (Andrew and Kravitz, 1974; Wahsner et al., 1996; Levitan et al., 1999). It is evident from this map that there are six basic associations, composed of four main mineral groups: smectite, illite, kaolinite and chlorite. (1) A chlorite-illite association is observed as two spots on the western slope of the North-Kara Rise. (2) A smectite-chlorite-illite association adjoins it from the west and elongates in submeridional direction not only on the western slope of the North-Kara Rise, but also in the eastern part of the bottom of the St. Anna Trough. (3) A chlorite-smectite-illite association (characterized by increased amount of smectite) is located in the western part of the trough and on its western slope; in the north it rises to the base of the Franz-Joseph Land Archipelago. (4) A chlorite-kaolinite-illite-smectite association with significant amount of kaolinite and a dominance of smectite occurs in the south-western part of the study region. In the direction to northwest, to the islands of the

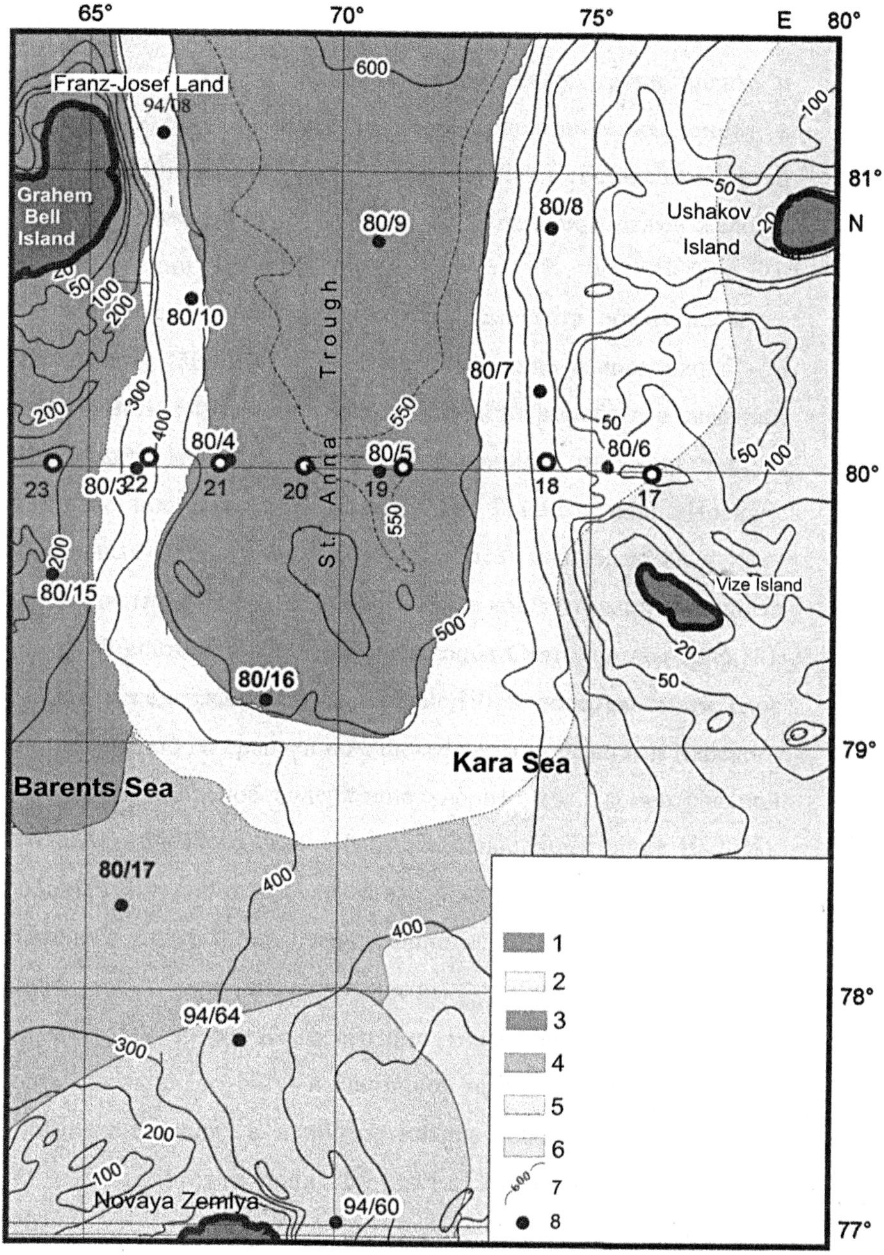

Fig. 10.5 Heavy-mineral assemblages in surface sediments from the St. Anna Trough (Levitan et al., 1999). 1 – black ore-epidote-clinopyroxene; 2 – epidote-clinopyroxene-black ore; 3 – epidote-black ore-clinopyroxene; 4 – black ore-amphibol-epidote-cliopyroxene; 5 – clinopyroxene-epidote-black ore; 6 – garnet-epidote-clinopyroxene-black ore; 7 – isobaths, m; 8 – geological stations

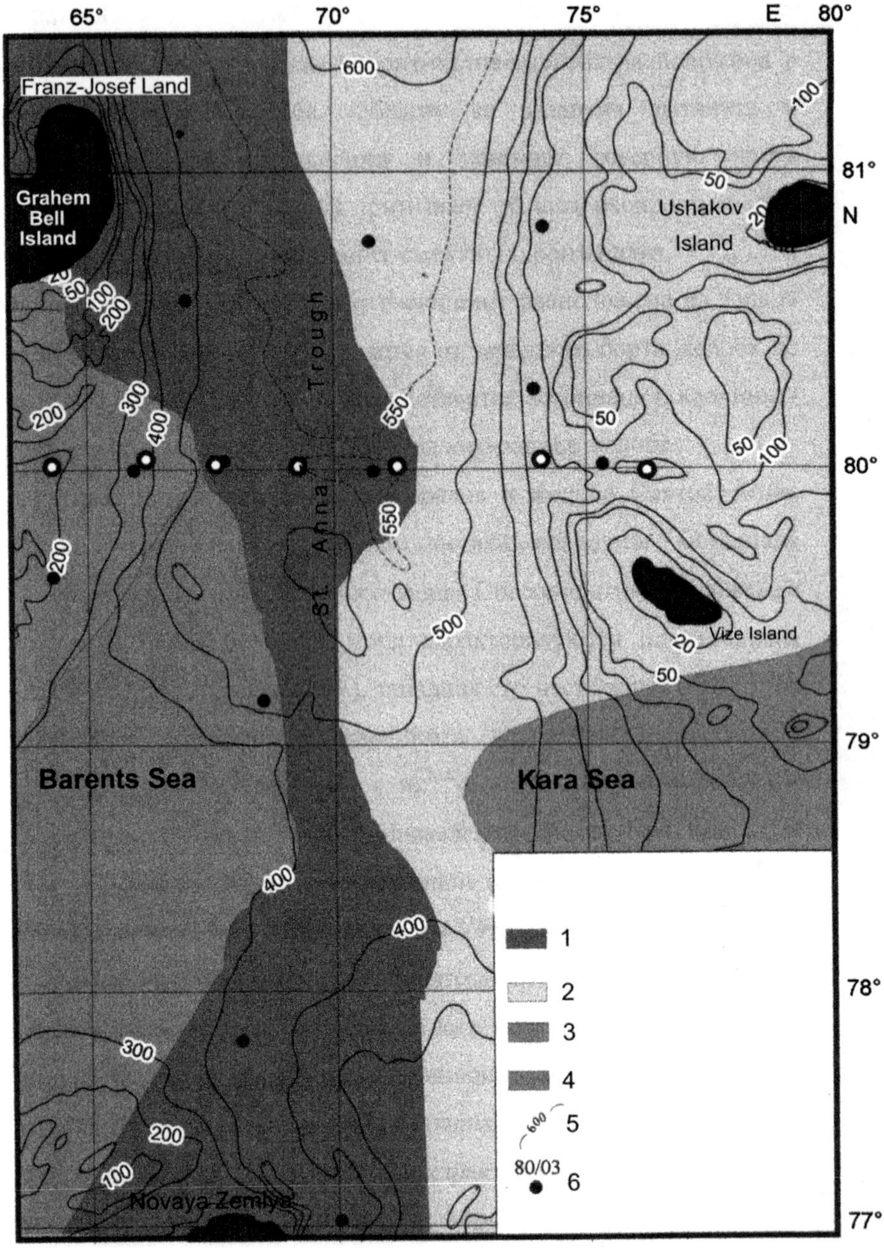

Fig. 10.6 Clay-mineral assemblages in surface sediments from the St. Anna Trough (Levitan et al., 1999). 1 – kaolinite-chlorite-illite-smectite; 2 – kaolinite-smectite-chlorite-illite; 3 – chlorite-illite-smectite-kaolinite; 4 – kaolinite-chlorite-smectite-illite; 5 – isobaths, m; 6 – geological stations

eastern part of Franz-Joseph Land, the associations consecutively changed to a (5) illite-kaolinite-smectite association and, further on, to a (6) illite-smectite-kaolinite association.

Discussion of Results

The facies zonality in surface sediments of the St. Anna Trough area was determined on the basis of lithological and grain-size composition data as well as mineral-composition data of fine sand, coarse silt and sub-colloidal (<2 mkm) fractions.

As for the fine-sand fraction, Franz-Joseph Land is the main source of quartz and rock fragments, whereas the North Kara Rise is the main source of feldspar. Enrichment of biogenic carbonate in the northern part of the trough is related to the action of Atlantic waters, which contributed to the development of the planktonic and benthic foraminifers. Supply of bivalve and barnacle detritus from the shoals of Franz-Joseph Land where they are a widely developed, plays a subordinate role. Enrichment of fine-sand fraction by rock fragments in the west and southwest of the studied region is likely to be connected with iceberg activity. Based on the position of the boundary between the second and third associations of the light minerals in surface sediments (Fig. 10.4), the (sub-) recent delivery of fine-sand material to the St. Anna Trough was probably more intensive from the Franz-Joseph Land rather than from the North Kara Rise.

The spindle-shaped aggregates of clay particles found in the 0.1–0.25 mm fraction of sediments of the straits between the islands of the Franz-Joseph Land are related to fecal pellets of zooplankton or zoobenthos. The formation of pellets is also described in old pack ice due to repeated cycles of melting and freezing (Pfirman et al., 1989). However, since old pack ice in the studied area does not exist, this ice hypothesis can be rejected. According to the pellet size and shape, their formation is most likely connected with the vital activity of macro-benthos.

Black ore minerals of the coarse silt fraction are known to increase towards the basic source provinces (Levitan et al., 1996). At the same time a specific mineral specialization is quite obvious as well: Novaya Zemlya serves as the basic source of epidote, Franz-Joseph Land associates with clinopyroxene supply, and the North Kara Rise delivers garnet and iron hydroxides. Hornblende is probably supplied from the northeastern part of the Barents Sea due to erosion of Mesozoic rocks, which are partially exposed on the surface of Admiralty Swell (Gurevich, 1995). An additional supply of clinopyroxenes by the Siberian Rivers into the studied area seems to be unlikely today, but may have been important during the last glacial and interglaciation cycles.

Based on the data by (Andrew and Kravitz, 1974) and (Wahsner et al., 1996) (Fig. 10.6), basic sources of chlorite and illite are the North Kara Rise and, to a certain degree, Novaya Zemlya. Kaolinite is supplied from the east of Franz-Joseph Land due to erosion processes of Triassic sandstones filled with kaolinite cement (Geological structure of the USSR and regularities of mineral deposits distribution, 1984).

The observed distribution of smectite is considerably difficult to explain. The continuous development of chlorite-illite association in the east and southeast of the studied area makes an active supply of smectite to the Kara Sea by the Siberian Rivers more unlikely (Levitan et al., 1996). It is sufficiently obvious that the area of increased abundance of smectite in the sediments between the kaolinite-dominated association in the west and the chlorite-illite association in the east can have been caused by weakening of the diluting influence of the corresponding clay-mineral supply in proportion to their removal from the source province. Furthermore, reduction in hydrodynamic activity in the center of the trough may play a substantial role, as indicated by the distribution of granulometric fractions.

There are three possible sources of smectite. First of all, it is the weathering crust of the Lower Cretaceous traprocks of Franz-Joseph Land, causing increased abundances of this mineral (>40%) in some areas in the straits between the islands (Nürnberg et al., 1995). Secondly, there are Upper Jurassic bituminous shales, also enriched by smectite (Elverhøi et al., 1995), which are outcropping in the region of the Admiralty Swell west of Novaya Zemlya (Gurevich, 1995). The Atlantic waters may act as the third source of smectite, which, according to the results of mapping the clay-mineral distribution in surface sediments of the eastern part of the Barents Sea (Nürnberg et al., 1995), are also enriched by this mineral. As noted earlier, the Atlantic waters penetrate St. Anna Trough, what is recorded by hydrological measurements. During the flow of waters to the east along the northern margin of Franz-Joseph Land an additional quantity of smectite could be delivered due to erosion processes of the weathering crust of Lower Cretaceous traprock. Finally, it is not possible to completely exclude the supply of some admixture of the "Siberian" smectite, although this source undoubtedly plays a minor role. According to Fig. 10.6, the bottom erosion of Upper Jurassic bituminous shales is apparently of important significance.

The investigations clearly showed that the mineral associations of recent sediments of the St. Anna Trough are mainly related to erosion processes of its surrounding rises: Franz-Joseph Land, Novaya Zemlya and North Kara Rise. The more remote sources, in particular, the ones located in the internal shelf delta areas of Pechora, Ob and Yenisei rivers, actually do not influence the composition of sediments of the trough.

Facies Zonality of Surface Sediments in the Eastern Kara Sea

Facies analysis of recent marine sediments (Murdmaa, 1987; and others) represents a basic tool for marine geologists studying lithology of bottom sediments. The facies structure of recent sediments in the eastern Kara Sea, strongly affected by Ob and Yenisei rivers, was already discussed in early publications by lithologists who studied this area back in 1930s–1960s (Belov and Lapina, 1961; Kulikov, 1961; Gorshkova, 1957; Kordikov, 1953; Ul', 1936). Traditionally, researchers paid a special attention to the facies analysis in many expeditions by VNIIOkeangeo, VNIIMorgeo, MMBI, MAGE, and AMIGE carried out in the 1960s–1980s. The data obtained have been summarized in the comprehensive work by V.I. Gurevich (1995). Facies indicators that were considered in all works listed above, mainly included sediment color, grain size distribution, mineral composition (heavy, light, and clay minerals), $CaCO_3$, TOC, opal, and heavy metal (Fe, Mn, and others) contents. We understand the facies concept of A. Gressly according to its description by Frolov (1995, p. 240): "Facies are distinguished *empirically* based on direct field observations ... according to *objective* lithological, paleontological, and other geological indicator*s*" (italics are marked by V. Frolov).

The next stage in the study of the facies structure of surface sediments from the eastern Kara Sea within a Russian-German cooperation with several joint expeditions of R/V *Polarstern* in 1995 and R/V *Academik Boris Petrov* in 1997, 1999, 2000, 2001, 2002, and 2003 (Matthiessen and Stepanets, 1998; Stein and Stepanets, 2000, 2001, 2002; Schoster and Levitan, 2003, 2004). The *Academik Boris Petrov* expeditions were carried out under the framework of the Russian-German SIRRO ("Siberian River Run-off") Project (see (Stein et al., 2003b) and numerous references therein). This stage was not only characterized by comprehensive investigations of sedimentary environment as a continuation of the work started by the P.P. Shirshov Institute of Oceanology during Cruise 49 of RV *Dmitry Mendeleev* in 1993 (Lisitzin and Vinogradov, 1995), but also by the study of many new facies indicators of bottom sediments, such as some biogenic remains (dinoflagellates, diatoms, and foraminifers), petrographic composition of the 63–125 mkm and >125 mkm fractions, sedimentation rates, and determination of a wide range of chemical elements using contemporary high-precision methods (X-ray fluorescence and instrumental neutron activation). Addition of new geological stations and geophysical profiles, as well as expansion of the study area relative to Cruise 49 of RV *Dmitry Mendeleev*, allowed us to identify a wider set of facies zones and subzones in the facies system of the studied area (Levitan, 2001, 2002). Based on own data and available publications (Kulikov, 1961; Levitan et al., 1994; Gurevich, 1995), we identified facies composition using a wide variety of indicators. The facies zonation of recent sedimentation in the study area reflects interaction in the land-sea system and includes successive transitions from fluvial to prodeltaic (estuaries of large rivers) and marine shelf facies.

Materials and Methods

Figure 10.7 and Table 10.4 shows the locations of geological stations obtained during cruises of RVs *Dmitry Mendeleev, Polarstern,* and *Academik Boris Petrov.* The study region (68°40′–78° N, 72°15′–92° E) includes the Ob and Yenisei downstream beds, the Taz riverbed (right tributary of the Ob River), the Ob Estuary, the Yenisei Bay, the Ob-Yenisei Shoal, and southern parts of the St. Anna and Voronin troughs. Three *Polarstern* stations are located east of this region near the Severnaya Zemlya archipelago (two stations in the Vilkitsky Strait and one station in the Voronin Trough).

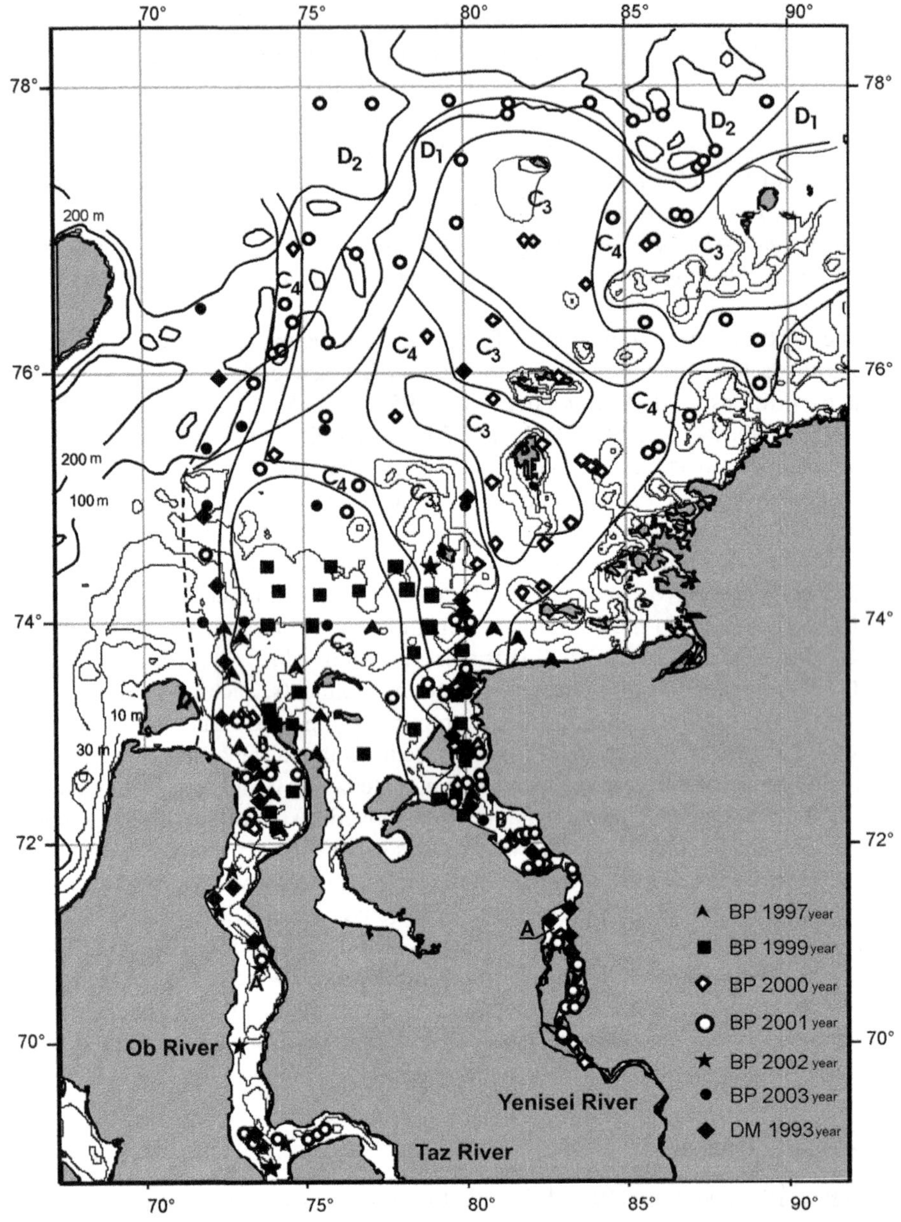

Fig. 10.7 Facies zonality and location of geological stations in the East Kara Sea (Levitan et al., 2005a). Cruises: BP – RV *Academik Boris Petrov*; DM – RV *Dmitry Mendeleev*. A–D – facies codes

In order to study recent sediments, we sampled the surface layer (0–1 cm) from box cores (in some cases, from Ocean–0.25 grab samples). Lithology of surface sediments collected during Cruise 49 of RV *Dmitry Mendeleev* is described by M.A. Levitan et al. (1994, 1996). Sediments obtained during cruises of RV *Polarstern* and RV *Academik Boris Petrov* were described in the following publications (Matthiessen and Stepanets, 1998; Rachor, 1997; Stein and Stepanets, 2000, 2001, 2002; Schoster and Levitan, 2003, 2004).

Grain-size analyses were performed in the Analytical Laboratory of the Shirshov Institute of Oceanology and the Alfred Wegener Institute. The results obtained were partially used in (Levitan et al., 1996; Müller and Stein, 1999; Levitan, 2001; Stein et al., 2004). In this paragraph we used the grain-size classification proposed in (Frolov, 1995).

The composition of the >63 mkm fraction is described in (Levitan, 2002; Levitan and Krupskaya, 2003). The composition of heavy minerals analyzed in the Institute of Oceanology is presented in several publications (Levitan et al., 1996, 1999; Levitan, 2001; Bourtman and Levitan, 2002). The composition of light minerals also analyzed at the Shirshov Institute of Oceanology, is given in (Levitan et al., 1998, and Levitan, 2001). Clay minerals studied at the Alfred Wegener Institute and the V.I. Vernadsky Institute of Geochemistry and Analytical Chemistry are described in (Shelekhova, 1998; Müller and Stein, 1999; Steinke, 2002; Krupskaya and Levitan, 2003; and Stein et al., 2004). Shelekhova's data of clay minerals in the <1 mkm fraction were recalculated for fraction <2 mkm using empirical coefficients given in (Shelekhova, 1998).

In order to study distribution patterns of characteristic chemical elements and compounds in surface sediments, we used results of routine wet chemical analyses (Kordikov, 1953; Nürnberg, 1996) and X-ray fluorescence analysis carried out in the German and Russian shore-based laboratories (Schoster and Stein, 1999), our unpublished results obtained in Vernadsky Institute of Geochemistry and Analytical Chemistry (I.A. Roshchina, supervisor), and neutron activation analyses (Levitan et al., 2002). Sea-water salts were not removed from the samples prepared for analyses.

Data on the distribution of palynomorphs (Matthiessen, 1999), diatoms (Polyakova, 2003), and benthic foraminifers (Khusid and Korsun, 1996), as well as calculation of sedimentation rates based on radiochemical methods, especially ^{137}Cs (Stepanets et al., 1999, 2000, 2003), are further very important for the facies analysis.

Facies Environments

We distinguish the following modern facies environments in the study region (Fig. 10.7): (A) the river environment (the Taz, Ob, and Yenisei downstream beds); (B) the estuarine environment (the Ob and Yenisei estuaries); (C) the inner shelf (the Ob-Yenisei Shoal); and (D) the outer shelf (the Voronin and St. Anna troughs).

The river environment is characterized by the following features: fresh river water (carefully controlled by the Niel Brown probe and instrumental salinity measurements from surface to bottom); several kilometers wide valleys (up to several tens of kilometers wide in the southern part of the Ob estuary); water depth up to 8–10 m; high current velocity (up to several decimeters per second); and rough bottom topography with numerous ledges of bedrock outcrops. The riverbed alluvium is virtually absent. As a rule, the thickness of loose sediments does not exceed several decimeters, and it reaches 2–3 m only at the confluence of Taz and Ob rivers. The river environment is characterized by strong seasonal variations in both sediment delivery and freshwater discharge into the Kara Sea: their major portions are related to the May–September period (Gordeev et al., 1996). The rivers are ice-covered during most part of the year.

Differences in geology of drainage areas significantly affect the sediment composition. The Ob and left Yenisei tributaries drain a huge region of the West Siberian Plate covered by Neogene–Quaternary friable clastic sediments whereas the right Yenisei tributaries erode the rock of the old East Siberian Craton (Levitan et al., 2005a). In the downstream area, erosion products of Permian–Triassic trap rock from the Putoran Plateau play an important role. The Taz River flows from the east to west across the Tydan watershed before its confluence with the Ob, and drains loose Quaternary sediments.

The estuaries of the Yenisei River (Yenisei Gulf) and Ob River (northern part of the Ob Bay) are traced in this work from the southernmost appearance of sea water to the strong hydrological front that marks the northern boundary of the mixing zone (Fig. 10.7). The latter is limited by the surface water salinity of 20‰ in both cases (Burenkov and Vasil'kov, 1995). Both estuaries are several tens of kilometers wide. The Ob estuary is up to 30 m deep and the Yenisei estuary is up to 37 m deep. The simplified water mass structure may be represented as two wedges. Salinity of the upper (relatively freshwater) wedge increases northward, whereas salinity of the underlying seawater wedge decreases southward. Current velocities in the estuaries are considerably lower than those in the downstream rivers (commonly, no more than a few centimeters per second). Strong temporal variations in current direction provoke an unstable position of southern boundaries of the estuarine environment in the water column. The bottom is commonly flat. Thickness of soft Holocene prodeltaic sediments may reach 20–30 m (Gurevich, 1995; Dittmers et al., 2003; Stein et al., 2003a, 2004). Narrow deep valleys or channels occur near estuarine coasts that serve (or served in the geological past) as transport routes for water and suspended matter flows. The thickness of Holocene sediments decreases sharply to a couple of meters near the coast (Dittmers et al., 2003).

Rivers serve as the main sources of sedimentary material. They deliver suspended matter, terrestrial plant (wood) remains, and freshwater flora. Coastal abrasion represents an additional source of sedimentary material. Ice-rafted material plays a minor role. Activity of brackish-water and marine fauna and flora, which becomes more important northward, is reflected in sediments. In the southernmost part of the estuarine environment, we can identify the proximal zone (B1) represented by the hydrodynamic barrier of the interaction between high-velocity river water flow and relatively slow estuarine currents. As a result of such interaction, increased amounts of rather coarse-grained sedimentary material are settled here. Sedimentation related to physicochemical and biological processes within the water column (Lisitsin, 1994) dominate in the northern distal zone (B2). When tracing the distal zone from the south to north, the precipitation of pelitic particles (due to change in the charge of clay minerals) gives way to the coagulation of colloids and the flocculation of dissolved Fe and organic matter (OM). Natural sorbents (supplied from rivers and newly formed) absorb significant amounts of elements dissolved in the seawater. The role of phytoplankton and bacterioplankton production progressively increases further to the north (Vedernikov et al., 1994). Up to 90% of the suspended matter discharged by rivers settles within the fresh water and seawater mixing zones ("marginal filter") (Lisitsin, 1994). First of all, this concerns silt- and sand-size particles.

The inner shelf environment involves a wide area of the Ob-Yenisei Shoal (down to a depth of 50–60 m) southward of the inner shelf edge. The shoal represents a flat (gently northward-sloping) submarine plain (C3). Its topography is locally complicated by island archipelagoes (Izvestiya TsIK, Academia Nauk, and others), shelf depressions, and submerged hydrographic network of the Ob and Yenisei rivers (C4). Bottom topography of the eastern inner shelf is rather poorly studied owing to hard ice conditions. However, it may be suggested that some small ridges consist of Pleistocene moraine deposits (Stein et al., 2002). Depth amplitudes commonly vary within several meters.

The water column is distinctly stratified (Burenkov and Vasilkov, 1994). In summer, the surface water mass is represented by relatively warm and freshened Kara Sea waters (their salinity increases northward up to 34.5‰). The intermediate water mass is formed by the mixing of surface and deep waters. The deep water is characterized by higher salinity and low temperature. Its genesis is related to cooling of the surface water during the autumn–winter formation of sea ice and the submergence of newly formed dense waters to lower levels. It is believed that northeastern directions of currents prevail (Moretzky, 1985) although the actual surface circulation patterns are complicated and highly variable. For example, the northwestern transport becomes very important in some periods (Burenkov and Vasilkov, 1994). Generally, velocities of surface and bottom currents are up to several centimeters per second. However, according to S.I. Muyakshin, the velocities may reach decimeters per second in latitudinal jet streams (Levitan et al., 1996).

The major part of the Ob-Yenisei Shoal represents an erosional or erosional-abrasional surface related to combined effects of recent uplift and erosion by bottom currents (Gurevich, 1995). Therefore, the thickness of soft sediments does not exceed a few decimeters in this area, or they are even missing in ferromanganese nodule fields, where bedrocks are exposed as Cretaceous (or Precambrian) sandstones and siltstones (Gurevich, 1995). Rare shelf depressions are commonly related to faults or their intersections (Explanatory note to tectonic map of the Barents and Kara Seas, 1998). They are filled with loose Quaternary sediments with a Holocene layer up to 5–8 m thick (Levitan et al., 2000). The system of submerged river valleys (channels) of various ranks (width up to 5–6 km) on the Ob-Yenisei Shoal is best described in (Musatov, 1989) and (Dittmers et al., 2008). In the downstream area, submerged Ob and Yenisei valleys cross the slope of the Ob-Yenisei Shoal and represent the outer shelf environment (Fig. 10.7). The channels are partially (in some cases, completely) filled with sediments. The thickness of Holocene deposits varies from several meters to 10 m (Dittmers and Schoster, 2004; Stein, 2001). Some channels represent sub-aqueous down cuttings barren of sediments (see (Dittmers et al., 2007) for synthesis).

The bottom erosion, which dominates in the major part of the Ob-Yenisei Shoal, promotes the washout and redeposition of the fine-grained clayey material in the submerged channels and holes that serve as natural sediment traps. The same channels, apparently, favor sediment transport in the nepheloid layer. In the inner shelf environment, a dominant proportion of the sedimentary material is supplied by bottom erosion of underlying sediments. The material of river discharge that overcomes the mixing zone also contributes to the sediment supply. It should be noted that erosion products of small islands and IRD serve as additional sources of material.

The outer shelf environment (Fig. 10.7) is confined to the transition from the Ob-Yenisei Shoal to deeper areas of the Voronin and St. Anna troughs (D1). This environment is also developed within the troughs (D2). The transition zone includes the northernmost part of the shoal near its edge, the steep slope, and the foothill. Both the slope and foothill are characterized by a rough paleotopography overlain by an almost continuous 3–7 m thick Holocene sediment cover. The depth difference between the shoal and foothill is 100–150 m. Environment D2 is characterized by a glacial topography of moraine deposits (Niessen and Dittmers, 2002) with decameter-scale rises that are locally overlain by fine soft sediments up to several decimeters in thickness. Considerably thicker young sediments are found in other areas of the troughs (Polyak et al., 1997; Rachor, 1997). The sedimentary material is derived from both the Ob-Yenisei Shoal and the local bottom erosion, whereas IRD plays a minor role.

Stratification of the water column in the outer shelf environment is similar to that on the inner shelf, but the warm intermediate Atlantic water appears in northern areas at a depth of 200–300 m (Polyak et al., 1997). The general hydrological situation and comparison with other regions allow us to assume a high hydrodynamic activity over the Ob-Yenisei Shoal edge and slope. However, direct current velocity measurements are absent. Surface-water salinity increases gradually northward reaching 34.7‰ at about 78° N.

Facies

River facies comprise riverbed alluvium and submarine deltaic sediments. They are characterized by gray or greenish gray color. Grain size of sediments varies from gravel and sand to silty mud depending on the hydrodynamic regime. The average grain size of river-sediment facies corresponds to sandy mud (Table 10.5). The southern part of the Yenisei Bay and the confluence of the Ob and Taz rivers contain slightly siliceous diatomaceous silty mud (Krupskaya and Levitan, 2003) containing more than 5% of opaline silica (Nürnberg, 1996).

The illite-smectite association of clay minerals with almost equal kaolinite and chlorite contents dominates in the <2 μm fraction of river sediments (Table 10.6). The comprehensive study of sediments in the Ob Bay showed that the Taz River is a local chlorite source. After its confluence with the Ob River, chlorite is gradually diluted by other clay minerals delivered by the Ob River (Krupskaya and Levitan, 2003). Sediments of the Yenisei River are enriched in smectite (to a lesser extent, in kaolinite) and depleted in illite relative sediments of the Ob River (Steinke, 2002; Stein et al., 2004).

Among light minerals of fraction 0.1–0.05 mm potassium feldspars and quartz dominate; Ratio quartz/feldspars is 3.3 (Table 10.7).

The association of epidote, black ore minerals, and clinopyroxene prevails among heavy minerals of this fraction. The clinopyroxene/epidote ratio is 1.7 (Table 10.8). The 63–125 mkm fraction and, especially, the >125 mkm fraction are strongly enriched in rock fragments and wood remains (Levitan, 2002). The sediments are characterized by high contents of SiO_2, Ba, Y, La, Ce, Nd, and Sm, but low contents of MnO, P_2O_5, Cr, and Eu (Table 10.9). The average TOC content varies within 1.0–1.5% and is less than 1.0% only in sandy sediments (Stein et al., 2004). The sediments contain abundant freshwater diatoms (Polyakova, 2003). Benthic foraminifers are absent (Khusid and Korsun, 1996).

River sediments are characterized by wide variations in their grain size, mineralogy, and chemistry. For example, the >125 mkm fraction at Station BP03-16 (southern part of the Yenisei Gulf) is totally composed of wood remains and muscovite flakes (Levitan et al., 2004b). Such a combination is not found in any of adjacent stations. Sediments taken at a distance of less than 1 km from each other are often represented by different lithological types.

It is undoubtedly interesting to compare specific features of sediments from each of the large rivers (Ob and Yenisei) in this region. According to our data, the following indicators are most typical for the material discharged by the Yenisei River: black ore minerals, wood remains in the sand fraction, high clinopyroxene/epidote ratio, high smectite/illite ratio, increased Ni/Al ratio and Fe content, and high magnetic susceptibility. Sediments of the Ob River are characterized by the absence of black ore minerals, high content of wood remains in the sand fraction, low smectite/illite ratio, high K/Al and Rb/Al values, and low magnetic susceptibility (Levitan, 2001; Levitan et al., 1996; Schoster and Stein, 1999; Stein et al., 2004).

Recent sediments are accumulating at a rate 350 cm/ky in the barrier zone of the Ob–Taz confluence area (Stepanets et al., 2003). However, sediment accumulation is not noted in the major part of the riverbed, because intense currents either wash out the fine fractions or erode the bottom of river valleys.

Sediments of estuaries (fresh and sea water mixing zones) demonstrate a specific composition. They are commonly represented by rather homogenous fine-grained olive gray or black clays (B2). These sediments are replaced by silty sandy mud only in the proximal (B1) facies (Table 10.5). In some cases, an admixture of

Table 10.5 Average grain-size composition (wt. %) of the recent facies [*Levitan et al.*, 2005a]

Facies	n	>10	10–7	7–5	5–3	3–2	2–1	1–0. 5	0.5–0.25	0.25–0.1	0.1–0.05	0.05–0.01	0.01–0.005	0.005–0.001	<0,001
A2	1	29.26	4.86	4.60	0.87	0.53	0.95	6.27	1.77	37.33	1.40	0.12	3.26	3.91	4.87
B1	8	0.00	0.00	0.00	0.00	0.19	0.32	1.42	10.71	15.08	11.89	2.42	8.24	17.85	31.88
B2	11	0.00	0.00	0.00	0.00	0.00	0.40	0.10	0.53	5.70	2.91	4.44	14.72	29.23	41.97
C3	13	0.00	0.00	0.00	0.00	0.00	0.43	0.13	1.19	19.17	12.47	3.26	10.14	20.76	32.46
C4	7	0.00	0.00	0.00	0.00	0.00	0.38	0.20	0.26	1.38	2.49	2.35	19.15	31.92	41.87
D1	1	0.00	0.00	0.00	0.00	0.00	0.00	0.35	0.15	1.24	2.28	0.45	14.13	30.69	50.71

Note: n – number of studied samples; grain size is given in mm.

Table 10.6 Average content of major clay minerals in different facies (fraction <2 mkm) (rel. %) [*Levitan et al.*, 2005a]

Facies	Illite	Smectite	Smectite/illite	Chlorite	Kaolinite
A (n=11)	24	46	1.9	16	14
B (n=30)	25	44	1.8	17	14
C3 (n=23)	22	47	2.1	16	15
C4 (n=16)	23	49	2.1	15	13
D1 (n=7)	30	44	1.5	16	10
D2 (n=4)	38	39	1.0	14	9

coarse-grained material is noted in the mud, reflecting an input of coastal abrasion products (Levitan et al., 2004). Sediments of the mixing zone virtually do not differ from riverbed deposits in terms of clay-mineral composition (Table 10.6). Light mineral associations of the 0.1–0.05 mm fraction in sediments of both proximal and distal mixing zones are almost identical. They differ from counterparts in alluvium only by somewhat lower quartz content and quartz/feldspar ratio. The K-feldspar content is very low in proximal sediments (Table 10.7). Against the similar background of dominant heavy-mineral associations, we observe noticeable differences between proximal and distal sediments: contents of clinopyroxene, hornblende, and garnet as well as the clinopyroxene/epidote ratio increase toward the northern boundary of the marginal filter, whereas proportions of epidote and black ore minerals decrease (Table 10.8). Wood remains and POLYCHAETA tubes prevail in the >125 mkm fraction (Levitan, 2002). The chemical composition of sediments in facies B2 is characterized by maximal concentrations of Al_2O_3, Fe_2O_3, TiO_2, MgO, CaO, Co, Sc, S, Eu, Tb, Yb, and Lu (Table 10.9). TOC content is commonly >1.5% (Stein et al., 2004), whereas the content of biogenic opaline silica ranges within 3–5% (Nürnberg, 1996). Freshwater algae, first of all diatoms, dominate among the biogenic remains. Benthic foraminifers are represented by the brackish-water *Elphidium clavatum–Haynesina orbiculare* assemblage (Khusid and Korsun, 1996).

Recent sedimentation rates are very high (440–1080 cm/ky) in depocenters of the mixing zone for both Yenisei and Ob River mouths (Stepanets et al., 1999, 2000). The combined analysis of data on sediments from the mixing zone (grain

Table 10.7 Average content of major light minerals in different facies (fraction 0.1–0.05 mm)(rel. %) [*Levitan et al.*, 2005a]

Facies	Quartz	Basic and intermediate plagioclases	Acid plagioclases	K-feldspars	Total of feldspars	Quartz/ feldspar ratio
A (n=8)	68.2	3.0	5.3	12.3	20.6	3.3
B1 (n=9)	55.4	6.8	8.4	5.8	21.0	2.6
B2 (n=13)	58.5	3.9	6.5	14.1	24.5	2.4
C3 (n=19)	65.8	2.5	3.9	18.3	24.7	2.7
C4 (n=12)	66.4	3.7	1.4	21.9	27.0	2.5
D1 (n=9)	59.0	0.5	7.1	13.7	21.3	2.8

Table 10.8 Average content of major heavy minerals in different facies (fraction 0.1–0.05 mm) (rel. %) [*Levitan et al.*, 2005a]

Facies	n	ClPx	Ep	Black ore minerals	HBl	Garnet	Ep/ClPx
A2	3	31.6	17.6	24.7	7.7	2.9	0.7
B1	9	37.2	13.0	25.7	8.2	3.0	0.4
B2	13	44.4	12.1	14.3	8.9	3.5	0.3
C3	19	31.0	14.5	21.7	8.3	6.0	0.6
C4	12	36.2	14.8	18.8	5.8	5.5	0.6
D1	9	21.9	23.5	19.8	5.1	8.0	1.1
D2	4	21.7	26.8	17.8	7.0	7.6	1.2

Note: n – number of studied samples; ClPx – clinopyroxene; Ep – epidote; HBl – hornblende

size, clay minerals, and chemical composition) and sedimentation rates lead us to the conclusion that the major mechanism of sedimentation in the estuarine facies was represented by the precipitation of huge amounts of clay minerals from suspension due to mixing of fresh and seawaters.

Bottom sediments of the inner shelf (Ob-Yenisei Shoal) differ considerably in terms of facies depending on the depositional environment. One can divide these sediments into the following two facies: (C3) areas between the valleys, i.e., buried plains, and (C4) valleys of paleo-rivers and shelf depressions.

Facies C3 is represented by relatively coarse-grained clastic sediments (various sands or muddy sands) and clastic-clayey sediments (sandy and sandy silty mud). In general, the grain-size composition corresponds to silty sandy clays (Table 10.5). The association of clay minerals in sediments of this facies zone virtually does not differ from that in sediments of the mixing zone (Table 10.6). However, the illite content increases and the smectite content decreases northward within the inner shelf (Shelekhova, 1998). The illite and kaolinite contents are higher, while the smectite content is much lower in the coastal zone off the Taimyr Peninsula (Müller and Stein, 1999). The quartz content and quartz/feldspar ratio noticeably increase in this facies zone (Table 10.7). We have to note in this context that typomorphic features of quartz from sediments of the Ob-Yenisei Shoal considerably differ from those in the facies described earlier (Levitan et al., 1998). In sediments of facies C3, quartz is derived from underlying Mesozoic bedrocks, whereas quartz transported by rivers from remote sources dominates in alluvium and mixing zone sediments. Sediments of this facies are relatively enriched in heavy minerals, such as black ore minerals, common hornblende, and garnet. However, the clinopyroxene concentration and clinopyroxene/epidote ratio are relatively lower (Table 10.8). Heavy mineral associations distinguished in the Ob-Yenisei Shoal (Levitan et al., 1999a, 1999b) reflect both local variations in provenance and fractionation of heavy minerals during their transport.

The >125 mkm fraction contains relatively few rock fragments. *POLYCHAETA* tubes predominate among biogenic remains off the Ob River mouth, whereas agglutinated benthic foraminifers are most abundant in the area off the Taimyr coast, and bivalve shells prevail in the major area of facies C3 (Levitan, 2002). The chemical composition of sediments is characterized by

Table 10.9 Average chemical composition of different facies [*Levitan et al.*, 2005a]

Facies	SiO_2	Al_2O_3	Fe_2O_3	TiO_2	MnO	MgO	CaO	Na_2O	K_2O	P_2O_5	Ba	Co	Cr	Ni	Sc	Sr	V	Y	Zn	Zr	S	Cu	Pb	Rb	La	Ce	Nd	Sm	Eu	Tb	Yb	Lu
A	74.97	13.7	7.43	0.87	0.19	2.17	2.08	2.32	2.52	0.36	650	21	111	50	15	210	140	27	93	310	0.04	120	85	77	42	72	26	6	1.25	1.14	4	0.57
	59.9	9.89	3.65	0.7	0.06	1.15	0.93	0.09	2.06	0.08	424	13	48	33	10	138	95	20	40	167	0.03	120	16	70	22	38	15	4	1.00	0.73	2	0.33
	68.01	11.94	5.8	0.79	0.18	1.53	1.33	1.52	2.26	0.2	508	17	65	42	12	173	118	25	70	235	0.04	120	53	74	33	58	22	5	1.00	0.95	3	0.43
	8	8	8	8	8	8	8	8	8	8	6	6	5	5	3	6	5	4	6	6	2	1	4	2	3	3	3	3	3	3	3	3
B2	63.52	14.68	9.97	1.15	1	4.62	3.92	7.67	2.68	0.5	550	39	123	90	24	450	196	28	108	201	0.23	180	73	101	34	64	27	7	2.02	1.20	3	0.54
	43.7	10.1	6.41	0.57	0.05	1.87	0.84	2.54	1.7	0.18	21	16	63	21	13	115	111	23	40	24	0.21	14	12	58	22	41	18	4	0.78	0.81	2	0.37
	53.73	12.77	8.05	0.82	0.42	2.96	1.95	4.26	2.04	0.28	364	24	75	46	18	214	153	25	80	141	0.22	53	25	79	27	49	21	5	1.51	0.97	3	0.46
	32	32	32	32	32	32	32	32	32	32	31	31	30	28	13	31	24	23	31	31	2	12	20	17	9	9	9	9	9	9	9	9
C3	83.5	13.57	9.87	0.88	0.89	3.4	2.69	6.28	2.83	0.33	790	27	124	330	21	765	200	31	90	450	0.26	130	93	79	30	55	23	6	1.56	1.00	4	0.60
	49.15	5.8	2.16	0.45	0.02	0.59	0.74	1.7	1.58	0.08	318	3	37	14	6	140	55	12	19	80	0.12	11	11	68	3	8	4	1	0.16	0.29	1	0.26
	67.27	10.12	5.05	0.66	0.34	1.83	1.51	3.32	2.18	0.18	516	13	52	59	12	243	109	21	46	236	0.14	38	43	80	29	43	17	4	1.12	0.80	2	0.40
	26	26	26	26	26	26	26	26	26	26	23	23	24	21	19	23	15	16	23	23	5	5	9	3	13	13	13	13	13	13	13	13
C4	78.53	16.39	11.75	0.87	1.58	3.35	2.59	6.54	3.22	0.61	735	34	112	350	20	335	211	28	150	412	0.27	300	79	93	30	59	25	7	2.52	1.13	4	0.65
	47	10.18	3.54	0.36	0.06	1.33	0.99	1.42	1.75	0.08	300	14	61	32	10	120	107	19	50	55	0.15	19	13	82	21	37	14	3	0.60	0.67	2	0.33
	56.16	12.51	7.61	0.7	0.59	2.51	1.43	4.27	2.38	0.25	400	23	72	74	15	203	176	24	85	169	0.22	109	45	83	28	52	21	5	1.37	0.96	3	0.45
	33	33	33	33	33	33	33	33	33	32	25	25	23	23	17	25	19	18	25	25	8	12	16	8	15	15	15	15	15	15	15	15
D1	76.9	13.9	9.36	0.76	2.63	3	1.65	7.85	2.5	2.42	594	32	87	53	14	230	211	27	85	184	0.24	190	91	–	53	95	34	8	1.53	1.23	4	0.57
	46.3	7.5	2.5	0.38	0.19	0.75	0.89	2.41	1.97	0.12	316	11	31	19	5	110	73	12	24	100	0.03	110	65	–	19	35	14	3	0.82	0.66	1	0.17
	57.17	11.32	6.94	0.62	1.17	2.24	1.23	4.66	2.29	0.42	466	21	40	40	10	182	158	19	59	152	0.17	162	74	–	30	54	21	5	1.21	0.90	2	0.40
	11	11	11	11	11	11	11	11	11	11	9	9	8	9	5	9	9	6	9	9	9	5	9	–	10	10	10	10	10	10	10	10
D2	61.92	14.97	8.36	0.88	2.35	3.25	1.73	6.23	2.86	0.29	546	36	92	60	15	201	226	25	94	200	0.21	–	86	–	35	63	22	5	1.50	0.92	2	0.41
	50.9	10.7	5.2	0.58	0.49	1.99	0.9	3.39	2.14	0.15	428	17	73	40	10	161	156	19	60	156	0.15	–	79	–	27	47	19	4	0.91	0.80	2	0.36
	55.84	12.69	6.76	0.71	1.13	2.46	1.34	4.83	2.4	0.21	484	23	84	49	12	189	188	22	75	180	0.18	–	81	–	30	54	21	5	1.19	0.87	2	0.38
	9	9	9	9	8	9	9	9	9	9	6	6	6	6	6	6	6	66	6	6	4	–	3	–	3	3	3	3	3	3	3	3

Note: Maximal, minimal, average values and number of samples are shown from top to bottom for every facies; oxides and sulphur are given in %, the rest of elements – in ppm; blank – no data.

very high SiO_2 contents and maximal concentrations of Ba, Sr, and Zr (Table 10.9), because the sediments have mainly siliciclastic composition dominated by quartz and feldspars. The TOC content is commonly <1.0% (Stein et al., 2004), while the concentration of biogenic opaline silica is negligible (Nürnberg, 1996). The concentration of freshwater diatoms rapidly decreases from the south to <20% in the northern area of facies C3 (Polyakova, 2003), while the dinoflagellate concentration increases in the same direction (Matthiessen, 1999). The benthic foraminifer fauna is dominated by the marine calcareous *Elphidium clavatum–Cassidulina reniforme* assemblage (Khusid and Korsun, 1996).

Facies C3 is almost universally represented by condensed sections. Therefore, it is impossible to use short-lived isotopes (^{137}Cs, ^{210}Pb, and others) for the determination of sedimentation rates. In general, one can suggest that the average sedimentation rate does not exceed a few centimeters per thousand years. Two ferromanganese nodule fields have been outlined in the northeastern and southwestern areas of facies C3 (Gurevich, 1995).

Facies C4 is represented by fine-grained clayey mud of olive gray or brownish gray color (Table 10.5). As noted above, the composition of clay minerals in these sediments is almost similar that in facies C3. This is quite natural, because clay minerals in both facies are mainly transported as suspended matter in a common surface water mass widespread over the area, and the minerals settle onto the bottom from the same water layer. The smectite content is slightly higher in the profile off the Yenisei River mouth as compared to that in the Ob River profile (Steinke, 2002; Stein et al., 2004). It is interesting that associations of light minerals of the 0.1– 0.05 mm fraction are also similar in both facies C3 and C4, except for a slightly higher content of sodium plagioclase and K-feldspar in facies C3 and C4, respectively (Table 10.7). In facies C4, the heavy-mineral association is characterized by very higher values of the clinopyroxene content and clinopyroxene/epidote ratio, relative to facies C3 (Table 10.8), possibly reflecting a certain contribution of riverine material. Sediments of facies C4 have a high magnetic susceptibility (Dittmers et al., 2003; Stein et al., 2004), which makes it possible to trace them in buried channels of the old Yenisei River. This fact also supports our suggestion. The content of the >125 mkm fraction is much lower than that in facies C3, but its composition is almost identical in both facies zones (Levitan, 2003).

The chemical composition of facies C4 is characterized by very high concentrations of Ni, Zn, and S, and relatively high contents of Al_2O_3, Fe_2O_3, MgO, K_2O, Co, Sc, V, Sm, Eu, Tb, Yb, and Lu (Table 10.9). This type of composition reflects very high contents of pelitic fractions (especially, clay minerals) and enrichment in iron oxides. The TOC content commonly exceeds 1.5% (Stein et al., 2004). The content of biogenic opaline silica is negligible (Nürnberg, 1996). Virtually the same assemblages of marine fauna are developed in both facies C4 and C3. According to radiocarbon datings (Levitan et al., 2000, 2004; Stein et al., 2003a, 2004), sedimentation rate in the inner shelf amounts to tens of centimeters per thousand years.

Bottom sediments of the outer shelf are represented by brown and dark brown very fine-grained clayey mud (Table 10.5). In some cases, it is slightly enriched in sand and silt on topographic highs. The sediments differ considerably from those of the inner shelf in their clay-mineral composition. Illite and smectite contents are almost equal (facies D2), and chlorite slightly prevails over kaolinite (Steinke, 2002; Stein et al., 2004). According to their quartz and feldspars contents, sediments of facies D1 are similar to those from the mixing zone depocenter (Table 10.7). Among heavy minerals, we observed a considerable decrease in the clinopyroxene content and clinopyroxene/epidote ratio, whereas the proportion of epidote and garnet increases, as compared to the inner shelf (Table 10.8). Differences between facies D1 and D2 in the heavy mineral composition are negligible. We noted an increased content of black ore minerals and rock fragments in the >125 mkm fraction without the consideration of biogenic material (Levitan and Krupskaya, 2003). In this fraction, *POLYCHAETA* tubes prevail among the biogenic remains in the eastern part of the facies zone, and agglutinated foraminifers dominate in the west (Levitan, 2002).

The chemical composition of sediments from the outer shelf is characterized by maximal contents of MnO_2, Na_2O, K_2O, P_2O_5, Cr, V, Cu, and Pb, whereas the REE concentration is rather low. The SiO_2/Al_2O_3 ratio is considerably lower in facies D1 as compared to facies D2. Differences for other elements and ratios are less significant. The TOC content commonly ranges within 1.0–1.5% (Stein et al., 2004). Content of biogenic opaline silica does not exceed 1.0% (Nürnberg, 1996). Marine diatoms and dinoflagellates dominate among the microflora remains (Matthiessen, 1999; Polyakova, 2003). Benthic foraminifers are mainly represented by relatively deep-water agglutinated marine species of the *Cristobrostomoides subglomosum–Tritaxis nana* assemblage (Khusid and Korsun, 1996). Based on radiocarbon datings (Polyak et al., 2002), facies D1 is characterized by sedimentation rates of several decimeters per thousand years. Sedimentation rates show much wider variations in facies D2 (Polyak et al., 1997).

Thus, outer shelf sediments differ considerably from inner shelf ones in mineral, chemical, and micropaleontological compositions. Some facies indicators point to a contribution of riverine and estuarine materials to sediments of the outer shelf. We believe that such a conclusion only apparently seems to be paradoxical. Taking into account the genesis and age of geological bodies exposed at the bottom surface, as well as the bottom topography, we may assume that remains of Late Pleistocene deltas and/or estuaries possibly occur here (Stein et al., 2004) and they contribute to the formation of recent sediments. Probably, the currently available data are insufficient for a more detailed subdivision of the outer shelf facies.

Processes of mineralogical and geochemical fractionation play an important role in the formation of the recent facies structure. Using data on the distribution pattern of heavy minerals and the chemical composition of sediments, the following conclusions are possible.

The Q-mode factor analysis of heavy minerals showed that hydrodynamic activity in the nepheloid layer, including the bottom erosion and transport in the bottom water by cold brines, is the main process controlling their distribution. The second process is related to the transport of floatable minerals, such as mica and chlorite, in the surface-water mass. Additional two processes are related to calm hydrodynamic conditions. We can suggest that one process reflects "biotransport mechanism" of heavy minerals for settling onto the bottom, whereas glacial (IRD) sedimentation is another process (Levitan et al., 1999a). The same statistical method allowed us to distinguish the following seven heavy mineral associations (Levitan et al., 1999b): (1) iron oxides/hydroxides + garnet + rutile + staurolite + tourmaline + Cr-spinel (highly resistant association from metamorphic rocks typical of sediments in the zone of intense reworking by bottom currents); (2) clinopyroxene + chloritoid (erosion products of Siberian trap rock mainly delivered by the Yenisei River into the Kara Sea); (3) orthopyroxene + anatase + spinel + andalusite (similar to the first association, but less resistant); (4) black ore minerals (concentrated in coastal areas or sites of intense bottom erosion due to high density); (5) biotite + chlorite (transported over long distances by surface currents); (6) leucoxene + epidote + apatite + sphene + disthene (derived from seafloor bedrocks represented by slightly metamorphosed granitoids and/or sedimentary rocks); (7) magnetite + common hornblende + actinolite/tremolite + glaucophane (erosion products of amphibolite-facies metamorphic rocks and, partially, of rocks metamorphosed at low temperature and high pressure).

Using X-ray fluorescence data (61 samples), we calculated the correlation between 20 elements at the confidence level of 0.95. The following associations of elements showed significant (0.60 or more) positive correlations: (1) Al–Ti–Mg–Fe–Co–Ni–Y–Zn–Zr (this association based on clay minerals, iron oxides/hydroxides, and absorbed trace elements, is typical of fine-grained sediments); (2) Si–Ba–Zr (association of quartz, feldspars, and zirconium minerals typical for sand and silt or for the clastic component of clastic–clayey sediments); (3) Mn–Co–V (the early diagenetic association with Co and V absorbed on manganese hydroxides); (4) Ca–Sr (biogenic association based on the isomorphic replacement of Ca by Sr in the biogenic calcite and aragonite related to foraminifer tests and bivalve and ostracod shells); (5) Fe–Sc–P (Sc is absorbed by Fe and incorporated into its compounds; phosphorus together with organic matter is concentrated in the same fine-grained Fe-rich sediments); and (6) Mg–Na (the most mobile elements concentrated in outer shelf sediments characterized by the highest degree of fractionation). Obviously, the sediments include some series of elements, in which each element is associated with several carrier-minerals. For example, cobalt correlates with Al, Fe, Mn, Mg, and P, related to different mineral phases, but the strongest correlation ($r = 0.82$) is registered with Fe.

Facies variability of recent sediments in the eastern Kara Sea is related to the superposition of many processes and factors. We believe that sediment sources, bottom topography, and the hydrological structure of each facies zone are among the leading factors. Processes of fractionation of sedimentary material during its transport and mixing in the course of sediment accumulation mainly control the facies characteristics.

Sediments of the river facies are mainly derived from drainage areas. The Ob and Yenisei drainage areas differ considerably in their geology. The relief of river valleys and generally high hydrodynamic level of sedimentation environment contribute to the grain-size distribution of river sediments. The river environment determines the development of freshwater phytoplankton and the absence of benthic foraminifers. Initial fractionation stages (including gravity, grain size, mineralogical, and geochemical fractionation) are responsible for the immaturity of clastic material, high contents of rock fragments and black ore minerals in the heavy fraction, the abundance of wood remains. Mixing of the pure riverine material with products of coastal abrasion and bottom erosion plays an additional role.

Sediments of the estuarine facies are essentially derived from the same sources that supply the river facies. However, a dramatic change in the sedimentation environment owing to the mixing of fresh water with sea water severely changed the facies characteristics of estuarine sediments. Intense ("avalanche") precipitation from suspension into sediment due to coagulation of fine fraction caused a major change in clay minerals. The transition of dissolved organic carbon and dissolved Fe into the colloidal or suspended form promotes the flocculation and transfer of corresponding compounds to bottom sediments. These processes are accompanied by the absorption of considerable amounts of trace elements from seawater. Marine planktonic microfossils and brackish-water calcareous benthic foraminifers appear in the sediments. Fractionation of the sedimentary material is still poorly developed. Coastal abrasion products also contribute to the sediment budget.

Recent sediments of the inner shelf facies are primarily derived from bedrocks and older sediments within the sedimentary basin, including those exposed on islands. Partially, the terrigenous material is also derived from river discharge that could escape the river water-seawater mixing zone (Stein et al., 2004), eolian input (Shevchenko, 2000; Shevchenko and Lisitsin, 2004), and IRD input. Intense bottom currents promoted the deposition of clastic and clastic-clayey sediments in areas between submerged river valleys of the Late Pleistocene-Early Holocene. It is possible that a part of these sediments represents relict or residual deposits. They contain material of various degrees of maturity. Mature well-sorted quartz sand occurs at some sites. The finest fractions are washed out, redeposited in the valleys, and supplemented with additional portions of pelite from estuaries that could escape the marginal filter barrier. Normal seawater salinity combined with shallow depths favor the dominance of marine phytoplankton and shallow-water calcareous benthic foraminifers among microfossil assemblages within this facies zone.

Sediments of the outer shelf facies are mainly related to erosion of ambient land and seafloor (Levitan et al., 1999a, 1999b). Eolian dust and IRD play an insignificant role. The significance of river discharge is also subordinate. Old fluvial sediments exposed on slopes of the Ob-Yenisei Shoal, rather than present-day river sediments, are more important in the outer shelf facies. Since the outer shelf facies of the Kara Sea is located in a distal zone relative to the major provinces (coast of the West Siberian Plain and Taimyr, as well as Ob and Yenisei River mouths), sedimentation rates were relatively low and redox potential was high in the upper layer of the sediments. Because of these processes and relatively high geochemical fractionation, sediments of the outer shelf facies are enriched in mobile elements, such as Mn, P, S, Cu, and others. Attenuation of the influence of modern riverine materials and growth of primary plankton production increased the proportion of marine organic matter in the total organic-carbon fraction, although terrestrial organic carbon remains still predominant (Fahl et al., 2003). Marine algae and relatively deep-water agglutinated benthic foraminifers predominate among biogenic remains.

In summary, we have demonstrated that recent lithofacies of the study region are controlled by several relatively independent processes related to changes in the hydrodynamic and hydrochemical structure of the water column, water depth, distance from the main sources of sediments, provenance, etc. These processes affect the fractionation and mixing of major components in the recent sediments. Data on facies environment and composition obtained from surface sediments (see above) may also give important information for interpreting ancient sediment data in terms of paleoenvironmental changes on land and in the oceans.

History of Sedimentation

Late- and Post-Glacial History of Sedimentation in the Eastern Part of the Barents Sea

There are several points of view among scientists regarding the development of Late Valday (Late Weichselian) glaciation east of the Barents Sea. There are "marinists" (Gritzenko et al., 1989), "moderate glacialists" (Pavlidis et al., 1990), and "glacialists" (Epstein, 1985; Gataullin and Polyak, 1992; Gataullin et al., 2001; Polyak et al., 1995). The "marinists" assume that a large LGM Barents Sea Ice Sheet never existed, and all diamictons are marine deposits. "Moderate glacialists" consider that several separated small glaciation centers related to the archipelagoes of islands and underwater rises, occurred east of the Barents Sea during the LGM. Above underwater depressions of different genesis there may have been floating ice shelves, which have not reached the seafloor. The view of glacialists proposing a large Barents-Kara Ice Sheet of sufficient thickness, which penetrated to the bottom not only on the rises but also in depressions, is supported by the data presented here.

During its formation, this LGM ice sheet has eroded older Quaternary sediments and, thus, LGM moraine deposits directly overlay Mesozoic bedrock. From our point of view, it is useful to map the boundary of the propagation of the LGM ice sheet, since large parts of the Pechora Sea and south of the Kara Sea have not been covered by glaciation at that time. It was possible to establish the position of this boundary in the Pechora Sea (Polyak et al., 2000a, 2000b; Gataullin et al., 2001) and also in the west and northwest of the Kara Sea (Polyak et al., 2000b; 2002; Stein et al., 2002). Here, we will concentrate on mapping and reconstructing the extent of the LGM glaciation in the eastern Kara Sea.

A comprehensive study of the LGM glaciation in the Barents Sea area was carried out by scientists of P.P. Shirshov Institute of Oceanology (Russian Academy of Sciences, Moscow) in the course of several cruises of RV *Academik Sergey Vavilov* in the central and eastern part of the Barents Sea in 1997–1998 (Murdmaa and Ivanova, 1999; Levitan et al., 2003a; Ivanova et al., 2002; Murdmaa et al., 2006).

Materials and Methods

25 sediment cores were retrieved on three transects by gravity corer and subsequently studied (Table 10.10, Fig. 10.8). The western north-south transect (Fig. 10.9) is located from the eastern Franz Victoria Trough to the southern end of the Central Deep. The northern transect is a west-east transect along 80° N (Fig. 10.10) through the central and southern inflows of the Franz Victoria Trough to the glacial trough located south of the island Slam. The eastern transect (Fig. 10.11) at 60° E. intersects the strait between Franz-Josef Land and Novaya Zemlya and is stretched to the south along the Western Novaya Zemlya Trough.

Sediment cores have been studied and described by I.O. Murdmaa, Yu. A. Bogdanov and M. A. Levitan onboard of the vessel. The intervals of sampling varied from 5 to 20 cm. The thickness of each sample was 1–2 cm. The grain-size analysis of six cores was carried out at the Analytical laboratory of P.P. Shirshov Institute of Oceanology (Russian Academy of Sciences). We accepted the boundaries between the gravel, the sand, silt and clay sizes as 1 mm, 0.05 mm, and 0.005 mm, respectively.

Table 10.10 Characteristics of the studied sediment cores from the central and eastern parts of the Barents Sea [*Murdmaa et al.*, 2006]

Station number ASV	Northern Latitude	Eastern Longitude	Water depth, m	Core length, m	Unit I thickness, m	Unit II thickness, m	Unit III thickness, m
841	71°56,6′	37°03,6′	290	1.26	0.30	0.40	0.56
858	75°50,6′	39°54,8′	312	3.56	2.07	0.49	1.00
861	76°43,0′	40°05,2′	196	2.94	0.86	0.44	1.64
875	79°56,6′	39°48,4′	315	4.56	0.35	0.41	3.80
877	79°55,0′	42°29,7′	358	4.20	0.10	0.30	3.80
880	79°55,5′	47°08,2′	388	5.08	3.10	0.94	0.90*
883	79°49,9′	53°01,4′	457	4.17	2.55	0.60	1.02
891	79°29,3′	59°06,3′	315	4.63	2.28	1.30	1.05
897	78°06,0′	59°56,2′	365	2.31	0.72	0.14	1.45
902	76°51,0′	59°58,6′	268	3.85	2.90	0.24	0.71
1006	71°07,4′	52°00,9′	182	4.45	4.45	–	–
1060	78°21,9′	49°50,4′	252	2.95	1.50	0.34	1.11
1157	70°35,0′	52°48,0′	169	4.86	4.40	–	–
1183	71°28,3′	40°48,3′	330	4.17	2.63	1.54	–
1190	73°12,4′	40°56,4′	316	2.76	0.23	=	2.53*
1200	75°54,4′	41°00,5′	308	3.60	2.25	0.47	1.88*
1212	78°30,6′	40°57,1′	289	2.04	0.12	0.48	1.44
1213	78°39,1′	41°04,0′	326	2.50	1.12	0.59	0.79
1217	75°37,6′	45°42,6′	313	2.12	1.03	0.27	1.09
1221	75°59,8′	42°47,9′	357	3.75	3.26	0.03	0.49*
1282	76°11,1′	43°29,9′	330	1.77	0.40	0.10	0.27
1283	75°34,8′	43°09,1′	328	2.24	2.07	0.17	–
1302	75°39,5′	46°09,7′	306	2.70	0.78	0.39	1.53
1310	75°06,1′	53°21,7′	228	4,20	3,89	0,31	–

Note: (∗) total thickness of unit III, thus the unit IV is penetrated; (–) unit has not been penetrated; (=) unit is omitted (hiatus)

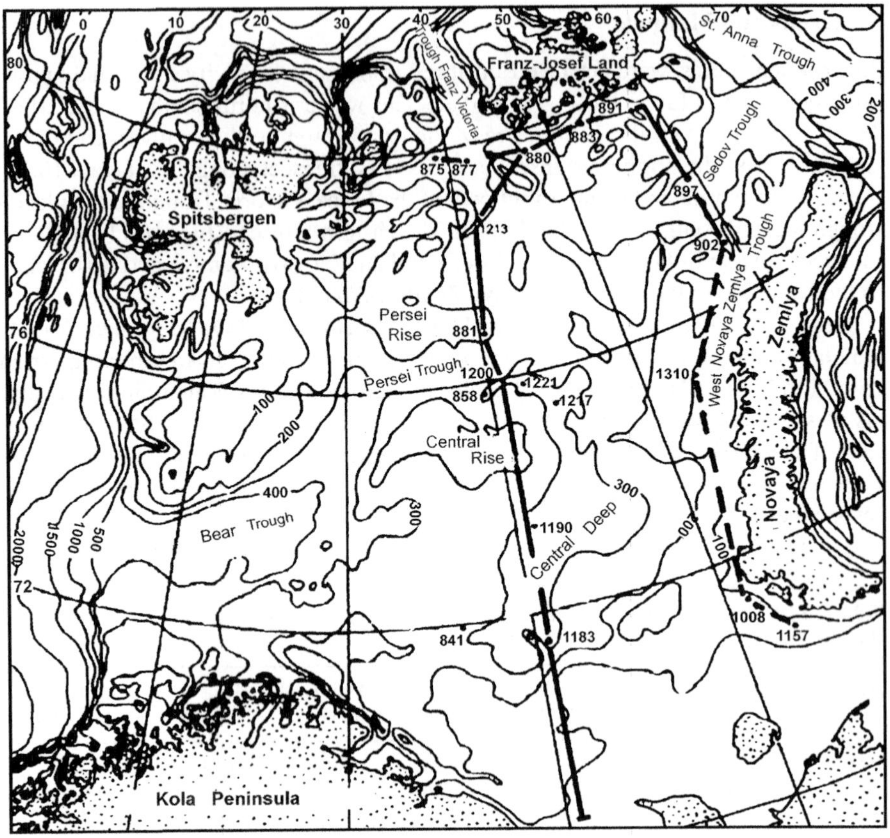

Fig. 10.8 Locations of sediment cores and lithological profiles of RV *Academik Sergey Vavilov* Expedition (Murdmaa et al., 2006)

Radiocarbon AMS ^{14}C datings are mainly obtained from samples of bivalve shells in Gif-sur-Yvette (France) under the management of J.-C. Duplessy (Table 10.11). Then, the corrected radiocarbon ages were transformed into calendar years according to (Stuiver and Reimer, 1993). Oxygen and carbon isotopes in planktonic foraminifer tests *N. pachyderma* (sin.) and benthic foraminifers *E. clavatum* and *M. barleeanus* were measured on mass spectrometer Finnegan MAT 251 in Gif-sur-Yvette (France), using standards NBS 19 and NBS 18 as internal standards.

For foraminifer analysis performed by E.V. Ivanova and T.A. Khusid, 2 cm thick samples were selected every 5 cm. All planktonic foraminifers were identified and quantified (calculation of abundances). 200 to 300 benthic foraminifer tests were identified and quantified in the >100 mkm fraction.

Results

Lithostratigraphy

Based on seismo-acoustic data complete sedimentary sections of Upper Valday (LGM)-Holocene sediments can only be obtained in the shelf depressions of the

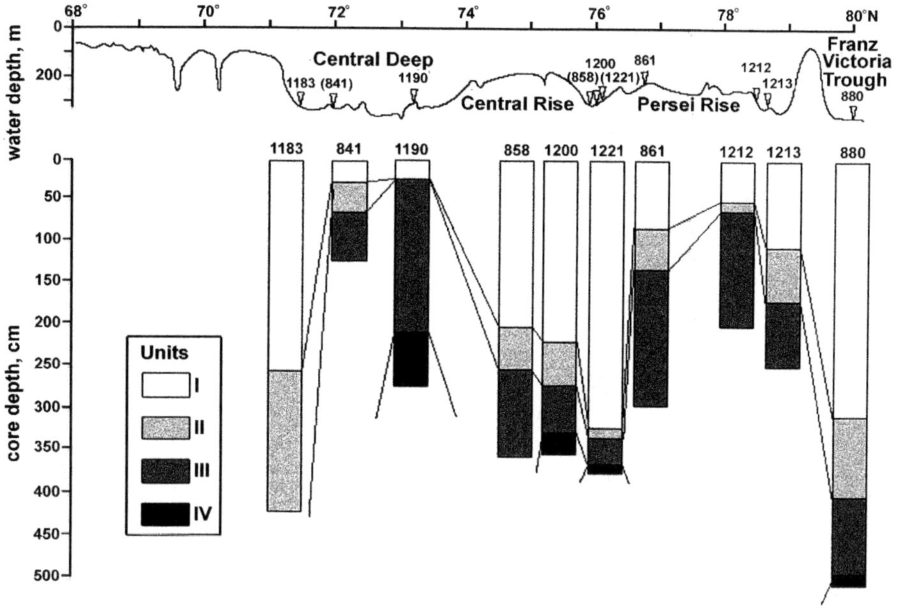

Fig. 10.9 Lithostratigraphic correlation of sediment cores along the western (40–41° E) transect. Location of sediment cores is shown on the water depth transect; for location see also Fig. 10.8 (Murdmaa et al., 2006)

Barents Sea (Gataullin et al., 1993; Polyak and Mikhailov, 1996). Thickness and degree of completeness of the sections of post-glacial sediments mainly depend on local bottom relief, but not on the water depth because the sediments may be redeposited by bottom currents in the narrow valleys and along the axial zones of the glacial troughs (Murdmaa and Ivanova, 1999; Gurevich, 1995).

In the studied sections it is possible to identify four lithostratigraphic units (Murdmaa et al., 2006), and their phases and thickness are greatly varied (see Figs. 10.9, 10.10 and 10.11).

Unit IV is the Upper Valday glacial (LGM) diamicton, recovered in the lowermost part (494–508 cm; 14 cm thick) of Core ASV 880 in the southeast of Franz Victoria Trough (Fig. 10.12). It is a dark gray, very dense diamicton with a chaotic mixture of fractions from large pebbles to finest pelite. The diamicton is characterized by very low water content (18%) and low pore-water salinity (about 13‰) (Bogdanov et al., 2001). Rare benthic and planktonic foraminifers are redeposited from older sediments. Based on these characteristics we suggest that the diamicton is probably a part of the continental moraine. Similar diamictons were recovered at cores ASV 1200 and ASV 1201 in the Persey Trough and sediment core ASV 1190 in the Central Deep.

Unit III is a proximal marine-glacial facies (phase of initial deglaciation). Sediments are gray and dark gray sandy-silty clays, which contain gravel and scattered pebbles. They differ from the underlaying diamicton by a more fine-grained composition, a more frequent presence of planktonic and benthic foraminifer tests,

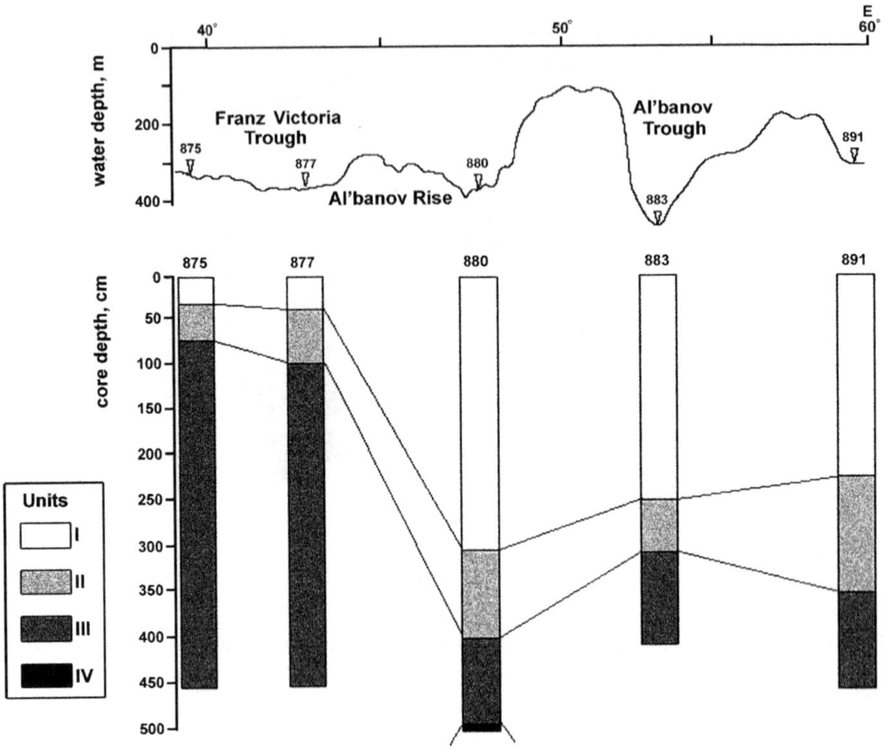

Fig. 10.10 Lithostratigraphic correlation of sediment cores along the northern (79°45′–79°57′ N) transect; for location see also Fig. 10.8 (Murdmaa et al., 2006)

and a salinity of pore waters more close to that of sea water (28–30‰) (Bogdanov et al., 2001). Massive unstratified sedimentary structures predominate, but weakly expressed layering may be present, developed by layers of pebbles or fragments of dense clays. Such interlayers were mainly observed near the top of the unit and are characterized by reverse gradation layering. The thickness of the unit varies from 0.9 to >3.8 m (see Table 10.10).

Unit II is a distal marine-glacial facies (main phase of deglaciation). On the existing seismo-acoustic profiles this unit is represented by a well-stratified sequence of 5 to 20 m in thickness. At some places, the thickness may reach even 100 m (Gataullin et al., 1993). At slopes and apexes of underwater rises the thickness decreases to 1 m and less, sometimes the unit is thinned out or only represented by a hardground layer. The unit is composed of fine brownish clays with a content of the <0.001 mm fraction of up to 74%, coarse sand, banded interlayers of sandy and silty clays of different nuances of gray and brown color, ferrous hardground, homogeneous interlayers of grayish silty clays with sandy lenses or lenticular layers. Furthermore, a distinct (to 34 cm) cyclicity of glacioturbidites occasionally occurred with a straight gradation showing basal folded layers of massive silty sands, that change into layered sandy-silty clays upward in the sedimentary section.

Table 10.11 Radiocarbon dates of samples from sediment cores ASV 880, 1200 and 1183 [*Murdmaa et al.*, 2006]

Lab number, Gif	Core depth, cm	Radiocarbon age, years	1 σ, years	Calendar age, cal. yrs. BP	Studied material
ASV 880					
99457	25	1550	60	1104	*Yoldiella lusida*
99385	46	2440	60	2082	*Yoldiella lenticula*
99386	166	6560	80	7081	*Yoldiella fraterna*
99458	216	7420	80	7866	*Yoldiella sp.*
99387	266	8830	90	9282	*Yoldiella fraterna*
99388	305	9150	90	9814	*Yoldiella fraterna*
ASV 1200					
99802	110	5810	70	6225	*Yoldiella sp.*
99803	120	6180	60	6607	Assemblage of benthic foraminifers
99804	165	6770	70	7300	*Yoldiella fraterna*
99805	195	9220	80	9981	*Yoldiella lusida*
99806	205	9450	80	10082	*Yoldiella sp.*
99807	210	9550	80	10331	*Yoldiella intermedia*
ASV 1183					
100083	218	9820	100	10707	*Yoldiella sp.*

Note: The radiocarbon dates of sediment core ASV 1157 are given below.

Unit I is a marine facies of shelf depressions. The sediments are distinctly different from the underlying sediments in terms of color (olive-gray color predominates), a more uniform, homogeneous and much more fine-grained structure with strong bioturbation (layered structures are not observed), a massive development of early diagenetic hydrotroilite, abundance and variety of foraminifers, shells of mollusks and tubes of polychaets, and a predominance of marine organic matter (to 80%) (Romankevich et al., 2000). The thickness of the unit varies from 5–6 m to several centimeters in the shelf depressions and is usually thinned out on the underwater rises.

In general, the arches of underwater rises at depths of <100 m are usually lacking of sedimentary cover, and outcrops of bedrocks or moraines are exposed here at the surface of the seafloor. Mixtures of boulders, pebbles, iron-manganese concretions, shells of mollusks and barnacles, mixed with clayey silt and silty sand, were discovered at several deep-water rises, for example, on the Murmansk, the Persey and Central rises. Coarse-grained material is enriched by IRD input, erosion of moraines or winnowing of fine fractions by bottom currents.

Grain-Size Composition

In general, the four sedimentary units are characterized by a thinning of upward sections with abrupt changes at their boundaries (Murdmaa et al., 2006). Only at

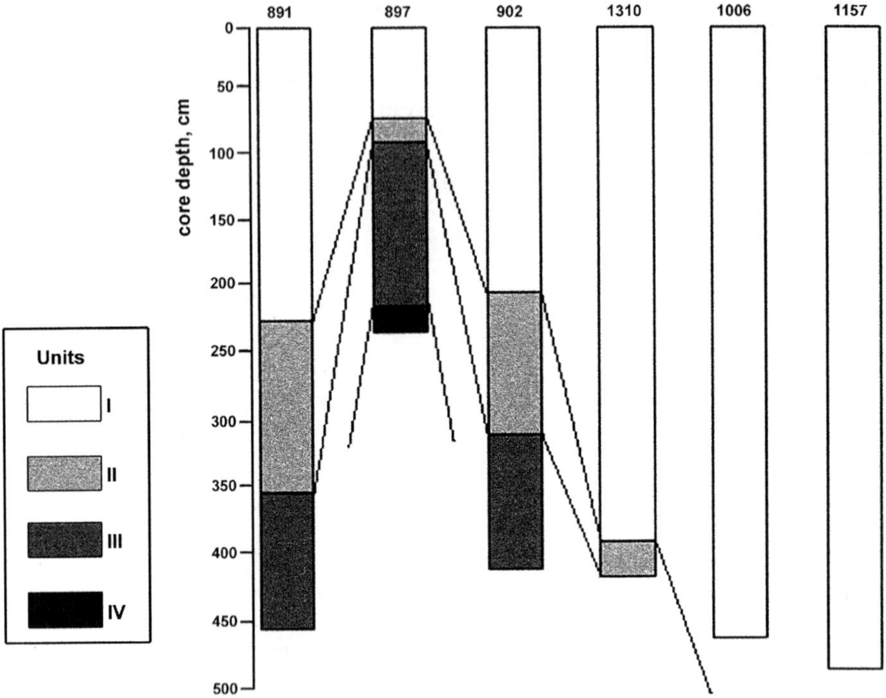

Fig. 10.11 Lithostratigraphic correlation of sediment cores along the eastern (60° E) transect; for location see also Fig. 10.8 (Murdmaa et al., 2006)

Core ASV 1183 (Fig. 10.13), the sediments of Unit I are coarser-grained compared to Unit II, presented by fine clayey silts. In some cores, for example, ASV 880 and ASV 1200, the basal layers of Unit I are enriched by sandy fractions.

The content of clay fractions varies from 30 to 60% in the diamicton matrix (Unit IV). Sediments of Unit III are granulometrically very close to the diamicton, but the content of sand is somewhat lower, and the clay content is somewhat higher. The silt content is approximately the same. It seems that the particles of >0.1 mm are composed of IRD in all units, and finer sand could be transported by low density flows or bottom currents. In terms of grain-size distribution, sandy clays of units III and II are very similar, except that the fine clays of Unit II are absent in Unit III. As already mentioned, the most fine-grained sediments generally predominate in Unit I. It is very interesting to note that the ratios of the 0.01–0.005 mm, 0.005–0.001 and <0.001 mm fractions in the sediments of the upper three units do not differ from each other, suggesting a common source and transport mechanism connected to melting of glaciers and extension of under-ice currents. Eolian delivery, as well as precipitation of fine fractions as a result of coastal abrasion, bottom erosion, remote river drainage, etc. have to be considered as additional basic sources and transport processes of sedimentary material.

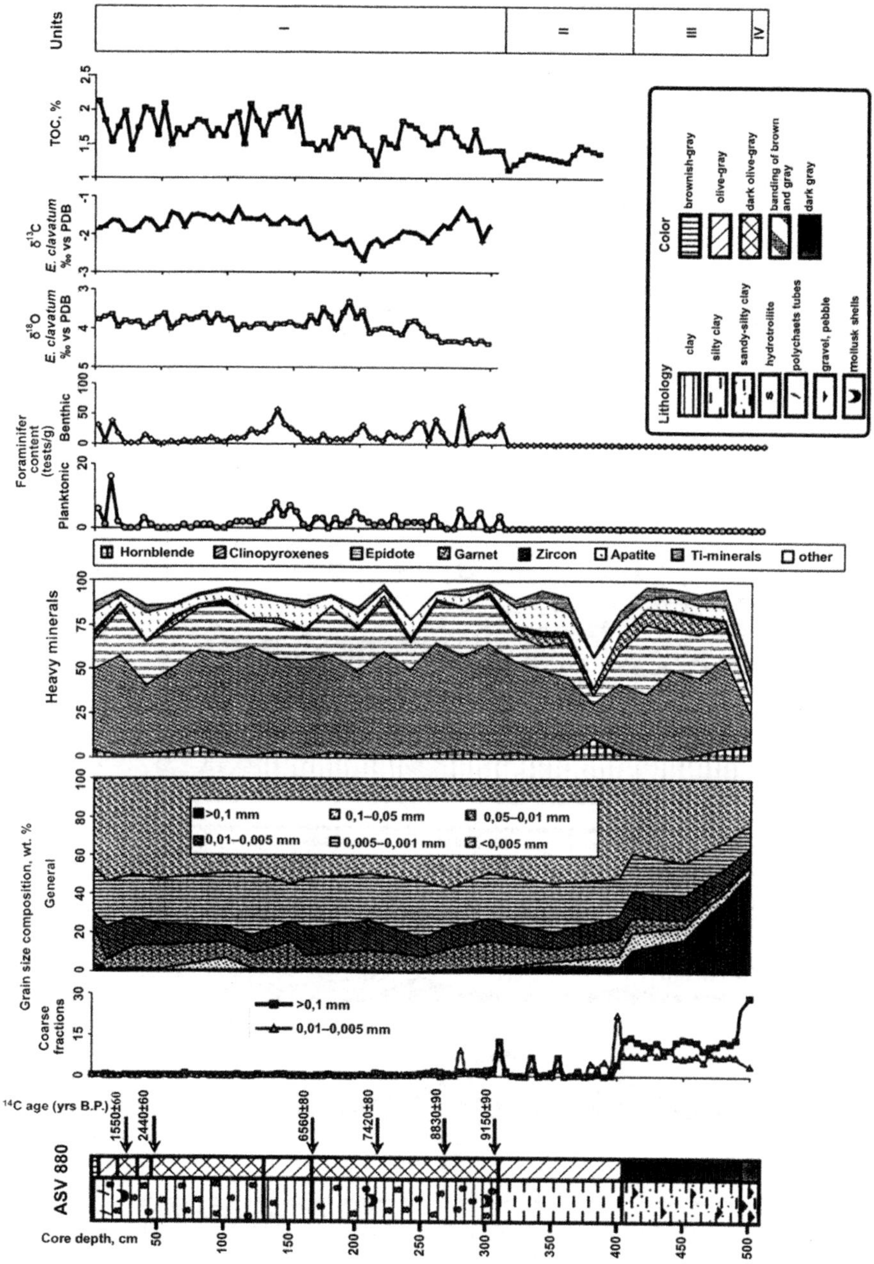

Fig. 10.12 Summary diagram of results of sediment core ASV880 (Murdmaa et al., 2006)

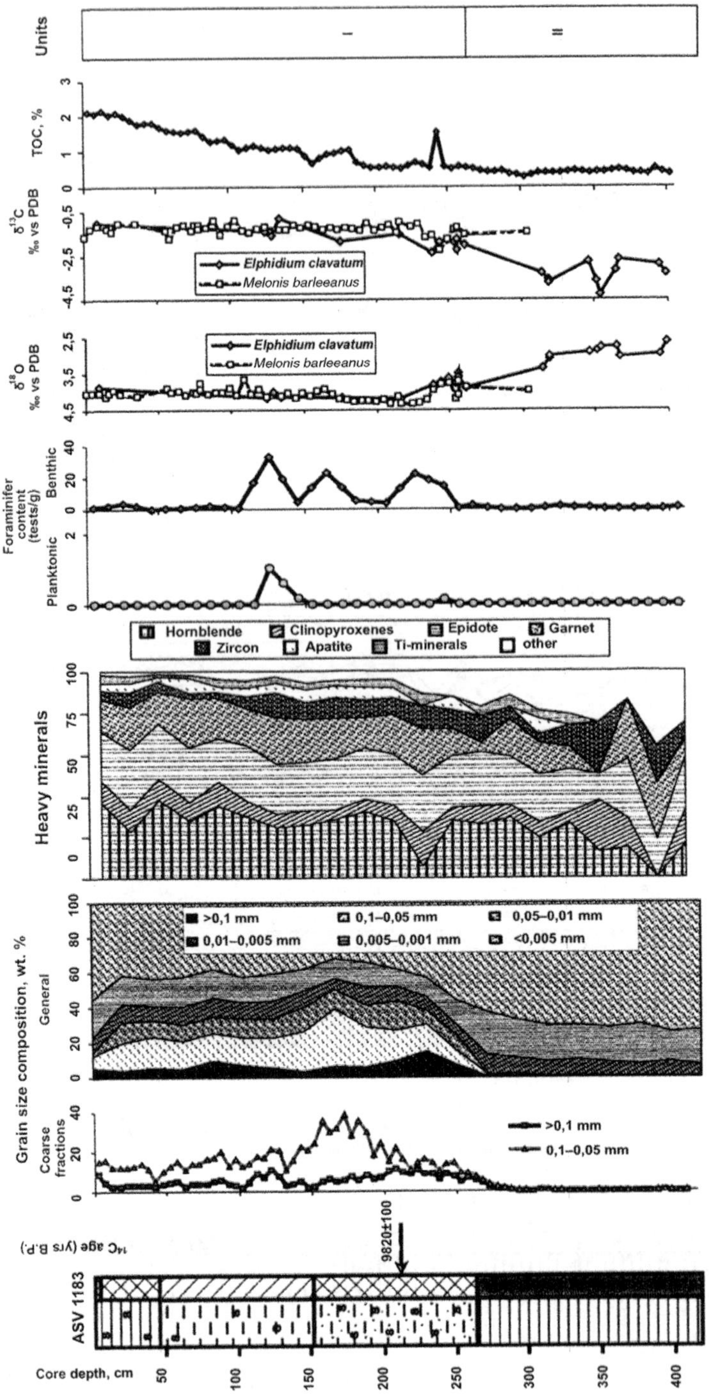

Fig. 10.13 Summary diagram of results of sediment core ASV1183 (Murdmaa et al., 2006). See legend in Fig. 10.12

Heavy Minerals

Heavy minerals are studied in four sediment cores of the western transect and in two cores of the eastern transect. In general, the results indicate the influence of regional source provinces with a continuous decrease of clinopyroxene abundance from the north to the south, supplied from Franz-Joseph Land, and by an increase of abundance of garnet and apatite, for which a basic source is the Baltic Shield (Murdmaa et al., 2006). Furthermore, numerous local variations were observed. Thus, the sediments of units IV-II Core ASV 1200 (Fig. 10.14) are characterized by an apatite-clinopyroxene-garnet association due to the supply from southern source provinces, and sediments of Unit I are characterized by a garnet-epidote-clinopyroxenes association as a result of mineral mixing by surface and bottom currents. Sediments of units IV-III of Core ASV 1190 are enriched in garnet, related to glacial erosion of the Kola Peninsula, while a hornblende-epidote-garnet association is described in the sediments of Unit I, typical for the southern part of the Barents Sea and related to riverine discharge (Levitan et al., 2003b). Near Novaya Zemlya, the sediments of the more northern Core ASV 1310 contain an epidote-garnet-hornblende association, which is supplemented by clinopyroxene in the south (Core ASV 1157).

Based on our data we can identify three basic terrigenous-mineralogical provinces in the 0.1–0.05 mm fraction: Southern-Barents (with a dominance of garnet, zircon, titaniferous minerals, apatite, mica, and hornblende); Novozemelsky (with a dominance of epidote); and North-Barents (with a dominance of clinopyroxenes and, partially, epidote).

Clay Minerals

Pilot studies of clay minerals in the <0.001 mm fraction were carried out in 8 sediment cores (Table 10.12).

An illite-kaolinite-chlorite association usually prevails among the clay minerals, and the content of smectite is usually low (Table 10.12). Units IV and III are characterized by very low contents of smectite (<10%), slightly lowered abundances of illite (18–39%) and very high abundances of chlorite and kaolinite (52–75% together) with a predominance of kaolinite. The specific similarity of the clay-mineral composition of units IV and III and the clay-mineral association of Unit I in the Russkaya Gavan fjord (Table 10.12) should be noted. This may be explained by the delivery of "glacial milk" from the adjacent outlet of the Shokalsky Glacier. In average, the content of smectite is steadily increasing in the sediments of Unit II, the illite concentration slightly increases and the total amount of chlorite and kaolinite is noticeably reduced.

The same tendency continues upward through the sedimentary section. Moreover, quite often the smectite content is not just increased in Unit I (compared to Unit II), but reaches its maximum in the uppermost layers of Unit I. In general, the composition of clay minerals in the studied sediment cores coincides with clay-mineral data determined in surface sediments from the same area (Nürnberg et al., 1995).

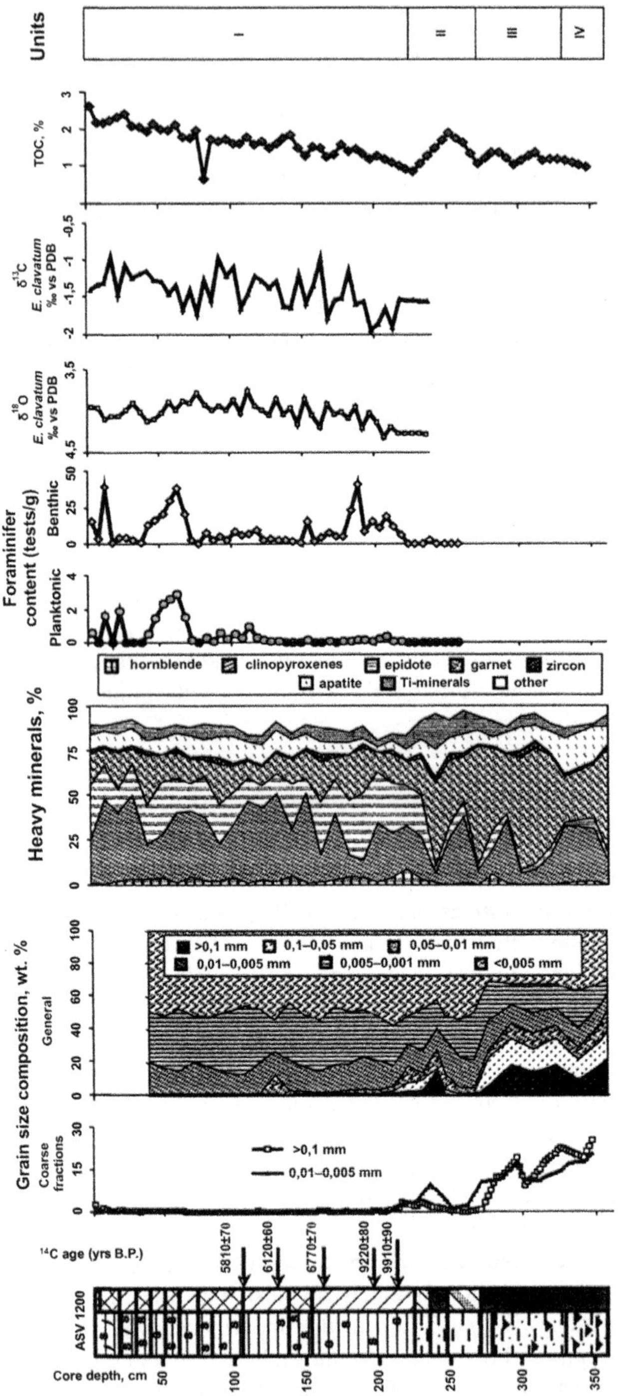

Fig. 10.14 Summary diagram of results of sediment core ASV1200 (Murdmaa et al., 2006). See legend in Fig. 10.12

Table 10.12 Content of clay minerals in the fraction of less than 0.001 mm in the eastern Barents Sea sediments (rel. %) [*Murdmaa et al.*, 2006]

Station number ASV	Interval, cm	Illite	Smectite	Chlorite	Kaolinite	Unit
880	0–3	22	30	26	22	I
	59–60	25	10	32	33	I
	3–5	44	12	22	22	I
	53–55	37	17	24	28	I
861	63–65	32	16	25	27	I
	78–80	31	8	31	30	I
	93–95	27	19	26	28	II
	103–105	30	6	32	32	II
	0–2	28	43	14	15	I
	13–15	42	26	17	15	I
	155–157	47	12	20	21	I
858	207–209	26	45	15	14	II
	257–259	26	22	25	27	III
	285–287	39	5	29	27	III
	330–332	35	1	36	28	III
	44–46	30	15	30	34	I
	189–191	28	9	32	31	I
1200	244–246	20	10	35	35	II
	304–305	18	7	35	39	III
	349–351	25	0	34	41	IV
	8–10	36	28	18	18	I
	30–32	50	19	15	16	II
841	100–102	57	4	20	19	III
	114–116	28	2	34	36	III
	119–120	28	8	30	33	III
	309–311	29	24	35	12	I
1310	349–351	34	6	45	15	I
	399–401	22	0	39	39	II
988	4–5	42	0	29	29	I
987	297–298	39	0	30	31	I
	582–583	42	5	27	26	I

Note: Sediment cores ASV 987 and ASV 988 are located in the fjord Russian Gavan'.

The absence of any correlation between the heavy-mineral and clay-mineral associations supports the hypothesis about the relative independence of the distributions of the sub-colloidal and fine-sand fractions due to different transport processes and accumulation of sedimentary material.

Foraminifers

Unit IV presents continental sequences and thus does not contain *in-situ* marine fauna. Only a few tests, i.e., well-preserved left-coiled *N. pachyderma* (Ehrenberg), were found in sediments of sediment core ASV 880, but according to a number of features they are related to reworked older sedimentary material.

A few foraminifers were found in Unit III, presented by Pleistocene and reworked Mesozoic benthic (mainly agglutinated) species. Their quantity does not exceed 0.01 tests per gram of sediment. Planktonic forms, the type of those found in the underlying Unit IV, are even more rare. Most parts of the sediments are barren in foraminifers. Nevertheless, the good preservation of the Pleistocene benthic foraminifer species *Elphidium spp., Cassidulina reniforme, Islandiella helenae/norcrossi,* and *Melonis barleeanus*, as well as the presence of well-preserved aragonitic pteropod shells of *Limacina helicina* (Murdmaa and Ivanova, 1999), indicate in-situ burial in the basin.

Sediments of Unit II contain reworked Mesozoic foraminifers and ostracod shells in all cores. In most intervals, sediments are barren (in particular, this is true for almost all cores of the northern and eastern transects). In comparison to Unit III, however, a wide variety of benthic and planktonic forms, especially in sediment core ASV 880, was observed. Together with the above-mentioned Arctic benthic foraminifers, the Atlantic forms *Pullenia bulloides, Cibicides wuellerstorfi, Cassidulina teretis* were also found. Besides the left-coiled planktonic foraminifer *N. pachyderma* (Ehrenberg) its right-coiled variety as well as *Turborotalia quinqueloba* are also present. The amount of foraminifers is still very insignificant and only exceeds the numbers of the underlying unit at some areas.

Foraminifers reach their maximum abundance and specific variety in the sediments of Unit I. Planktonic forms in a more or less noticeable quantity are only present in sediments from those core locations, which are influenced by Atlantic waters (for example, sediment core ASV 880). The content of benthic forms sometimes reaches 41 and even 63 tests per gram of sediment. The preservation of carbonate tests strongly varies; moreover their strongest dissolution is noted for the southeastern part of the studied area (cores ASV 1183, 1157, 1310), where barren intervals are encountered. In general, for the central and eastern part of the Barents Sea the variable distribution of the foraminifer fauna is characteristic in terms of preservation, quantity, and associations. In the north, the Lower, Middle and Upper Holocene sediments are noticeably differed from each other. Moreover, the Middle Holocene is characterized by the maximum abundance and specific variety of fauna, with notable abundance of the Atlantic boreal forms.

Oxygen and Carbon Isotopes

Oxygen and carbon isotopes were measured in carbonate tests of benthic foraminifers in cores ASV 880, 1157, 1183 and 1200, whereas measurements of tests of planktonic foraminifers *N. pachyderma* sin. were only carried out for cores ASV 880 and 1200.

All isotopic data confirm that Unit I is of Holocene age. Small variations ($<1‰$) in the isotopic composition of oxygen are probably caused by both changes in the water temperature and its isotopic composition, which can be related to fluctuations of the glacial ice-sheet volume, by local melting of glaciers, by the inflow of riverine freshwater, and by entering of Atlantic waters. For further details we refer to (Duplessy et al., 2001, 2005).

Discussion of Results

All data obtained within this study give important information on the glacial ice-sheet history, the deglaciation and the ongoing influence of marine accumulation during the LGM-Holocene time interval. There is no question regarding the development of Unit IV during times of maximum continental glaciation (LGM). Clay-minerals data clearly point to the delivery by glacier. It is clear that the development of Unit III occurred under following conditions: (1) onset of decay of the Barents Sea Ice Sheet, (2) penetration of seawater from the north into the deepest shelf depressions, and (3) active development of icebergs and initial development of lateral flows of sediment material from the slopes, gradually released from ice shelf rises in the depression. Probably, the developed sequences can be named as "sea moraine". According to seismostratigraphic data, units IV and III correspond to the glacial seismostratigraphic complex proposed by (Gataullin and Polyak, 1977). I.O. Murdmaa and E.V. Ivanova (1999) believe that sediments of Unit III should be assigned to proximal glacial-marine formations of the initial deglaciation stage. According to opinion of M.A. Levitan, the stratified diamicton of Unit III has a greater petrographic similarity to underlaying diamicton of Unit IV than to the overlaying sediments of Unit II. Therefore, Unit III should be appointed to the final glacial stage in this area, in particular, to the substage of the initiation of the Barents Sea Ice Sheet disintegration.

Not much is known about the exact age of the boundary between the two units (II and III). It can be diachronic, if we consider the map of the boundaries of glacial formations of different age on the Barents Sea seafloor (Landvik et al., 1998). Then, the age may be between 12 and 15 ^{14}C ka for the larger part of the studied region and <12 ka for the Novaya Zemlya area.

Apparently, there are no doubts about the fact that Unit II has been developed under conditions of active deglaciation, with sedimentation predominantly under sea-ice cover. Thus, turbidites were found in Unit II of Core ASV 1060, sediments of mud flows are described in the core ASV 1183, sediments of grain flows were described in Core ASV 1310, and sediments of debris flows were found in Core ASV 1200. Furthermore, it is obvious, that many structures have been developed due to accumulation from nepheloid flows. For the larger part of the studied region, the transition from Unit III to Unit II probably occurred at about 13.3 ka (Polyak and Mikhailov, 1996). Primary production at this time, although somewhat increasing in comparison to the period of underlaying unit, remained very low.

In 2004 M.A. Levitan participated in cruise 63 of RV *Professor Shtokman* in the central and western (Norwegian) part of the Barents Sea together with Russian scientists from the P.P. Shirshov Institute of Oceanology and Norwegian scientists from the Bjorness Paleoclimatic Center in Bergen. Geological and geophysical studies showed that in this area of the Barents Sea it is possible to identify similar units as those described above for the eastern part of the sea. Preliminary results of these studies are shown in Table 10.13.

Table 10.13 Location of sediment cores and thickness of the individual units (by I.O. Murdmaa and M.A. Levitan)

Station number PSh	Northern Latitude	Eastern Longitude	Water depth, m	Core length, cm	Thickness of the Unit I, cm	Thickness of the Unit II, cm	Observed thickness of the Unit III, cm
5138	73°35′14	20°59′34	501	277	277	–	–
5140	74°26′96	27°45′72	397	56	19	–	37
5141	75°29′00	30°19′19	372	173	111	42	20
5142	75°40′76	31°03′16	355	61	33	3	25
5147	76°04′54	32°56′88	314	253	46	203	4
5150	78°09′77	13°45′37	431	204	101	93	10
5151	78°20′83	15°12′78	249	437	351	86	–
5154	78°03′04	29°40′96	319	71	–	29	42
5155	78°50′53	37°38′33	301	319	297	22	–
5157	78°55′80	41°52′87	461	457	>187	–	–
5159	71°21′65	22°38′81	418	505	505	–	–

Note: Blank means lack of sediments.

Holocene Sedimentation History in the Southern Novaya Zemlya Trough

The Late Quaternary history of the Russian part of the Barents Sea and its southeastern segment – the Pechora Sea - has been actively studied in recent years using AMS ^{14}C dated sediment cores (Polyak and Solheim, 1994; Lubinski et al., 1996; Polyak et al., 2000; Duplessy et al., 2001), which allow to substantially deepen and develop previously existing ideas. Nevertheless, high-resolution studies of the Holocene sedimentation history in the studied area are still very rare. As a result, problems such as the identification of sediment sources, the degree of the Holocene Climate Optimum expression in sediments from the Pechora Sea, the history of cold brines, and estimates of paleoproductivity have not been examined in detail so far. Furthermore, a number of the problems, connected with the extension of Late Valday glaciation (LGM), still cause a vigorous discussion (Epstein and Chistyakova, 2005; Svendsen et al., 1999; Bauch et al., 2002).

Methods

On board of RV *Academik Sergey Vavilov*, M.A. Levitan retrieved sediment core ASV 1157 (70°33'N, 52°48'E; water depth is 169 m) in the central section of the Southern Novaya Zemlya Trough (Fig. 10.15).

Recent Sedimentation Environment

The Southern Novaya Zemlya Trough is an elongated (120 x 25 km on the isobath 140 m) west-northwest directed depression in the Pechora Sea with steep slopes and flat bottom, located to the south of the Southern Island of Novaya Zemlya. It is

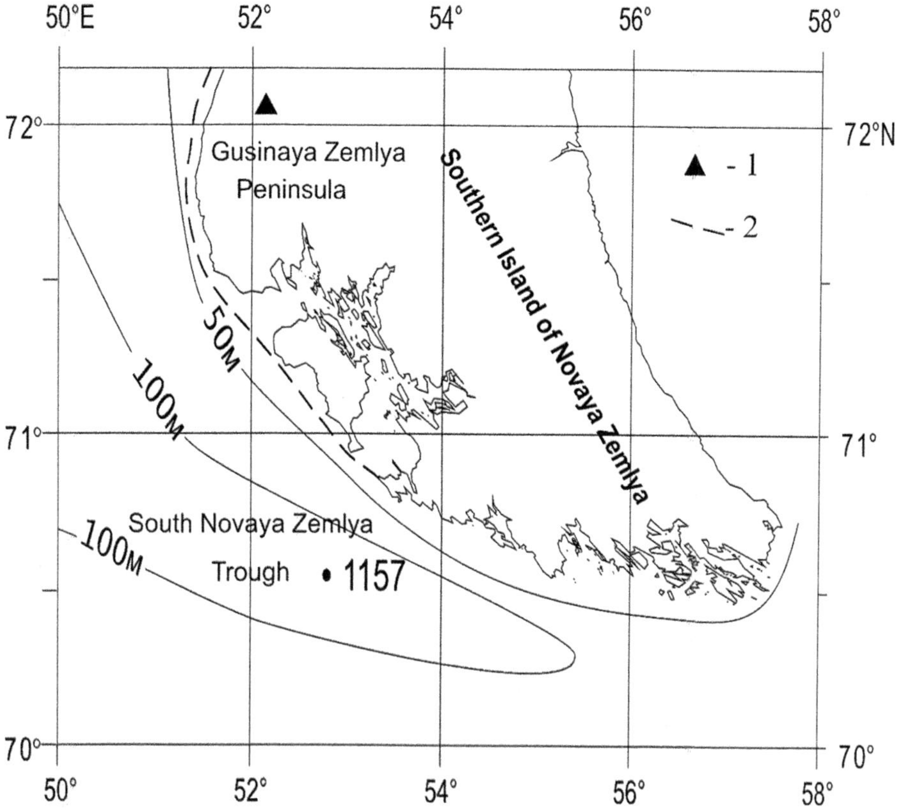

Fig. 10.15 Location of sediment core ASV 1157 in the South Novaya Zemlya Trough (Levitan et al., 2004). 1 – location of sedimentary section with peat layer with an age of 15310 years (Serebryanny and Malyasova, 1998); 2 – southern part of moraine of Admiralty complex (Gataullin and Polyak, 1977)

filled with 5–10 meter thick Holocene sediments, up to 75–80 m thick Upper Valday (upper Pleistocene) marine-glacial sediments and a 20–25 m thick diamicton (interpreted as basal moraine) of the Late Valday (LGM) age, which overlay (together with older diamictons?) the Jurassic-Triassic bedrock (Gataullin et al., 2001).

The Admiral moraine complex shown in Fig. 10.15 relates to the youngest glacial formation of the Barents Sea, since the radiocarbon age of the overlain moraine sediments is about 8 ka (Landvik et al., 1998), i.e., the complex has been developed, most likely, during the Younger Dryas. A narrow shallow (<60 m of water depth) shelf zone is located to the north of the trough between its northern boarder and the coastline of the Southern Island.

At the coastline of the island, Lower Paleozoic, weakly metamorphosed (greenstone phase) detrital rocks are cropping out, which are overlain by red colored siliciclastic rock and basalts of Triassic age (Geological structure of the USSR and regularities of mineral deposits distribution, 1984). Thermo-abrasion of the coast and river extensions of small creeks and their channels from the Southern Island are the main mechanisms of sedimentary delivery into the trough in recent time (Khasankaev, 1978). Some sediments can also be supplied from the south, from the shallow part of the Pechora Sea. According to heavy-mineral and clay-mineral data in surface sediments from the Pechora Sea it is established that the Southern Island is the main source of material for the present-day sedimentation of the trough (Levitan et al., 2000a).

It is no doubt that a leading role in transport of sediments belongs to processes, which take place at the underwater slopes of the trough. Furthermore, in the considered case we assume the existence of longitudinal bottom currents, which wash out fine fractions and redeposit bottom sediments along the long axis of the trough. The measured rate of bottom currents comprises <10 cm/s (Gurevich, 1995).

At the end of August 1998 the following stratification of water layer was registered: (1) upper layer of relatively warm (1.0–5.5°C) and weakly-salt waters (30.8–33.3‰), 23 m thick; (2) very sharply pronounced thermocline and halocline of several meters in thickness; (3) intermediate water mass with temperatures of <–1.4°C and salinities of 33.3–34.5‰, 70 m thick; 4) cold (<–1,4°C) and saline (34.5–34.9‰) bottom waters which are likely to correspond to cold brines developed under ice formation during fall times (Levitan et al., 2003b).

It is important to note that in the studied area there are no signs of the presence of Atlantic waters. The nearest jet stream, which contains transformed Atlantic waters, the so-called Kanin-Kolguev Current, passes 250 km to the west from the Southern Novozemelsky Trough (Levitan et al., 2003b). The sea surface above the trough is covered with ice more than half of the year, from fall to spring. The average annual primary production is very low (<100 mg C/m^2 per day) (Romankevich and Vetrov, 2001). This is mainly caused by the development of phytoplankton at the sea-ice edge.

Results

The length of sediment core ASV 1157 is 486 cm (Fig. 10.16). The sediments are soft, fine, yellowish-gray in upper 6 cm and olive-gray, olive and black below (Levitan et al., 2003a). IRD was not found in the above sediments. Rare tubes of polychaets with a length of up to 5 cm and with a diameter of 2 mm are noted from the surface to the depth of 140 cm. Throughout the entire core bivalve shells of *Macoma calcarea, Tridonta borealis, Leda sp.* and *Portlandia arctica* of up to 3 cm and their detrite (3–5 mm) were found. The most abundant numbers of shells are noted at intervals of 13–15, 38, 50–55, 93–95, 165–173, 255, 268–284, 309–324, 354, 372–373, 400, 410, 440 and 485–486 cm.

The core is characterized by a distinct layering, caused by different color of sediments and by their structure and grain-size composition (see Fig. 10.11). Figure 10.16 clearly demonstrates that the sediments are generally bioturbated. The most intensive bioturbation is noted at intervals of 6–49, 78–109, 129–258, and 379–440 cm. The layering caused by the alteration of sediment colors is developed in the interval of 440–470 cm. This structure repeatedly was described in sediments, which

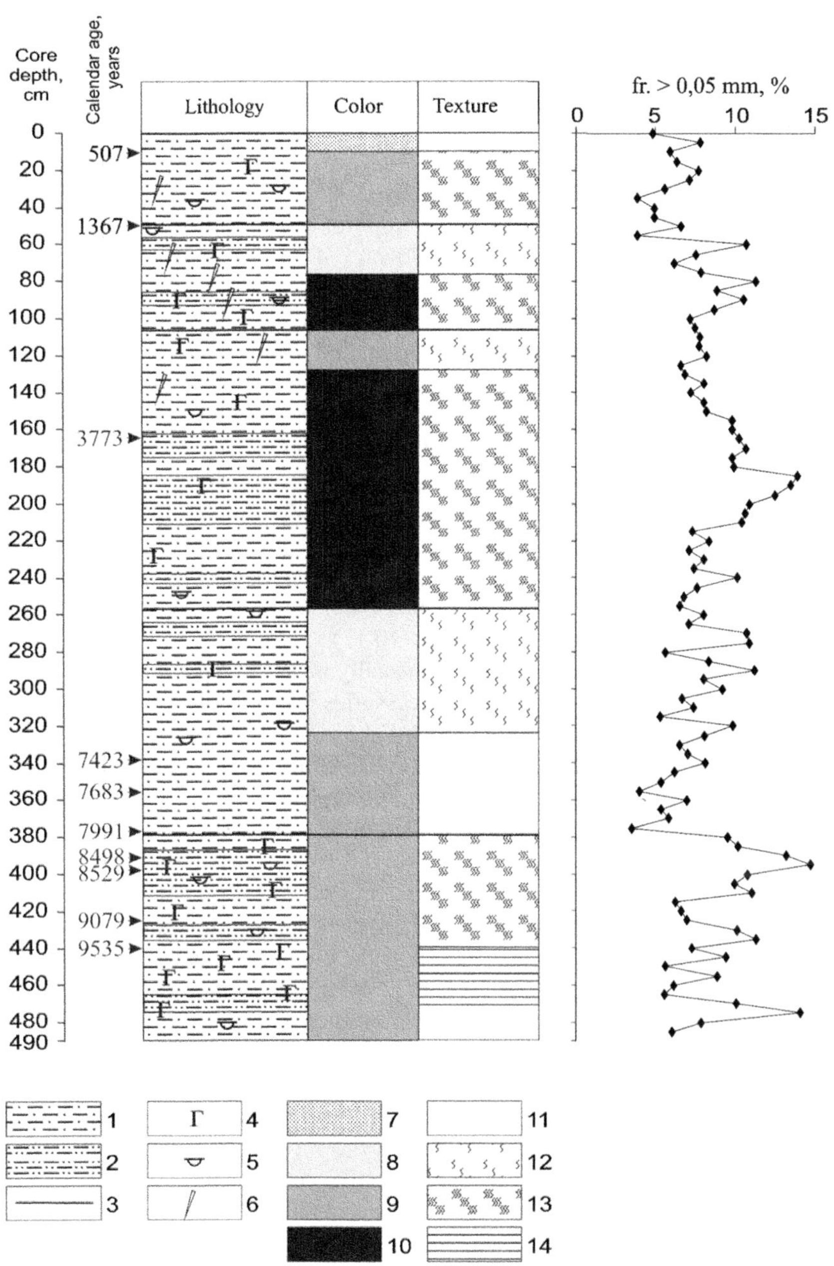

Fig. 10.16 Lithology and amount of the >0.05 mm fraction (in %) in sediment core ASV 1157 (Levitan et al., 2004). 1 – silty clay, 2 – silty clay with sand admixture, 3 – boundaries of layers with different textures, 4 – hydrotroilite, 5 – bivalves shells, 6 – polychaets tubes; 7–10 – sediment color: 7 – yellowish-gray, 8 – olive-gray, 9 – olive, 10 – black; 11–14 – texture: 11 – massive, 12 – with weak and intermediate bioturbation, 13 – with intensive bioturbation, 14 – laminated. *Arrows* show calendar age of mollusks shells

were accumulated during deglaciation in the Barents Sea (Murdmaa and Ivanova, 1999). Starting from 10 cm and down to the bottom, abundant black spots and scattered patches of hydrotroilite were found in the sediments

In accordance with the granulometric classification of V.T. Frolov (1995), the dominating sediments are aleurite-clayey silts. Sandy admixtures of up to 15% (see Fig. 10.16) were observed at some spots. Based on the distribution of the >0.05 mm fraction, there are eleven layers (from top to bottom): 0–20, 20–60, 60–90, 90–150, 150–185, 185–205, 205–380, 380–415, 415–470 and 470–486 cm. On a larger scale it is possible to reduce these layers to the following four main intervals: 0–60, 60–185, 185–380 and 380–486 cm. The first interval is characterized by sand concentration of 4–8% (in average 6%), the second one by 6–11% (in average 9%), the third one by 4–14% (in average 8%, with maximum 14% at 180 cm and with a distinct drop of the sand content downward), and the fourth one 6–14% (in average 10%, with a maximum of 14–15% at 390 and 475 cm). The very sharp boundaries of the isolated layers and horizons by granulometric composition should be noted.

The recent sediments of the Southern Novaya Zemlya Trough belong to the Southern-Barents type of terrigenous-mineralogical provinces by composition of heavy minerals (Gurevich, 1995). There are about 40 mineral species found in Core ASV 1157. The content of only four of them, i.e., clinopyroxenes, epidote, garnet and black ore minerals, exceeds 10% (Fig. 10.17).

Iron oxides, hornblende and rutile are usually present in amount of several percentages. The content of other mineral accessories is negligibly small. Recent sediments of the Pechora Sea are enriched by hornblende, epidote, garnet, and black ore minerals, and to a less extent by zircon (Levitan et al., 2003b). Rutile and clinopyroxenes, as a rule, are absent. Garnet is a more typical mineral for the western part of the Pechora Sea, and epidote is the characteristic mineral for the eastern province. Triassic basalts of the Southern Island serve as the source of clinopyroxenes for the sediments of the Southern Novaya Zemlya Trough.

Examination of the distribution of a number of heavy minerals and their variability downward the sedimentary section of the core reveals a number of interesting special features. In general, the relationship of the four main minerals changes weakly. However, the clinopyroxene/epidot ratio clearly divides the core into two parts: in the upper part (0–200 cm) it varies from 0.7 to 2.0 (in average 1.4), whereas in the lower part (>200 m) typical values vary from 0.3 to 1.7 (in average 0.9) (see Figs. 10.16 and 10.17). An explicit increase of the epidote content is obvious in the interval below 460 cm.

In the same interval there is a noticeably sharp increase in the epidote/garnet ratio (see Fig. 10.17). The defined confinement of such "floating" minerals as hornblende and muscovite to the interval of 220–380 cm is brought to notice. If we examine the primary results of mineralogical analyses, not converted to the determinable grains, then we could clearly see an obvious connection with the same horizon of the maximum of the content of grains with hematite coating (see Fig. 10.17).

The study of TOC distribution (Fig. 10.18) allows to identify two units, each of which, in turn, consists of two subunits. The first (from top to bottom) unit (0–125 cm) is characterized by relatively high TOC (in average 1.75%). The upper

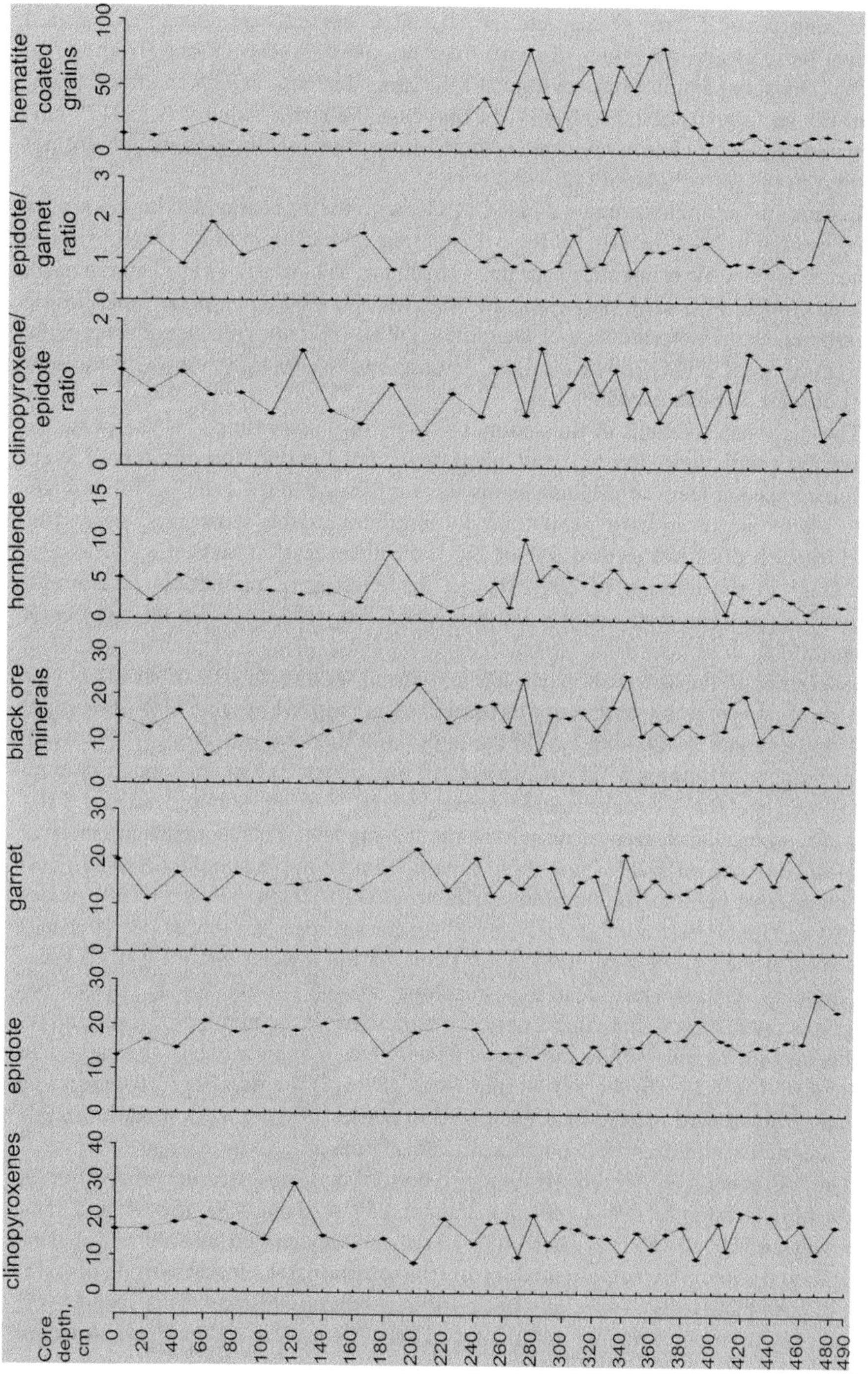

Fig. 10.17 Distribution of heavy minerals (in %) and their ratios in sediment core ASV 1157 (Levitan et al., 2004)

subunit (0–48 cm) contains high TOC concentrations (1.5–2.0%), while for the lower one (48–125 cm) is characterized by TOC contents increasing downward against the background values of sharp fluctuations of concentrations (from 0.7 to 2.2%). The second unit differs in terms of the clear decrease of TOC content downward the section (from 2.2 to 0.6%). In this case the upper subunit (125–425 cm) is characterized by clearly noticeable fluctuations of concentrations, which are not obvious in the lower subunit (425–486 cm).

In general, sediments contain 2–4% $CaCO_3$, with two exceptions of higher values reaching 8 and 13% (Fig. 10.18). If we do not take into account these two maximum values, it is possible to identify four units (from top to bottom): (1) 0–180 cm, with a stable content of $CaCO_3$ between 2 and 4%; (2) 180–400 cm, with a weak inrease of carbonate content to the base of the unit; (3) 400–460 cm, with a clear drop in the concentration of $CaCO_3$ downward the section; and (4) 460–486 cm, with more or less constant carbonates values.

The carbonate content in the sediments is mainly determined by the presence of bivalves and their detritus, and planktonic and benthic foraminifers. Layers with increased content of mollusk shells were indicated in the lithological description. There are almost no planktonic foraminifers in the sediments; as a rule, their fraction does not exceed 1% of the total number of foraminifers in a sample. Only in the interval of 300–307 cm the number of planktonic foraminifer *N. pachyderma* sin. tests sharply increases, reaching 15–40% of total number of foraminifers.

According to the distribution of benthic foraminifer tests in 10 g of sediment (see Fig. 10.18) it is possible to identify three units (from top to bottom): 1) 0–300 cm; 2) 300–460 cm and 3) 460–486 cm. In the upper unit the tests are strongly dissolved. Their number composes 5–25 (in average 10) pieces per 10 g in the subunit interval 0–110 cm and 0–13 (in average 5) pieces per 10 g in the subunit interval 110–300 cm, where the degree of dissolution is the highest. The second unit (interval 300–460 cm) can be divided into three subunits. In the upper subunit (300–375 cm) the number of benthic foraminifers varies from 0 to 50 (in average to 30) pieces per 10 g. The highest number of 20 to 90 (in the average 50) pieces per 10 g was observed in the middle subunit (375–400 cm). In the lower subunit (400–460 cm) the quantity of tests again decreases, reaching 5 to 65 (in average 25) pieces per 10 g. It is interesting to note that a noticeable increase in the number of foraminifers in this unit (in comparison to the first unit) only led to an increase of the carbonate content of 1.0–1.5%. In the lowest unit (460–486 cm) the number of foraminifers increases downward the section, reaching 130 pieces per 10 g at the bottom; in this case, an increase of carbonate content was not observed.

The following five horizons (from top to bottom) are separated in composition of the benthic foraminifers: 0–70 cm, 70–258 cm, 258–420 cm, 420–450 cm and 450–486 cm (see Fig. 10.18). The horizon 0–70 cm is characterized by a normal-marine assemblage with many agglutinated forms (this additionally supports strong dissolution of carbonate tests), especially of *Spiroplectammina biformis*. This species was found only in minor amounts in the interval 20–40 cm, where *Elphidium clavatum* dominates.

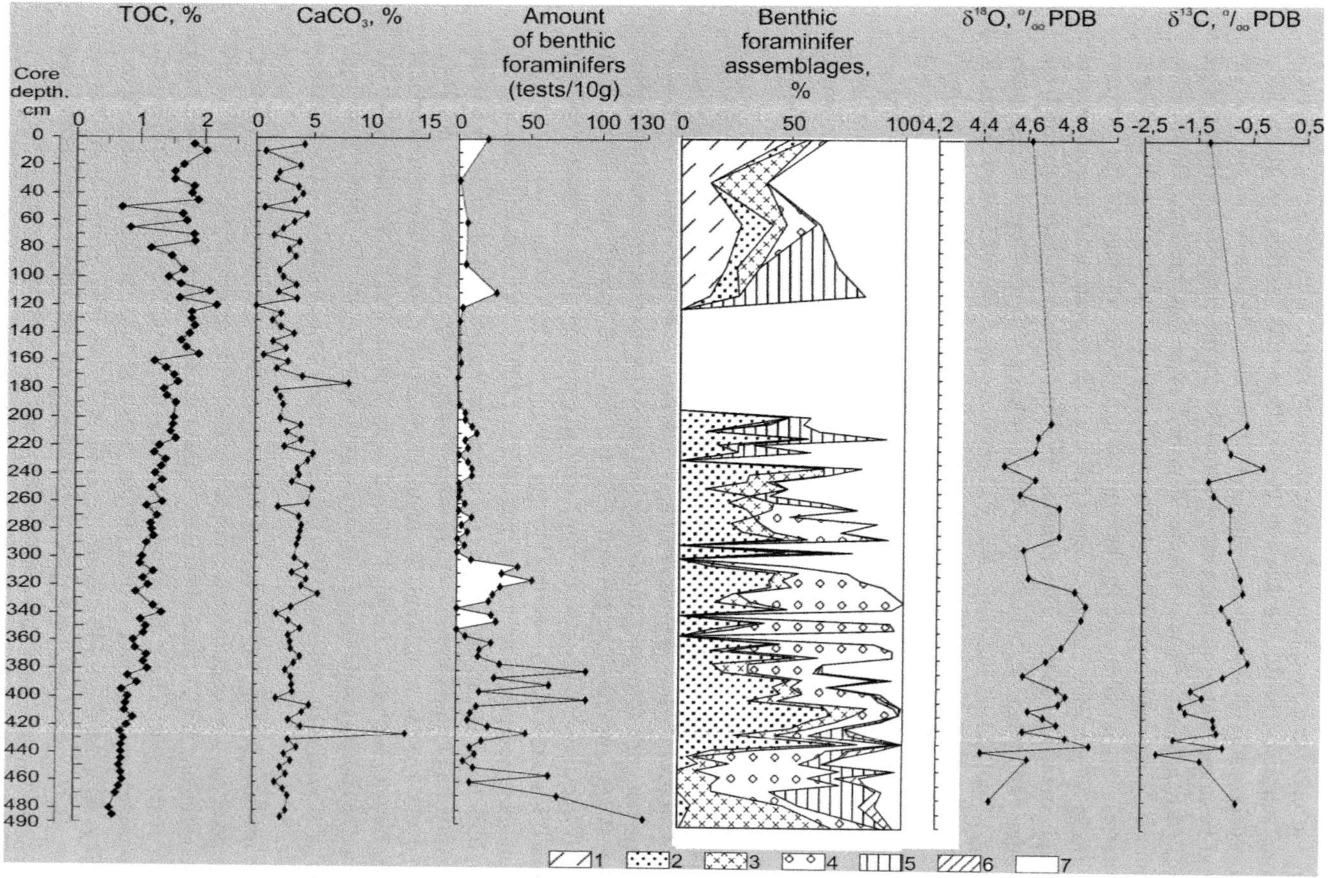

Fig. 10.18 Distribution of TOC, $CaCO_3$, benthic foraminifers, and oxygen and carbon isotope in *N. labradoricum* tests of the sediment core ASV 1157 (Levitan et al., 2004). 1–7 – species of benthic foraminifers: 1 – *Spiroplectammina biformis*, 2 – *Nonion labradoricum*, 3 – *Elphidium clavatum*, 4 – *Cassidulina reniforme*, 5 – *Haynesina orbiculare*, 6 – *Pyrgo williamsoni*, 7 – other species

Two forms of benthic foraminifers prevail in the interval of 70–258 cm: *Hainesina orbiculare* is a shallow form, which maintains strong distillation, and *Nonion labradoricum* is a relatively deeper-water form, which dwells in waters of normal salinity (Khusid and Korsun, 1996). Thus, this assemblage points to a certain redeposition of sedimentary material from the "shoulders" of the trough. It should be noted that in the interval of 120–190 cm the number of foraminifers is so insignificant that it could not be identified as an assemblage. The benthic foraminifers *N. labradoricum, Cassidulina reniforme, E. clavatum* (with dominant *N. labradoricum*) predominate in the normal-marine assemblage of the horizon 258–420 cm which may support somewhat lower temperature of bottom waters (Hald and Steinsund, 1996). The interval of 420–450 cm is characterized by a foraminifer assemblage in which the relative role of *N. labradoricum* decreases from top to bottom of the core, and *C. reniforme and H. orbiculare* increase. The presence of *Pyrgo williamsoni,* an euryhaline and almost euritherm benthic form, plays an important role. In general, the foraminifer assemblage has been accumulated under less saline conditions and lower temperatures in comparison with the overlaying interval. Finally, the lowermost interval (450–486 cm) is characterized by a benthic foraminifer assemblage dominated by *E. clavatum, H. orbiculare, C. reniforme, P. williamsoni*, suggesting a proximal glacial-marine environment.

Oxygen and carbon isotopes were studied in tests of *N. labradoricum* and *E. clavatum*. Due to the uneven distribution of *E. clavatum* in the sedimentary section the data given below are restricted to *N. labradoricum* (see Fig. 10.18). Based on isotope data, the following three intervals (from top to bottom) could be distinguished: 0–200 cm, 200–430 cm and 430–486 cm. The identification of the first interval is questionable because the only available data are from to the top and the bottom of this interval. The second interval from top to bottom is characterized by heavy oxygen isotopes (with values from 4.50 to 4.89‰ relative to PDB) and an uneven lightening of carbon isotopes (in the range from 0.30 to 1.80‰ relative to PDB). Maxima of light oxygen isotope values of are observed at intervals 230, 284–308, 380 and 420 cm. In general, fluctuations of carbon isotopes are anticorrelated with the oxygen isotopes. The third interval (only represented by two data points) is characterized by very light oxygen isotope values (see Fig. 10.18), related to a large supply of glacial melt water into the bottom waters, and distinctly heavier carbon isotopes.

The results of AMS ^{14}C datings of bivalves are listed in Table 10.14. At first, the obtained ages were corrected considering a marine reservoir age of 400 years (Duplessy et al., 2001), then they were converted into calendar years following (Stuiver et al., 1998). These data show that during accumulation of the entire 440 cm thick sedimentary sequence sedimentation rates did not change, reaching about 45 cm/ky. At a depth of 440 cm an age is of 9535 calendar years before present (cal. yrs. BP). We do not have data of sedimentation rate down to the bottom of the section. For the Barents Sea, substantially higher sedimentation rates have been repeatedly described for sediments of deglaciation stage, compared to overlaying marine sediments (Murdmaa and Ivanova, 1999; Polyak and Solheim, 1994; Polyak

Table 10.14 Age of bivalve mollusk shells from sediment core ASV 1157 [*Levitan et al.*, 2003b]

Lab number, Gif	Core depth, cm	^{14}C age, years	1 σ, years	Corrected ^{14}C age, years	Calendar age, cal. yrs. BP	Mollusk species
99 635	13	840	60	440	507	*Macoma calcarea*
99 636	50	1900	60	1500	1367	*Tridonta borealis*
99 637	165	3910	70	3510	3773	*Macoma calcarea*
99 638	338	6890	90	6490	7423	*Leda sp.*
99 639	355	7290	90	6890	7683	*Macoma calcarea*
99 640	378	7600	80	7200	7991	*Macoma calcarea*
99 641	393	8130	90	7730	8498	*Macoma calcarea*
99 642	398	8160	90	7760	8529	*Macoma calcarea*
99 643	423	8570	100	8170	9079	*Portlandia arctica*
99 644	440	8970	90	8570	9535	*Macoma calcarea*

et al., 1995; Hald et al., 1996). For interval 0–440 cm J.-C. Duplessy compiled the following equation:

$$t = 335 + 20.83\,h,$$

where t is calendar age (in years) and h is core depth (in cm) (Levitan et al., 2003a).

Discussion of Results

All presented results will be discussed versus calendar ages. In Table 10.15, the Holocene scale of Blitt-Sernander is given with ages of boundaries between stages (Merkt, 1999).

Based on this time scale, most intensive bioturbation is noted between 9500 and 8230 (i.e. at the very beginning of marine conditions), 5709 and 3022 (into the Subboreal Stage), 2605 and 1960, and 1360 and 460 cal. yrs. BP (i.e. into the Subatlantic Stage).

Table 10.15 Calendar age (cal. yrs. BP) of Holocene stages by Blitt-Sernander scale [*Merkt*, 1999]

Stages	Beginning	End
Subatlantic	2600/2400[1]	Now
Subboreal	6200/5700[1]	2600/2400[1]
Atlantic	9200	6200/5700[1]
Boreal	10640	9200
Preboreal	11560	10640

Note: [1] – depending on region

The duration of layer accumulation characterized by different content of the >0.05 mm fraction varies from 415 to 3645 years, moreover values on the order of 420 and 740 years are repeated several times. The boundaries between four basic stages of grain-size composition change are related to the intervals 8250, 4190 and 1585 cal. yrs. BP. The first stage has an Early Holocene age and is characterized by accumulation of relatively coarse-grained sediments. The same phenomenon is described for the Early Holocene of the almost entire eastern part of the Barents Sea (Murdmaa and Ivanova, 1999).

The extremely sharp (engaged of approximately 300 years) thinning of sediments is noted at the boundary of the first and second stages (at a level of 8250 cal. yrs. BP). By the way, this boundary also clearly coincided with the boundary between the layer of intensive bioturbation and the overlying layer with massive structure (see Fig. 10.16). Furthermore, during the second stage (from 7900 up to 4190 cal. yrs. BP) there is a steady increase in the sizes of sediment grains, which led toward the end of the stage of sediment accumulation of actually the same composition as at the end of the first stage. During the third stage (from 4190 up to 1585 cal. yrs. BP) the gradual (uneven) accumulation of thinner sediments is noted. Finally, during the last stage (from 1585 cal. yrs. BP to the present) an insignificant coarsening of sedimentary material is observed. In general, the layers with intensive bioturbation differ somewhat in terms of coarser-grained composition (see Fig. 10.16). Probably, this is explained by granulometric separation as a result of the vital activity of the digging organisms of zoobenthos.

If a change in granulometric composition of separate layers is possible to explain by the action of such local mechanisms as the activity of zoobenthos, the supply of slope material or a change in the speed of bottom currents, then the noted stages of granulometric composition changes, which are significant in terms of time period, require another explanation. From the beginning of the Holocene to about 6000 cal. yrs. BP, the relative uplift of continental blocks surrounding the Barents Sea probably depended on the combination of global eustatic sea-level rise (Fairbanks, 1989) and regional glacio-isostatic ascending motions (Forman et al., 1999).

After 6000 cal. yrs. BP, when the sea level became relatively stabilized, the glacio-isostasy has begun to play a dominant role. It is of interest that in the Laptev Sea, located far beyond the limits of the zone of intensive Late Valday glaciation (LGM) (Svendsen et al., 1999), the very sharp reduction of sedimentation rates was noted precisely after 6000 cal. yrs. BP (Bauch et al., 1999). The processes of glacio-isostatic rise occurred unevenly, with periodic delays, caused by changes in rheological properties of lithosphere and by adaptation to the new environment. Thus, for the Northern Island of Novaya Zemlya a stop of glacio-isostatic motion was observed from 4400 to 2500 cal. yrs. BP (Forman et al., 1999), and for Kola Peninsula from 8000 up to 4500 cal. yrs. BP (Corner et al., 1999).

It is possible to assume that during such a stop finely dispersed terrigenous material was supplied into the sedimentation basin, while upon intensification of the ascending motions more coarse-grained sediment was added. The presence of outcrops of peat with the radiocarbon age of 15 ka (Serebryanny and Malyasova, 1998), in the west of the Southern Island (in the northern part of the peninsula of the Goose

Land) (see Fig. 10.15), most likely suggests a small thin ice sheet on the island during the Late Valday (LGM). Hence, weak variations in the rates of glacio-isostatic rise, which were incapable to influence changes of sedimentation rates, and which were registered only in the grain-size composition of the Southern Novaya Zemlya Trough sediments.

At the same time it is not possible to eliminate the specific effect of climatic (paleo-glaciological) changes to the grain-size composition. From this point of view, it is logical to assume a development of the layered unit in the lower part of the sedimentary section due to year-round existing sea ice, as typical for deglaciation stage. Then, accumulation of relatively coarse-grained sediments developed after the accumulation of the layered unit, can be explained by the disappearance of a closed sea-ice cover above the trough, caused by the Early Holocene warming.

The accumulation of very fine-grained sediments with the above mentioned massive structure, beginning at 8250 cal. yrs. BP and lasting about 300 years, probably correlates with the known Holocene short-term temperature drop at about 8.2 cal. yrs. BP, which may have caused the development of a more closed sea-ice cover in the studied region. In this case, the absence of bioturbation may suggest a sharp reduction in primary production. The nature of this sea-ice cover is not clear yet: was it "normal" sea ice, or did a floating glacier exist as the peripheral part of the Southern Novozelsky glacial cupola? Then, at 4190 cal. yrs. BP (i.e. at the end of the Atlantic stage and the beginning of Subboreal stage) the grain size of sediments was increasing in connection with the development of fine layers of seasonally formed sea ice.

The period from 4190 to 1585 cal. yrs. BP probably corresponds to the "interglacial" stage of the Southern Island development, which was expressed in thinning of sediments because of the development of a more distinct seasonal sea-ice cover in the trough area. The last 1600–1500 cal. yrs. BP were possibly warmer which led to the recent climatic environment with the development of seasonal sea ice of insignificant extension. The glacier was not only finally retreated from the north of the Pechora Sea, but also generally melted at the surface of the Southern Island of Novaya Zemlya. Accordingly, bottom sediments have become somewhat coarser-grained. For testing this hypothesis (see Fig. 10.15), more detailed studies of the terrestrial sections of Holocene sediments of the Southern Island are needed.

The study of mineral composition showed that sediments, which have been accumulated at the end of deglaciation (up to approximately 10000 cal. yrs. BP), are characterized by the increased content of epidote and by high epidote/garnet and low clinopyroxene/epidote ratios. In total these data indicate an increased role of the southern sources, connected with the eastern part of the Pechora Sea (Levitan et al., 2003b), against the background of the dominant role of the Southern Island of Novaya Zemlya as the basic source of terrigenous material. A certain (substantially smaller) sediment input from the southern sources (indicated by the clinopyroxene/epidote ratios) continued up to 4500 cal. yrs. BP (i.e., the middle part of the Subboreal Stage).

Noteworthy is the period from 8250 to 4920 cal. yrs. BP, during which a significant quantity of grains with iron coatings were supplied into the bottom sediment. This period sufficiently well coincides with a second change in the grain-size composition described above. Probably, this phenomenon is caused by the accumulation of "coated" grains due to strengthening of chemical weathering during the Middle Holocene stage, because in the south of the Southern Island the climate optimum occurred during the Atlantic Stage approximately from 8.2 to 6.3 ka (Serebryanny and Malyasova, 1998), i.e., from 8700 to 6600 cal. yrs. BP.

A slightly increased content of hornblende and micas in the sediments of this age was most likely caused by the widespread development of outer-delta silts in the Pechora Sea during the Middle Holocene (Levitan et al., 2000; Polyak et al., 2000a), and partial by the supply of minerals from these coeval silts into the trough. The decrease of hornblende and muscovite content in the Lower and Upper Holocene sediments of the trough, which were accumulated at the time when marine sands were spread to the south of the trough, support these assumptions (Polyak et al., 2000a).

The TOC content continuously increased from the beginning of the considered time period until about 2940 cal. yrs. BP. Then, up to 1335 cal. yrs. BP a decrease in TOC occurred, followed again by an increase. Regretfully, no data on the organic-carbon source, i.e., marine versus terrigenous, are available yet in the investigated sediments. Thus, a general predominance of marine organic carbon over terrigenous organic carbon is assumed, based on the data from surface sediments of the Barents Sea (Romankevich and Vetrov, 2001), and processes of organic-carbon mineralization during diagenesis were considered.

Thus, increased primary production, which controlls to a considerable degree the TOC supply into the bottom sediments, is most likely restricted to the more recent epoch, i.e., the late part of the Subatlantic Stage. For the period from 2940 to 1335 cal. yrs. BP, a certain decrease of paleoproductivity due to neoglaciation is assumed. The noted decrease in TOC content in older sediments, however, should not only be explained by changes in paleoproductivity, but, first of all, by diagenesis. Intervals of intensive bioturbation in older sediments may suggest episodes of increased activity of burrowing zoobenthos, probably triggered by increased paleoproductivity.

In sediments accumulated at the end of the deglaciation period, low carbonate contents of 2.4 to 3.0% are noted. After the end of deglaciation period until 8670 cal. yrs. BP (i.e., approximately prior to the beginning of the Atlantic Stage), a steady increase of the background $CaCO_3$ content up to 5.0% is observed. At 8670 cal. yrs. BP, the carbonate content sharply decreases to 2.0% and stays more or less stable until 7420 cal. yrs. BP (middle of the Atlantic Stage), when within 200 years the $CaCO_3$ content increased to 5.5%. Later, at 4085 cal. yrs. BP (middle of Subboreal Stage) a clear decrease of the carbonate content to 1.8% was observed. Then, up to the recent stage the $CaCO_3$ content remained stable. It seems that precisely within the period from 7210 to 4085 cal. yrs. BP, i.e., in the Middle Holocene, the most intensive dissolution of biogenic carbonates occurred. The reason for this phenomenon was probably an increased paleoproductivity, which led to the accumulation of high TOC concentrations in the sediments and, therefore, to an intensive formation of CO_2 due to diagenetic processes at the sediment/bottom water boundary, that also contributed to the dissolution of carbonates.

Support for this interpretation is coming from the zones of strongest dissolution of benthic foraminifers in the surface sediments of the Kara Sea occurring in sediments most enriched in TOC (Hald and Steisund, 1996). In the eastern and southern Kara Sea (Levitan et al., 2000a; Ivanova, 2002) and in the Franz Victoria Trough (Duplessy et al., 2001) intensive dissolution of foraminifers was observed, starting at 6.9 ka in both areas (7.3 cal. yrs. BP). An increase in paleoproductivity implied from different parameters, started at 7.0 ka (7.3 cal. yrs. BP) in the Fram Strait area (Levitan et al., 2000a; Hald and Aspeli, 1997; Wollenburg et al., 2001), during the Middle Holocene in the St. Anna Trough (Hald et al., 1999), and at about 7.2 ka (7.6 cal. yrs. BP) in the eastern part of the Central Arctic Ocean (Stein et al., 1994a). In all these studies, this phenomenon of increased paleoproductivity was related to an intensification of advection of warm and salty Atlantic waters into the Arctic Ocean. These results are the basis for the interpretation of the relationship between increased paleoproductivity and increased dissolution of biogenic carbonates in the sediments of the Southern Novaya Zemlya Trough.

Peak numbers of planktonic foraminifers were registered for the time period from 6730 to 6584 cal. yrs. BP (at the end of the Atlantic Stage), related to a short-term migration of the Kanin-Kolguev Current to the east that occurred only 200–350 years after the Holocene climate optimum in the Franz Victoria Trough in the north of the Barents Sea (Duplessy et al., 2001). Earlier, from 8670 to 8250 cal. yrs. BP (at the beginning of the Atlantic Stage), the well-expressed maximum in the number of benthic foraminifers is probably connected with the known short-term temperature drop at 8200 cal. yrs. BP (Merkt, 1999).

The same phenomenon was observed in the sediments from the Yermak Plateau, where maximum values planktonic and benthic foraminifers often occurred in adjacent layers, because planktonic foraminifers are abundant under conditions of increased paleoproductivity. The latter caused high concentrations of dissolved CO_2 which lead to the strong dissolution of benthic foraminifers (Wollenburg et al., 2001). Therefore, the maximum numbers of benthic foraminifers characterize conditions of increased, but not maximum paleoproductivity.

Data on benthic foraminifers may indicate the stages of changing characteristics of bottom-water mass. In the Boreal Stage (up to 9710 cal. yrs. BP) glacial-marine conditions with very cold and partially freshened bottom waters were found. During the period from 9710 to 9083 cal. yrs. BP this water clearly became more saline and warm. From 9083 to 5710 cal. yrs. BP (i.e., during the Atlantic Stage) bottom waters had a normal salinity, but were somewhat colder than today. Later, conditions close to the recent environment existed. In the period between 5710 and 1795 cal. yrs. BP the supply and redeposition of some benthic foraminifers occurred, that dwelled at higher bathymetric levels, in the fresher and warmer water mass, on the bottom of the trough.

According to oxygen isotope data on benthic foraminifers, the inflow of glacial meltwaters into the bottom water layers ceased about 9300 cal. yrs. BP. A sharp isotope shift to 0.5‰ occurred for approximately 110 years at the boundary of the Boreal and Atlantic stages. Further isotopic trends support a change in the characteristics of bottom-water mass toward lighter oxygen-isotope composition. Taking into account the recent T-S parameters of bottom waters, in particular, the temperature of bottom-water mass near the freezing point, it can be assumed that changes in fresh-water supply played a much more important role. It is thought that due to brine formation some part of the light oxygen isotope could be seized from the surface waters and transferred to the bottom.

If this is true, maxima in brine formation occurred at 9084, 8250, 6751–6251 and 5126 cal. yrs. BP. These maxima had durations from a few hundred to 1.2 thousand years. It has to be mentioned that these conclusions are only related to the part of the section, for which isotopic data are available. The Holocene thermal "marine" optimum, noted for the Franz Victoria Trough at 7.85–6.90 cal. yrs. BP (Duplessy et al., 2001), was not observed in the sediments of the Southern Novozemelsky Trough, where the Atlantic waters have probably never reached the bottom during Holocene times.

The short-term increase in the number of planktonic foraminifers mentioned above is possibly related to the irregular migration of the Kanin-Kolguev Current to the east. The noted differences in the isotopic composition of bottom-water masses during the period of deglaciation and during the following marine stage suggest the existence of a constant sea-ice cover above the trough during deglaciation stage, which changed into an ice regime with a seasonal sea-ice cover at about 9300 cal. yrs. BP, because only a seasonal ice cover allows the formation of cold brines.

There is a certain interest in the identification of the boundary between sediments of deglaciation stage and strictly marine sediments. Sediments of the deglaciation stage differ in terms of the presence of lamination structure, high epidote content, high epidote/garnet ratios, very low and stable TOC concentration, a specific fauna of benthic foraminifers, and a characteristic oxygen-isotope and carbon-isotope composition of foraminifers. As shown above, the position of the boundary is differed for 45 cm core depth according to different parameters. The boundary between both facies was probably caused by climatic changes. Here, benthic foraminifers and their isotopic composition are the most sensitive indicators of climate change in our study. Consequently, the deglacial stage ended at 430 cm core depth, i.e., about 9300 cal. yrs. BP. From the number of different litho-geochemical parameters the most clearly given boundary is reflected in the distribution of TOC. The remaining parameters appeared to change somewhat earlier, i.e. approximately near 10000–9500 cal. yrs. BP. This difference is probably caused by the specific independence of each of the parameters and, correspondingly, by different reaction rate to the proceeding climatic changes.

Because it follows from data given above that the Southern Island of Novaya Zemlya served as the basic supplier of sedimentary material for the trough during the Holocene, it is possible to assume that all sediments delivered from the island southwards was accumulating in the natural sedimentary trap presented by the trough. Consequently, the average sedimentation rate of 45 cm/ ky (or <0.5 mm/year) is approximately equal to the average rate of the island uplift during the Holocene.

Taking into account seismo-acoustic data (Gataullin et al., 2001), the uplift rate was probably about 1.0 mm/year, which corresponds to the very low rates of glacio-isostatic rise (Trifonov, 1999). This supports an insignificant extension of a glacial ice sheet on the island.

Thus, glacial-marine sediments found in the lowermost part of core ASV 1157, were accumulated at the end of the deglaciation stage, i.e., after the Younger Dryas phase of glaciation. These sediments are characterized by a glacial-marine assemblage of benthic foraminifers and an oxygen-isotope composition indicating an active delivery of melt waters. A sea-ice cover existed throughout the year. The Southern Island of Novaya Zemlya was the basic source of sedimentary material, however, there was a slight influence of southern sources (i.e., the area of the Pechora Sea) (Fig. 10.19).

Then, with the onset of the marine stage of sedimentation conditions characterized by a seasonal sea-ice cover, production of cold brines, and variable intensity of paleoproductivity were developed. During the Atlantic Stage, a warmer climate causing a noticeable displacement of the timber line and tundra in the European part of Russia to the north (MacDonald et al., 2000) and a change from a psychrophilic tundra vegetation to a more hemophilic flora on the Southern Island of Novaya Zemlya (Serebryanny and Malyasova, 1998), led to the partial degradation of permafrost of the Timan-Pechora lowland and an increase in water supply by the Pre-Pechora River (MacDonald et al., 2000).

Outer-delta clayey silts were developed almost all over the entire area of the recent Pechora Sea, up to the southern edge of the studied trough. Therefore, part of the sediments entered the trough due to the delivery of these silts. Simultaneously with the climate warming, chemical weathering on the Southern Island was strengthened (see Fig. 10.19). Beginning at approximately 7000 cal. yrs. BP, intensive advection of warm and saline Atlantic waters into the Arctic Ocean contributed to the increase of primary production in the West Arctic and, as a result, to intensive dissolution of the carbonate tests of benthic foraminifers (see Fig. 10.19).

Then a gradual temperature drop led (1) to the retreat of the timber line and tundra to the south during the Subboreal Stage between 4500 and 3200 cal. yrs. BP, representing the beginning of neoglaciation (MacDonald et al., 2000), (2) to a decrease of the Pre-Pechora River discharge, (3) to a substitution of outer-delta silts by marine sands, (4) to a weakening of the influence of southern sources on sedimentation processes in the trough, and (5) to an increase of physical weathering.

The Southern Island which main relief features were developed prior to the Holocene, has undoubtedly served as the main source of material for sediment accumulation in the Southern Novozemelsky Trough during the entire Holocene period (see Fig. 10.19). The uneven glacio-isostatic rise during the Holocene coinciding with paleo-glaciological changes, occurred in four stages that were reflected in the grain-size composition. Moreover, sedimentation rate remained constant at 45 cm/ky. Conditions for layer formation have changed about every several hundred years, apparently reflecting fluctuations of bottom-current rates in the trough, delivery of material by gravitational processes from the slopes, and other possible factors.

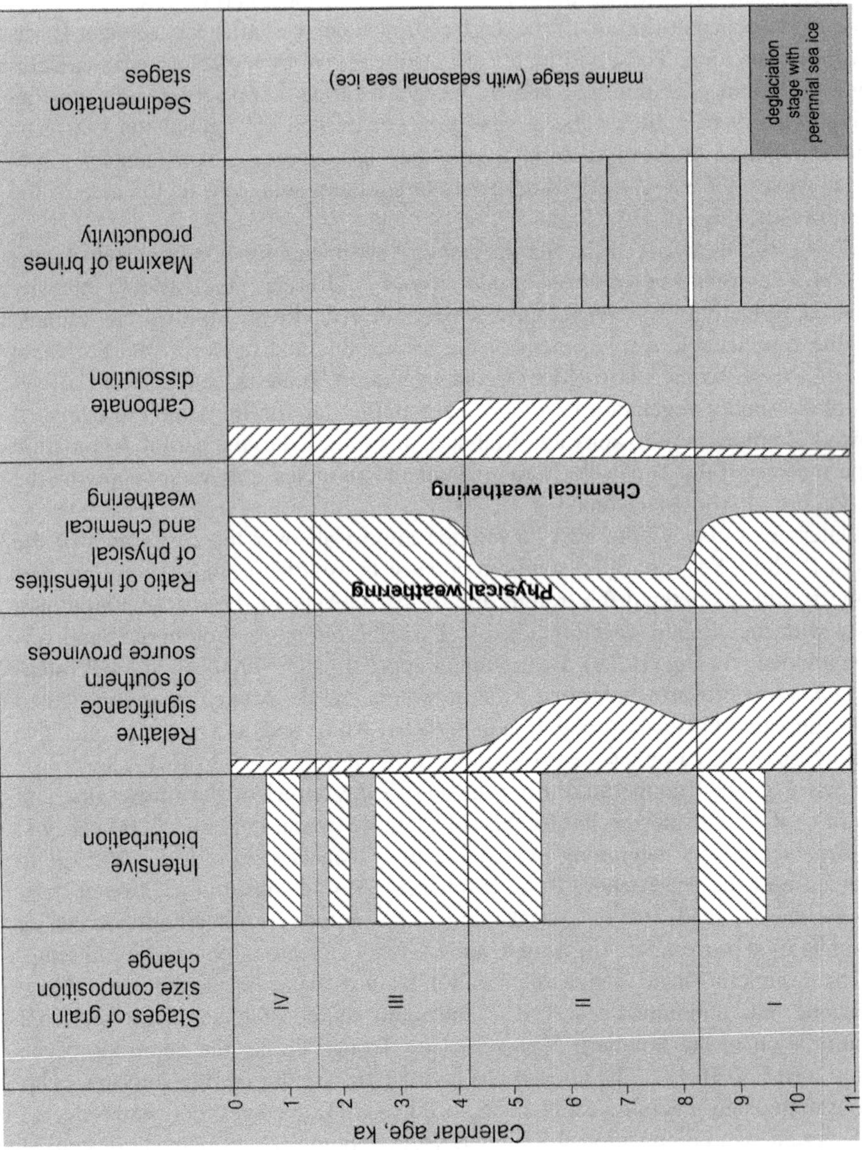

Fig. 10.19 Changes of some sedimentation parameters in the South Novaya Zemlya Trough during Holocene times (Levitan et al., 2004)

History of Sedimentation in the Pechora Sea During the Late Pleistocene and Holocene

Material and Research Results

The present paragraph is mainly based on data published by (Levitan et al., 2003b; Mel'nikov and Spesivtzev, 1995; Gataullin et al., 2001; Levitan et al., 2000; Polyak et al., 2000a). These studies demonstrated that marine-geological data of Quaternary and, in particular, Holocene sediments of the Pechora Sea allowed to identifiy two basic types of sedimentary sections. The first type is extended in the extreme northwest of the Pechora Sea, and the second type is located in the southeast. The position of the boundary between both types of sections is shown in publications (Gataullin et al., 2001; Polyak et al., 2000a).

Upper Valday and Lower-Middle Valday glacial diamictons divided and overlain by non-glacial Upper Quaternary sediments (Polyak et al., 2000a), were recovered in the northwest of Kolguyev Island in marine bore holes. In the south-eastern part of the sea, only Lower-Middle Valday glacial formations revealed were recovered in bore holes. The Late Valday and the period of deglaciation are not represented in the bore holes due to a hiatus. The sections of Upper Quaternary sediments on Vaygach Island and on the north of the Pechora lowland (up to the line of Markhidy) are of the same type as described above (Mangerud et al., 1999). The Lower-Middle Valday sections with an erosional unconformity overlays bedrock of Late Cretaceous age.

The section typical for the southeast of the Pechora Sea was recovered in a number of closely spaced bore holes B–210–218 (Fig. 10.20) with the following lithologies (from bottom to top): gray terrigenous sandstones and siltstones of Late Cretaceous age (125–115 m); diamicton of Early-Middle Valday age (115–65 m); gray clayey silt and silty clays with fauna indicating an open-sea environment typical for interglacial periods (65–37 m) (Okuneva, 1991); layered silts which were accumulated in an outer-delta environment adjacent to the coast (37–28 m); Holocene marine sands, divided into an interval of 15–5 m of silty clays and silts on top (Polyak et al., 2000a).

It is especially important that all varieties of Quaternary sediments (except the diamicton) are dated by AMS ^{14}C method. Only one dating (37.1 ka), which is likely to be too young, is from the diamicton. The gray silty clays are dated to 39.4–28.3 ka, the layered silt – to 28.3–23.6 ka, and the lower sand unit to 9.7 to 9.1 ka (Polyak et al., 2000a). The middle Holocene silty clays have an age of 8.2 to 5.0 ka (Tarasov et al., 2000; Polyak et al., 2000a), and the upper unit of the Holocene sand can be distinguished by a series of values of younger than 5.0 ka (Table 10.16) (Levitan et al., 2000; Polyak et al., 2000a). Furthermore, the age of the bottom of the Holocene "blanket" sand decreases from north to south.

The sedimentary sections recovered by gravity cores at the Prirazlomnoe and Varandey sites (for location see Fig. 10.20) consist of (from top to bottom): Upper Holocene sand and "transitional" layer; Middle Holocene "black clays" and "gray clays"; Lower-Middle Valday (?) diamicton. According to seismo-acoustic data, the Holocene sediments contain much gas in some areas, which sometimes escapes into

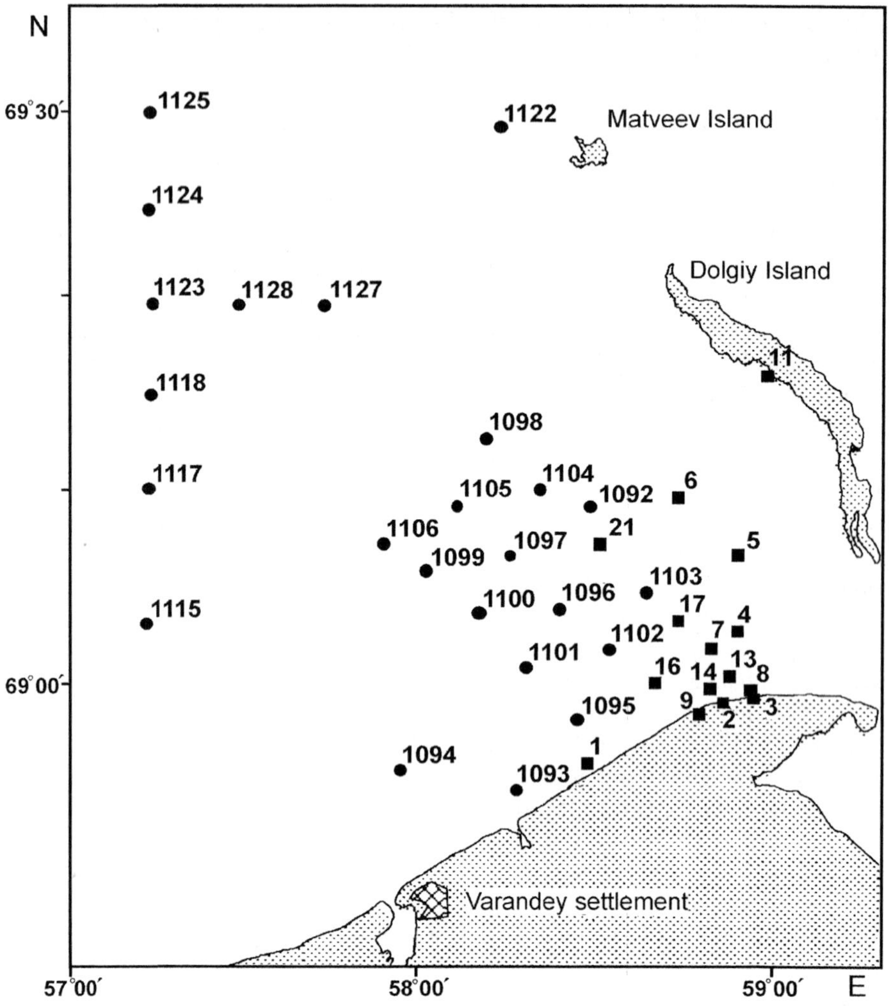

Fig. 10.20 Location of geological stations in the Pechora Sea

the bottom water through pockmarks (e.g., Station 1123) (Levitan et al., 1999). The grain-size characteristics of the sediments (with the exception of the "transitional" layer) are given in Table 10.17, and some characteristics of their chemical composition and physical properties are presented in Table 10.18.

Based on the data of Table 10.17, fine and fine-grained sands prevail in the sandy unit. Among the "black clays", sandy-clay and clayey-sandy silts are most common, and the "gray clays" are characterized by finer composition with a predominance of sandy-silt and silty clays. The matrix of the diamicton resembles the composition of "black clays". Sedimentation rates in the sandy unit vary from 26.9 to 149.1 cm/ky (Levitan et al., 2000) and rapidly change within the sedimentation basin.

Table 10.16 Radiocarbon age of bivalve mollusk shells *Astarta borealis* and *Macoma calcarea* from the "blanket" sand of the Pechora Sea studied in 13-th cruise of RV *Academik Sergey Vavilov* [*Levitan et al.*, 2000], with changes

Station ASV	Horizon, cm	Lab number of the sample	Radiocarbon age, years	Corrected age, years
1095	5–10	IOAN–4206	540±100	140
1100	75–87	IOAN–4201	3630±160	3230
1100	100–110	IOAN–4203	3630±180	3230
1115	70–80	IOAN–4204	950±170	550
1117	120–126	IOAN–4202	2950±210	2550
1124	75–84	IOAN–4205	2390±100	1990

Note: Radiocarbon dates are received by benzol method in P.P. Shirshov Institute of Oceanology, RAS by V.M. Kuptsov; the species determinations of bivalve mollusks are made by N.V. Kucheruk; the location of stations is shown at Fig. 10.12.

Data on grain-size and chemical sediment composition (see Table 10.18) suggest that the "black" and "gray" clays were formed in an outer-delta environment. Moreover, "black clays" were probably developed closer to the northern limitation of the mixing zone of river and sea water under conditions of increased primary production.

Coarse-grained material of different degree of roundness is present in the sandy as well as the clay units presented by dark gray solid sandstones, siltstones, argillites and clayey shales. The sizes of clasts vary from the gravel to the pebbles (up to 6×4×2 cm in size). A fragment of brownish-black peat 8×6×4 cm was found at Station 22. The clastic content increases towards the coastline. A significant part of this coarse-grained material probably enters the seafloor due to coastal abrasion and melting of coastal shore ice.

The sandy and clayey units also contain shells of mollusks (bivalves); *Astarta borealis* and *Macoma calcarea* predominate. Their sizes increase towards the coast. Shells of gastropods were found more rarely. Mollusks were not found in the diamicton.

A diamicton was observed in six cores at the Varandey site and, probably, outcrops at the seafloor in the area of Station 1092. The diamicton is very solid or semi-solid and has a gray and dark gray color as well as a massive homogeneous structure. Angular or weakly rounded fragments of rock (including of dark gray limestone) of different size (with length up to several centimeters) occur in rare abundance (a few percentages of the rock volume in the core). The matrix is mainly composed of mictite with a predominance of sand or silt fractions (see Table 10.17). Its physical characteristics (in particular, the very high shear strength) suggest an Early-Middle Valday age.

For most of the records, the boundary between the diamicton and the overlaying "clayey" unit probably represents a hiatus. At stations 23, 1095 and 1102, this boundary is presented as a fine layer containing weakly rounded fragments of gravel to pebble size, which may indicate erosional processes in the uppermost part of the diamicton.

The "transitional layer" is composed of a chaotic mixture of lenses of pure sand and lenses of underlaying "black clays". Moreover, the latter are characterized by a irregular shape and a strongly dismembered surface. They could fall into the described layer only immediately after accumulation of the "black clays" layer (or almost simultaneously), when compaction of these sediments has not just started. All sedimentological and radiocarbon data (see Table 10.16) suggest a geologically instant accumulation of the "transitional layer", possibly under conditions of a very high hydrodynamic activity of bottom waters (benthic storm?) at the base of sandy bars. This interpretation would contradict the assumption proposed by L.V. Polyak (oral communication) about the Middle Valday age of the described clayey-silt sediments ("black and gray clays").

Table 10.17 Grain-size composition (wt. %) of the Pechora Sea Quaternary sediments [*Levitan et al.*, 2003b]

Station No.	Horizon, cm	>10	10–7	7–5	5–3	3–2	2–1	1–0.5	0.5–0.25	0.25–0.1	0.1–0.05	0.05–0.01	0.01–0.005	0.005–0.001	<0.001	Lithotype
								Sandy unit								
1096	5–10	–	–	–	–	–	–	0.04	0.70	39.24	48.47	1.89	9.66	–	–	Small and fine-grained sand
1097	20–30	–	–	–	–	–	0.04	0.05	9.29	76.79	7.72	0.58	5.53	–	–	Small-grained sand
1098	110–120	–	–	–	–	–	–	0.05	0.05	56.60	38.80	1.28	3.22	–	–	Fine and small-grained sand
1100	20–30	–	–	–	–	–	–	–	4.35	6.37	83.11	0.62	5.55	–	–	Fine-grained sand
1102	105–110	–	–	–	–	–	–	0.12	3.60	39.13	26.40	7.09	2.38	5.52	15.76	Silty-clayey. sand
2	10–20	–	–	–	–	–	0.23	0.12	1.60	70.64	21.47	1.45	4.49	–	–	Fine and small-grained sand
12	40–50	–	–	–	–	–	–	0.04	1.29	59.80	31.65	3.16	4.06	–	–	Fine and small-grained sand
23	18–25	18.99	8.87	5.74	9.90	7.29	9.35	3.37	11.97	17.19	4.97	0.84	0.92	–	–	Mixed-grained sand with gravel and pebble
								"Black clays"								
1097	140–150	–	–	0.40	–	–	–	0.10	0.80	19.88	10.34	9.24	9.44	20.87	28.93	Clay-sandy silt
1098	50–60	–	–	–	–	–	0.08	–	0.16	7.25	3.62	14.03	9.38	25.84	39.64	Sandy-clayey silt
1099	95–105	–	–	–	1.44	–	0.34	0.17	0.42	15.83	24.05	13.97	4.91	12.11	26.76	Clayey-silty sand
1101	60–70	15.66	–	0.17	0.75	1.24	1.49	0.83	0.99	11.18	15.66	8.95	8.37	16.90	17.81	Clay-sandy silt with pebble
1103	40–50	–	–	–	0.40	0.24	0.40	0.16	0.80	6.46	8.70	19.23	7.82	11.57	44.22	Sandy-silty clay
1104	90–100	–	–	–	–	–	–	0.35	0.27	5.92	20.41	16.96	13.96	17.40	24.73	Clay-sandy silt
1104	115–125	–	–	–	–	–	–	0.17	0.42	5.68	3.39	5.17	6.27	22.63	56.27	Silty clay
1105	30–40	–	–	–	–	–	–	–	0.38	21.13	7.07	14.34	6.79	18.36	31.93	Sandy-clayey silt
1106	30–40	–	–	–	–	–	–	–	0.24	1.90	15.04	41.73	0.24	16.46	24.39	Sandy-clayey silt

Table 10.17 (continued)

"Gray clays"																
13	20–30	–	–	–	–	1.00	0.80	0.51	1.31	10.67	9.14	9.80	8.56	19.45	38.76	Sandy-silty clay
13	80–90	–	–	–	0.66	0.50	0.33	0.33	0.66	5.74	5.41	10.40	11.06	21.05	43.86	Sandy-silty clay
14	70–80	–	–	–	–	–	–	0.58	1.50	2.98	3.15	8.75	12.27	20.97	49.80	Silty clay
15	40–50	–	–	–	–	–	–	–	–	0.57	1.41	12.87	15.35	31.61	38.19	Clayey silt
23	35–40	–	–	–	–	–	–	0.67	0.53	4.47	8.07	18.93	12.27	17.86	37.20	Sandy-clayey silt
Diamicton																
1095	80–90	–	–	–	–	–	0.15	0.07	0.45	22.80	18.39	11.81	4.48	5.01	36.84	Silty-clayey sand
1096	95–105	–	–	–	–	–	–	–	–	0.15	1.30	22.76	12.72	22.76	40.31	Clayey silt
1096	120–130	–	–	–	–	–	–	–	–	0.21	5.76	12.09	10.12	28.04	43.78	Clayey silt
1102	150–160	–	–	–	–	–	–	0.07	0.29	5.56	13.40	21.16	8.35	14.49	36.68	Sandy-silty clay
1104	140–150	–	–	2.89	1.44	–	5.77	5.77	2.02	14.79	14.43	20.20	4.40	7.22	21.07	Clayey-silty sand
15	70–80	–	–	–	–	–	–	–	–	0.29	2.13	14.57	13.98	21.92	47.11	Clayey silt
23	35–40	–	–	–	–	–	–	0.67	0.53	4.47	8.07	18.93	12.27	17.83	37.20	Sandy-clayey silt

Note: Size of fractions is given in mm; blank – lack of fraction; grain-size analyses were performed in the Analytical laboratory of P.P. Shirshov Institute of Oceanology, RAS by V.P. Kazakova.

Table 10.18 Arithmetic mean values of TOC and $CaCO_3$, and the physical properties values of the basic Quaternary sedimentary units [*Levitan et al.*, 2003a]

Unit	Sandy	Transitional	"Black clays"	"Gray clays"	Diamicton
TOC (wt. %)	0.44	0.56	0.78	0.60	0.54
$CaCO_3$ (wt. %)	0.78	1.27	3.22	3.05	2.51
Water content (wt. %)	22.0	30.0	46.0	33.0	23.0
Bulk density (g/cm^3)	2.03	1.89	1.71	1.84	2.01
Shear strength (kPa)	–	33.0	18.0	39.0	126.0

Note: TOC and $CaCO_3$ were determined by N.P. Tolmacheva in P.P. Shirshov Institute of Oceanology, RAS on the express carbon-analyzer An-7529; physical properties were determined by A.V. Kondratenko (VNIIOkeangeo).

In general, the relationship between the layers and units in the coastal zone of the southern Pechora Sea described above, is complex due to strong variability (Levitan et al., 2000) and frequently occurring hiatuses. A geological model of the structure of the studied Quaternary sediments is shown in Fig. 10.14. To the north, in the area of the Prirazlomnoe site, however, the geological situation becomes more stable, the top of diamicton is sharply immersed (Mel'nikov and Spesivtzev, 1995), and the thickness of all Holocene layers noticeably increased.

History of Sedimentation

The Late Quaternary sedimentary history in the Pechora Sea reflects a complex interaction of climate change (i.e., changes between glacial-deglacial-interglacial stages and Holocene interglacial variability), glacio-isostasy and neotectonic motions, which led to repeated changes in the position of sea level, variations in the delivery of river discharge, and fluctuations of the hydrodynamic regime in the sedimentatary basin.

In first half of Middle Valday (45–60 ka) the Pechora Sea was occupied by a continental glacier, which originated from the Kara Sea through the Vaygach Island and south of the Southern Island of Novaya Zemlya (Larsen et al., 1999; Polyak et al., 2000a). In particular, the carbonate rock fragments found in the described diamicton confirm this assumption. After the end of this glaciation period, the diamicton underwent erosion in the territory north of the Pechora lowland and recent coastal zone. In the more northern open-sea areas silty-clayey muds were accumulated, which changed into layered outer-delta silt facies at about 30 ka as a result of sea-level fall and, possibly, strengthening of delivery of river discharge.

Furthermore, the period of sea-level fall was associated with accumulation under terrestrial conditions of the Late Valday glacier, which extended to the northwest of the recent Pechora Sea from the Barents Sea and the Southern Island of Novaya Zemlya (Polyak et al., 2000a). The southeastern part of the study area was most likely to be eroded at this time under conditions of alluvial plain. The paleo-hydrological reconstructions indicate that during the Late Valday the average annual discharge of the Pechora River reached about 220 km^3 (Sidorchuk et al., 1999) in comparison with the recent value of 130 km^3.

Based on the data obtained from the southern Barents Sea (Polyak et al., 1995), deglaciation began approximately 13.5 ka and ended about 10 ka (i.e., earlier than in the Southern Novaya Zemlya Trough where deglaciation was dated to about 9.3 cal. kyrs. BP (Levitan et al., 2003a)). In the southeast of the study area, the sediments of this age were either eroded in the course of the beginning of the sea-level rise or not accumulated at all.

The marine transgression started in the Early Holocene (10–8 ka), related to the global eustatic sea-level rise (Fairbanks, 1989). This process was probably more important than the glacio-isostatic rise of the seafloor, which was apparently very small due to the relatively small extent and thickness of the Upper Valday glacial ice sheet in the studied area. Under such conditions the basal unit of marine sands was accumulated in the Pechora Sea.

The noticeable changes occurred in the Middle Holocene (8–5 ka). At the end of this period, eustatic sea-level rise ceased (Fairbanks, 1989) and the climate of the adjacent land became much warmer and more moist (MacDonald et al., 2000). Humidity increased due to both atmospheric precipitations and degradation of permafrost. Under these conditions, river discharge significantly increased, and marine sediments have changed into outer-delta silt and silt-clay deposits in most part of the Pechora Sea (with exception of the Southern Novozemelsky Trough, where a marine environment existed).

In the Late Holocene, i.e., after the end of the climatic optimum, the climate of the surrounding land became cooler and drier. Possibly, just at this time (after 5 ka) the complex of the insular barrier Gulyaevskie Koshki was formed, which blocked most of the solid discharge of the Pechora River in the Pechora Bay. As a result, the river discharge entering the remaining sea area was sharply reduced, and gradually

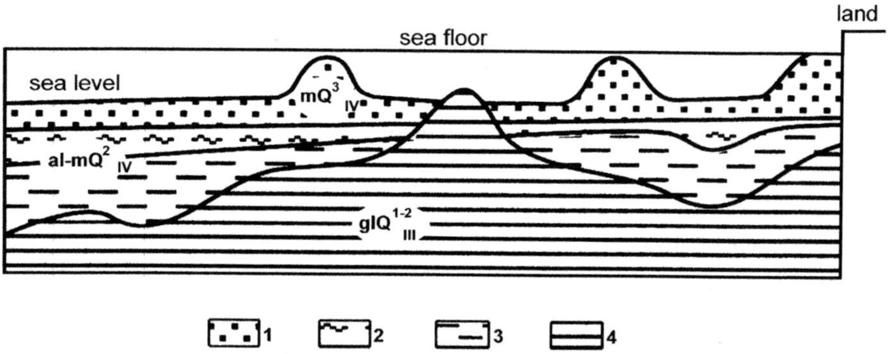

Fig. 10.21 Geological model (profile) of Quaternary deposits in the south-eastern Pechora Sea (Levitan et al., 2003b). 1 – sand; 2 – "black clay"; 3 – "gray clay"; 4 – diamicton

(earlier in the north and later in the south) outer-delta fine sediments changed into a blanket of marine sands. Their thinning upward through the sedimentary section was probably caused by a decrease in the rate of glacio-isostatic rise (neotectonic uplift?) of the Pechora Sea seafloor (Musatov, 1989) and a weakening of bottom erosion, rather than by sea-level rise and subsequent weak increase of river discharge (although the latter can not be excluded). The basic features of recent sedimentation were probably established during the Subatlantic Stage of the Holocene.

As already mentioned, there are other publications which dispute a number of the conclusions presented above about the geological history of the Pechora Sea during the Valday and Holocene periods (Epstein and Chistyakova, 2005). These authors reinterpreted the same data from bore holes from the shelf and the seismo-acoustic profiles obtained in work by AMIGE, which was used also in publications (Gataullin et al., 2001; Polyak et al., 2000a). There is an idea about the presence of an Upper Valday moraine and an older one in the central and southwestern parts of the Pechora Sea. Subsequently, the Upper Valday moraine underwent intensive glacio-erosion and wash-out in some areas during the catastrophic descent of dammed lakes.

Our model presented in Fig. 10.21 for a number of areas of the Pechora Sea still seems to be true even with the new interpretation mentioned above. It should be noted that O.G. Epstein and I.A. Chistyakova (2005) did not give any new radiocarbon datings, they have only re-interpreted several geophysical profiles. At the same time, their scientific opponents used a whole series of original datings and an extensive network of profiles. Certainly new detailed, well-dated sedimentary records are needed to solve the problem of glaciation history in the area of the Pechora Sea.

Light Fraction Mineralogy of the Upper Quaternary Sediments from the Saint Anna Trough and Its Paleoceanographic Interpretation

The present paragraph is mainly based on research conducted by (Levitan and Kukina, 2002). Three sediment cores (PL 94–08, PL 94–60, PL 94–64) recovered during the *RV Professor Logachev* cruise in 1994 (Fig. 10.4) were selected for the comprehensive research by specialists from several Russian and German institutions. The main results of this joint research were published in a scientific monograph (Stein et al., 1999). Data of the light fraction mineralogy in the three cores mentioned above, however, have not been included in the monograph because these data had been produced much later.

Materials and Methods

The three cores were opened and described at AWI. Grain-size distribution was also analyzed at AWI applying the Atterberg technique (Andreeva et al., 1999). A special study showed that results obtained for same samples using the Atterberg technique and the combined decantation/pipette method by V.P. Petelin (1967) are very similar (Alekseeva and Svalnov, 2000). Trask coefficients were calculated using the method of moments (Romanovsky, 1977). We only used normalized entropy values (Hr) to evaluate sediment sorting (Table 10.19).

The 0.25–0.125 mm fraction was selected for the mineralogical analysis and divided into the heavy and light subfractions in the AWI laboratory using sodium polywolframat heavy liquid (2.87 g/cm^3). Constant slides were then prepared mounting the light fraction into Canada balsam (index of refraction 1.54). Roundness of quartz grains was determined in the 0.5–0.25 mm fraction using a 5-grade scale (Khabakov, 1946), from angular grains with sharp edges (grade 0) to perfectly rounded grains with evenly smooth surface (grade 4). To determine the average roundness, we estimated the morphology of at least 100 grains. The number of grains in each group was then multiplied by its grade number, the products were summarized, and the sum was divided by the total number of measured grains.

In homogeneous sediments, samples for analysis were selected in intervals of 5–10 cm. In laminated sequences, samples were taken from each layer. N.A. Kukina performed the grain-size study together with colleagues from AWI (R. Stein), VNIIOkeangeo (A.A. Krylov, T.V. Ponomarenko), and IO RAS (M.V. Bourtman). The light fraction study was also carried out by N.A. Kukina.

Table 10.19 Degree of sediment sorting according to normalized entropy [*Romanovsky*, 1977]

Degree of sediment sorting	Normalized entropy (*Hr*)
Perfectly sorted	0–0.1
Well sorted	0.1–0.25
Moderately sorted	0.25–0.5
Poorly sorted	0.5–0.75
Non sorted	0.75–0.9
Absolutely nonsorted	0.9–1.0

Results

The Upper Quaternary sediments recovered in the three studied cores from the Saint Anna Trough are represented by three lithostratigraphic units: glacial, glaciomarine, and marine (Polyak et al., 1997). The oxidized surface layer that some authors distinguish as an individual unit (Krylov, 2000) was included into the upper unit of marine sediments. Glacial sediments are represented by dense, massive, gray or dark gray diamicton that contains rock fragments up to 10 cm in size and rare sand lenses. In their genesis, these sediments are thought to belong to a basal till (Polyak et al., 1997). The glaciomarine unit is characterized by distinct lamination. It comprises intercalated brown and reddish brown buried oxidized layers of stiff clay; gray and greenish gray sand layers (with graded bedding in the lower part of the unit), gray silty clay with a layered structure, and some other lithological varieties. The unit of marine sediments mainly consists of bioturbated olive gray clay with hydrotroilite. It becomes yellowish gray in the upper part of the section. According to published radiocarbon datings (Polyak et al., 1997; Hald et al., 1999), the glacial unit is likely accumulated during the Last Glacial Maximum prior to 13.3 ka (corrected ^{14}C age); the glaciomarine unit spans the time interval from 13.3 to 10 ka; and the marine unit is deposited after 10 ka. The thickness of glaciomarine and marine units in the St. Anna Trough increases from the north to the south (Andreeva et al., 1999; Levitan et al., 1999). AMS ^{14}C dates on bivalve shells and benthic foraminifer tests are listed in Table 10.20.

Core PL 94–08 (Fig. 10.23) recovered the following lithostratigraphic sections (from top to bottom): (1) the marine unit (0–55 cm) is represented by brown silty clay of various color shades; (2) the glaciomarine unit (55–105 cm) comprises intercalated greenish gray and dark olive silt with minor sand and silty sand; and (3) the glacial unit (105–185 cm) consists of a reddish brown and dark gray diamicton with a fine sand layer at 113–120 cm.

The average grain-size composition of the marine unit is as follows (wt. %): 0.94% gravel (>2 mm); 5.3% sand (2.0–0.63 mm); 41.27% silt (0.63–0.002 mm); and 52.49% clay (<0.002 mm). The upper part of the unit shows a more fine-grained composition (higher clay content) as compared to the lower part. The average grain-size composition of the glaciomarine unit is 0.69% gravel; 12.5% sand; 53.88% silt; and 32.93% clay. The upper part of the unit (55–60 cm) is enriched in sand and gravel, and contains less silt than the lower part. The clay content is almost equal throughout the section. The average grain-size composition of the glacial unit

Table 10.20 Radiocarbon dating of the sediment sample from the St. Anna Trough

Sediment core	Core depth, cm	Corrected ^{14}C age, years	Reference
PL 94–08	40	9560	Kolstad, 1996
PL 94–60	87	2180	Hald et al., 1999
PL 94–60	137,5	3135	Hald et al., 1999
PL 94–60	254	5025	Hald et al., 1999
PL 94–60	304	5440	Hald et al., 1999
PL 94–60	395	7060	Hald et al., 1999
PL 94–64	360	9560	Kolstad, 1996

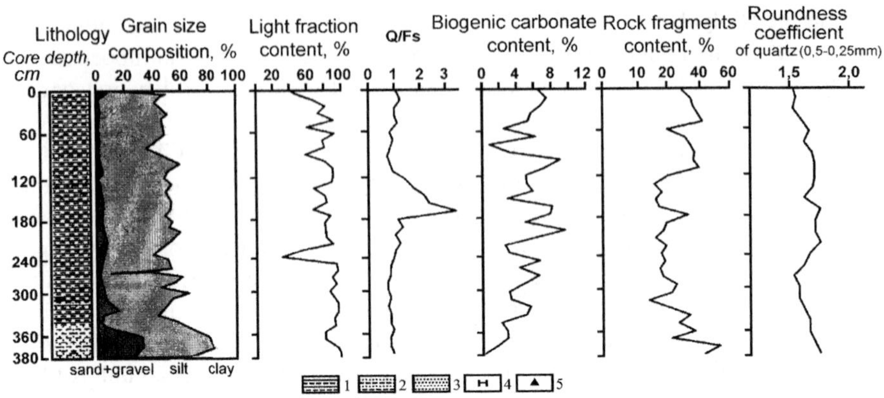

Fig. 10.22 Lithology of sediment core PL94-60 (Levitan and Kukina, 2002). 1 – silty clay; 2 – silt; 3 – sand; 4 – hydrotroilite; 5 – IRD

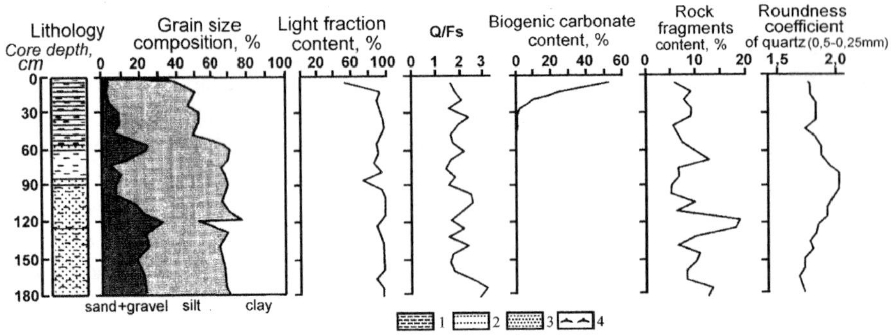

Fig. 10.23 Lithology of sediment core PL94-08 (Levitan and Kukina, 2002). 1 – silty clay; 2 – silt; 3 – sand; 4 – IRD

is 3.77% gravel; 18.7% sand; 44.07% silt; and 33.46% clay. The gravel content increases downward within the section. Sand and silt concentrations show irregular variations along the core, whereas clay fraction content is rather constant.

Normalized entropy values in the marine unit range from 0.37 to 0.51 indicating moderate sorting of sediments. Corresponding values for glaciomarine sediments vary from 0.52 to 0.68 showing their poor sorting. Glacial sediments are characterized by maximum normalized entropy values from 0.6 to 0.7, and are thus even less sorted than glaciomarine sediments. Therefore, sediments become finer and better sorted upward in the core section, with the transition from glacial through glaciomarine to marine ones. The sand and gravel contents decrease, and the clay content increases. According to their grain-size distribution, the marine sediments may be classified as silty clay, whereas clayey silt predominates among the glaciomarine sediments, and glacial sediments are represented by sandy clayey silt (mictite).

The heavy-fraction content demonstrates minor variations along the section ranging within 5–8%. Distribution patterns of the following four light-fraction parameters are described in the section: quartz/feldspar (Q/Fs) ratio, biogenic carbonate content, concentration of rock fragments, and rounding coefficient of quartz grains. The Q/Fs ratio does not show any significant changes downward in the section within the interval 0–160 cm, varying around a mean value of 2. Further below, in the glacial sediments, it increases up to 3.0. The biogenic carbonate content decreases downward within the interval 0–30 cm from 53% to 0%, and carbonate is absent at all below. The concentration of rock fragments in the studied fraction of marine and glaciomarine sediments varies within 5–10%, except for a sandy layer at 60 cm, where it reaches 13%. The content of rock fragments is higher in the glacial unit reaching 18% at a depth of 120 cm. The average rounding coefficient of quartz grains from marine sediments is 1.84 (i.e., quartz is poorly rounded). Quartz is more rounded in glaciomarine sediments, and the rounding coefficient value reaches 2.3 in the lower sandy layer. The rounding coefficient decreases downward in the glacial unit, and it equals 1.6 at the bottom of the core.

Core PL 94–60 (Fig. 10.22) recovered the following lithostratigraphic sections (from top to bottom): (1) dark gray and gray silty clay with plant remains, *POLYCHAETA* tubes, and hydrotroilite inclusions (0–308 cm); (2) dark brown silty clay intercalated with sandy silt layers, which dominate in the interval 345–382 cm and contain rare quartz and chert fragments, 1–5 cm in size (308–382 cm). According to radiocarbon dates (Table 10.20, both subunits have a Holocene age. The discrepancy between the depth of core bottom in our description and the location of the oldest dated sample in Table 10.20 appeared, because several cores with different total lengths were recovered at this station.

The average grain-size composition of the upper gray colored subunit contains 0.15% gravel; 1.9% sand; 43.74% silt; and 54.21% clay. Fine-grained mud of the lower subunit is similar to the upper one in its grain-size distribution, but coarser sediments are characterized by the following content of size fractions: 6.79% gravel; 21.04% sand; 47.74% silt; and 24.43% clay. Normalized entropy values in the interval 0–340 cm range from 0.34 to 0.51 corresponding to moderately sorted sediments whereas the values vary between 0.69 and 0.73 below indicating poor sorting. Therefore, sediments become coarser and less sorted in the lower part of the section. In general, the fine-grained sediments are attributed to silty clay, and the coarse-grained ones are named as sandy clayey silt with gravel and pebbles.

The heavy-fraction content decreases from 20% at the core top to 3–4% at the base of the section. The Q/Fs ratio value is about 1 throughout the section except for the interval 120–180 cm, where the value is commonly higher reaching 3.0 in one of the samples. The biogenic carbonate content is generally low. An uneven downward decreasing trend from 6–8% of carbonate in the upper part of the core to 0% at the level of 380 cm is rather apparent. The content of rock fragments slightly decreases downward from 25–30% to 15% within the interval 0–315 cm, and considerably increases up to 40%, near the core bottom. The rounding coefficient of quartz grains slightly increases downward in the core from 1.55 to 1.8. The data obtained allow us to suggest that reworking products of glacial sediments contribute to the composition of the lower (mictite) part of the section. We realized that diamicton is exposed

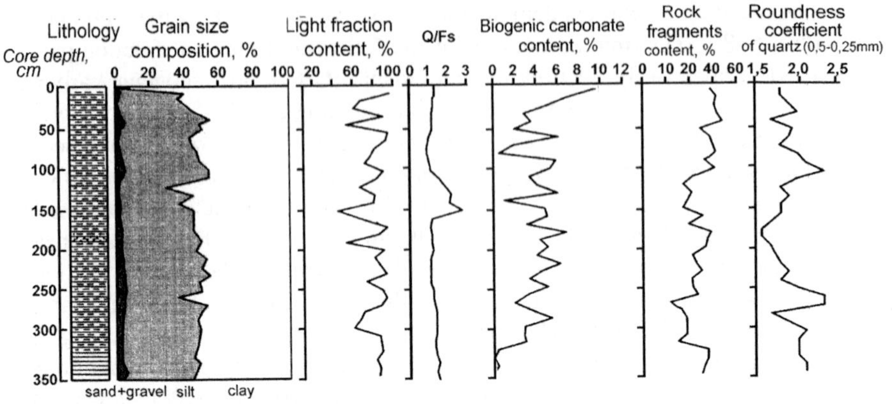

Fig. 10.24 Lithology of sediment core PL94-64 (Levitan and Kukina, 2002). See legend at Fig. 10.23

on trough slopes as small ridges (side moraines?), and that products of its reworking were found in glaciomarine sediments recovered by cores PL 94–07 and PL 94–08 (Levitan et al., 1999; Polyak et al., 1997).

Core PL 94-64 (Fig. 10.24) recovered a rather uniform section of dark gray silty clay with hydrotroilite inclusions and rare bivalve-shell fragments. Fine layers of olive brown and greenish brown silty clay occur sporadically. Sediments within the interval 170–180 cm are slightly enriched in sand. Based on a single radiocarbon date (Table 10.20), and by comparison with the lithology of the nearby well-dated core PL 94–67 (Polyak et al., 1997; Hald et al., 1999), the section described here belongs to the Holocene.

The amount of gravel is commonly <1%. In most samples, the sand content ranges from 1% to 35%, the silt content varies within 37–52%, and the clay fraction makes up 46–60% of the sediment. Average values are 0.88%, 2.65%, 42.87%, and 53.6%, respectively. Normalized entropy values range from 0.30 to 0.49 corresponding to a moderate sorting of sediments.

The heavy-fraction content demonstrates rather wide variations, with an average value of about 20%. The Q/Fs ratio is rather uniform throughout the section ranging within 1.0 and 1.3. Higher values occur in the interval 105–150 cm reaching a maximum of 2.9 at 150 cm. The biogenic carbonate content is generally low, and it shows a gradual downward decreasing trend from 8 –10% in the surface layer to zero at a depth of 350 cm. The concentration of rock fragments decreases downward, from 30% in the surface layer to 18% at 315 cm, and then increases sharply to 28%. The rounding coefficient of quartz grains is variable, but a clear downward increasing trend is observed from 1.8 at the surface to 2.1 at the core bottom.

Discussion

As noted in the previous chapter, data on the light fraction of glaciomarine and glacial sediments were only obtained from Core PL 94–08. Combining these data with published descriptions of many cores raised from the St. Anna Trough (Tarasov et al., 2000) and with additional analytical results, however, we are able to come to several conclusions about paleoceanographic and paleoclimatic environments of the St. Anna Trough area.

The diamicton is characterized by abundant reworked Quaternary foraminifers with an age beyond the limit of ^{14}C dating, i.e. >35 ka (Hald et al., 1999) as well as by Cretaceous palynomorphs (data by A.Yu. Sharapova, and personal communication by S.A. Korsun, 2000). Heavy minerals are mainly represented by pyrite, siderite, and black ore minerals (Levitan et al., 1999). The majority of these minerals are probably reworked from Cretaceous rock underlaying the diamicton. The minerals mentioned above make up >50% of the heavy fraction in the Cretaceous sandstone from Vise Island located on the North Kara Rise (Kosheleva and Yashin, 1999). Clay minerals show higher kaolinite and lower illite contents in comparison to the overlaying glaciomarine and marine sediments (Krylov, 2000). According to kerogen microscopy data, coal particles and other terrestrial organic remains predominate in organic matter composition of the diamicton (Boucsein, 2000). The terrigenous origin of dispersed organic matter is also indicated by high C/N ratios and low HI index values (Stein and Knies, 1999). All these data are evidence for a terrestrial glacial (bottom moraine) origin of the diamicton. The ice sheet probably reached the trough bottom, where Cretaceous rock and older Quaternary sediments served as main sediment sources. Sporadic water streams (possibly under-ice ones?) only appeared at the end of glaciation, when sand interbeds and lenses were deposited.

Sea water penetrated into the trough at the beginning of the deglaciation at 13.3 ka. Melting of the ice sheet resulted in a simultaneous distinct freshwater discharge indicated by light oxygen-isotope values of benthic foraminifer tests (Hald et al., 1999; Polyak et al., 1997). The presence of marine organic matter in sediments may reflect the development of water exchange with the Central Arctic Ocean (Boucsein, 2000; Stein and Knies, 1999). Rare *in-situ* planktonic and benthic foraminifers including species related to episodic Atlantic Water penetration into the St. Anna Trough also appear (Hald et al., 1999). The thickness of the ice sheet successively decreased during the deglaciation. Permanent sea-ice cover existed at the beginning of deglaciation. Pulsating sub-ice sedimentation predominated with alternating instantaneous deposition from gravity flows and slow accumulation of fine-grained sediments from the nepheloid layer that led to accumulation of buried oxidized interbeds. We believe that a better roundness of the quartz grains in turbidite sand interbeds is an evidence of active hydrodynamics during their accumulation, but not a result of changed quartz source or longer transport. Based on IRD layers enriched in coarse rock fragments and containing poorly-rounded quartz grains, the ice cover disappeared episodically during the late deglaciation, thus allowing a drift of icebergs. The decreased thickness of the ice sheet led to sediment supply from both trough slopes. Changes in light-, heavy-, and clay-mineral associations indicate a significant sediment supply from Novaya Zemlya and the North Kara Rise slopes. Indeed, clinopyroxenes, black ore minerals, epidote, and amphiboles begin to dominate among heavy minerals. However, layers with a heavy-mineral association characteristic for the diamicton also occur sporadically (Levitan, et al., 1999; Polyak et al., 1997). The illite and smectite proportions increase sharply among the clay minerals, and the kaolinite content decreases (Krylov, 2000).

The grain-size distribution of the marine Holocene sediments in three studied cores is quite similar. The sediments are represented by moderately sorted silty clay, except for the lower part of Core PL 94-60, where a layer of coarser sediments occurs. Taking into account the location of this core in a deep arm-like depression extending northward of Novaya Zemlya (Fig. 10.4), we suggest that this layer was probably formed by downslope gravity processes. Available data on the composition of this layer support such a suggestion. Therefore, generally uniform sediments accumulated at the Saint Anna Trough floor throughout the Holocene. Sedimentation rates considerably decreased from south to north.

Rather constant quartz to feldspar ratios in all three cores suggest that the relative contributions from the main provinces to the sediments supply into the Saint Anna Trough have not changed during the last 10 ka. Only an episode of increased sediment supply from glacial sediments exposed on slopes of a valley near Novaya Zemlya affected the Q/Fs ratio in cores PL 94–60 and PL 94–64 between 2.5 and 4 ka.

A significant trend of increasing content of rock fragments and decreasing of roundness of sand-size quartz grains upward in the Holocene section from the southern part of the trough are interpreted as an evidence for a decrease of mean annual sea-ice cover and an intensification of iceberg rafting related to a general warming during the Holocene. A higher content of rock fragments in sediments from the lower part of Core PL 94–64 (>5.2 ka) coincides with an increase in the median diameter of the surrounding sediments, explained by local slope processes.

The distribution of biogenic carbonate in the light fraction of the Holocene sediments is especially interesting. In Core PL 94–08, biogenic carbonate is represented by both planktonic and benthic foraminifer tests which appear in the middle part of the Holocene and rapidly increase in abundance upward in the section. Foraminifer analysis carried out in the nearby Core PL 94–07 (Hald et al., 1999) revealed the same foraminifer-distribution trend, as that found in our core. Distribution patterns of the benthic species *Cassidulina teretis* related to Atlantic Water, are also similar. Unfortunately, the age control is insufficient for both cores. We can only suggest that the increase in abundance of both benthic and planktonic foraminifers related to intermediate Atlantic-water inflow, started sometimes in the middle Holocene. This suggestion does not contradict our data on an increase in the number of foraminifer tests upward in the section of Holocene sediments from the Yermak Plateau that started 7 ka (Levitan et al., 2000). However, such a trend is much weaker in sediments from the Franz Victoria Trough (Lubinski et al., 1996) likely related to a local hydrological situation. As the Atlantic-water advection is highly important for the Arctic Ocean climate, this problem needs further investigation. In the southern part of the Saint Anna Trough (cores PL 94-60 and PL 94-64) biogenic carbonate is represented by bivalve shell fragments. Upward increasing trends of biogenic carbonate noted in the Holocene sections of both cores possibly reflect intensified advection of the transformed Atlantic Water (Barents Sea Water) into the southern part of the trough. This led to an increased primary production which, in turn, caused an increase in the zoobenthos biomass.

The volume of ice sheets further decreased during the Holocene that at a certain stage glacier melt-water discharge lost its leading role in terrigenous sediment supply. More remote sources were increasingly involved in sediment delivery, including the inner regions of the Kara Sea, as shown by the higher clinopyroxene content among the heavy minerals in the southeastern part of the trough (Krylov, 2000). According to mineralogical data (Levitan et al., 1999), the relative importance of Franz-Josef Land as a source province increased gradually during the Holocene. This is likely a result of considerably wider land area released from the ice sheet on Franz-Josef Land and of its higher average topographic level, as compared to the North Kara Rise. The productivity of the basin increases as reflected not only by biogenic carbonate and hydrotroilite abundances, but also by the organic-matter composition (Boucsein, 2000; Stein and Knies, 1999) as well as by widespread bioturbation structures. The change in dominating sedimentation mechanisms at the transition from deglaciation to Holocene regimes led to a more important role of long-distance sediment transport by surface and bottom currents, as it is expressed by both sediment sorting and quartz-grains roundness. In general, sedimentation rates decreased during the Holocene. Higher sedimentation rates prior to 8 ka possibly reflect a more abundant sediment supply from Siberian rivers (Polyak et al., 2000, 2002).

Thus, the evolution of sedimentation in the Saint Anna Trough during the last approximately 15 ka was related to the history of climatic changes. Three types of sedimentation can be distinguished: glacial, glaciomarine, and marine. A specific set of sedimentation parameters existed at each evolution stage including basin type, extension of sea ice and glacier ice, mechanisms of sediment erosion, transport, and accumulation, provenance, relationships with other sedimentary basins, and primary production.

During the glacial stage of sedimentation, i.e., during the Late Weichselian, the ice sheet reached several hundreds of meters of thickness, and directly contacted the trough floor. Basal and side moraines formed under terrestrial conditions. The trough floor mainly composed of Cretaceous clastic rocks served as a major material source. Areas of material erosion, transport routes, and accumulation sites were located very close to each other.

During the deglaciation (13.3–10.0 ka), the thickness of the ice sheet decreased, and seawater began to flow under ice into the trough from both north and south. Main episodes of intense sedimentation were related to the ice-sheet melting and to pulsating delivery of huge freshwater masses into the basin where they mixed with seawater. Source provinces occupied gradually more and more area on trough slopes including side moraines. Primary production was increasing slowly and very unevenly. During the final deglaciation stage, sea ice cover disappeared at times, and icebergs drifted within the aquatory.

The relative role of sea ice reduced considerably when deglaciation terminated at 10 ka. The thickness of pack ice likely did not exceed several meters. Settling of suspended matter from the nepheloid layer became a dominating sedimentation mechanism. Biological transport mechanisms developed being related to a sharp increase in primary production, especially during the second half of the Holocene. The geography of source provinces expanded considerably, and the relative contribution of Franz-Joseph Land to the sediment supply increased. Relations with sedimentary basins in the south (Barents and Kara seas) and in the north (sediment exchange with the Central Arctic) became closer. Distances between sediment sources and sites of accumulation reached their maximum.

Many aspects of the described succession of geological events need a more careful geochronological control and a better time resolution. Due to the lack of precise datings, further studies in the Saint Anna Trough are still needed.

Holocene History of Yenisei River Discharge

River discharge in the Arctic Ocean has a strong impact on the sea-ice formation, water-mass stratification, thermohaline circulation, and other parameters related to the Arctic climate. The Yenisei River accounts for nearly one-fourth of the annual river discharge into this ocean (Aagaard and Carmack, 1989). Therefore, the Yenisei discharge history is very important for understanding the Arctic paleoclimate.

The study of the late Pleistocene-Holocene history of the Kara Sea was restricted for a long time due to the absence of a reliable geochronological database. In the past, studies were mainly based on lithostratigraphic (Kosheleva and Yashin, 1999) or ecostratigraphic (Levitan et al., 1994) models. More recently, sedimentological, geochemical, and micropaleontological studies using radiocarbon AMS ^{14}C datings for age control were carried out (Levitan et al., 2000; Hald et al., 1999; Polyak et al., 1997, 2000, 2002; Stein et al., 2002, 2003a, 2004). As a main result, all these publications indicate that the early Holocene sedimentation rate was significantly higher in most part of the Kara Sea, related to enhanced suspended-matter discharges by the Ob and Yenisei rivers.

The chemical and mineral compositions of bottom sediments of the Kara Sea were described in several publications (Belov and Lapina, 1961; Gurvich et al., 1994; Gurevich, 1995; Schoster and Stein, 1999; Siegel et al., 2001a, 2001b; Levitan et al., 1996, 2002; Steinke, 2002; Stein et al., 2004). However, works devoted to the evolution of the chemical and mineral compositions of bottom sediments of the Kara Sea based on a reliable geochronological database and linked to the Holocene sedimentation history are still very rare. The present paragraph provides the first more detailed insight into this issue. Particular attention is given to the Yenisei runoff history based on lithological, mineralogical, and geochemical data. Micropaleontological indicators are also discussed on the basis of literature data. Concerning the short-term variability in Yenisei river discharge through Holocene times, the best AMS ^{14}C-dated sedimentary record including high-resolution data of lithology, magnetic susceptibility, grain-size distribution, palynomorphs, diatoms, terrigenous biomarkers, and terrigenous organic-carbon input, was obtained from sediment core BP99-04/7 (73°24.9 N, 79°40.5 E; water depth 32 m) recovered in the northern Yenisei Estuary (Kraus et al., 2003; Stein et al., 2003a, 2004).

Materials and Methods

We studied bottom sediments from cores BP00–23/7 (73°28.5′ N, 79°51.3′ E; water depth 33 m) and BP00–7/6 (74°39.5′ N, 81°08.5′ E; water depth 38 m) recovered from the Yenisei transect during the R/V *Academik Boris Petrov* cruise within the joint Russian-German SIRRO project in 2000 (Stein and Stepanets, 2001). Results of the macroscopic description and comprehensive investigation of smear-slides (Stein and Levitan, 2001) were supplemented with data on the water content and wet bulk density obtained by weight analysis (Kodina et al., 2001). The grain size analysis was carried out by the combined method of V.P. Petelin (1967) in the Analytical Laboratory of the Shirshov Institute of Oceanology under the supervision of V.P Kazakova. In order to allow a correlation with published grain-size data from the same area (e.g., (Stein et al., 2004)), grain sizes of 0.05 and 0.002 mm were accepted as the sand-silt (silt) and silt-clay (pelite) fraction boundaries, respectively. Names of lithotypes of sediments in the Russian version of this paragraph (Levitan et al., 2007) are given according to (Frolov, 1995).

Results of the radiocarbon AMS ^{14}C analysis at the facility of Arizona State Univerisyt of bivalve mollusk (*Portlandia arctica*) shells are given in Table 10.21. In order to carry out the subsequent calculations, the obtained dates were corrected for the reservoir age of 440 year (Stein et al., 2002) and transformed into calendar years according to (Stuiver et al., 1998). All dates in the present paragraph are given in the calendar age scale (cal. yrs. BP).

The chemical composition of sediments was analyzed at the Vernadsky Institute of Geochemistry and Analytical Chemistry by the INAA method described in (Kolesov, 1994). In addition, concentrations of several elements were determined by the ICP-AS and atomic absorption methods (Sedykh et al., 2000). Characteristics of the studied sediments are supplemented by results of semi-quantitative determination of benthic foraminifer population (Ivanova, 2001) and the analysis of clay- and heavy-mineral assemblages carried out by V.V. Krupskaya and M.V. Bourtman, respectively. The clay minerals were analyzed in the <0.001 mm fraction, while the heavy minerals were examined in the heavy subfraction (separated by bromoform with n = 2.89 g/cm^3) of the 0.1–0.05 mm fraction.

Table 10.21 Radiocarbon age of the bivalve *Portlandia* shell samples [*Levitan et al.*, 2004]

Lab. no.	Sediment core	Core depth, cm	Weight, mg	$\Delta^{13}C$	^{14}C age, years
AA44370	BP00–07/6	240	77	–2,76	4185±43
AA44371	BP00–07/6	645	108	–2,86	7482±50
AA44372	BP00–23/7	137	79	–4,04	8607±50
AA44373	BR00–23/7	290	36	–5,00	8891±62

Results and Discussion

Sediment Core BP00-23/7

In the Gulf of Yenisei the following subfacies zones of recent sediments (from south to north) can be distinguished: subaerial delta, submarine delta, proximal marginal-filter zone, depocenter, and distal marginal-filter zone (Fig. 10.25). Core BP00–23/7 is located in the distal zone.

Lithological and Grain-Size Composition

Sediments of Core BP00–23/7 include the following lithostratigraphic units (from top to bottom):

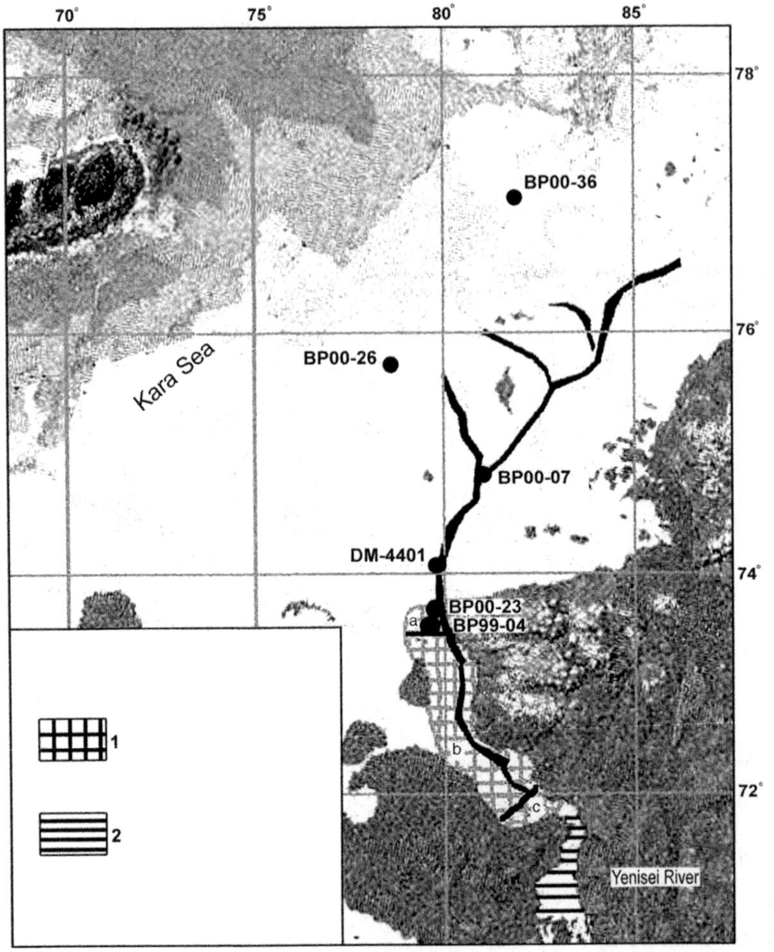

Fig. 10.25 Location of Yenisei Transect and geological stations (Levitan et al., 2004). 1 – marginal filter (*a* – distal part, *b* – depocentre, *c* – proximal part), 2 – delta

Unit 1: 0–85 cm. Olive-gray to dark olive-gray and black sandy-silty clay with a coarse sandy-silty clay (hereafter, coarse silt) interlayer in the middle part. Two wood fragments (1–3 cm) were detected at 0–25 cm and a bivalve mollusk fragment was found at 24 cm. At 25 cm and further down-core, black spots of hydrotroilite and bioturbation of low and medium intensities are typical. The sediments have the following average composition: clay minerals 50%, quartz 25–30%, and feldspars 10–15%. Rock fragments and heavy minerals account for as much as 5–7% of the sediment in the upper 25 cm and do not exceed 3–5% below. Diatoms and plant remains account for <5%. Benthic foraminifers occur in rare amounts (2–5%).

The sand fraction (Fig. 10.26) varies form 13 to 20% in the upper 60 cm of the section and abruptly drops to 8% below. In contrast, the silt content increases from 27–30% in the upper interval to 34% in the lower part of the section. The clay content remains relatively stable with 52–57%. The water content drastically decreases from 52 to 40% in the upper 25 cm of the section and increases to 50% in the lower interval. The wet bulk density gradually increases down-core from 1.20 to 1.40 g/cm^3.

Thus, Unit 1 can be subdivided into the upper (0–25 cm) and lower (25–85 cm) subunits.

Unit 2: 85–335 cm. Olive-gray silty clay. Black sediment layers, 2–4 thick, occur at 144–159 cm and 183–192 cm core depth. Bivalve mollusk detritus was found at 88, 120, 137, and 290 cm core depth. Unit 2 is characterized by abundant hydrotroilite patches and coatings and moderate to strong bioturbation. In com-

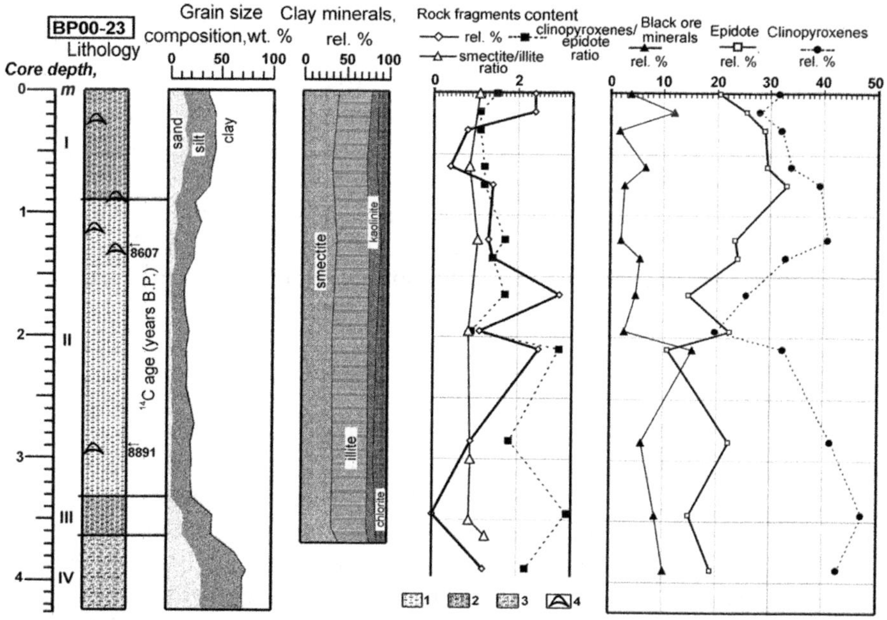

Fig. 10.26 Lithology and mineral composition of sediment core BP00-23/7 (Levitan et al., 2004). I–IV – lithostratigraphic horizons. *Arrows* show radiocarbon dates. 1 – silty clay; 2 – sandy-silty clay; 3 – clayey-sandy silt and silty sand; 4 – shells of bivalve mollusks

parison to Unit 1, Unit 2 is depleted in quartz and enriched in clay minerals and plant remains. The amount of benthic foraminifers remains rare at 90–230 cm and increases to a content of 5–10% in the lower interval.

Based on grain-size distribution, two intervals (85–175 and 175–335 cm core depth) can be identified. The first interval is characterized by a down-core decrease of the sand and silt contents (8–1 and 36–28%, respectively) and an increase of the clay content (57–72%). In the lower interval, the sand content varies from 0.7 to 2.1%, the total silt content increases from 30 to 38%, and the clay content decreases from 70–72% to 61%. At 230 cm core depth, the decrease of the clay content becomes more prominent, probably related to the increase of foraminifer remains. In the upper interval, the water content is quite stable (50–52%), and the wet bulk density decreases from 1.40 to 1.20 g/cm^3 at 130 cm, increases to 1.47 g/cm^3 at 130–140 cm, and remains stable downward the section. In the lower interval, the water content remains 50–52% up to the level of 230 cm. Below this level, the water content at first drastically decreases to 42% and then rises to 50%. The wet bulk density is 1.47 g/cm^3 and increases to 1.50 g/cm^3 at 280 cm. Based on these data, Unit 2 can be divided into three sub-units (85–175, 175–230, and 230–335 cm).

Unit 3: 335–374 cm. Olive-gray slightly bioturbated sandy–silty clay and coarse silt with a lower content of clay minerals. The concentration of rock fragments and heavy minerals (including black ore minerals) is distinctly increased. Benthic foraminifers are rare. The grain-size distribution is as follows: sand fraction 14–16%, silt fraction 33–38%, and clay fraction 50%. The water content decreases down-core from 50 to 24%. The wet bulk density is 1.47 g/cm^3.

Unit 4: 374–427 cm. Olive-gray clayey-sandy silt and clayey-silty sand (at the base) with a distinct horizontally-bedded structure of interlayers of 1–2 cm in thickness. The clay-mineral content is very low (5–10%) at the core base. The rock-fragment content increases to 25%. The concentration of heavy minerals, primarily black ore minerals, is drastically increased. The sediments contain abundant plant remains and fresh-water diatom frustules. Benthic foraminifers are absent.

The sand content is 32% in silty interlayers and 44–47% in sandy interlayers. The silt content is 50 and 24–27%, respectively. The clay content is as much as 26–31%. The water content and wet bulk density are up to 30% and 1.75 g/cm^3, respectively, at the core base.

Mineral Composition

As is evident from Fig. 10.26, the clay-mineral assemblage is rather monotonous in the <0.001 mm fraction: smectite 34.2–44.3% (average 38.3%), illite 35.6–41.8% (average 39.5%), kaolinite 6.8–13.5% (average 9.9%), and chlorite 8.1–15.7% (average 12.3%). The average smectite/illite ratio is 1.0 (0.9 in the majority of cases), while the chlorite/kaolinite ratio is 1.2. The smectite/illite ratio slightly increases at the core base and in the surface layer. A chlorite–smectite–illite assemblage prevails in the sediments.

In contrast, the heavy-mineral fraction (0.1–0.05 mm) demonstrates a distinct trend (Fig. 10.26): the clinopyroxene/epidote ratio and the average clinopyroxene content decrease downward the section. The down-core decrease of the content of black ore minerals is less distinct. Like the smectite/illite ratio, the clinopyroxene/epidote ratio increases in the surface layer and the lower half of the core. The content of iron hydroxides and grains with an iron coating increases in the middle part of Unit 2. The surface layer is significantly enriched in basaltic hornblende, enstatite-hypersthene, titanite, fluorite, andalusite, muscovite, sillimanite, and rock fragments.

Chemical Composition

The distribution of chemical elements and their ratios in the studied sediments show a very characteristic signature for the following five types ("sand", "silt", "clay", "feldspar", and "epidote"). In the "sand" type, the La/Yb ratio rises in the upper and lower intervals relative to the middle interval (Fig. 10.27). This can be explained by the geochemical behavior of LREE and HREE and the grain-size distribution of the sediments. The "silt" type is marked by a more or less stable concentration of Zr, Cd, Ni, and Ti, as well as Ce/La, Eu/Sm, and La/Sm ratios (Fig. 10.27). Some of these elements are concentrated in the silt. The silt content is less variable than the sand or clay content.

The "clay-type" distribution (with higher concentrations in the middle interval and lower ones in the upper and lower intervals) is inherent to Al, Fe, Mn, P, U, Mo, Cu, Zn, Co, Cr, Th, Sc, Yb, and Lu (Fig. 10.27). The Al and Fe distribution deserves a special attention. Evidently, Al is mainly concentrated in the crystal lattice of clay minerals. The Fe content increases from 2.0–2.5% in sediments of Unit

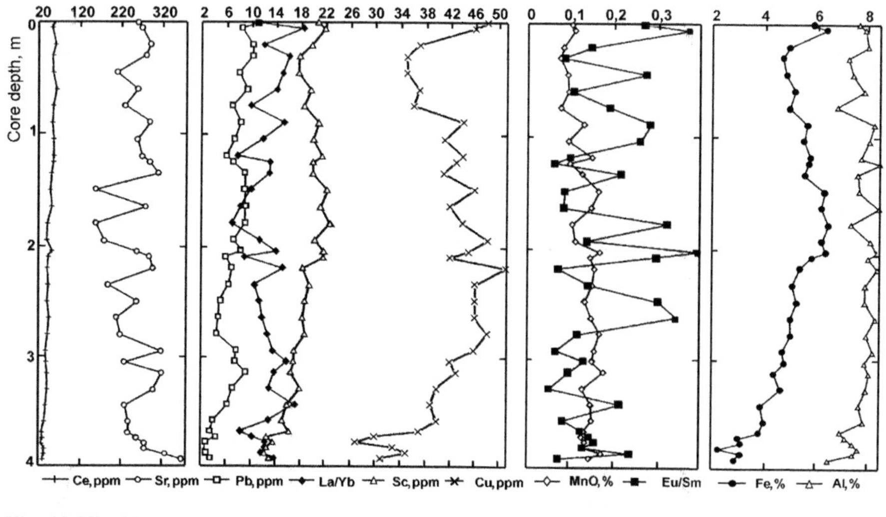

Fig. 10.27 Chemical composition of sediment core BP00-23/7 (Levitan et al., 2004)

4 to 6.0–6.5% in the middle interval of Unit 2. The latter interval is characterized by rather high contents of iron hydroxides and grains with a ferruginous coating (~20%). Because the Al concentration reaches maximum values in the entire Unit 2, the Fe concentration lag (relative to Al) should be attributed to the phenomenon observed in the present-day mixing zones of fresh and sea waters, i.e., the shift of maximum Fe concentrations toward the depocenter where Fe is extensively transferred from the true dissolved state to the colloidal state and precipitated as colloidal particles and organic iron compounds. Such processes take place in the present-day surface water of the Yenisei Estuary at a salinity of 5‰ (Dai and Martin, 1995). In contrast, Al concentrated in clay minerals, already start to precipitate in the proximal marginal-filter zone at the beginning of the mixing of fresh and sea waters. This process continues up to the frontal-filter zone. The major part of minor elements of the studied assemblage is associated with clay minerals or unaltered iron compounds. Therefore, the behavior of minerals can also explain the behavior of minor elements. In this context, it should be mentioned that the Sc behavior exactly mimics the Fe distribution (Fig. 10.27). Another significant process is the coagulation of dissolved organic matter simultaneously with the transformation of iron forms (Artem'yev, 1993).

The "feldspar-type" distribution only includes Sr that continuously increases downward the section (Fig. 10.27). This phenomenon can probably be explained by the predominance of fluvial matter at the core base. According to (Levitan et al., 1998), the majority of feldspars (first of all, basic and intermediate plagioclases) are derived from the Yenisei River runoff. The carbonate content does not exceed a few percents in the studied sediments. Hence, the major portion of Sr is concentrated in crystal lattices of feldspars (particularly potassium varieties) that prevail in the sediments (Levitan et al., 1998).

Finally, the "epidote-type" distribution is characterized by concentrations of Hf, Pb, Cs; La, Ce, Nd, Sm, and Eu. Their concentrations decrease downward the section (Fig. 10.27). This can be explained by changes in the sedimentary environment (Fig. 10.28) from the submarine delta (Unit 4) to the proximal marginal-filter zone (Unit 3), depocenter (Unit 2), and distal marginal-filter zone with the participation of products of coastal abrasion of the Taimyr Peninsula (Unit 1). Indeed, such behavior of REE during the transition from the alluvial facies to the coastal-marine facies is well known and described earlier (Balashov, 1976). Based on the study of REE behavior, we revealed that the basal sediments (370–400 cm) are closely related to the Siberian trap basalts, while sediments of Unit 1 are similar to platform sedimentary rock (Table 10.22).

Sedimentation Rates and Mass Accumulation Rates

Based on the radiocarbon dates (cal. kyrs. BP) coupled with their linear interpolation and extrapolation, the age and sedimentation rate of Unit 2 are estimated to be 9.1–9.5 cal. kyrs. BP and 475 cm/ky, respectively. The sedimentation rate of Unit 1 is 10.5 cm/ka. Taking into account the grain-size distribution of the sediments of Unit 1, it is suggested that the sedimentation rate is decreasing to a certain extent upward

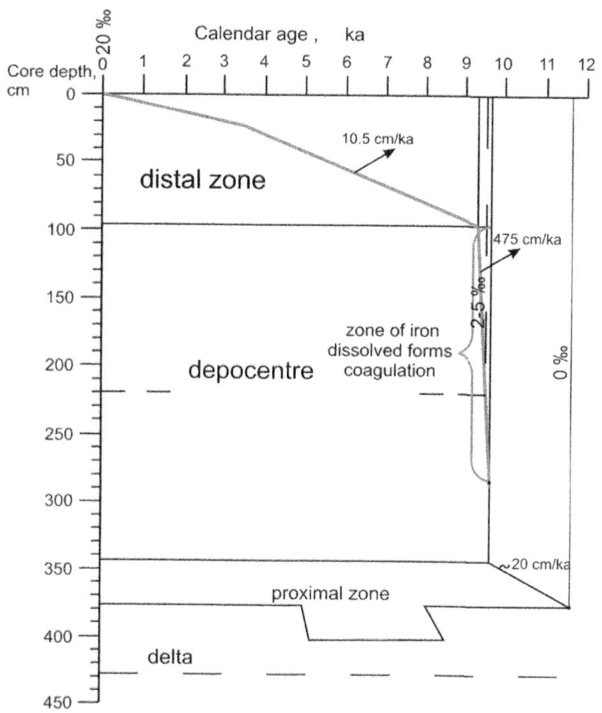

Fig. 10.28 Age model and environmental evolution of sediment core BP00-23/7 (Levitan et al., 2004)

the section. Assuming the radiocarbon age for boundary of units 3 and 4 at 10 cal. kyrs. BP (Polyak et al., 2000, 2002; Stein, 2001), the sedimentation rate in Unit 3 is equal to ~20 cm/ky. The sedimentation rate is unknown for Unit 4. Using data

Table 10.22 REE content (ppm) in sediments of the Yenisei marginal filter [*Levitan et al.*, 2004]

	Bottom sediments in Core BP00-23/7			
Elements	Interval 0–80 cm (n = 7)	Interval 370–400 cm (n = 6)	Trap basalts of the Putoran Plateau, Mokulaev formation (Lightfoot et al., 1990)	Platform sedimentary rock (Balashov, 1976)
La	30.0	17.3	6.9	35.5
Ce	56.1	32.0	16.5	67.0
Nd	23.2	13.2	11.2	33.0
Sm	6.53	3.3	3.21	6.7
Eu	1.19	0.47	1.11	1.24
Tb	0.95	0.62	0.65	1.0
Yb	2.28	2.86	2.36	2.95
Lu	0.37	0.25	0.36	0.45

on water content and wet bulk density, mass accumulation rates of ~8.2 $g/cm^2/ky$ (Unit 1), 357.5 $g/cm^2/ky$ (Unit 2), and 20.6 $g/cm^2/ky$ (Unit 3) were calculated.

Sedimentation History

Based on the data described above, the basal section of Core BP00-23/7 includes submarine delta-facies sediments accumulated in a fresh-water zone of the Yenisei River about 11.3 cal. kyrs. BP (Unit 4). This interpretation is also supported by the absence of benthic foraminifers in these sediments. In the present-day sediments of the studied region, benthic foraminifers exactly disappear in the submarine delta-facies sediments where the bottom water salinity reaches zero (Khusid and Korsun, 1996). Based on the mineral and chemical composition, the relatively coarse-grained fraction of the sediments of Unit 4 is primarily composed of erosion products of the Putoran Plateau basalts and its weathering crust.

Unit 3 includes the relatively fine-grained sediments of the proximal marginal-filter zone (Levitan, 2001). The zero isohaline line of surface water existed in this area at 11.3 cal. kyrs. BP. Subsequently, the salinity gradually increased to 4% at about 9.5 cal. kyrs. BP. The hydrodynamic barrier in the mixing zone of active currents of the submarine delta with the calm sea-water mass resulted in the preferential precipitation of the relatively coarse-grained fraction and mass accumulation rates of ~20.6 $g/cm^2/ky$.

The depocenter formed in a few hundred years in the Yenisei Estuary area. A large volume of clayey material contained in the water suspension of the mixing zone settled to the seafloor. When the salinity of the surface water reached 5‰ approximately 9.4 cal. kyrs. BP, this material was supplemented with a large amount of Fe, organic matter, and minor elements. During this period, i.e., the formation of Unit 2, the mass accumulation rate reached very high values of 357.5 $g/cm^2/ky$. Such dramatic events could only be provoked by the vigorous and rapid increase in Yenisei River discharge as a result of the Early Holocene warming (MacDonald et al., 2000) and the consequent active degradation of the permafrost. Numerous examples of high sedimentation rates in Early Holocene records from the Kara Sea are described in (Polyak et al., 1997, 2000, 2002; Stein et al., 2002, 2003a, 2004).

After 9.1 cal. kyrs. BP the sedimentation rate drastically decreased mainly due to the reduction of clay fractions. At the same time, the sediments were relatively enriched in coarser fractions. Because the role of river run-off diminished, the silty–sandy material delivery was mainly related to products of coastal abrasion of the adjacent Taimyr Peninsula that was composed of Lower Paleozoic clastic-clayey sediments of the greenstone facies (Geological structure of the USSR and regularities of mineral deposits distribution, 1984). Based on mineralogical data, a relative increase of the contribution of river runoffs is only assumed for the uppermost interval of Unit 1, the precise age of which is unknown. The mass accumulation rate of Unit 1 is in average 8.2 $g/cm^2/ky$, and the salinity of surface water in the studied area has increased from ~10 to 20‰.

Distal sediments have an insignificant thickness in Core BP00–23/7. Therefore, we cannot describe in detail the history of river discharges during the last 9.2 ka for this area. However, results of the study of sediments in the adjacent sediment core BP99–04/7 (Fig. 10.25) provide such an opportunity (Stein et al., 2002, 2003a, 2004; Polyakova and Stein, 2002, 2004). In this core, the boundary between the proximal marginal-filter zone and depocenter is dated to 9.25 cal. kyrs. BP. The significant attenuation of magnetic susceptibility, probably related to a decrease of the concentration of ferromagnetic minerals delivered by erosion of the Putoran trap basalts, was recorded at 8.74 cal. kyrs. BP (Stein et al., 2003a, 2004). The significant decrease in sedimentation rate by a factor of 2 took place at 8.0 cal. kyrs. BP.

Sediment Core BP00–7/6

Core BP00–7/6 is located slightly north of Core BP00–23/7 in a sediment-filled channel (Stein and Stepanets, 2001), one of the river valleys of the hydrographic network of the old Yenisei River on the Kara Sea shelf (Fig. 10.25).

Lithology and Grain Size Composition

The studied core can be subdivided into the following lithostratigraphic units (from top to bottom):

Unit 1: 0–510 cm. Silty clay with single clayey sediment interlayers (Fig. 10.29). Dark olive-gray and dark gray sediments prevail in the 0–365 cm interval. Dark gray and black sediments predominate in the lower interval. Sediments of the 0–25 cm interval have a homogeneous structure. They are moderately or strongly bioturbated in the lower interval. Hydrotroilite patches and coatings are universally developed in the section. Bivalve shells were found at 237 and 270 cm. The sediments are composed of clay minerals (50–75%), quartz (25–50%), and feldspars (5–10%). The content of heavy minerals and rock fragments does not exceed 5%. Locally, rock fragments may increase to 8–10% (e.g., at 49 cm), and plant remains were found (e.g., at 325 and 430 cm). The content of benthic foraminifers is generally in the range of 5–10% and decreases to 2–5% below 430 cm core depth.

The average grain-size distribution is rather monotonous and stable: sand fraction 2%, silt fraction 30%, and clay fraction 68%. Water content decreases from 70% at the surface to 50% at the base of Unit 1. Intermediate maxima were recorded at 100, 200, and 350 cm core depth, and a prominent minimum was observed at 150 cm. The wet bulk density inversely correlates with the water content and increases from 1.15 g/cm^3 at the surface to 1.35 g/cm^3 at the base. Wet bulk density fluctuations are similar to those of the water content.

Unit 2: 510–665 cm. Dark gray and black silty-sandy clay with rare clayey sediment interlayers and single sandy-silty clay strata (525 and 568 cm). The sediments are moderately to strongly bioturbated. Bivalve shells were found at 544 and 644 cm. The composition of the sediments is similar to that of Unit 1. The coarser-grained interlayers, however, are depleted in clay minerals and enriched in rock and

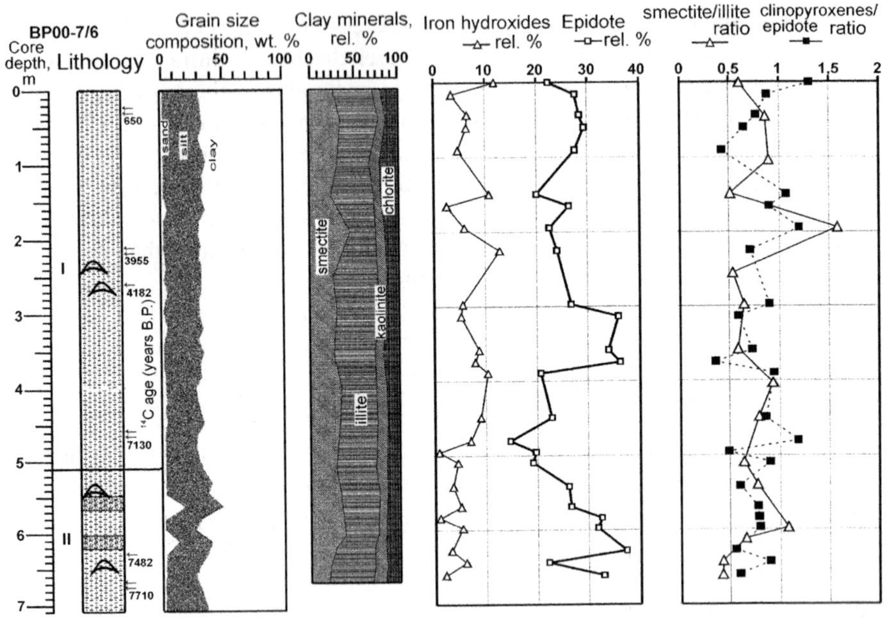

Fig. 10.29 Lithology and mineral composition of sediment core BP00-07/6 (Levitan et al., 2004). I–II – lithostratigraphic horizons. Radiocarbon dates are shown by single arrows for sediment core BP00-07/6 and double arrows for sediment core BP00-07/7 (Stein et al., 2002). See legend at Fig. 10.26

heavy-mineral fragments. Plant remains are absent. The content of benthic foraminifers is generally 2–5%, increasing to 5–10% only below 610 cm core depth.

The grain-size distribution is characterized by a strong variability typical of layered sediments. The composition of fine silt layers is approximately similar to that in Unit 1. The coarser-grained variety is enriched in sand (up to 17%) and silt (up to 39%) and depleted in clay (51%). Both sand and silt fractions display an independent distribution mode in the core. The water content decreases to 45% downward the section with some fluctuations. The wet bulk density remains more or less at the previous level and weakly responds to changes in the grain-size distribution.

Mineral Composition

The clay minerals are mainly represented by a kaolinite–chlorite–smectite–illite assemblage (Fig. 10.29). The composition is as follows: smectite 21.6–45.2% (average 30.5%), illite 28.5–50.4% (average 42.8%), kaolinite 8.0–14.9% (average 10.9%), and chlorite (12.9–21.2%) (average 15.8%). The average smectite/illite and chlorite/kaolinite ratios are 0.7 and 1.5, respectively. The smectite/illite ratio increases to values >1.0 at 195 and 600 cm and reaches a value of 0.9 at 45–110 and 405 cm. A minimum value of 0.4 is typical for the lower part of the core. In general, the smectite/illite ratio is slightly higher in the interval 0–510 cm relative to the underlying interval.

The heavy-mineral composition shows a distinct trend of an up-section increase of the pyroxene/epidote ratio (Fig. 10.29). Furthermore, this ratio is >0.8 at 650, 590, 500–520, 320, 145–210, 60, and 0–20 cm. In general, intervals of high pyroxene/epidote ratios coincide with high smectite/illite ratios. Concentrations of clinopyroxene, hydrotroilite, black ore minerals, titanite, and chlorite show a similar although less distinct increasing trend. The epidote content decreases upward the section. Sediments of Unit 2 are commonly enriched in aegirine-augite and rock fragments relative to Unit 1, which is slightly enriched in tourmaline, Cr-spinel, and volcanic glass. Other minerals are distributed irregularly.

Chemical Composition

The chemical composition of the sediments is characterized by a very low variability within the whole sedimentary section. We can clearly identify the existence of one single province and accumulation of chemical elements in similar constraints. We have recognized two groups of elements that mainly differ in terms of the type of material that absorbed the elements from seawater. Elements of the first group (Sc, Co, Cr, Sb; La, Ce, Nd, and Sm) were absorbed by iron oxides and hydroxides. Like Fe, these elements are characterized by a slight down-core decrease of concentrations (Fig. 10.30). The behavior of elements of the first group is retained in the black sediments below 365–400 cm, but their amplitude of variation is less relative to that in the overlaying gray sediments. Thus, the hydrotroilite content, which is

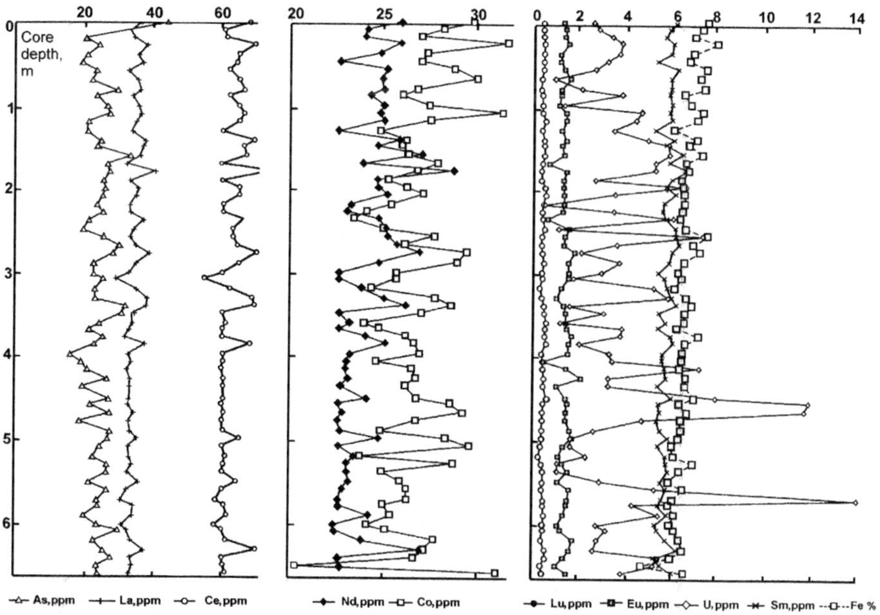

Fig. 10.30 Chemical composition of sediment core BP00-07/6 (Levitan et al., 2004)

responsible for the black color, has a certain impact on the chemical composition of the sediments. Elements of the second group (Th, U, Ba, Zr, Zn, Se, As; Eu, Tb, Yb, and Lu) were mainly absorbed by clay minerals. Exactly this type of sorbent distribution was reported for rare earth elements by (Gurvich et al., 1980). In general, these elements are characterized by a stable average value throughout the section (Fig. 10.30), except for As and U. The As content drastically increases in surface sediments, probably related to anthropogenic contamination (Siegel et al., 2001a, 2001b). The U concentration is marked by rare maxima possibly related to the presence of organic matter in individual interlayers (Fig. 10.30).

Sedimentation and Mass Accumulation Rates

Calculations of sedimentation rate in Core BP00–23/7 are based on the following two assumptions. First, our observations revealed that the gravity corer probably did not recover the uppermost 15 cm interval of fine clayey sediment. Therefore, the thickness of the uppermost interval, for which sedimentation rates were calculated, was increased by 15 cm. Second, the study of magnetic susceptibility of additional cores recovered at the same station BP00–07 (Dittmers et al., 2001) revealed that the sedimentary sections recovered at these cores can be correlated one to one. Therefore, we assume that radiocarbon dates determined in Core BP00–07/7 (Stein et al., 2003a, 2004) can be used for chronology and calculation of sedimentation rates at our Core BP00-07/6 (Fig. 10.29). They showed that the sedimentation rate was very high (789.5 cm/ky) at 8.1–7.9 cal. kyrs. BP (i.e., the time of Unit 2 accumulation), resulting in a very high mass accumulation rate of 525.0 g/cm^2/ky. These extremely high values are based on the virtually same ages of sediments at 510 cm (7.9 cal. kyrs. BP) and 645 cm (7.9 cal. kyrs. BP). If these ages are correct, this requires a short-lived event of sediment supply (focusing) difficult to explain in this shallow-water environment. Based on sediment accumulation data determined in near-by cores, R. Stein et al. (2003a, 2004) suggest that the very high accumulation rates in this area are caused by a marginal-filter setting, i.e., the marginal filter was located more to the north at that Early Holocene time interval. M.A. Levitan et al. (2004) propose that this event could be related to slump. Further data on well-dated sediment cores are needed to solve this problem. After 7.9 cal. kyrs. BP, the sedimentation and mass accumulation rates were in the range of 57.6–93.4 cm/ky and 31.6–47.6 g/cm^2/ky, respectively, until 268 cal. yrs. BP when they increased again to 171.6 cm/ky and 72.1 g/cm^2/ky, respectively.

Sedimentation History

The Ob-Yenisei shallow-water zone, which hosts Station BP00–07/7, is a late Pleistocene flooded alluvial plain. Its geological evolution during the late Pleistocene-Holocene was mainly controlled by the global eustatic sea-level rise (Fairbanks, 1989) and the subordinate irregular supply from the Yenisei River runoff. The sea-level rise promoted a successive southward migration of the coastal zone and the consequent migration of the Yenisei marginal filter (Stein et al., 2003a, 2004).

It is possible to identify the suspended matter discharge of the Yenisei River based on indicators of fluvial materials in the sediments, such as wood remains (Levitan, 2001) and high clinopyroxene/epidote and smectite/illite ratios as well as high magnetic susceptibility values (Levitan et al., 1996; Dittmers et al., 2003; Stein et al., 2003a, 2004). The heavy-mineral composition shows that the river impact is best reflected in the clinopyroxene and iron hydroxide contents and, primarily, the clinopyroxene/epidote ratio (Fig. 10.31). Based on plant remains and the mineral composition of the sediments, river runoff was most prominent between 7.5 and 6.2 and between 3.3 and 2.4 cal. kyrs. BP. Correspondingly, runoff was reduced between 6.2 and 3.3 and between 2.4 and 1.4 cal. kyrs. BP. Based on the data from Core BP00-07/7, it seems to be that during the last 1.4 cal. kyrs. BP, the Yenisei River discharge began to increase again. In this context, however, it should be mentioned that, based on the high-resolution record of Core BP99-04/7, R. Stein et al. (2004) suggests a decreased Yensei River discharge during the last about 2.4 cal. kyrs. BP. At present, river discharge is very significant and reflects the global warming in the XX–XXI centuries.

Our data match the magnetic susceptibility (Stein et al., 2003a) as well as ostracod and mollusk (Taldenkova et al., 2003) data from Core BP00–07/5. There is also a partial correlation with the oxygen and carbon isotope data determined in ostracod shells (Simstich et al., 2003), and a weak correlation with benthic foraminifer data from this core (Ivanova, 2003). Interestingly, foraminifer data point at the same time of Atlantic waters influence on the Kara Sea shelf as in Franz Victoria Trough (Duplessy et al., 2005).

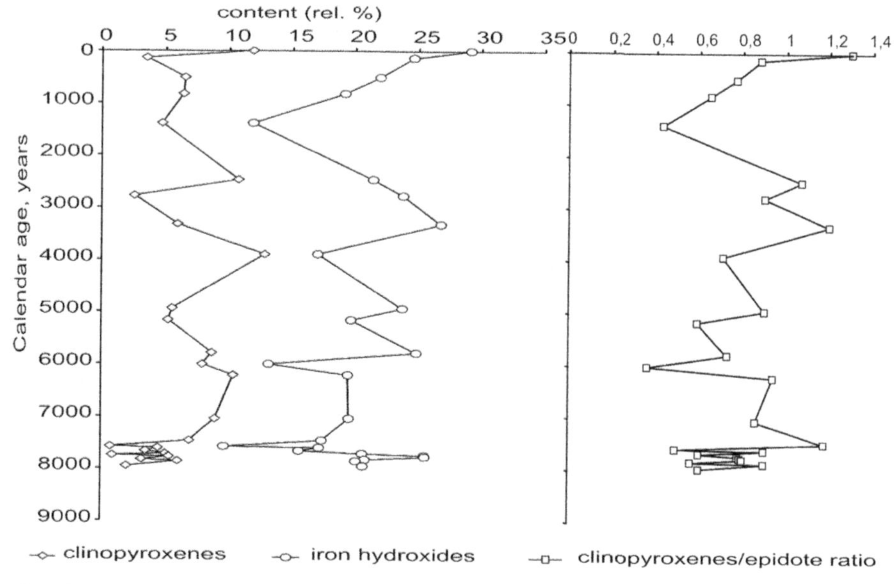

Fig. 10.31 Age distribution of clinopyroxenes, Fe-hydroxides, and clinopyroxene/epidote ratio in sediment core BP00-07/6 (Levitan et al., 2004)

The idea about slump origin of Unit 2 is supported not only by radiocarbon datings but also the lack of wet bulk density changes in this Unit and specific grain-size and mineral composition.

Thus, the studied southern Kara Sea sediments accumulated during the last 7.9 cal. kyrs. BP in the course of the Holocene transgression. The transport of pelitic particles from the surrounding seamounts by bottom currents and proportion of the Yenisei River runoff that escaped the marginal filter, served as the main sources of sedimentary material. As shown above, the relationship between these sources repeatedly changed during the Holocene period. In general, the available data suggest that the BP00–07/7 area remained beyond the marginal filter even during the most active delivery of Yenisei material.

Sediment core DM–4401 described in (Polyak et al., 2002b) is located in a channel, a tributary of the Pro-Yenisei River, between sediment cores BP00–07/6 and BP00-23/7 (Fig. 10.25). Our calculation of sedimentation rates and mass accumulation rates for this core show that these parameters were equal to 26.7 cm/ky and 22.6 g/cm^2/ky, respectively, during the period from 0 to 3.27 cal. kyrs. BP; 17.2 cm/ky and 17.6 g/cm^2/ky, respectively, during the period from 3.27 to 7.92 cal. kyrs. BP; and 120.6 cm/ky and 106.1 g/cm^2/ky, respectively, during the period from 7.92 to 8.28 cal. kyrs. BP. The latter rates probably continued down to about 9.4 cal. kyrs. BP. The lowermost part of the section down to the base at 415 cm is composed of coarser-grained sediments with very distinct features of river influence (high clinopyroxene contents and clinopyroxene/epidote ratios, high concentrations of plant remains and fresh-water diatom frustules, and light oxygen-isotope composition of benthic foraminifers). We assume higher sedimentation rates and a marginal-filter setting for this portion.

Thus, we have recorded a strong influence of river runoff at Core DM–4401 during the early Holocene (until 9.4 cal. kyrs. BP?). All the data suggest that this influence diminished during the period up to 7.2 cal. kyrs. BP and then slightly increased up to the level of ~1.5 cal. kyrs. BP. Data uppermost period are contradictory. Data on heavy minerals indicate an appreciable decrease of the Yenisei runoff contribution, while the diatom data suggest the contrary effect, i.e., increase of the Yenisei contribution (see also (Polyakova and Stein, 2004; Stein et al., 2004) for discussion). This may be caused by different mechanisms of transport of heavy minerals on the one hand, and the diatoms, on the other hand. In the first case, one can expect a preferential transport of heavy minerals in the nepheloid layer of the bottom-water mass (Levitan et al., 1996); in the second case the preferential transport took place in the surface water. This assumption is supported by our data on Ni/Al ratios in the sediments of Core DM-4401. The Ni/Al ratio can be used as facies indicator for Yenisei River runoff, as shown in surface sediments from the Kara Sea including the Ob and Yenisei estuaries (Schoster and Stein, 1999; Schoster et al., 2000) and primarily reflects the composition of the pelitic (clayey) fractions mainly transported in surface waters. Based on these assumptions, we found that the Ni/Al ratio is $>12.4 \times 10^{-4}$ in the sediments younger than 3.27 cal. kyrs. BP and varies from 10.1 to 11.0×10^{-4} in the older sediments (down to the core base).

History of Holocene Sedimentation along the Yenisei Transect in the Southern Kara Sea

The studied cores belong to two large facies zones of shelf sedimentation, i.e., the mixing zone of fresh and sea waters in the Yenisei Estuary (Yenisei marginal filter) and the old river valley of a tributary of the Pre-Yenisei River in the Ob-Yenisei shallow-water zone. Based on our data, each of these facies zones is characterized by a specific Holocene sedimentation history (Levitan et al., 1996). We have reconstructed the following historical–geological succession of submarine deltaic sediments in the marginal filter area: proximal sediments, depocenter sediments accumulated at avalanche rates, and distal sediments with contribution of coastal abrasion products. The major portion of the geological body of the marginal filter in the study area was formed within a very short time span (9.5–9.1 cal. kyrs. BP).

Based on various facies indicators (proxies), the following periods of strong influence of river runoff in the northern Core BP00-07/6 area were identified: 7.5–6.2, 3.3–2.4, and 1.4–0 cal. kyrs. BP. The basal core section probably represents a part of the submarine slump (see discussion above).

In general, the Holocene sedimentation history of the southern Kara Sea is related to synchronous processes of global eustatic sea-level rise (Fairbridge, 1989; Stein et al., 2003a, 2004), subsidence compensated by sedimentation in the Yenisei marginal filter (Gurevich, 1995), and fluctuating Yenisei River runoff. Concept of the variable basin depth (Stein et al., 2002) suggests that the Atlantic Water did not penetrate the study area.

The comparison of all available data on mass accumulation rate of fluvial material of the Yenisei transect (see Fig. 10.25) reported by L. Polyak et al. (2002) and R. Stein et al. (2002, 2003a, 2004) with data in the present research shows a north-to-south rejuvenation trend of the marginal-filter (more precisely, its depocenter) boundary with the overlying marine shelf sediments: 11.0 cal. kyrs. BP in Core BP00–36/4, 9.8 cal. kyrs. BP in Core BP00–26/4, 9.4 (?) cal. kyrs. BP in Core DM–4401, 9.1 cal. kyrs. BP in Core BP00–23/7, and 8.8 cal. kyrs. BP in Core BP00–04/7 (sediments of cores BP00–07/5, BP00–07/6, and BP00–07/7 did not reach this boundary?). This is caused by the southward Flandrian transgression related to the global post-glacial eustatic sea-level rise. The transgression rate was approximately 146 km/ky (probably, up to several hundred km/ky) during the Early Holocene in the northern Kara Sea and decreased to 22 km/ky during the later period (within the present-day marginal filter zone). This difference in transgression rate could be linked to the deepening of the northern areas of the Barents and Kara seas during the deglaciation and Early Holocene as a result of glacio-isostasy (Lubinskii et al., 2001). The southern Kara Sea did not undergo such deepening. At the same time, the differences can also be explained by a higher rise rate of the global sea level between 11.0 and 9.5 cal. kyrs. BP relative to the time interval of 9.5–8.0 cal. kyrs. BP (Fairbanks, 1989). We believe that the latter reason is more probable, because this process is global. However, one cannot rule out the simultaneous impact of both factors.

The subsequent fluctuations of the Yenisei River runoff were commonly characterized by one or more order of magnitude lower relative to the Early Holocene ones. Depending on the nature and mechanism of transport, various indicators of fluvial material indicate different periods of intensification or attenuation of Yenisei River runoff in each of the studied cores.

Holocene History of Ob River Discharge

Ob River discharge makes up to more than 12% of the modern annual total river discharge into the Arctic Ocean (Gordeev et al., 1996). Its Holocene history has not been discussed yet in scientific publications. Here, we present some new results related to the Ob River discharge and obtained within the Russian-German Kara Sea Project "SIRRO" ((Stein et al., 2003b) for background).

Materials and Methods

This study is mainly based on sedimentological, geochemical, and micropaleontological investigations of two sediment cores (BP03-07 and BP03-19) recovered and described during the 2003 cruise of R/V *Academik Boris Petrov* (Schoster and Levitan, 2004). These cores were located to the north of the mixing zone of fresh and sea water in a submerged river valleys of the Proto-Ob River (facies zone C4 according to (Levitan et al., 2005a)) along a submeridional Ob transect, which was already studied during cruise 49 of the R/V *Dmitry Mendeleev* in 1993 (Levitan et al., 1994). Data on the location of the two sediment cores are given in Table 10.23 and Fig. 10.32. Furthermore, data on sediment Core DM 4397 located further to the north along the same transect (facies zone D1, according to (Levitan et al., 2005a)) and used for comparison with the new cores, are shown as well. The data from the Ob cores are discussed later together with some cores from the Yenisei River transect (see below).

The selection of sites for sediment coring from R/V *Academik Boris Petrov* was performed using a new parametric ATLAS PARASOUND III seismo-acoustic system, which was subjected to certification tests (Dittmers and Schoster, 2004). A sequence of finely stratified sediments was detected in the PARASOUND profiles at both sites, reaching up to 4 m at Station BP03-07 and up to 9.5 m at Station BP03-19.

The lithology of bottom sediments was described by M.A. Levitan and V.V. Krupskaya (Levitan et al., 2004; Schoster and Levitan, 2004). Physical

Table 10.23 Location of studied sediment cores of the Ob River transect

Sediment core	Northern Latitude	Eastern Longitude	Water depth, m	Core length, cm
BP03-19	73°59,99′	73°07,81′	34	672
BP03-07	75°33,61′	73°07,64′	108	390
DM-4397	76°00,00′	72°02,30′	150	384

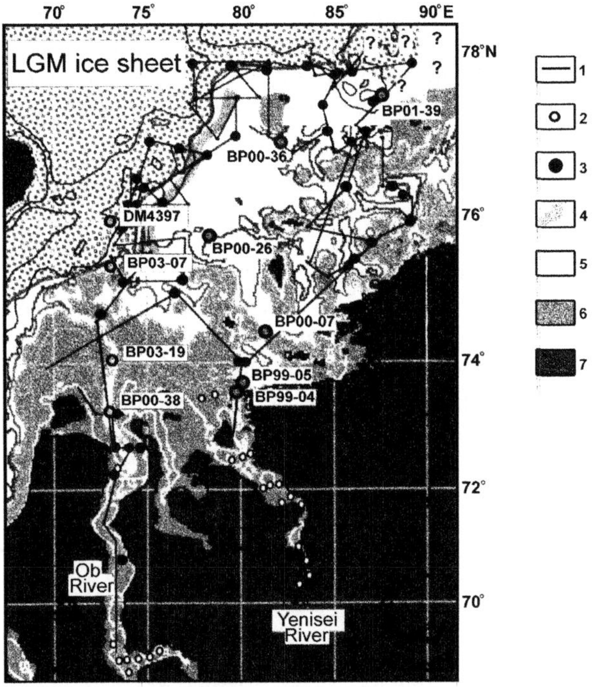

Fig. 10.32 Map of the study area of the "*Academik Boris Petrov*" Expedition 2001 in the inner Kara Sea and Ob and Yenisei estuaries, presenting locations of sampling sites and sediment echograph profiles (Stein and Stepanets, 2002). Based on the profiling results, the extent of the south-eastern margin of the Last Glacial Maximum (LGM) ice sheet and the LGM paleo-river channels were reconstructed (from (Stein et al., 2002)). Further sediment cores from the Ob Transect investigated in this study, were added. 1 – geophysical lines; 2, 3 – sediment cores: 2 – with radiocarbon dates, 3 – without age determination; 4 – paleo-Ob valley; 5– 7 – land: 5 – 11 ka ago, 6 – 9 ka ago, 7 – recent

properties were investigated by L.N. Vlasova, and mollusk shells were identified by E.A. Frolova (Levitan et al., 2004). Grain-size, chemical, and radiocarbon analyses were carried out at onshore laboratories. The sand-silt and silt–clay boundaries were taken to be 0.05 and 0.005 mm, respectively.

Table 10.24 Results of the radiocarbon analysis of mollusk shells

Lab no.	Sample No.	Radiocarbon age, years	Corrected age, years	Calendar age, cal. yrs. BP	Mollusk species
Poz–11062	BP03–07/248 cm	8990±40	8550	8985	*Yoldia hyperborea*
Poz–11063	BP03–07/44 cm	6010±40	5570	5925	*Macoma moesta*
Poz–11064	BP03–19/108 cm	595±30	155	recent	*Macoma moesta*
Poz–11065	BP03–19/440 cm	8250±50	7810	8290	*Musculus niger*
Poz–12677	BP03–19/143 cm	595±30	155	recent	*Portlandia arctica*
Poz–12678	BP03–19/608 cm	8560±50	8120	8580	*Portlandia arctica*

Six shells of marine bivalve mollusks were analyzed at the Poznan Radiocarbon Laboratory (Poland) by accelerator mass spectrometry under the supervision of T. Goslar. The results are listed in Table 10.24. As reservoir age, 440 years were used, and the AMS ^{14}C ages were transferred into calendar years according to (Stuiver et al., 1998).

Sediment Core BP03-19

The core consists of two lithostratigraphic units.

Unit I. The interval of 0–152 cm is composed of fine clayey mud (in average <10% sand and <15% silt), occasionally with fine-grained sand lenses of 0.5–1.0 cm thickness, and a 12 cm thick layer of silty-sandy clay at the base (Fig. 10.33). The uppermost 0.5 cm thick layer shows a dark olive brown color, whereas very dark gray or, occasionally, dark olive gray colors are predominant in the deeper layers. Intact shells of bivalves were encountered at levels of 10 and 108 cm. Black

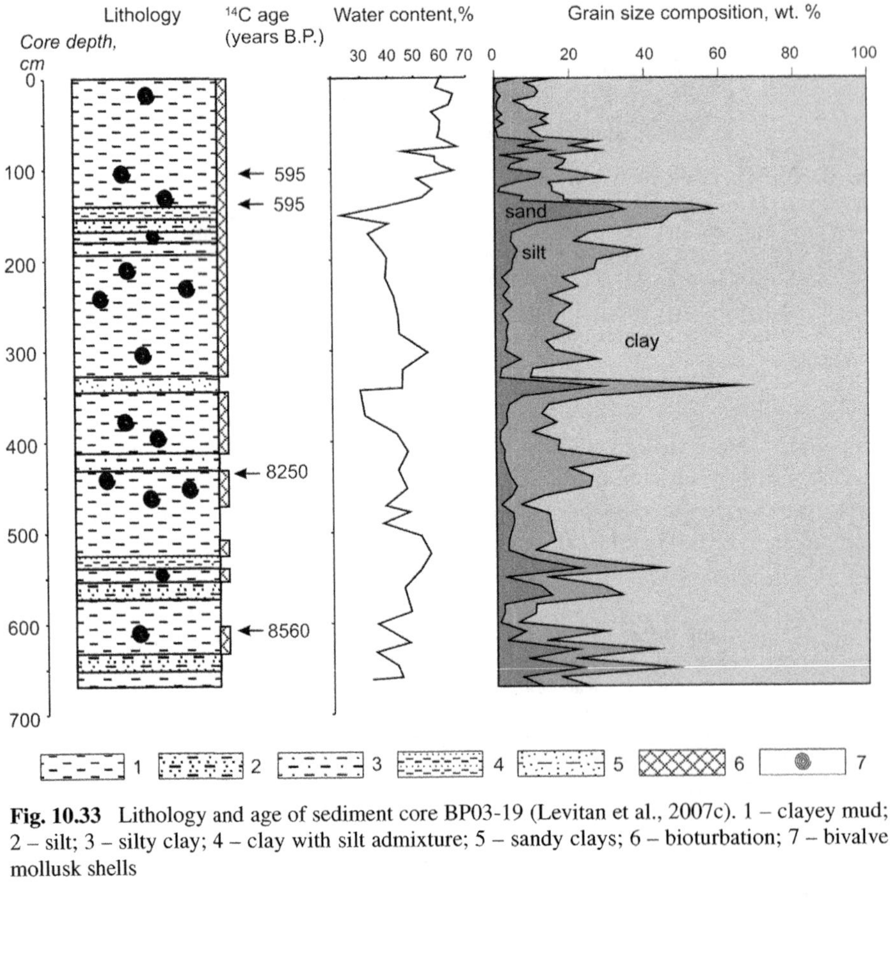

Fig. 10.33 Lithology and age of sediment core BP03-19 (Levitan et al., 2007c). 1 – clayey mud; 2 – silt; 3 – silty clay; 4 – clay with silt admixture; 5 – sandy clays; 6 – bioturbation; 7 – bivalve mollusk shells

spots, streaks, and lenses of hydrotroilite were found below 12 cm. Accumulation of hydrotroilite sometimes caused the black color of the sediment. Moderate bioturbation, more intense below 100 cm, was typical. The water content is 55–65% in the upper 100 cm and decreases abruptly at deeper levels to 22% at the boundary to the underlaying Unit II (Fig. 10.26). The wet bulk density increases gradually from 1.28 g/cm^3 to 1.40 g/cm^3 at 125 cm and, then, increases abruptly up to 1.70 g/cm^3 (Levitan et al., 2004). Such a distribution of physical properties may suggest a hiatus at the boundary between the two lithological units, which is represented by a weakly developed hardground.

Unit II. The interval of 152–672 cm is dominated by dark gray and black clayey mud layers, with subordinate layers and lenses of silty-clayey muds and silty-sandy or sandy-silty clays of the same color (Fig. 10.26). The coarser units are usually no more than a few centimeters thick. The clayey muds usually contain $<$10% sand and 8 to 22% silt. The sediments enriched in silty and sandy fractions contain 11–34% sand and 18–33% silt (Fig. 10.26). Bivalve shells occur throughout the entire section. At a level of 605–608 cm, an extensive accumulation of *Portlandia arctica* with detritus and shells of the gastropod *Criptonatica sp.* was found. The most extensive bioturbation and diagenetic mottling resulting from the development of hydrotroilite aggregates, were described in the interval between 150 and 385 cm. The lower part of the horizon (below 343 cm) displays the development of a distinct fine-layered structure composed of lithological varieties with different shades of gray and black colors, massive or bioturbated structures, and finer or coarser structures. On average, the water content varies around 40% (from 30% in the coarsest sediments to 50% in the finest clayey muds) (Fig. 10.26), and the average wet bulk density is about 1.60 g/cm^3 (ranging from 2.00 g/cm^3 in layers enriched in silty and sandy fractions to 1.40 g/cm^3 in clayey muds).

Chemical Composition

The chemical composition of the two units is strongly different. Unit I is enriched in Na, Mn, Al, Fe, P, and Ba, but depleted in Si, Ca, Ti, S, Rb, Sr, and Zr compared to Unit II (Fig. 10.34). In general, the chemical composition of Unit I resembles the composition of modern facies C4 (filling the channels of an old river network) (Levitan et al., 2005a). Unit II shows a complex relationship of features observed in the modern facies of the rivers (Na, Cr, Mn, Fe, V, Co, and Ni) and mixing zones (P, to some extent, Si, Al, and Ba). The correlation analysis of various oxides, elements, and grain-size fractions (Tables 10.25, 10.26 and 10.27) allowed us to distinguish the following associations in Unit I at a confidence level of 95%: (1) Na_2O and S confined to the 0.50–0.25 mm fraction (i.e., sulfide sulfur and sodium from plagioclases are predominant); (2) MgO, $A1_20_3$, P_2O_3, K_2O, TiO_2, Cr_2O_3, MnO, FeO, V, and, to some extent, Ni confined to the subcolloidal fraction; and (3) SiO_2, CaO, Rb, Sr, and Ba confined to the sand and 0.05–0.01 mm fractions. These correlations are less obvious in Unit II, where the following associations were observed: (1) Na_2O, K_2O, Ni, and Ba confined to sands; (2) MgO, $A1_20_3$, TiO_2, MnO, FeO, P_2O_3, S, and V confined to the subcolloidal fraction; (3) SiO_2, CaO, Rb, Sr, Zr, and Ba confined

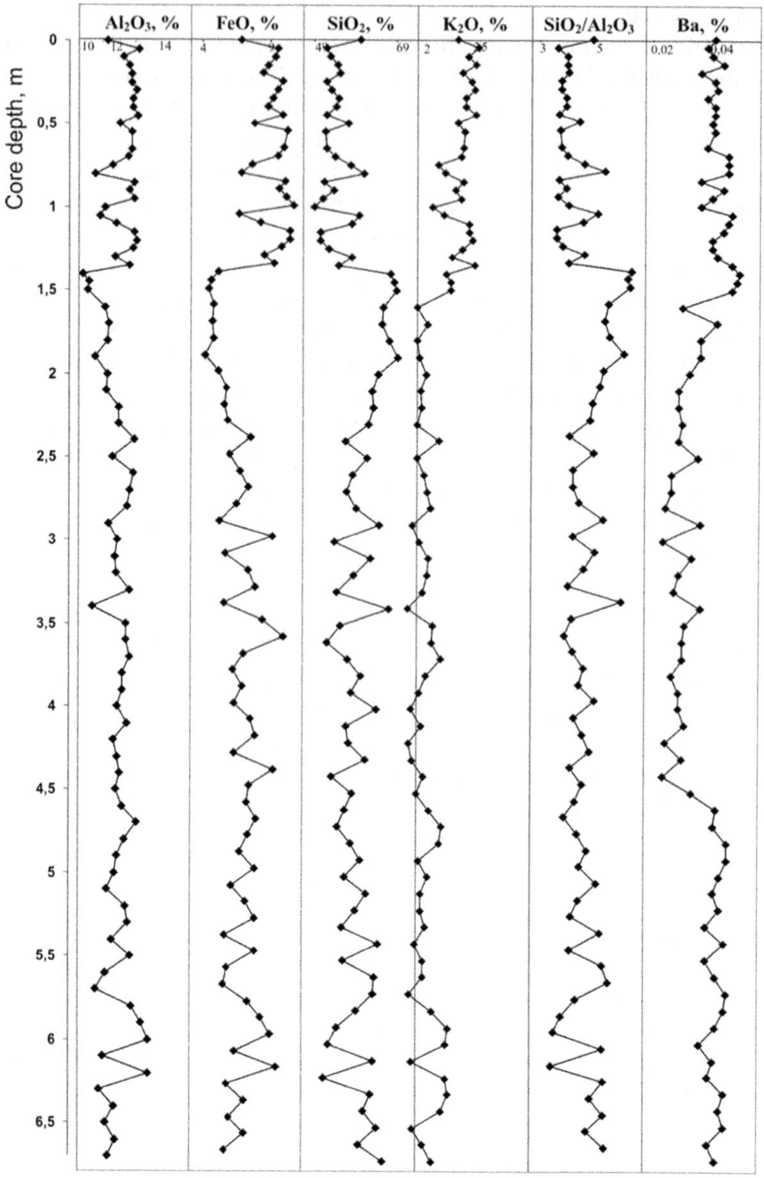

Fig. 10.34 Distribution of chemical elements in sediment core BP03-19 (Levitan et al., 2007c)

to sands and silts (Sr significantly substitutes for Ca in large fragments of mollusk shells); and (4) Cr_2O_3 confined to the <0.001 mm fraction.

The available radiocarbon datings (Table 10.23) show that Unit I was most likely accumulated relatively recently as a result of a submarine slumping. Between 440

Table 10.25 Correlation matrix of sediment core BP03-19 (upper horizon)

n=27	Na_2O	MgO	Al_2O_3	SiO_2	P_2O_5	K_2O	CaO	TiO_2	Cr_2O_3	MnO	FeO	S	V	Ni	Rb	Sr	Zr	Ba	1–0,5	0,5–0,25	0,25–0,1	0,1–0,05	Sand	0,05–0,01	0,01–0,005	Silt	0,005–0,001	<0,001	Clay
Na_2O	1.00	–0.02	–0.17	–0.33	–0.51	–0.21	0.02	–0.16	–0.01	–0.04	0.18	0.65	–0.10	–0.16	–0.30	–0.19	–0.18	–0.32	–0.09	0.55	–0.22	–0.18	–0.18	–0.15	0.09	0.06	–0.28	0.27	0.16
MgO	–0.02	1.00	0.87	–0.87	0.32	0.66	–0.67	0.54	0.47	0.37	0.88	0.01	0.89	0.19	–0.32	–0.88	–0.88	–0.37	–0.47	–0.14	–0.29	–0.73	–0.62	–0.63	–0.16	–0.23	–0.20	0.59	0.52
Al_2O_3	–0.17	0.87	1.00	–0.71	0.37	0.78	–0.71	0.74	0.53	0.35	0.72	–0.24	0.84	0.31	–0.07	–0.79	–0.75	–0.37	–0.61	–0.43	–0.32	–0.76	–0.68	–0.69	–0.18	–0.26	0.07	0.48	0.52
SiO_2	–0.33	–0.87	–0.71	1.00	–0.21	–0.37	0.63	–0.39	–0.44	–0.40	–0.97	–0.40	–0.80	0.03	0.60	0.94	0.94	0.54	0.52	–0.29	0.35	0.75	0.65	0.62	0.09	0.18	0.36	–0.66	–0.53
P_2O_5	–0.51	0.32	0.37	–0.21	1.00	0.17	–0.23	0.08	0.12	0.58	0.35	–0.40	0.50	–0.07	–0.33	–0.25	–0.27	0.10	–0.10	–0.33	0.13	–0.09	–0.01	0.02	0.27	0.26	–0.02	–0.03	–0.04
K_2O	–0.21	0.66	0.78	–0.37	0.17	1.00	–0.57	0.58	0.46	0.07	0.39	–0.43	0.53	0.52	0.26	–0.53	–0.45	–0.09	–0.38	–0.59	–0.18	–0.56	–0.48	–0.46	–0.25	–0.30	0.23	0.30	0.41
CaO	0.02	–0.67	–0.71	0.63	–0.23	–0.57	1.00	–0.40	–0.47	0.02	–0.68	0.01	–0.66	–0.44	0.11	0.81	0.62	0.17	0.79	0.20	0.20	0.64	0.52	0.56	0.21	0.28	0.03	–0.59	–0.59
TiO_2	–0.16	0.54	0.74	–0.39	0.08	0.58	–0.40	1.00	0.59	0.34	0.33	–0.28	0.55	0.46	0.20	–0.43	–0.36	–0.37	–0.53	–0.41	–0.52	–0.69	–0.71	–0.70	–0.37	–0.45	0.32	0.15	0.29
Cr_2O_3	–0.01	0.47	0.53	–0.44	0.12	0.46	–0.47	0.59	1.00	0.24	0.44	–0.02	0.45	0.13	–0.01	–0.50	–0.40	–0.14	–0.43	–0.09	–0.23	–0.53	–0.45	–0.46	–0.48	–0.51	–0.15	0.35	0.30
MnO	–0.04	0.37	0.35	–0.40	0.58	0.07	0.02	0.34	0.24	1.00	0.40	–0.01	0.48	–0.26	–0.50	–0.29	–0.37	–0.38	–0.19	0.01	–0.23	–0.37	–0.33	–0.15	0.32	0.27	–0.20	–0.14	–0.23
FeO	0.18	0.88	0.72	–0.97	0.35	0.39	–0.68	0.33	0.44	0.40	1.00	0.35	0.84	–0.04	–0.64	–0.96	–0.96	–0.44	–0.52	0.25	–0.23	–0.69	–0.55	–0.54	–0.03	–0.11	–0.39	0.63	0.48
S	0.65	0.01	–0.24	–0.40	–0.40	–0.43	0.01	–0.28	–0.02	–0.01	0.35	1.00	–0.06	–0.45	–0.63	–0.30	–0.30	–0.39	–0.12	0.95	–0.11	–0.14	–0.10	–0.04	0.08	0.07	–0.40	0.25	0.09
V	–0.10	0.89	0.84	–0.80	0.50	0.53	–0.66	0.55	0.45	0.48	0.84	–0.06	1.00	0.21	–0.37	–0.82	–0.81	–0.17	–0.52	–0.16	–0.18	–0.60	–0.49	–0.53	0.06	–0.02	–0.16	0.39	0.33
Ni	–0.16	0.19	0.31	0.03	–0.07	0.52	–0.44	0.46	0.13	–0.26	–0.04	–0.45	0.21	1.00	0.43	–0.11	0.01	0.19	–0.32	–0.54	–0.26	–0.34	–0.35	–0.33	–0.21	–0.24	0.38	0.12	0.29
Rb	–0.30	–0.32	–0.07	0.60	–0.33	0.26	0.11	0.20	–0.01	–0.50	–0.64	–0.63	–0.37	0.43	1.00	0.48	0.55	0.31	0.19	–0.63	–0.02	0.27	0.15	–0.01	–0.29	–0.27	0.53	–0.27	–0.06
Sr	–0.19	–0.88	–0.79	0.94	–0.25	–0.53	0.81	–0.43	–0.50	–0.29	–0.96	–0.30	–0.82	–0.11	0.48	1.00	0.93	0.45	0.67	–0.18	0.26	0.76	0.63	0.61	0.11	0.19	0.25	–0.66	–0.58
Zr	–0.18	–0.88	–0.75	0.94	–0.27	–0.45	0.62	–0.36	–0.40	–0.37	–0.96	–0.30	–0.81	0.01	0.55	0.93	1.00	0.51	0.45	–0.16	0.28	0.72	0.61	0.62	0.05	0.13	0.30	–0.67	–0.56
Ba	0.32	0.37	0.37	0.54	0.10	0.09	0.17	0.37	0.14	0.38	0.44	0.39	0.17	0.19	0.31	0.45	0.51	1.00	0.36	0.25	0.65	0.68	0.73	0.57	0.19	0.26	0.02	0.41	0.41

Table 10.26 Correlation matrix of sediment core BP03-19 (lower horizon)

n=55	Na_2O	MgO	Al_2O_3	SiO_2	P_2O_5	K_2O	CaO	TiO_2	Cr_2O_3	MnO	FeO	S	V	Ni	Rb	Sr	Zr	Ba	1–0,5	0,5–0,25	0,25–0,1	0,1–0,05	Sand	0,05–0,01	0,01–0,005	Silt	0,005–0,001	<0,001	Clay
Na_2O	1.00	–0.47	–0.42	0.30	0.10	0.62	–0.13	–0.51	–0.21	0.11	–0.27	–0.41	0.24	0.50	–0.19	0.24	0.30	0.70	0.09	–0.04	0.76	0.56	0.68	–0.26	0.26	0.05	–0.44	–0.29	–0.44
MgO	–0.47	1.00	0.89	–0.91	0.29	0.14	–0.29	0.45	0.18	0.36	0.83	0.42	0.39	–0.03	–0.10	–0.79	–0.84	–0.54	–0.05	0.19	–0.58	–0.66	–0.69	–0.38	–0.57	–0.58	0.06	0.70	0.73
Al_2O_3	–0.42	0.89	1.00	–0.85	0.39	0.23	–0.34	0.61	0.06	0.56	0.77	0.24	0.54	0.13	–0.26	–0.68	–0.74	–0.35	–0.10	0.01	–0.61	–0.67	–0.71	–0.39	–0.69	–0.67	–0.03	0.79	0.79
SiO_2	0.30	–0.91	–0.85	1.00	–0.45	–0.23	0.32	–0.34	–0.16	–0.54	–0.96	–0.31	–0.55	0.03	0.37	0.83	0.90	0.38	0.02	–0.16	0.48	0.60	0.61	0.43	0.56	0.60	0.05	–0.70	–0.69
P_2O_5	0.10	0.29	0.39	–0.45	1.00	0.13	–0.25	0.19	0.23	0.80	0.53	–0.19	0.73	0.24	–0.73	–0.24	–0.40	0.39	0.27	–0.13	–0.16	–0.10	–0.12	–0.36	–0.30	–0.38	–0.29	0.36	0.27
K_2O	0.62	0.14	0.23	–0.23	0.13	1.00	–0.38	–0.14	–0.16	0.37	0.21	–0.11	0.46	0.48	–0.07	–0.28	–0.22	0.34	–0.09	0.07	0.35	0.11	0.19	–0.44	–0.20	–0.35	–0.32	0.18	0.07
CaO	–0.13	–0.29	–0.34	0.32	–0.25	–0.38	1.00	–0.04	0.06	–0.28	–0.37	0.03	–0.34	–0.18	0.08	0.59	0.42	0.10	0.08	–0.06	–0.02	0.12	0.08	0.15	0.13	0.17	0.17	–0.20	–0.15
TiO_2	–0.51	0.45	0.61	–0.34	0.19	–0.14	–0.04	1.00	0.00	0.30	0.27	0.07	0.21	0.11	–0.11	–0.16	–0.11	–0.24	–0.04	–0.08	–0.66	–0.44	–0.55	–0.06	–0.51	–0.39	0.18	0.47	0.54
Cr_2O_3	–0.21	0.18	0.06	–0.16	0.23	–0.16	0.06	0.00	1.00	–0.02	0.22	0.26	0.12	–0.21	0.03	–0.19	–0.27	–0.07	0.09	0.14	–0.21	0.14	0.05	0.00	0.11	0.08	0.04	–0.08	–0.07
MnO	0.11	0.36	0.56	–0.54	0.80	0.37	–0.28	0.30	–0.02	1.00	0.59	–0.22	0.74	0.17	–0.70	–0.33	–0.43	0.32	0.10	–0.19	–0.18	–0.18	–0.20	–0.41	–0.41	–0.49	–0.18	0.43	0.38
FeO	–0.27	0.83	0.77	–0.96	0.53	0.21	–0.37	0.27	0.22	0.59	1.00	0.26	0.58	–0.10	–0.47	–0.85	–0.90	–0.30	–0.03	0.14	–0.41	–0.49	–0.50	–0.44	–0.46	–0.53	–0.01	0.59	0.59
S	–0.41	0.42	0.24	–0.31	–0.19	–0.11	0.03	0.07	0.26	–0.22	0.26	1.00	–0.21	–0.36	0.35	–0.33	–0.33	–0.55	–0.05	0.29	–0.36	–0.29	–0.33	–0.03	–0.17	–0.14	0.21	0.20	0.28
V	0.24	0.39	0.54	–0.55	0.73	0.46	–0.34	0.21	0.12	0.74	0.58	–0.21	1.00	0.44	–0.62	–0.37	–0.47	0.43	0.08	–0.10	–0.06	–0.14	–0.12	–0.52	–0.46	–0.57	–0.35	0.49	0.38
Ni	0.50	–0.03	0.13	0.03	0.24	0.48	–0.18	0.11	–0.21	0.17	–0.10	–0.36	0.44	1.00	–0.16	0.16	0.09	0.51	–0.03	–0.15	0.16	0.05	0.08	–0.32	–0.31	–0.37	–0.49	0.31	0.14
Rb	–0.19	–0.10	–0.26	0.37	–0.73	–0.07	0.08	–0.11	0.03	–0.70	–0.47	0.35	–0.62	–0.16	1.00	0.09	0.26	–0.44	–0.07	0.11	0.02	0.08	0.07	0.31	0.24	0.32	0.26	–0.29	–0.21
Sr	0.24	–0.79	–0.68	0.83	–0.24	–0.28	0.59	–0.16	–0.19	–0.33	–0.85	–0.33	–0.37	0.16	0.09	1.00	0.87	0.51	0.12	–0.21	0.26	0.42	0.41	0.35	0.31	0.38	–0.07	–0.43	–0.46
Zr	0.30	–0.84	–0.74	0.90	–0.40	–0.22	0.42	–0.11	–0.27	–0.43	–0.90	–0.33	–0.47	0.09	0.26	0.87	1.00	0.41	0.05	–0.16	0.37	0.54	0.53	0.32	0.44	0.46	0.06	–0.59	–0.57
Ba	0.70	–0.54	–0.35	0.38	0.39	0.34	0.10	–0.24	–0.07	0.32	–0.30	–0.55	0.43	0.51	–0.44	0.51	0.41	1.00	0.10	–0.26	0.52	0.51	0.55	–0.24	0.07	–0.06	–0.45	–0.15	–0.31

Table 10.27 Correlation matrix of sediment core BP03-07

n=39	Na_2O	MgO	Al_2O_3	SiO_2	P_2O_5	K_2O	CaO	TiO_2	Cr_2O_3	MnO	FeO	S	V	Ni	Rb	Sr	Zr	Ba	1–0,5	0,5–0,25	0,25–0,1	0,1–0,05	Sand	0,05–0,01	0,01–0,005	Silt	0,005–0,001	<0,001	Clay
Na_2O	1.00	0.21	0.46	–0.26	–0.09	0.26	–0.20	0.27	–0.09	–0.19	0.14	–0.01	0.20	0.04	0.01	–0.13	–0.06	–0.17	0.09	0.01	0.12	–0.21	0.10	0.06	0.04	0.06	0.37	–0.17	0.02
MgO	0.21	1.00	0.80	–0.79	0.18	0.61	–0.23	0.70	0.05	0.30	0.86	–0.18	0.80	0.53	–0.63	–0.31	–0.75	–0.63	0.33	0.29	0.18	–0.66	–0.39	0.35	0.30	0.38	0.39	–0.23	–0.02
Al_2O_3	0.46	0.80	1.00	–0.76	0.10	0.64	–0.22	0.63	0.02	0.01	0.66	–0.18	0.64	0.46	–0.27	–0.44	–0.43	–0.50	0.16	0.19	0.15	–0.56	–0.38	0.19	0.22	0.25	0.42	–0.18	0.03
SiO_2	–0.26	–0.79	–0.76	1.00	–0.19	–0.63	0.08	–0.48	0.11	–0.35	–0.80	0.14	–0.57	–0.20	0.26	0.55	0.47	0.44	–0.30	–0.26	–0.36	0.44	0.11	–0.08	–0.21	–0.20	–0.31	0.27	0.09
P_2O_5	–0.09	0.18	0.10	–0.19	1.00	0.12	0.02	0.30	–0.13	0.33	0.27	–0.01	0.12	–0.07	0.08	–0.26	0.07	–0.03	0.16	–0.19	0.07	–0.18	–0.09	–0.18	0.03	–0.06	0.22	0.00	0.10
K_2O	0.26	0.61	0.64	–0.63	0.12	1.00	–0.16	0.33	0.08	0.01	0.67	–0.56	0.68	0.42	–0.26	–0.56	–0.41	–0.35	0.29	0.37	0.52	–0.34	0.03	0.11	0.28	0.26	0.19	–0.39	–0.26
CaO	–0.20	–0.23	–0.22	0.08	0.02	–0.16	1.00	–0.09	–0.09	0.15	–0.19	0.30	–0.33	–0.29	0.19	0.08	0.13	0.06	–0.24	–0.29	–0.18	–0.14	–0.17	–0.18	–0.33	–0.33	0.03	0.41	0.38
TiO_2	0.27	0.70	0.63	–0.48	0.30	0.33	–0.09	1.00	–0.03	0.19	0.46	0.07	0.45	0.35	–0.43	0.01	–0.43	–0.28	0.03	0.05	–0.17	–0.76	–0.54	0.31	–0.02	0.13	0.47	0.17	0.36
Cr_2O_3	–0.09	0.05	0.02	0.11	–0.13	0.08	–0.09	–0.03	1.00	–0.18	0.00	0.04	0.05	0.17	–0.20	0.07	–0.19	0.17	0.07	–0.02	–0.21	0.15	–0.60	0.07	–0.12	–0.06	–0.06	0.11	0.06
MnO	–0.19	0.30	0.01	–0.35	0.33	0.01	0.15	0.19	–0.18	1.00	0.44	0.21	0.08	–0.10	–0.10	–0.25	–0.20	0.01	0.03	0.04	0.33	–0.18	–0.03	–0.27	0.14	–0.02	–0.16	0.03	–0.05
FeO	0.14	0.86	0.66	–0.80	0.27	0.67	–0.19	0.46	0.00	0.44	1.00	–0.35	0.83	0.47	–0.53	–0.53	–0.70	–0.63	0.34	0.35	0.48	–0.44	–0.05	0.22	0.41	0.41	0.24	–0.45	–0.29
S	–0.01	–0.18	–0.18	0.14	–0.01	–0.56	0.30	0.07	0.04	0.21	–0.35	1.00	–0.61	–0.46	0.24	0.23	0.21	0.28	–0.18	–0.26	–0.48	0.00	–0.36	–0.23	–0.49	–0.48	0.11	0.56	0.54
V	0.20	0.80	0.64	–0.57	0.12	0.68	–0.33	0.45	0.05	0.08	0.83	–0.61	1.00	0.72	–0.75	–0.19	–0.80	–0.68	0.30	0.40	0.26	–0.46	0.00	0.54	0.51	0.63	0.29	–0.52	–0.32
Ni	0.04	0.53	0.46	–0.20	–0.07	0.42	–0.29	0.35	0.17	–0.10	0.47	–0.46	0.72	1.00	–0.66	0.09	–0.63	–0.43	0.02	0.21	0.03	–0.29	–0.13	0.54	0.57	0.67	0.04	–0.38	–0.32
Rb	0.01	–0.63	–0.27	0.26	0.08	–0.26	0.19	–0.43	–0.20	–0.10	–0.53	0.24	–0.75	–0.66	1.00	–0.36	0.92	0.53	–0.20	–0.31	0.16	0.43	0.09	–0.75	–0.36	–0.62	–0.22	0.26	0.13
Sr	–0.13	–0.31	–0.44	0.55	–0.26	–0.56	0.08	0.01	0.07	–0.25	–0.53	0.23	–0.19	0.09	–0.36	1.00	–0.10	0.19	–0.28	–0.14	–0.72	–0.05	–0.13	0.50	–0.10	0.15	–0.04	0.33	0.27
Zr	–0.06	–0.75	–0.43	0.47	0.07	–0.41	0.13	–0.43	–0.19	–0.20	–0.70	0.21	–0.80	–0.63	0.92	–0.10	1.00	0.61	–0.25	–0.33	–0.01	0.46	0.11	–0.68	–0.40	–0.61	–0.27	0.36	0.19
Ba	–0.17	–0.63	–0.50	0.44	–0.03	–0.35	0.06	–0.28	0.17	0.01	–0.63	0.28	–0.68	–0.43	0.53	0.19	0.61	1.00	–0.26	–0.29	–0.18	0.40	0.07	–0.40	–0.29	–0.40	–0.36	0.37	0.16

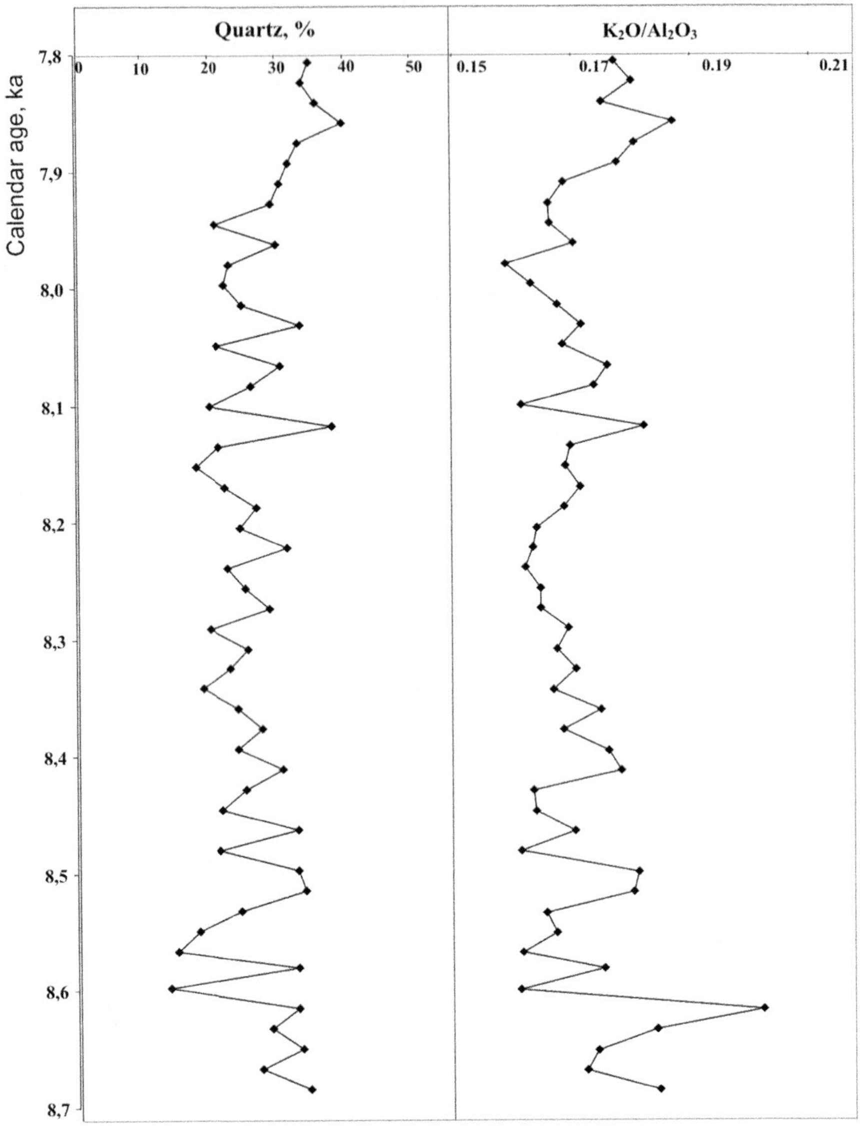

Fig. 10.35 Age distribution of quartz and K_2O/Al_2O_3 ratio in sediment core BP03-19 (Levitan et al., 2007c)

and 608 cm, the sedimentation rate was 569 cm/ka, which is within the range of modern sedimentation rates in the depocenter of the mixing zone of fresh and sea waters in the southern Kara Sea (Levitan et al., 2005a). The average mass accumulation rate of sediments in Unit II is 546 g/cm^2/ky. Extrapolation of the sedimentation rate yields an age of 7.79 cal. kyrs. BP for the upper boundary of the second layer,

8.69 cal. kyrs. BP for the base of Unit II, and 9.18 cal. kyrs. BP for the start of sediment filling of the channel (based on PARASOUND profiling, the thickness of filling was estimated to be about 9.5 m). Taking into account that there was a certain time lag between the beginning of channel incision by river currents and the beginning of its sedimentary filling, the initial accumulation of the river valley (channel) may have started at about 11–10 cal. kyrs. BP.

According to K. Dittmers (2006), (probably all) paleochannels of the meander system in the southern part of the Kara Sea were developed during Last Glacial Maximum, and (partly) filled with sediments after 15 cal. kyrs. BP. There seems to be a considerable hiatus between the age of the end of meandering/channel incision and the age of the basal sediments. This hypothesis, however, has still to be proven by direct evidence from well-dated sediment cores.

Smear slides revealed that the sediments are almost free of remains of siliceous organisms and, consequently, of biogenic opal (Levitan et al., 2004). As previously shown by M. Levitan et al. (1998), the Ob River is a major source of quartz and potassium feldspars in the sediments of the Kara Sea. It is reasonable to suppose that the SiO_2/Al_2O_3 ratio as well as the quartz content, calculated after E. Gurvich (Levitan et al., 1998) as $SiO_2 - 2.55\ Al_2O_3$, are indicators of the sediment discharge by the Ob River. For potassium feldspars K_2O/Al_2O_3 ratio has the same significance. K/Al and Rb/Al ratios are the good proxies of Ob riverine discharge (Schoster and Stein, 1999).

Figure 10.35 shows the distribution of quartz content and K_2O/Al_2O_3 in Unit II of the core. It is obvious that this ratio decreases upward the section from about 8.7 to 8.2 cal. kyrs. BP, which can be related to a decrease of riverine sediment input. This trend includes several periods of maximum sediment input by the Ob River at 7.85, 8.12, 8.42, 8.5, 8.58, 8.62, and 8.68 cal. kyrs. BP.

Sediment Core BP03-07

Lithostratigraphy

Core BP03-07 consists of one lithological unit. This unit is presented by intercalated sandy-silty and silty-sandy clays with rare fine lenses and layers of silty sand (with sand and silt contents of about 30 and 20%, respectively). The content of clay fraction generally ranges between 40 and 60%; occasionally some higher occurs (Fig. 10.36). The uppermost centimeters are composed of dark brown sediments, whereas the rest of the section is dominated by olive gray and dark olive gray colors. Polychaets were found in the upper 30 cm. Detritus of mollusk shells (bivalves and, occasionally, gastropods) occurs throughout the entire section down to 390 cm, whereas intact shells are more rare. It is interesting to note that the shells of *Yoldia hyperborea* which lives in the upper sub-littoral zone in water depths <20 m (data E.A. Frolova (Levitan et al., 2004)) were only found between 200 cm and the base of the core. Hydrotroilite and relatively strong bioturbation were determined down to a core depth of 85 cm but not below this level. A hydrogen sulfide odor was noted below 150 cm core depth. Within the interval 175–320 cm, single crystals

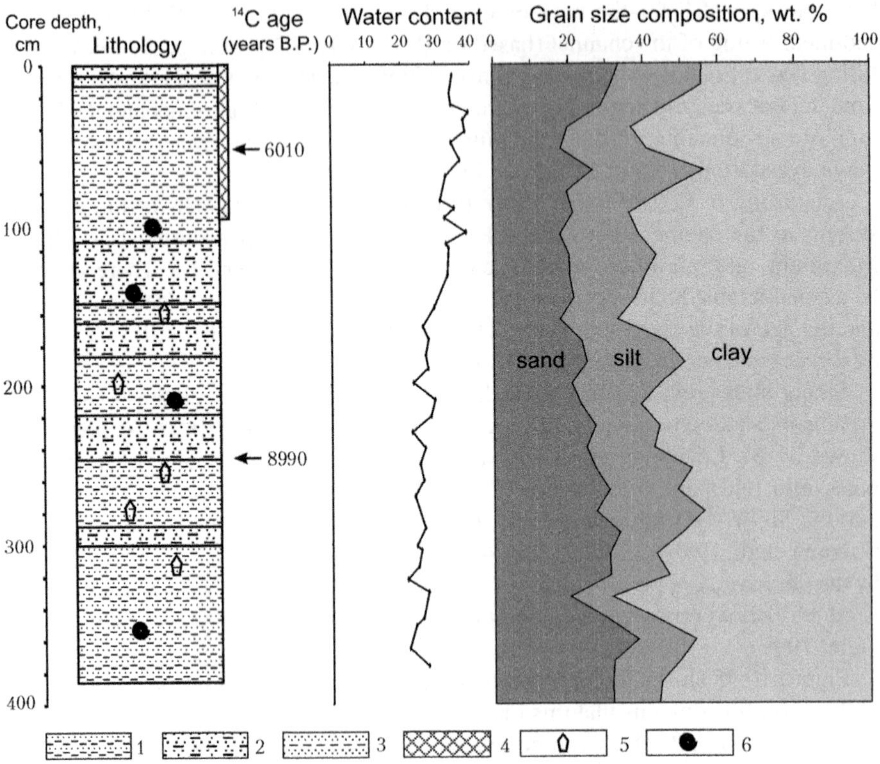

Fig. 10.36 Lithology and age of sediment core BP03-07 (Levitan et al., 2007c). 1 – silty clay; 2 – clay with silt admixture; 3 – sandy clays; 4 – bioturbation; 5 – ikaite; 6 – bivalve mollusk shells

and aggregates (up to 10 cm long) of ikaite of brown and honey-yellow colors were detected at levels of 175, 191, 199, 260–270, 280–285, 300–305, 315, and 320 cm. Ikaite was repeatedly described in sediment cores from the Yenisei transect (Levitan et al., 1994; Kodina et al., 2003), but this is the first finding of ikaites in the Ob transect. In general, the number of coarser-grained layers is increasing downward in the section.

Chemical Composition

Most major elements (Na, Mg, Al, Si, P, Ti, Fe, and Co) in the sediments correspond to those of modern river sediments and show some similarity to the proximal subfacies of the mixing zone (Levitan et al., 2005a). A number of elements (K, Mn, Fe, Zr, and Ba) display a more complex distribution. The concentrations of S and Sr are unexpectedly high, whereas those of Ni and Rb are very low. Correlation analysis (Table 10.27) revealed in principle the same element associations as in Core BP03-19. The following associations can be reliably distinguished: (1) MgO, Al_2O_3, K_2O, TiO_2, FeO, MnO, V, and Ni confined to two grain-size fractions, i.e.,

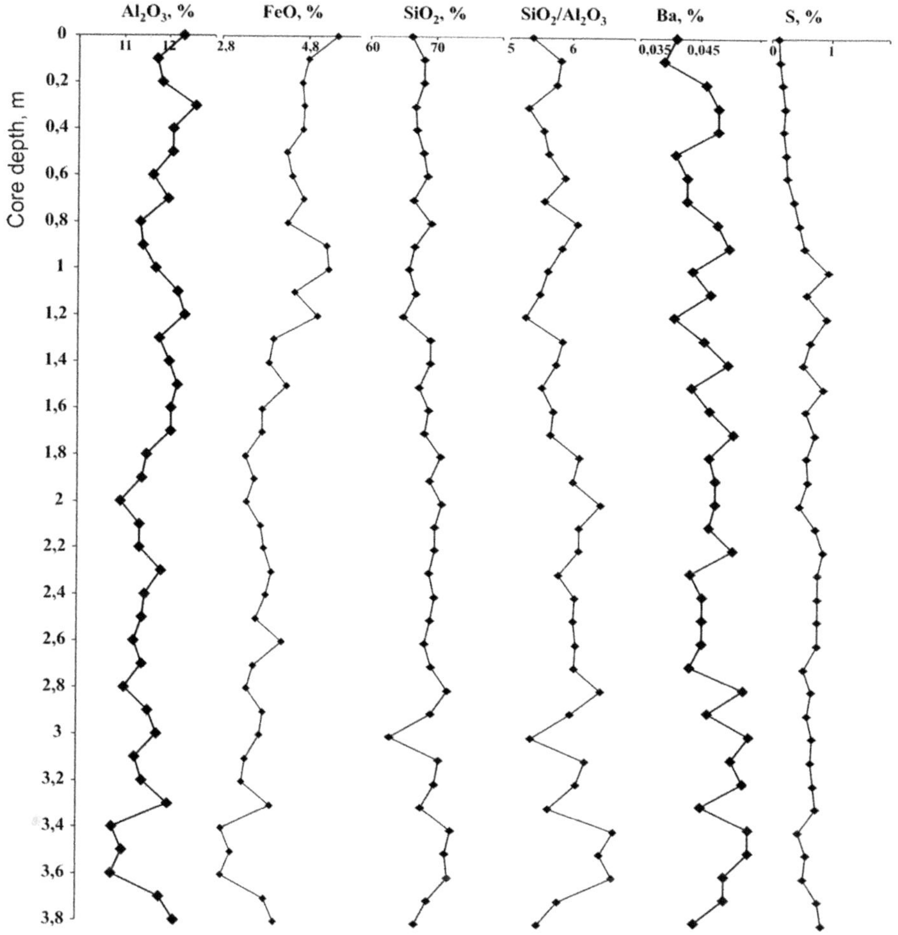

Fig. 10.37 Distribution of chemical elements in sediment core BP03-07 (Levitan et al., 2007c)

the 0.005–0.001 mm fraction and (for Fe) the sandy-silty fraction; (2) SiO_2, Sr, Zr, Ba, and Rb confined to the 0.1–0.05 mm fraction; (3) P_2O_5 partly correlating with TiO_2 and MnO; and (4) CaO, Cr_2O_3, and S showing no correlation with other elements; sulfur is definitely confined to the subcolloidal fraction.

In agreement with the lithologies distribution, the concentrations of most elements correlated with the sandy-silty fraction (for instance, SiO_2 and Ba), and the SiO_2/Al_2O_3 ratio increases from top to bottom of the section (Fig. 10.37). This trend becomes obvious at different depths for different elements (for example, at 125 cm for Si, at 115 cm for Sr, and at 305 cm for Zr). The enrichment of sulfur downward in the section is probably related to the parallel increase in TOC.

The available radiocarbon ages allowed to calculate the average sedimentation rate as 67 cm/ky. The average sediment mass accumulation rate is 94 $g/cm^2/ky$. An

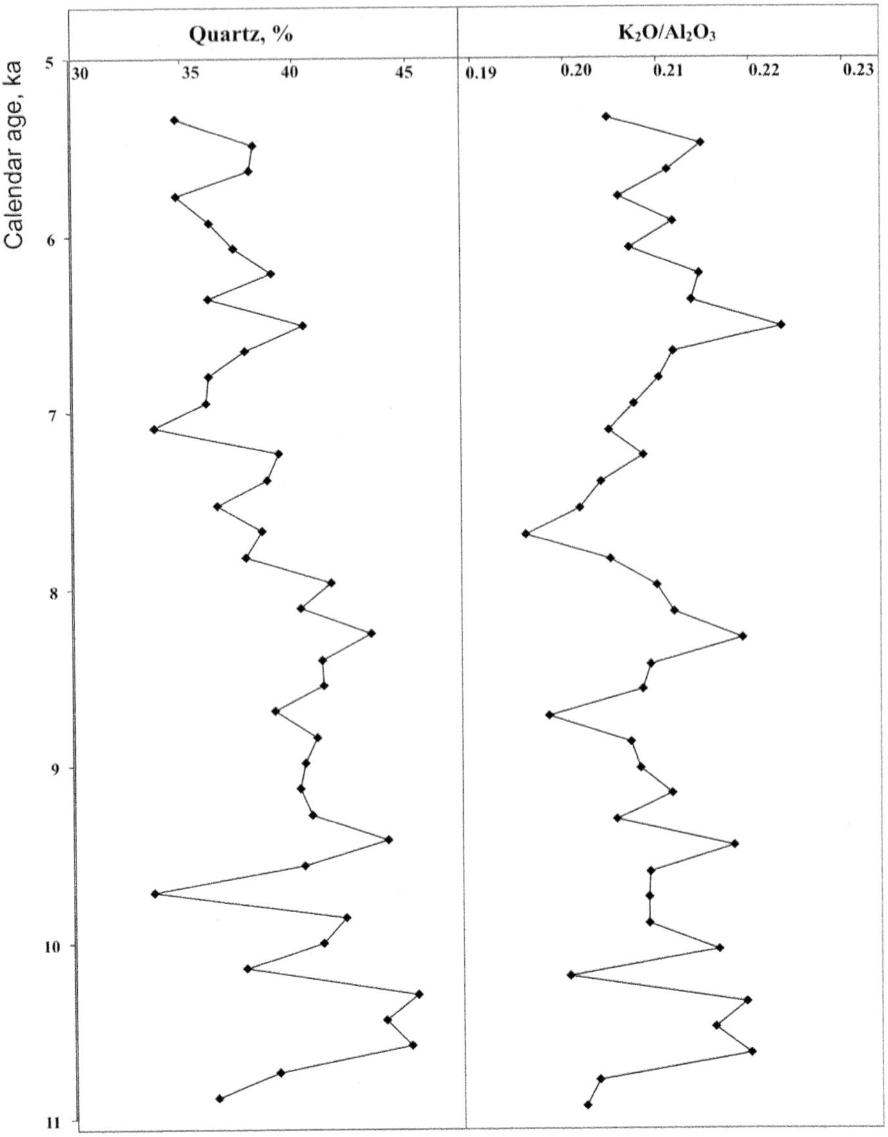

Fig. 10.38 Age distribution of quartz and K_2O/Al_2O_3 ratio in sediment core BP03-07 (Levitan et al., 2007c)

age of 5.27 cal. kyrs. BP was obtained for the top of the section, 11.1 cal. kyrs. BP for the base of the section, and 11.25 cal. kyrs. BP for the oldest sediments of the channel. Thus, the ages of the generation of Pro-Ob channels are approximately the same at the two stations and correspond to the beginning of the Holocene.

An irregular increase in the quartz content with increasing sediment age is clearly obvious from Fig. 10.38. The peaks of K_2O/Al_2O_3 practically coincide with quartz peaks, although the trend of increase in this ratio downward in the section is much less pronounced than that for quartz.

Based on the K_2O/Al_2O_3 ratios and quartz contents, the following periods of maximum sediment discharge by the Ob River can be distinguished: 6.5, 7.2, 8.3, 9.6, 9.8–10.0, and 10.3–10.7 cal. kyrs. BP (Fig. 10.38; remember that only the lower part of the Holocene was represented by the sediment core). The first two maxima of Ob discharge correspond to the periods of elevated Yenisei discharge at 7.5–6.2 cal. kyrs. BP, whereas those older than 7.9 cal. kyrs. BP and, especially, older than 9.4 cal. kyrs. BP correspond to the maximum sediment discharge (Levitan et al., 2004).

If compare the maxima of river discharge recorded in cores BP03-07 and BP03-19, it becomes obvious that the values are much higher in Core BP03-19, resulting in a much higher time resolution between samples. The time period between adjacent maxima of sediment discharge may be several tens to several hundreds of years. Considering the sampling density and lithological description, especially for the lower part of Unit II Core BP03-19, some of the maxima may even only be a few years apart.

In the diagram of hydrolysate module ($A1_2O3$ + TiO_2 + Fe_2O_3 + FeO + MnO)/SiO_2 (HM) versus TA – total alkalinity (Na_2O + K_2O) (Fig. 10.39; for background of approach see (Yudovich and Ketris, 2000)), the sediments of both lithological units I and II at Core BP03-19 show identical HM values (from 0.25 to 0.50) mainly corresponding to normal hydrolysate and, in part, low hydrolysate siallites (i.e., silty-clay sediments and clayey silts). The sediments of Core BP03-07 show HM values from 0.2 to 0.3 corresponding to miosilites (in our case, clayey-detrital rocks enriched in quartz). The sediments are even more clearly distinguished by TA values, which increase from Unit II of Core BP03-19, through Core BP03-07, to

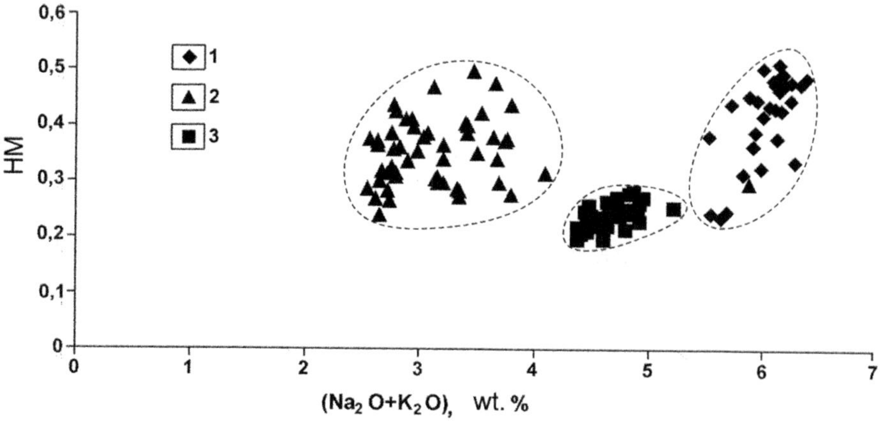

Fig. 10.39 Correlation diagram of HM-TA (see text for explanation) for sediments of the Ob Transect which allows to separate sediments belonging to different facies (Levitan et al., 2007c). 1 – upper unit of core BP03-19; 2 – lower unit of core BP03-19; 3 – core BP03-07

Table 10.28 Chemical composition (wt. %) of sediment cores of Ob River transect

Sediment core		Na_2O	MgO	Al_2O_3	SiO_2	P_2O_5	K_2O	CaO	TiO_2	Cr_2O_3
BPO3-07 (n=39)		2.08–2.57 2.21	1.28–2.00 1.71	10.78–12.58 11.68	63.63–72.9 68.94	0.06–0.093 0.077	2.35–2.59 2.44	0.95–1.96 1.22	0.646–0.773 0.728	0.01–0.023 0.013
BPO3-19	0–1.52 м (n=27)	3.11–4.12 3.57	1.89–3.05 2.53	11.58–14.22 13.42	49.14–61.21 54.27	0.113–0.348 0.258	2.27–2.63 2.50	0.81–1.38 0.94	0.71–0.81 0.77	0.018–0.024 0.022
	1.52–6.7 м (n=55)	0.423.48 1.07	1.47–2.64 2.13	10.82–14.84 12.88	51.77–69.53 60.60	0.055–0.285 0.153	2.08–2.41 2.23	0.99–2.47 1.26	0.65–0.94 0.80	0.014–0.036 0.019
Sediment core		**MnO**	**FeO**	**S**	**V**	**Ni**	**Rb**	**Sr**	**Zr**	**Ba**
BPO3-07 (n=39)		0.047–0.27 0.073	2.87–5.48 4.05	0.12–0.98 0.64	0.0098–0.0172 0.0132	0.0051–0.0083 0.0065	0.0038–0.0103 0.0058	0.026–0.040 0.032	0.015–0.030 0.020	0.039–0.054 0.047
BPO3-19	0–1.52 м (n=27)	0.096–0.594 0.263	6.7–10.64 9.12	0.19–2.29 0.38	0.0137–0.0184 0.0167	0.0041–0.0066 0.0053	0.0063–0.0088 0.0080	0.010–0.023 0.013	0.009–0.021 0.012	0.038–0.049 0.044
	1.52–6.7 м (n=55)	0.053–0.433 0.145	4.15–9.98 6.73	0.16–2.78 0.70	0.0091–0.0161 0.0121	0.0032–0.0061 0.0046	0.0073–0.0104 0.0090	0.008–0.036 0.020	0.014–0.030 0.020	0.024–0.052 0.038

Note: Maximum and minimum values are shown in the numerator, average – in denumerator; n – number of samples.

Unit I of Core BP03-19 (Fig. 10.39). This trend is interpreted here as a reflection of the facies transition from a marginal filter environment through a marine-fluvial environment to a fully marine environments (with a seaward increase in the mobility of alkalis, especially Na, during geochemical differentiation). The established chemical features can be clearly illustrated by comparing the average compositions of the two units of Core BP03-19 and Core BP03-07 sediments (Table 10.28).

Sediment Core DM-4397

Core DM-4397 was recovered not far to the north of Core BP03-07 (Table 10.23) on the slope of the Ob-Yenisei Shoal and is related to the facies zone D1 of the outer shelf (Levitan et al., 2005a). The lithology was described in (Levitan et al., 1994); and the data on benthic foraminifers, mineral composition, oxygen and carbon isotopes of calcareous foraminifers, and two radiocarbon datings were published in (Polyak et al., 2002b). L.L. Demina et al. (2006) described the distribution of bulk concentrations and speciation of Fe, Mn, Zn, Cu, Co, Cr, Ni, and Pb in Core DM 4397.

Two main lithostratigraphic units, 0–190 cm (Unit I) and 190–384 cm (Unit II), can be distinguished. Unit I is composed of olive gray (except for the oxidized upper brown layer) marine silty-clayey (0–70 cm) and clayey (70–190 cm) mud with characteristic bioturbation structures and hydrotroilite streaks. Unit II consists of greenish gray silty-clayey mud with layers of black silt and silty sands in the middle part (250–280 cm) and pebble massive sands at the base of the unit. A bivalve shell was found at a depth of 275 cm.

The total number of benthic carbonate foraminifers increases downward from 0 to 40 per gram of sediment, then decreases abruptly at a depth of 235 cm to 10 individuals per gram, increases again to 40 at 260 cm, and declines to zero at the base of the sediment core. The number of marine forms proper (*C. reniforme*) increases upward from the bottom of the section to 240 cm core depth, whereas the number of forms living in the zone of mixing of fresh and sea waters (*E. clavatum*) decreases toward 190 cm core depth. The oxygen isotopes of *E. clavatum* tests become enriched at 160 cm downward, especially from a depth >210 cm, whereas the carbon-isotope composition becomes heavier in the same direction, especially below 230 cm core depth. The content of clinopyroxene increases significantly below 230 cm core depth. In general, the sediments show a downward increase in the bulk content of Fe, Zn, Cu, Co, Cr, and, to some extent, Pb starting in average at about 190 cm.

At 190 cm and 280 cm core depth, ages determined on mixed foraminifer assemblages, are 8.15 cal. kyrs. BP and 9.8 cal. kyrs. BP, respectively. Thus, the age of the boundary between the marine horizon and the horizon showing significant river influence, is 8.15 cal. kyrs. BP. The calculated sedimentation rates are 23.3 cm/ky for Unit I (assuming that the section terminated by modern sediments was not disturbed) and 54.4 cm/ky for Unit II. The lower 4 cm of the core are probably composed of purely riverine sediments (as suggested from lithology and the complete absence of foraminifers). Then, the age of their top is approximately 11.6 cal. kyrs. BP, which approximately corresponds to the base of the Holocene (similar to the two sediment cores described above). Thus, river influence decreases regularly from bottom to top in the section, approximately to the level of 8.15 cal. kyrs. BP. The sediment discharge of the Ob increased definitely only within the interval 9.25–9.80 cal. kyrs. BP. This is in agreement with the data from Core BP03-07 and the sediment discharge of the Yenisei River.

Discussion

Based on the data described above, we can make some important conclusions related to the history of the sediment discharge of the Ob. First, the cores described here probably represent sediments from a single channel of the Proto-Ob, which originated at the Pleistocene-Holocene boundary. This suggestion is in agreement with data from core BP00-38, which was recovered north of the zone of mixing of Ob and sea waters and for which three radiocarbon age determinations were published (Stein et al., 2003a, 2004). Second, the geochemistry of Early Holocene sediments provides compelling evidence that the sediments of the zone of mixing of fresh and sea waters are replaced from south to north along the Ob transect by marine-fluvial and fluvial-marine facies. Third, it should be noted that marine-fluvial sedimentation ceased in the channel of this river during the Middle Holocene. Taking into account that similar processes in the northern part of the Ob-Yenisei Shoal were related to the channels of the Proto-Yenisei (cores BP00-36/4 and BP01-39/2; (Stein et al., 2003a)), it can be supposed that the Middle Holocene glacio-isostatic uplift of the crustal segments that were covered during the Late Valday (Late Weichselian) by the Kara Sea Ice Sheet (Stein et al., 2002) may have also affected part of the adjoining regions of the Ob-Yenisei Shoal. Fourth, there is a quite clear trend of decreasing river influence during the early Holocene, probably markedly disturbed during the stage of 9.80–9.25 cal. kyrs. BP. Fifth, there are arguments to suggest that the Ob and Yenisei underwent similar histories of changes in river discharge during the Holocene, not surprisingly because their watersheds are located in very similar climatic conditions.

Noteworthy are the results by K.V. Kremenetski et al. (2003) on the timing of the beginning of peat bog accumulation in West Siberia. These authors used statistical data on radiocarbon dating of peats and correlated the main peak in the intensity of accumulation with the period from 11 to 7 cal. kyrs. BP (with a maximum at 10 cal. kyrs. BP). It should be noted that this age interval is also close to the period of maximum sediment discharge by the Yenisei and elevated discharge of river material through the channel of the Proto-Ob. This similarity could be related to the elevated humidity caused primarily by intense thawing of permafrost in the watershed areas (Kremenetski et al., 2003).

Therefore, it is probable that diagenetic processes of ikaite crystallization, which were described in detail by (Kodina et al., 2003), were related to Lower Holocene sediments formed in paleo-river channels just northward from mixing zones of fresh and sea water with sedimentation rates ca. dozens cm/ky, strongly influenced by river discharge in paleochannels.

Chapter 11
Eastern Arctic Seas

The East-Arctic seas include the Laptev Sea, the East Siberian Sea and the Chukchi Sea. All three marginal seas are characterized by the following features: severe ice regime, shallow water depths of relatively flat bottom in which channels of a paleo-hydrological network were developed, absence of glacial ice sheets during the last climatic cycle, and well-expressed distribution of permafrost in the southern part of the seas. The Laptev Sea is separated from the Kara Sea by the Severnaya Zemlya Archipelago, and from the East Siberian Sea by the Novosibirsk Islands. The East Siberian Sea is separated from the Chukchi Sea by Wrangel Island and is connected with the Bering Sea through the Bering Strait.

Recent Sedimentation Environment

The Laptev Sea

The sea ice is present year-round in the Laptev Sea, except polynyas and the delta of the Lena River. The sea has a size of 498 thousand km^2, a volume of 24 thousand km^3, and an average water depth of only 48 m (Jakobsson et al., 2004). Annual riverine freshwater discharge is 738 km^3, from which 523 km^3 are discharged by the Lena River. Suspended-matter supply by the rivers reaches 28.6 million t/yr (Gordeev and Rachold, 2004). Coastal abrasion supply of sedimentary material equals 58.4 million t/yr in the Laptev Sea (Grigorjev et al., 2004), and an eolian delivery is 0.298 million t/yr (Shevchenko and Lisitzin, 2004). The Laptev Sea is the most important Arctic sea in terms of sea-ice delivery into the Arctic Ocean, reaching about 670 km^3/yr (Eicken, 2004). Average primary production is the lowest one among the Arctic seas and composes 25–40 g of C/m^2/yr (Sakshaug, 2004).

During summer times, winds of northern directions predominate. The tide and tidal fluctuations of sea level, together with active wave activity, contribute to active litho-hydrodynamics of recent sedimentation, to the displacement of fine pelite material, supplied by rivers, and also, with coastal abrasion, to the accumulation of bars and sandbars (Axenov et al., 1987). Despite the very monotonous grain-size

M.A. Levitan, Y.A. Lavrushin, *Sedimentation History in the Arctic Ocean and Subarctic Seas for the Last 130 kyr*, Lecture Notes in Earth Sciences 118,
DOI 10.1007/978-3-642-00288-5_11, © Springer-Verlag Berlin Heidelberg 2009

composition of the clayey silts, in which the content of the <0.001 mm fraction may exceed 70%, the mineral composition of recent sediments in the western and eastern parts of the sea is various. The Khatanga River flowing into the western part of the sea, drains the Putoran Plateau, composed by Permian-Triassic basalts (Explanatory note to tectonic map of the Barents and Kara Seas, 1998). Thus, the sediments of the western part of the Laptev Sea are characterized by the increased content of black ore minerals, clinopyroxenes, and smectite. Garnet is supplied by the Anabar River, which drains the Anabar crystalline massif (Behrends et al., 1999; Peregovich et al., 1999). The rivers Lena, Omoloy, Jana that mainly drain the Verkhoyanskiy sedimentary complex and supply minerals such as amphiboles, illite and chlorite, flow into the central and eastern part of the sea.

Thus, surface sediments from the eastern Laptev Sea have a very different mineralogical signature in comparison to the western Laptev Sea (Behrends et al., 1999; Peregovich et al., 1999; Rossak et al., 1999; Stein and Korolev, 1994), and mineralogical proxies can be used to distinguish between sea-ice sediments transported in the Polar and Siberian branches of the Transpolar Drift, originating from the eastern and western Laptev Sea, respectively (e.g., (Dethleff et al., 2000)).

The East Siberian Sea

By its climatic and ice conditions the East Siberian Sea is considered to be the most severe of all Eurasian marginal seas. The East Siberian Sea has a size of 987 thousand km^2, a volume of 57 thousand km^3, and an average water depth of only 58 m (Jakobsson et al., 2004). Annual riverine freshwater discharge is about 233 km^3, from which 122 km^3 are discharged by the Kolyma River. Suspended-matter supply reaches about 25.15 million t/yr (Gordeev and Rachold, 2004). Annual coastal abrasion input of sedimentary material into the East Siberian Sea is 66.5 million t (Grigorjev et al., 2004). Eolian delivery gives 0.589 million t/yr (Shevchenko and Lisitsin, 2004). 150 km^3/yr of ice enters from the basin into the central Arctic (Eicken, 2004). Primary production is similar to that determined for the Laptev Sea (Sakshaug, 2004).

The bottom of the sea is covered by fine-grained sediments as in the Laptev Sea. The flooded channels of rivers like Kolyma, Indigirka and smaller ones play a significant role in bathymetry of the East Siberian Sea. They are covered by the finest clayey silts, which may contain up to 60–70% of clay (Naugler, 1967). The composition of heavy-mineral associations is sufficiently uniform: amphiboles, clinopyroxenes, and epidote, garnet prevail (in these estimates by (Naugler, 1967) the black ore are excluded from the calculation). Near the Novosibirsk Islands and in the extensions of the Indigirka River, garnet and zircon play a relatively important role, and in the extensions of the Kolyma River the increased content of clinopyroxenes is observed. Toward the east, orthopyroxene becomes dominant.

The association of apatite and tourmaline seems to be typical for Indigirka River input. Near Wrangel Island an increased quantity of epidote and garnet is described

(Naugler, 1967). Among the light minerals quartz, potassium feldspar and plagioclases prevail (they are listed in decreasing order). The content of plagioclases in the recent sediments slightly increases to the east from the mouth of the Kolyma River, which indicates an increase of basic rock input. Among the clay minerals, a chlorite-illite association absolutely predominates (Naugler, 1967).

The Chukchi Sea

In the bottom relief of the Chukchi Sea noticeably differs from the other two East-Arctic seas. In the outer shelf and in the upper part of the continental slope there is a unique morphostructure, the Northwind Ridge, a marginal plateau. In total, the Chukchi Sea has an area of about 620 thousand km^2, a volume of about 50 thousand km^3, and an average water depth of 80 m (Jakobsson et al., 2004). Annual river discharge into the sea (without Alaska) equals 20.4 km^3 and suspended-matter discharge reaches 0.7 million t/yr (Gordeev and Rachold, 2004).

Annual coastal abrasion supply of solid sedimentary material is about 70 million t (Grigorjev et al., 2004). Eolian sediment supply is no more than 0.371 million t/yr (Shevchenko and Lisitsin, 2004). The export of sea ice into the Arctic Ocean is $<20\ km^3/yr$ (Eicken, 2004). Lowered formation and export of sea ice and increased summer sea-surface temperature (Pavlidis, 1982), and related increased primary production (from 20 to >400 g $C/m^2/yr$) (Sakshaug, 2004), are caused by the inflow of warm, saline and nutrient-rich Pacific Ocean water, entering through the Bering Strait. This also results in the development of large areas of recent diatom-bearing (14% of biogenic opal) silts in the central (deep-water) part of the Chukchi Sea, a unique phenomenon for the Arctic Ocean (Pavlidis, 1982).

Besides the diatom-bearing silts, a wide variety of lithotypes is observed in surface sediments of the Chukchi Sea (Fig. 11.1), caused by both regional special features of bottom relief and hydrodynamic particularities (Pavlidis, 1982). Here, the presence of channels such as the Chukchi Trough and the paleo-Hope valley is worth noting. Well-sorted sands with a predominance of the fine-sand fraction were developed in the accumulation areas of the inner shelf (to water depths of 35–40 m). The gravel-pebble badly sorted sediments with noticeable admixture of coarse-grained sand prevail in the abrasion areas. The clayey muds predominate in the central and outer parts of the shelf. The banks located here, for example the Herald Bank, are covered with sands.

The composition of clay minerals is sufficiently uniform: illite comprises $>60\%$, chlorite varies between 10 and 20%, and kaolinite and smectite do not reach $>10\%$ of the <2 mkm fraction (Wahsner et al., 1999).

History of Sedimentation

The history of sedimentation of the East-Arctic seas has been studied extremely unevenly. The Laptev Sea was intensively investigated, especially within a major

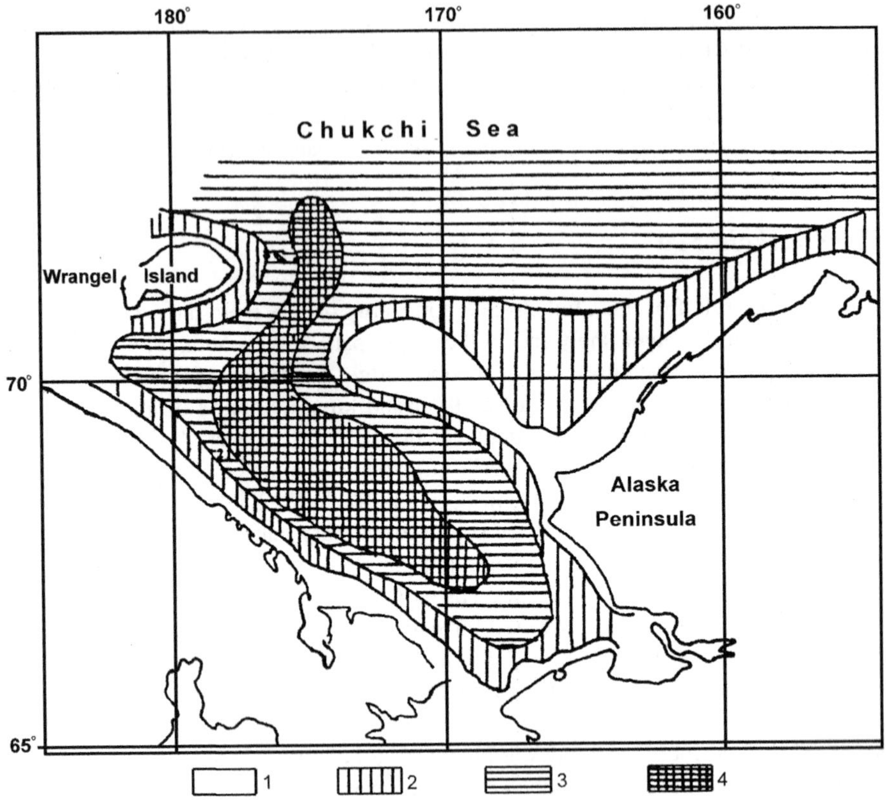

Fig. 11.1 Distribution of biogenic opal (wt. %) in surface sediments of the Chukchi Sea (Pavlidis, 1982). 1 – <1; 2 – 1–5; 3 – 5–10; 4 – >10

joint Russian-German research project, i.e., the "Geo-system Laptev Sea" project (Kassens et al., 1999). In the East Siberian Sea, on the other hand, no detailed data on sediment cores are available at all. For the Chukchi Sea, only discontinuous records on the history of shelf sedimentation, which are mainly based on foraminifer analysis, are known (Saidova, 1982).

History of Sedimentation in the Laptev Sea During the Late Weichselian to Holocene by Geophysical and Geochemical Data

Geophysical Data

In the latest reconstructions of the largest glaciations in northern Eurasia, including the shelf sea areas (Svendsen et al., 2004), most scientists supports the view about the absence of a marine ice sheet in the Laptev Sea, the local development of glacial cupolas on the islands of Severnaya Zemlya, and a restricted mountain-valley glaciation in the Central Taimyr during the Late Weichselian. The wide development of cryogenic complexes with vein ice ("edoma") in the south of the Laptev Sea and its coastal areas indicates the presence of extensive alluvial-lake plains with active participation of eolian material during MIS 2 and also rejects the assumption about the development of a major ice sheet in the area of the Laptev Sea at this time (Arkhipov, 2000).

Parasound geophysical studies (4 kHz) in the region of Severnaya Zemlya, Vilkitsky Strait, the Khatanga-Anabar depression and on the western continental slope of the Laptev Sea allowed to obtain important data about the development of the region during the last climatic cycle (Kleiber et al., 2001). These authors were able to separate four seismo-stratigraphic complexes (SSC) that are characterized by its geometry, internal structure, distribution and other parameters. From top to bottom SSC I to IV were distinguished. SSC-I has an integumentary shape, acoustically transparent structure, is developed everywhere, and conformly overlap the underlying complexes. Its general thickness of 1.0–2.5 m on the continental slope may increase to 18 m in local depocenters. Based on numerous radiocarbon datings this complex has a Holocene age.

SSC-II forms several wedge-shaped depocenters along the edge of the shelf, where its thickness may reach 100 m, but in other investigated regions does not exceed 2–3 m. The fine acoustic stratification is a special characteristic of this complex. Small transparent lenses are sometimes included. The described depocenters are probably previous fans of the rivers Anabar, Khatanga, Olenek, etc. (Kleiber et al., 2001). From our point of view, their origin in lakes dammed by the ice barrier from the north cannot be excluded as well. In accordance with radiocarbon datings and general considerations, SSC-II is approximately of MIS 2 age coinciding with a low sea-level standing.

SSC-III is very similar to the overlaying complex, it only differs from the latter by the absence of wedge shapes and lower gradients of thickness (from 2 m in the lower part of the continental slope to 35 m at the shelf edge). The thickness increases to 50 m at some areas in relief depressions of the underlying complex. The sediments of SSC-III were probably supplied by rivers at times of – in comparison to SSC-II – higher sea level. Based on indirect conclusions and confirmed by the presence of *Gephyrocapsa spp.* in the slope sediments, it is possible to conclude that the time of accumulation of this complex was MIS 3.

SSC-IV is mainly composed of transparent sediment lenses, interpreted as debris-flow (Kleiber et al., 2001) forming deep-water fans at the continental slope and rise, and related to the advance of an ice sheet during the MIS 4. Based on most reconstructions, the MIS 4 glaciation in the area of the Laptev Sea was more severe than during MIS 2 (Kleiber et al., 2001; Svendsen et al., 2004).

Geochemical Data

Organic-geochemical data may be used to characterize the geological situation at the continental slope of the Laptev Sea. As an example, data from sediment cores PS2474–3 and PS2741–1 are presented here (Stein et al., 2001). The first core is located at the upper continental slope off the Khatanga-Anabar channel (77°40.2′ N,

118°34.5′ E, water depth is 1494 m), the second one is located to the northeast of Severnaya Zemlya at the middle continental slope (77°40.2′ N, 118°34.5′ E, water depth is 2400 m). The age models of both cores are mainly based on oxygen-isotope stratigraphy, confirmed by radiocarbon datings for the upper part of the section of Core PS2474-3. In both cores maximum values of TOC content, sedimentation rate, and mass accumulation rate were determined for the Holocene time interval (Fig. 11.2; (Stein and Fahl, 2004)).

The predominance of terrigenous organic matter is suggested from hydrogen index values (Stein et al., 2001). Some increases in the TOC content, sedimentation

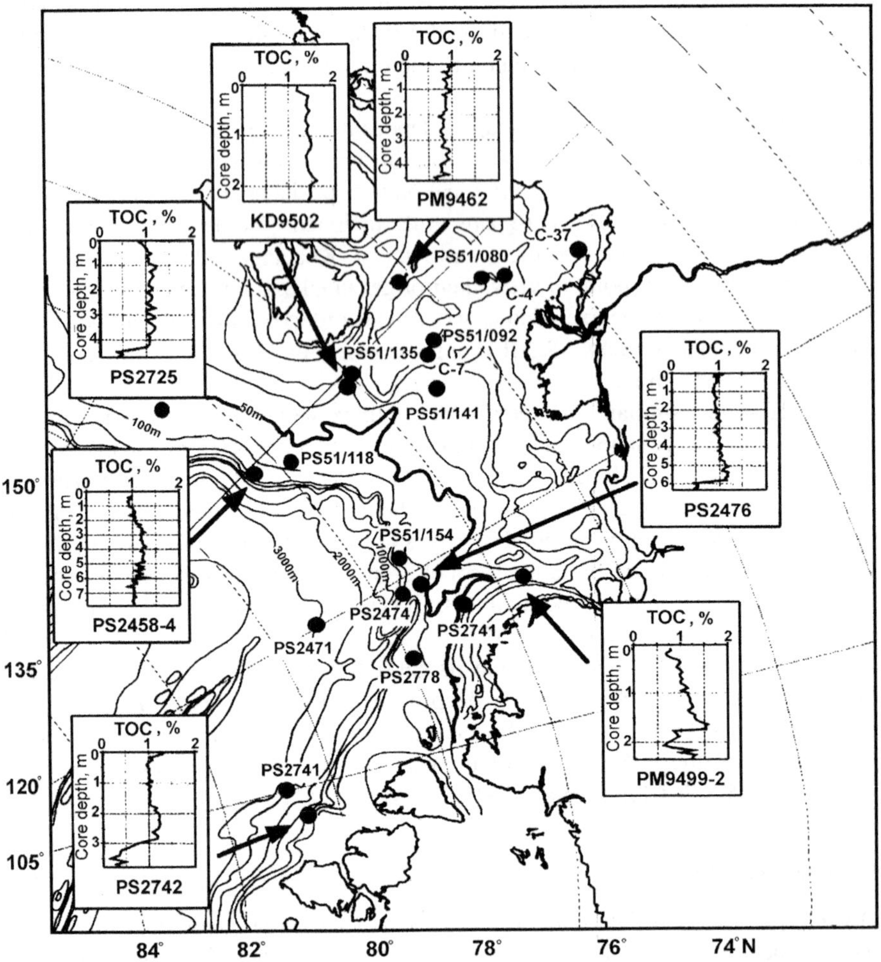

Fig. 11.2 TOC distribution in sediment cores from the Laptev Sea continental slope (Stein and Fahl, 2004)

rate and mass accumulation rate also occurred in the middle part of MIS 3, the bottom of MIS 5 and in MIS 6.

The organic-geochemical data point to the absence of an ice sheet during MIS 2 in the Laptev Sea (Stein et al., 2001). During MIS 6, dark organic-carbon rich interlayers are related to the carbon-rich sediments located between North Taimyr and the Lena River delta and transported by glacial erosion to the continental slope of the western Laptev Sea. Occasional enrichment of organic carbon in the various periods of MIS 3 and MIS 5 are most likely related to the inflow of Atlantic waters through the Fram Strait.

Holocene History of the Lena and Other Rivers Discharge in the Laptev Sea

In the following, the Holocene and pre-Holocene history of river discharge into the Laptev Sea, is briefly described, mainly based on the studies (Bauch et al., 1999, 2001) and some others (Stein and Fahl, 2000; Polyakova et al., 2005; Taldenkova et al., 2005). Within these studies, 14 sediment cores located in the upper part of the continental slope, in the outer, central and inner shelf (see Table 6.1, Fig. 11.3) and well-dated by 119 AMS ^{14}C determinations on bivalve shells, were investigated. In most of the cores, sedimentological and mineralogical data as well as TOC distribution and isotope data of the organic matter were determined.

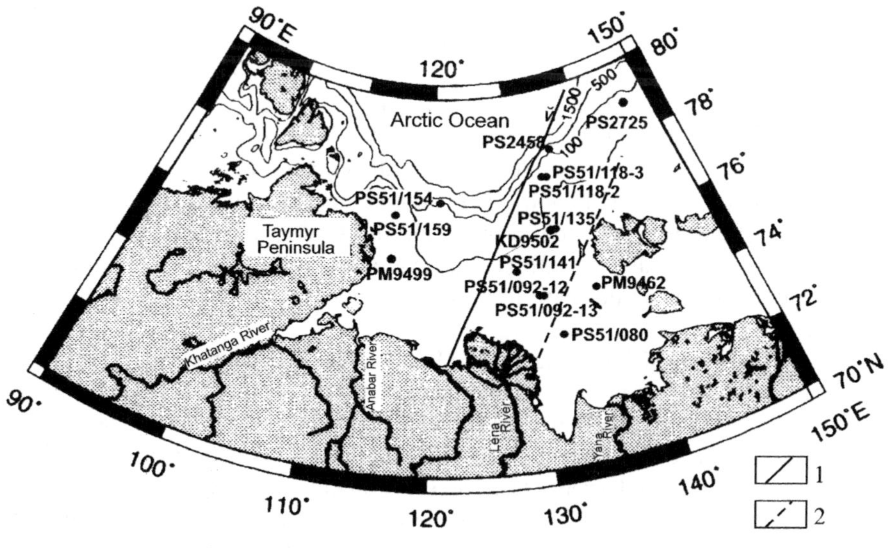

Fig. 11.3 Location of studied sediment cores in the Laptev Sea (Bauch et al., 2001). 1 – transform fault (Explanatory note to tectonic map of the Barents and Kara Seas, 1998); 2 – boundary between complete and uncomplete cross-sections of the Holocene

Core PS 51/154 located at the continental slope northwest of the Laptev Sea contains the oldest sediments of the studied cores with an age of 15.9 cal. kyrs. BP. Numerous plant remains and flakes of mica were found, and also vivianite – diagenetic phosphate – was described, which suggested to be formed under lake conditions in organic-carbon-rich bottom sediments. These data coincide rather well with data from SSC-II described above. It seems to be that this core has penetrated a fan or sediments of a pre-glacial lake, connected with the output of the Khatanga – Anabar channel to the continental slope.

At the base of Core PM 9499 located at the southern end of this channel river (delta?) sediments with layered structure, enrichment of mica (due to input from the Byrranga mountains in Taimyr), remains of vegetation, sandy-silt fractions and ikaite were described. The age of these sediments is approximately 12.3 cal. kyrs. BP. A sharp drop in sedimentation rate upward the sedimentary section was observed at 11.1 cal. kyrs. BP in the north and at 7.1 cal. kyrs. BP in the south.

A similar north-south tendency of rejuvenation in age of the boundary is noted for the joint channel Jana-Lena, where sediments with high sedimentation rates and strong river influence are overlain by younger predominantly marine sediments. The basic reason for this rejuvenation, as in the Yenisei region, is related to the Flandrian transgression, which started in connection with deglaciation after the Last Glacial Maximum.

The transgression in the Laptev Sea has occurred unevenly. The highest position of sea level is registered at 11.1, 9.8 and 8.9 cal. kyrs. BP, and corresponded to the isobaths –50 m, –43 m and –31 m, respectively. The recent sea level was approximately reached at about 5 cal. kyrs. BP. According to paleontological data, i.e., the basic groups of benthic organisms, however, the recent environment was established in the channel of the paleo-Jana River at about 10.3 cal. kyrs. BP and in the channel of the paleo-Lena River at about 8.2 cal. kyrs. BP (Taldenkova et al., 2005).

Based on diatom, dinoflagellate and planktonic foraminifer data, the influence of Atlantic waters for this region has appeared at about 10.7–9.2 cal. kyrs. BP (Polyakova et al., 2005; Taldenkova et al., 2005). In Core PS2725 from the outer Laptev Sea shelf area, a noticeable drop in sedimentation rate occurred at 9.6 and 5.3 cal. kyrs. BP (Stein and Fahl, 2000). In a number of cores taken in the paleo-Lena channel on the inner shelf, a pronounced drop in fresh-water diatom inflow was recorded at 7.0 cal. kyrs. BP (Polyakova et al., 2000), almost contemporaneously with a significant decrease in average sedimentation rate at 7.3 and 7.1 cal. kyrs. BP.

Using all data available so far for the Laptev Sea, H. Bauch et al. (2001) found that the earliest Holocene drop in sedimentation rate occurred between 11.1 and 9.0 cal. kyrs. BP. The latter boundary coincides with the end of main deglaciation in the Nordic Seas and the establishment of normal marine environment in the West-Arctic Seas. The second strong decrease of sedimentation rate, also confirmed by organic geochemistry and diatom data, occurred in the region at about 7.2 cal. kyrs. BP and is most likely connected with a distinct decrease of river discharge.

A careful examination of all the radiocarbon datings of Laptev Sea sediment cores (Bauch et al., 2001) showed that only cores from the channels in the inner shelf in the southeastern part of the sea have continuous sedimentary sections throughout the Holocene. At the same time, almost all the remaining cores have more or less complete Lower Holocene sedimentary sections and clearly expressed sharp reductions in sedimentation rate or even hiatuses in the younger sections. According to the data by (Stein et al., 2002), similar records were also obtained in some areas in the north of the Ob-Yenisei Shoal at the end of the Yenisei transect, in south of which a hiatus has not been frequently registered (Levitan et al., 2004; Stein et al., 2002, 2003a).

In the case of the Ob-Yenisei Shoal it was noted in the previous chapter that sediment cores with preserved Lower Holocene sedimentary sections and large hiatuses in later sedimentation are located along the northern and north-western margins of this structure, near the boundaries of the Kara Sea Ice Sheet (MIS 2). It is logical to assume that glacio-eustatic motions, which certainly occurred during the removal of ice load, affected the relief of adjacent structures, especially in the boundary zone.

In the Laptev Sea, where an influence of an ice sheets was limited during the last climatic cycle, the reason for hiatuses formation at the shelf area in Holocene was fundamentally different. In this case, we do not exclude an influence of the Gakkel Ridge propagating on the continental margin and the outer shelf of the Laptev Sea. The analysis of Fig. 11.3, at which is shown location of cores with the radiocarbon datings mentioned above, leads to several conclusions. All studied cores with hiatuses are located within upper continental slope, the outer and central shelf (north-westward from dashed line) in contrast to the southeastern area of the sea (inner shelf), where undisturbed whole Holocene cross-sections have been found. The bottom relief between the shelf edge and the boundary between the inner and central shelf areas (dashed line) is an anomalously hilled surface (I.G. Avenarius, personal communication, 2005). It seems to be that propagation of the Gakkel Ridge (including the end of the Holocene) led to its impact to the continental margin of the Laptev Sea. This impact was expressed in the movements along the transform fault (Fig. 11.3) and the activation of the ancient Yenisey-Khatanga depression (development of a specific bottom relief, increased seismicity, possible underwater slumps). All these mechanisms could contribute to the hiatuses in sedimentation processes. The continuous Holocene sedimentary sections, represented by different genetic types of continental and marine sediments, have been preserved in the cores beyond the limits of the activization zone.

Organic Geochemical Data About Sedimentation History Along the Continental Slope of the East Siberian Sea During the Last Climatic Cycle

Due to the absence of the well-dated sediment cores from the shelf of the East Siberian Sea it is necessary to study material from the continental margin in order to obtain at least indirect data on the history of this sea. In this context, we briefly present results from a study of two sediment cores recovered from the continental

Table 11.1 Sedimentation rate (cm/ky) and bulk mass accumulation rate (g/cm^2/ky) at the continental slope of the East Siberian Sea [*Stein et al., 2001*]

MIS	PS 2763–7	PS 2761–10
Sedimentation rate		
1	12.6	2.0
2	10.2	3.4
3	13.6	4.2
4	–	3.3
5	–	3.8
6	–	2.5
Bulk mass accumulation rate		
1	10.59	1.98
2	10.42	3.02
3	13.22	4.43
4	–	3.37
5	–	3.83
6	–	2.55

Note: Blank – lack of sediments

slope of the East Siberian Sea (Stein et al., 2001): Core PS2761–10 (81°11, 5′N, 150°29, 0′E; water depth 2640 m; core length 620 cm) and Core PS2763–7 (80°16, 9′N, 150°26, 1′E; water depth 1597 m; core length 740 cm). Sedimentation rate and mass accumulation rate data are given in Table 11.1 (for age model see (Stein et al., 2001)).

TOC data of these two cores show a significant TOC enrichment in sediments of MIS 1 and of the lower part of MIS 5 as well as in single layers in MIS 3 (Core PS2763–7) and MIS 6 (Core PS2761–10). The distribution of hydrogen index values point to an increase of marine organic matter in the sediments of MIS 3 and especially MIS 5. The behavior of all examined parameters indicates an absence of an ice sheet on the shelf of the East Siberian Sea during MIS 2 and suggests the delivery of terrigenous organic matter by rivers from the shelf into the deep-water basin. Phases of increased marine organic-matter input are explained by increased primary production during warm (interstadial and interglacial) intervals, which is related to intensified advection of Pacific waters (Stein et al., 2001).

Preliminary Data About Accumulation of Diatom-Bearing Clayey Silts at the Chukchi Sea Shelf

In Section "Recent Sedimentation Environment" we discussed a phenomenon in recent sediments of the Chukchi Sea: a restricted area of diatom-bearing clayey silts in the middle part of the shelf. For sediment core PSh 40 (69°30′N, 171°E; water depth 51 m, core length 350 cm) some analytical data are listed in Table 11.2.

The whole Core PSh 40 is composed of diatom-bearing clay silts, with low contents of Fe, Mn, Ti, and $CaCO_3$ and increased TOC contents. Regretfully, there are no radiocarbon datings available, and the stratigraphy of the core is only based

Table 11.2 Grain-size (wt. %) and chemical composition (wt. %) of diatom-bearing silty clays in sediment core PSh 40 at the Chukchi Sea shelf [*Aksenov et al.*, 1987]

Horizon, cm	1–0.1 mm	0.1–0.01 mm	0.01–0.001 mm	<0.001 mm	Fe	Mn	Ti	SiO_2 am.	TOC	$CaCO_3$
0–15	0.71	4.08	53.64	41.57	4.24	0.03	0.31	10.15	1.62	1.25
15–20	0.69	4.04	58.82	36.45	n.d.	n.d.	n.d.	n.d.	n.d.	n.d.
50–60	0.25	2.43	54.02	43.40	4.46	0.02	0.31	9.79	1.65	1.23
100–110	1.04	1.34	57.11	41.01	4.34	0.04	0.33	9.35	1.40	1.30
150–160	0.40	2.38	54.69	42.03	4.40	0.04	0.32	9.85	1.55	1.31
200–210	3.08	3.25	65.07	30.78	4.18	0.03	0.33	12.86	1.43	1.28
250–260	0.43	1.66	55.81	42.20	4.11	0.02	0.36	11.79	1.40	1.25
300–310	0.26	0.64	59.53	39.57	4.69	0.03	0.34	7.63	1.68	1.84
340–350	0.36	0.72	59.27	40.35	4.80	0.03	0.37	7.95	1.59	1.23

on foraminifer data (Saidova, 1982). According to these data, the upper 15 cm of sediments were formed during the Subatlantic Stage, the interval 15–120 cm – during the Subboreal Stage, the interval 120–215 cm – during the Atlantic Stage, and the interval 215–350 cm – during the Boreal Stage of the Holocene. Agglutinated foraminifers prevail in the Subatlantic Stage, suggesting strong dissolution of carbonates.

Based on the data of Table 11.2, recent sediments and sediments from the transitional zone between the Boreal and the Atlantic stages are mostly enriched in biogenic opal. Taking into account all available data, the second interval was probably developed during the Holocene Climatic Optimum in this region, i.e., between 9.5 and 8.9 ka along the coast of the Chukchi Sea (Anderson et al., 2002), or during the time interval 9.5–5.5 ka in accordance with data of oxygen isotopes in planktonic foraminifers, or 8.0–5.5 cal. kyrs. BP according to data of oxygen isotopes in benthic foraminifers (de Vernal et al., 2005).

The Atlantic Stage was identified by the total content of benthic foraminifers (up to 3200 tests/50 g of sediment in comparison to the average value of 500–600 tests/50 g of sediment above and below the interval) and increased concentrations of such species as *Busselainusitata* and *Protelphidium orbiculare*. Furthermore, the abundance of *Nonionellina labradorica* is increased, although this parameter is only of marginal importance. There is a general similarity between foraminifer distribution in the Chukchi Sea and in the Barents Sea in Holocene sediments (Pogodina, 1994).

In conclusion, we may suggest that most likely the enrichment of biogenic components in Arctic and Subarctic sediments are related to stages or episodes of increased temperatures and penetration of warm, saline and nutrient-rich Atlantic or Pacific waters into the Arctic Ocean. This topic will be discussed in more detail in the following chapter.

Chapter 12
Seas of the Eastern Subarctic

Recent Sedimentation Environment

The Bering Sea and Sea of Okhotsk known as the marginal seas of the Pacific belong to the eastern Subarctic, where sea ice plays a significant role in the hydrological regime. According to (Grosswald, 1998), both seas and their drainage areas were covered by large ice sheets during the Last Glacial Maximum. This view, however, was recently disputed by many researchers (Axenov et al., 1987) and data presented below will give other possible paleogeographic scenarios.

The Bering Sea is isolated from the Arctic Ocean by the shallow Bering Strait. It consists of two large parts: the northern (shelf) and southern (deep-water) areas. The southern part is isolated from the Pacific Ocean by the Komandor and Aleutian islands. In the surface-water circulation of the deep-water area the Bering cyclonic circulation gyre plays an important role (Ohtani, 1991). In general, the Bering Sea receives 12.5 million m^3/s water from the Pacific Ocean, and 11.7 million m^3/s water are leaving the sea. Primary production ranging between 150 and 500 g $C/m^2/yr$, is very high on the shelf, along its edge and on the periphery of the sea, while for the oceanic part typical values are between 50 and 100 g $C/m^2/yr$ (Sakshaug, 2004).

The recent clastic, clastic-clayey sediments and clayey muds are developed on the shelf: Fig. 12.1 presents our lithology data from the Navarin area in the western parts of the shelf, obtained in 1982 during the Cruise 29 of RV *Dmitry Mendeleyev*. For the recent to Subatlantic shelf sediments, mass accumulation rates of TOC are <0.25 g $C/m^2/ky$, and at the base of the Holocene these values may exceed 1.25 g $C/m^2/ky$ (Fig. 12.2). Mass accumulation rates of TOC are mainly related to mass accumulation rates of terrigenous matter. During the Holocene the input of terrigenous material onto the shelf was so high that any specific biogenious sediments (sapropels or diatomaceous oozes) were not accumulated. In contrast to the shelf, our data suggest that already during the Holocene diatomaceous mud was deposited everywhere in the deep water part of Bering Sea (Table 12.1), and overall mass accumulation rates of terrigenous material were several orders of magnitude lower than on the shelf.

The Sea of Okhotsk, the southernmost region of occurrence of sea ice in the North Pacific, is located in the transition zone from the continent to the ocean. Here,

M.A. Levitan, Y.A. Lavrushin, *Sedimentation History in the Arctic Ocean and Subarctic Seas for the Last 130 KYR*, Lecture Notes in Earth Sciences 118,
DOI 10.1007/978-3-642-00288-5_12, © Springer-Verlag Berlin Heidelberg 2009

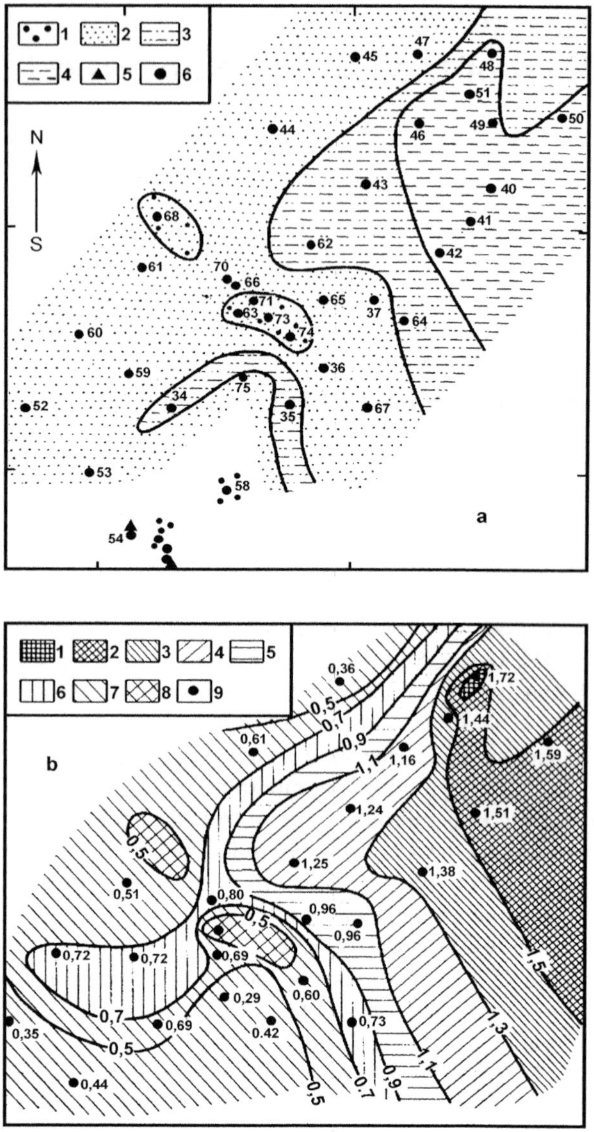

Fig. 12.1 Lithology of the Navarin Basin Subatlantic sediments (the Bering Sea north-western shelf) and distribution of TOC (Marina et al., 1985). *a* – lithology: 1 – gravel and pebble, 2 – sand, 3 – silt, 4 – clayey mud, 5 – outcrops of hard rock, 6 – geological stations; *b* – TOC content (wt. %): 1 – >1.7, 2 – 1.7–1.5, 3 – 1.5–1.3, 4 – 1.3–1.1, 5 – 1.1–0.9, 6–0.9–0.7, 7 – 0.7–0.5, 8 – <0.5, 9 – geological stations

one issue of controlling mechanisms of climatic changes related to history of sedimentation in this area is of special interest: the evolution of atmospheric circulation, changes in the thermohaline circulation or a combination of both mechanisms.

The Sea of Okhotsk has been studied by marine geologists for many years, especially by Russian scientists (Gorbarenko, 2004 and numerous references therein). During recent years, further research was carried out by scientists from Japan, South Korea, China, Germany, USA and other countries. For the earlier studies, the length

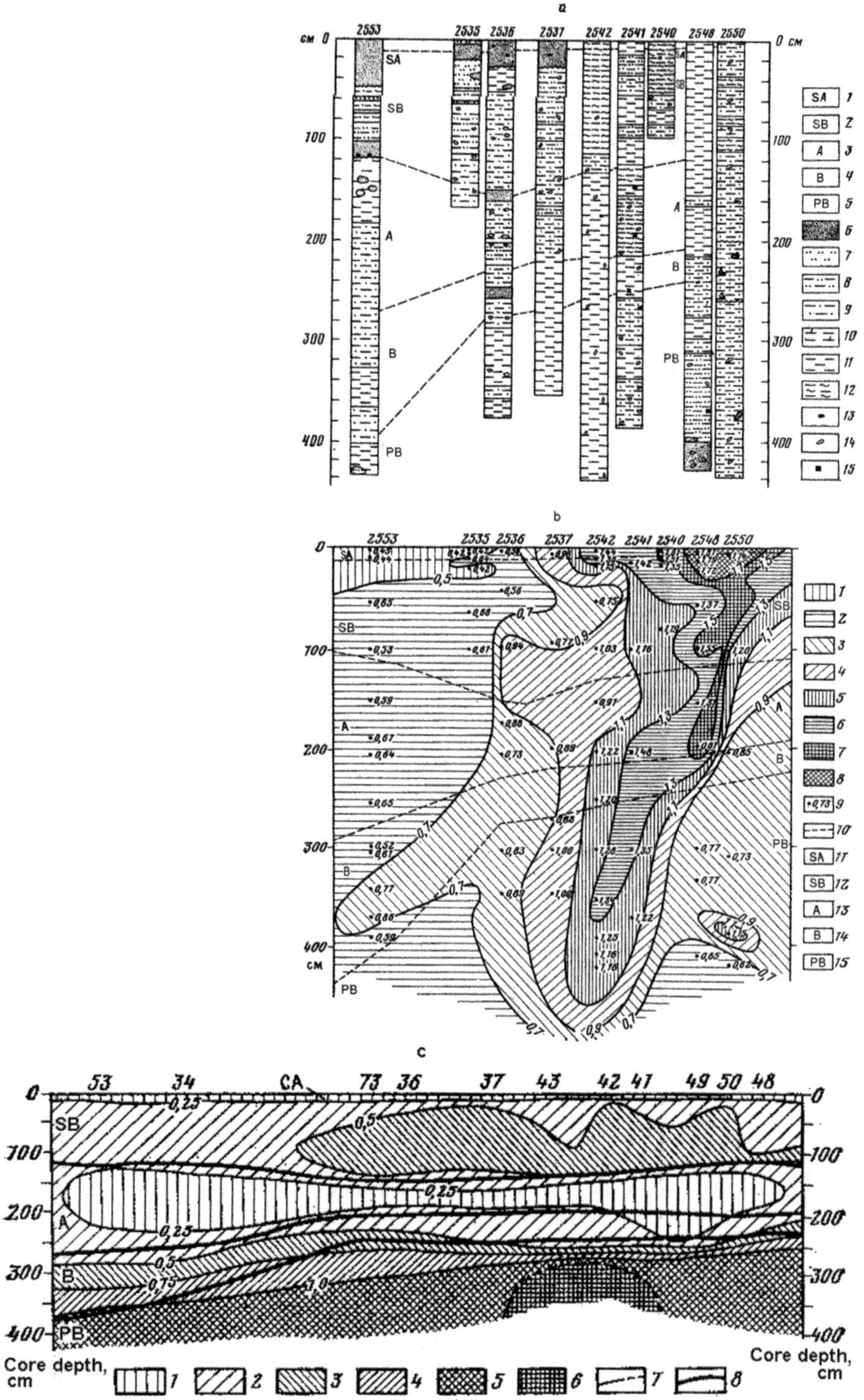
a
2553 2535 2536 2537 2542 2541 2540 2548 2550
SA
SB
A
B
PB
SA 1
SB 2
A 3
B 4
PB 5
b
2553 2535 2536 2537 2542 2541 2540 2548 2550
c
53 34 CA 73 36 37 43 42 41 49 50 48
Core depth, cm
Core depth, cm

Table 12.1 Thickness of Holocene diatom oozes in the Bering Sea according to M.A. Levitan (29-th cruise of RV *Dmitry Mendeleev*)

Sediment core DM	Geological feature	Northern Latitude, N	Longitude	Water depth, m	Core length, cm	Thickness of Holocene diatom ooze, cm
2533	Komandor Basin	59°34,6′	175°47,0′E	3700	405	62
2578	Komandor Basin	60°39,6′	177°28,0′E	3280	498	360
2579	Komandor Basin	60°13,4′	177°52,7′E	2895	462	176
2580	Komandor Basin	58°42,3′	169°30,3′E	3055	213	213
2594	Shirshov Ridge	56°56,4′	169°53,0′E	1780	472	92
2597	Komandor Basin	56°43,0′	169°31,6′E	3840	546	15
2598	Aleutian Basin	56°10,0′	174°55,7′W	3875	274	30
2599	Aleutian Basin	56°13,3′	174°12,4′W	3860	120	8
2600	Aleutian Basin	56°10,0′	178°05,3′W	3790	174	168

of available sediment cores was mainly restricted to a few meters, and only more recently the use of a new piston corer allowed to obtain cores with lengths of >10 m (Nürnberg and Tiedemann, 2004). The evolution of different sedimentary mechanisms (including sea ice transport) under changing conditions during Pleistocene glacial-interglacial cycles was intensively studied for the first time by the participants of an international RV *Marion Dufresne* expedition carried out in 2001 within the IMAGES program (Bassinot et al., 2003; Levitan, 2003). In the course of this cruise, three sediment cores (MD01-2413, -2414 and -2415) were recovered in the international waters of the central region of the northern Okhotsk Sea (Fig. 12.3). Cores MD01-2413 and MD01-2415 were investigated by an international group of scientists from Germany (GEOMAR, Kiel and AWI, Bremerhaven) and Russia (GEOKHI and IO RAN, Moscow; and TOI, Vladivostok), and Core MD01-2414

Fig. 12.2 (continued) Lithological transect through the Navarin Basin (*a*) and TOC distribution (*b, c*) (Marina et al., 1985). *a*: 1–5 – stratigraphic horizons: 1 – Subatlantic, 2 – Subboreal, 3 – Atlantic, 4 – Boreal, 5 – Preboreal; 6–12 – lithology: 6 – sand, 7 – silty sand, 8 – silt, 9 – silty clay, 10 – marly mud, 11 – clayey mud, 12 – diatom mud, 13 – pebble, 14 – shell, 15 – pyrite. *b*: 1–8 – TOC (wt. %) distribution: 1 – 0–0.5, 2 – 0.5–0.7, 3 – 0.7–0.9, 4 – 0.9–1.1, 5 – 1.1–1.3, 6 – 1.3–1.5, 7 – 1.5–1.7, 8 – >1.7, 9 – TOC values; 10 – boundaries of stratigraphic horizons, 11–15 – stratigraphic horizons. (see Fig. 12.2a) *c*: 1–6 – TOC mass accumulation rates (g/cm^2/ka): 1 – <0.25, 2 – 0.25–0.50, 3 – 0.50–0.75, 4 – 0.75–1.00, 5 – 1.00–1.25, 6 – >1.25; 7 – lines of equal mass accumulation rates; 8 – boundaries of stratigraphic horizons

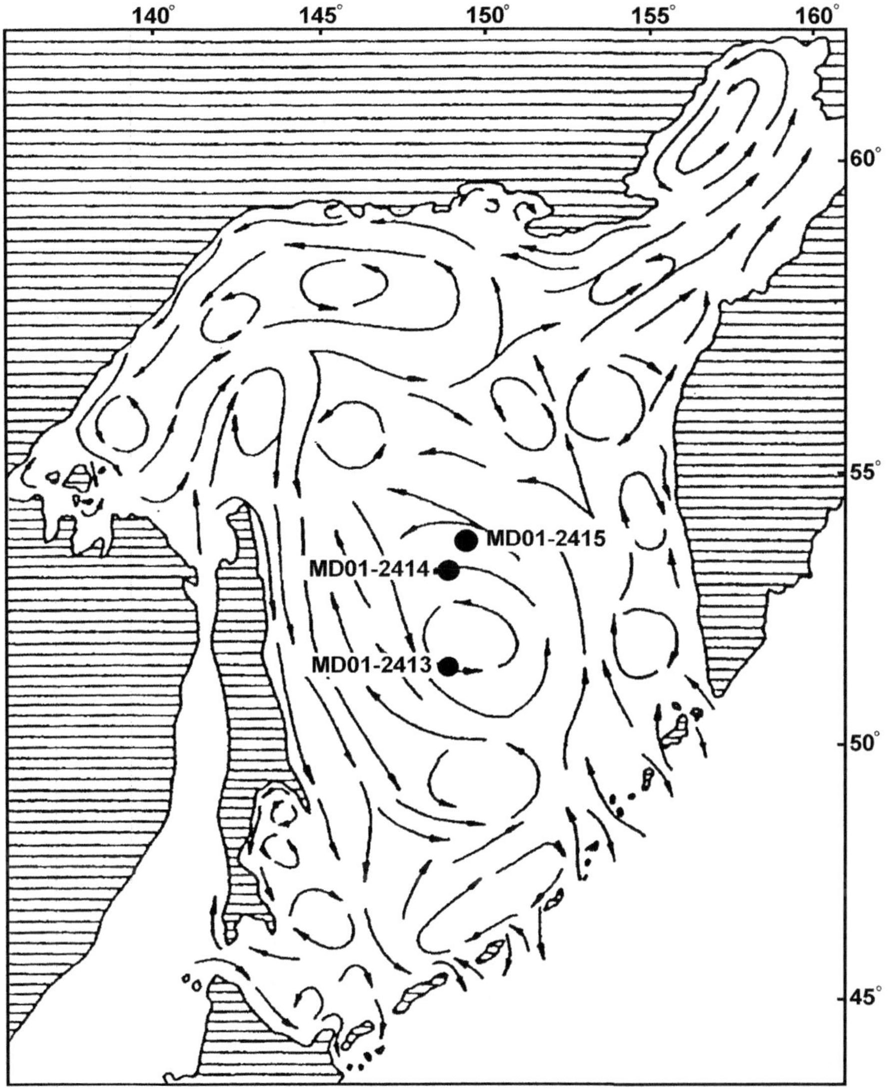

Fig. 12.3 Scheme of surface-water circulation (Chernyavsky et al., 1993) and location of RV *Marion Dufresne* sediment cores in the Sea of Okhotsk

was studied by scientists from Taiwan. The long-lasting attempt to organize deep-water drilling in this area, however, was unfortunately not succeeded yet.

The Sea of Okhotsk is surrounded by West Kamchatka, the Okhotsk volcanic belt, Sakhalin and the Kuril Islands. Like the Bering Sea, it is also divided into a shallower northern part (in which the water depths, however, are not typical for a shelf, reaching in average 1000 m with two deep depressions (TINRO and Deirugin) and a number of underwater rises) and a deep-water southern part with the deepest depression of the Sea of Okhotsk – the Southern Kuril Basin. In this sea, as in the Bering Sea, a distinct sea-ice cover is developed starting in November and lasting until June. The Sea of Okhotsk is mostly covered by sea ice (80%) with a thickness of 0.5–1.0 m during March (Aota and Uematsu, 1989).

The Alaska Current plays the most important role in the hydrology of the northern part of the Pacific Ocean. It flows to the north along the periphery of North America and, reaching the Alaska coast, sharply changes the direction to the west. Flowing along the Aleutian bank it transfers relatively warm Pacific waters into the Eastern Hemisphere in the water layer of 200–400 m. In this case a substantial part of the Alaska Current arrives in the Bering Sea through the deep-water straits of the Aleutian Islands. These waters, partially freshened in proportion to cooling and circulation, enter the Pacific Ocean again through the deep Kamchatka Strait as the powerful Kamchatka Current. A minor part of the Pacific Ocean waters passes through the Bering Strait into the Chukchi Sea. After flowing along the East Kamchatka coast, part of the waters of the Kamchatka Current enters the Sea of Okhotsk through a number of straits (especially, the deepest one – Krusenstern Strait) (Chernyavsky et al., 1993). Figure 12.3 shows a detailed diagram of the circulation of surface waters in the Sea of Okhotsk. It is demonstrated, that the Western Kamchatka and Eastern Sakhalin currents form a cyclonic circulation. As a result, terrigenous material supplied by the Amur River (about 14 $t/km^2/yr$) (Gorbarenko, 2004) is practically not transported into the center of the sea. In the northern branch of this circulation gyre several flows are isolated, while in the southwestern part the warm Soya Current transports subtropical waters, entering from the Sea of Japan through the La Perouse Strait. The annual primary production in the Sea of Okhotsk is approximately 100–200 g C/m^2, decreasing toward the pelagic direction (Gorbarenko, 2004).

Most waters of the Bering Sea and the Sea of Okhotsk originate from the Pacific Ocean. The characteristic property of hydrology of the Sea of Okhotsk is the development of so-called dichothermal cold waters at depths of 50–150 m with temperatures $<0°C$, which are formed during the winter cooling of surface water (ice formation) and remain throughout the whole year. At present, the direct ventilation of waters in the Sea of Okhotsk, stimulated by the formation of sea ice on the northern shelf and by the inflow of salt waters with the Soya Current from the Sea of Japan, reaches depths of 400–600 m. At depths of 750–1300 m, the warmer Pacific waters with positive temperatures of 2.0–2.5°C are located. Below, the temperature is reduced to 1.8°C at the bottom (Dobrovolsky and Zalogin, 1982).

Waters of the northwestern part of the Pacific Ocean, the Bering Sea and the Sea of Okhotsk are known to be the final component of the global salinity conveyor belt (Broecker, 1991). Therefore, these water masses are the oldest waters of the World Ocean and have the highest content of nutrients. This circumstance explains such a high primary production of these areas. The intermediate waters of the Sea of Okhotsk, saturated in oxygen, flow into Pacific Ocean with a velocity of 2.7 Sv and considerably contribute to the formation of intermediate waters in the northern part of the Pacific.

The detailed study of the recent sediments of the Sea of Okhotsk showed that only in the vicinities of the basic source provinces of terrigenous material (West Kamchatka, Sakhalin and others) the terrigenous or diatom-bearing sediments are developed (Gorbarenko, 2004). Holocene diatomaceous ooze with varying amounts of biogenic opal was found in the remaining part of the sea. Their thickness is usually 20–40 cm, increasing to 190–220 cm toward the boundary to the Southern Kuril Basin (Gorbarenko, 2004).

In the surface sediments of the Sea of Okhotsk in the area of Core MD01-2415, smectites clearly predominate among the clay minerals, supplied from the south of West Kamchatka and Kuril Islands (Kurnosov and Murdmaa, 1978). Their content usually exceeds the summary concentration of illite, kaolinite and chlorite by a factor of two. In this case, the basic sources of illite are only located to the north of the studied region, increasing towards to the northern branch of the described circulation gyre. Kaolinite and chlorite enter the Sea of Okhotsk from the northern and western continental regions (Kurnosov and Murdmaa, 1978). Rare-earth elements in the sediments of the Sea of Okhotsk, as far as we know, were not studied yet.

History of Sedimentation

History of Sedimentation in the Deep-Water Part of the Shirshov Ridge (Bering Sea) During the Last Three Marine-Isotope Stages

In the northern relatively shallow part of the Shirshov Ridge, clastic-clayey sediments and terrigenous clayey muds are exceptionally predominant in the sedimentary sections of our studied cores DM 2587-2593 (Table 12.1). The sedimentary section at Core DM 2594 from a water depth of 1780 m, however, is very different. The lithological characteristics of this core are described by M.A. Levitan (including macro-description and examination of smear-slides). Grain-size and heavy-mineral analyses performed by V.P. Kazakova. Analyses of the biogenic components ($CaCO_3$, TOC, and diatom content), mass accumulation rates of TOC, oxygen-isotope composition in foraminifer tests *N. pachyderma* sin. and *U. auberiana* (in ‰ to PDB), and also three radiocarbon datings were published by S.A. Gorbarenko (1996).

Lithostratigraphic Characteristics

From top to bottom, the following lithostratigraphic units were separated in the sedimentary section of Core DM 2594:

Unit I (0–92 cm): Clayey diatom ooze of olive color, cheese-like structure, uniform, with high water content. The content of <0.005 mm fraction always exceeds 70%. The concentration of the 0.01–0.005 mm fraction usually varies between 18 and 28%. Clay minerals dominate in the sediments (30–35%), different diatoms (30–35%), quartz-feldspar clastics (to 30%). Furthermore, green and brown hornblende, pyroxene, globules of pyrites, brown and colorless volcanic glass occur in varying abundances. The content of the total heavy fraction is very uneven and changes from 1.56 to 8.64%. Black ore minerals (28–33%), rhombic pyroxene (19.5–33%), and hornblende (16.4–32%) predominate among the heavy minerals. The content of monocline pyroxene sometimes increases to 12.9%. Acid volcanic glass (20–34%), indeterminable remains of rock and minerals (most likely, volcanic ash) in a quantity to 30%, basic and intermediate plagioclases (from 8 to 23%) and remains of carbonate organisms predominate among the light minerals. Small detritus of bivalves was only found in very minor amount.

Unit II (92–145 cm) is characterized by very fine clayey terrigenous mid of light gray color and homogenous structure, high water content and viscosity. The content of <0.005 mm fraction always exceeds 70%. The concentration of the 0.01–0.005 mm fraction usually varies between 10 and 28%. The sediments are characterized by abundant clastics (about 40%), represented by basaltoids, cherts, quartz, green and colorless volcanic glass, pyroxene, hornblende, pyrite and other. About 50% of the sediment is composed of clay minerals, and the remainder consists of several percentages of carbonate (Bivalve, Bryozoa, Echinoderma, benthic foraminifers) and siliceous (in essence, diatoms) organisms. The content of total heavy fraction is quite stable: 3.15–3.70%. Its composition strikingly differs from the composition of Unit I: hornblende (14–16%), monocline pyroxene (19–24.5%), and epidote (21–26%). Sometimes, pyrite plays a leading role. Quartz (28–41.3%), indeterminable rock remains (17.6–24%), and certain amount of acid plagioclases and by places (in the lower part of the section) abundant remains of biogenic carbonates (up to 37.5%).

Unit III (145–472 cm) is represented by sediments very similar to those of the overlaying Unit II. Gradually, they are compacting downward the section. At 216–229 cm, lenses of pinkish-brown volcanic ash of slightly more coarse-grained composition were observed. Pebbles of dark fine-crystalline rocks with a size of up to 2.5 cm were found at 248 cm and 427–430 cm core depth. Clay fraction predominates in grain-size composition, sometimes with some increased content of silt admixture. Volcanic ash contains clay (69.49%), silt (24.13%), and sand (5.38%). According to smear-slide-analysis data, very fine (clay) clastic component (up to 60%) prevails, presented by clastics of rock, quartzose cherts, light and heavy transparent minerals, volcanic glass, and pyrites. 30–35% of sediment are composed of clay minerals and their aggregates. Practically there are no remains of carbonate and siliceous organisms.

The amount of the heavy fraction varies between 2.5 and 3.8%. Hornblende (12.6–17.3%), epidote (11.3–23.2%), and monocline pyroxene (16.6–33.2%) predominate among the heavy minerals. Pyrite was found in relatively high abundances, ranging between 9.5 and 29.6%. Quartz (>32%, sometimes up to 45.6%) and indeterminable mineral grains (22–43%) prevail in the sediment composition. Occasionally, abundances of potassium feldspar increase up to 11.8%. In one sample (185–195 cm), a high content of biogenic carbonates of 25% is noted. Volcanic glass is mainly distinguished by its composition in the light fraction: there are 68.6% of acid glass and 12.3% of indeterminable grains and fragments (also ashes?). As mentioned above, the sediments of Unit III are lithologically and mineralogically very similar to the sediments of Unit II, but they contain somewhat more of fine clastics ("glacial milk"?) and less biogenic components. We assume that sediments of Unit II and Unit III were predominantly accumulated from sea ice, and water exchange with other basins was reduced to a minimum.

Stratigraphy and Several Paleoceanographic Proxies

For Core DM 2594, three radiocarbon datings were carried out, and corrected AMS ^{14}C dates look as follows: 9.07 ka at 82 cm, 11.2 ka at 111 cm, and 19.3 ka at 235 cm (Fig. 12.4). It means that sedimentation rates are 9.04, 13.62, and 15.3 cm/ky, respectively. Such way, Younger Dryas sedimentation rates were 1.5 time higher than later, and MIS 2 sedimentation rates were slightly higher than in MIS 1. Oxygen-isotope stratigraphy supports the radiocarbon datings: the boundary between MIS 1 and MIS 2 is located at 145 cm, the boundary between MIS 2 and MIS 3 at 282 cm. Based on TOC and $CaCO_3$ data, it possible to subdivide MIS 1 into two substages: 0–92 cm and 92–145 cm.

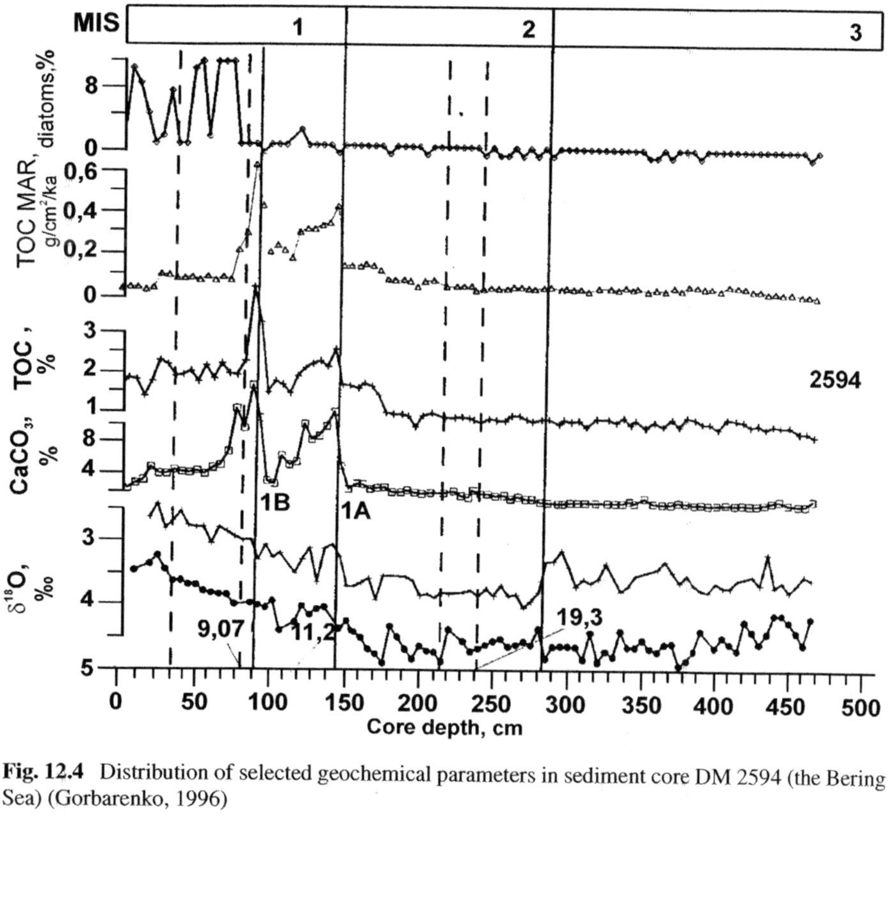

Fig. 12.4 Distribution of selected geochemical parameters in sediment core DM 2594 (the Bering Sea) (Gorbarenko, 1996)

In Unit I (corresponding to the first substage 0–92 cm) the content of diatom frustules changes from 1 to 12×10^4, and concentrations and mass accumulation rates of $CaCO_3$ and TOC are relatively high reaching maximum values at the base of Unit I, probably corresponding to termination 1B (Fig. 12.4). A similar evolution of the carbon record is also obvious for the second substage of MIS 1, where the maximum values corresponds to termination 1A; diatoms, however, are actually absent in the lower part of Unit I. Within MIS 2 and MIS 3 there are essentially no sedimentological changes, but oxygen isotopes of planktonic and benthic foraminifer become slightly lighter at the MIS 2 /MIS 3 boundary, probably reflecting somewhat increased temperatures of water masses at that time.

Thus, it is assumed that during MIS 1 the mass accumulation rates of biogenic components were somewhat higher than in MIS 2 and MIS 3, and mass accumulation rates of terrigenous substance, on the contrary, were lower. This can be explained by an increase in primary production during the Holocene, which may have resulted from a change in circulation system, an increase in water exchange with the surrounding basins and also a decrease in terrigenous matter input at this time, when the suspended-matter discharge of the Yukon has moved from the Bering Sea to the Arctic Ocean. Mineralogical data on Core DM 2594 remarkably illustrate changes in the source provinces at the MIS 1/MIS 2 boundary, when the Bering Strait was opened.

Data on both heavy and light minerals allowed to assert that as the source province for Unit I served in essence the volcanic islands of the Aleutian island arc, and the source provinces for the terrigenous material of Unit II and Unit III, most likely, were located on the large land masses, which frame the Bering Sea – Chukotka and Alaska. Abundance of the rock fragments in the heavy and light fractions were probably in favor of the massive inflow of icebergs from Alaska. The presented data from the shelf of Bering Sea in the area of the Navarin site completely coincide with the proposed concept.

History of Sedimentation in the Northern Sea of Okhotsk During the Last 1.1 Ma

Materials and Methods

Sediment core MD01-2415 (53°57.09′ N, 149°57.52′ E; water depth 822 m, length 46.23 m) was most thoroughly investigated in order to reconstruct the paleoclimatic and paleoceanographic history in the Northern Sea of Okhotsk. The core represents a hiatus-free sequence of sediments spanning the time interval from 1.1 Ma to recent (Nürnberg and Tiedemann, 2004; Matul, 2005).

The detailed layer-by-layer lithological description of the core was carried out onboard and supplemented by smear-slide description and core photographs. Determination of color reflectance of sediments were accomplished manually by a MINOLTA camera system (Bassinot et al., 2003). Various physical properties (primarily magnetic susceptibility and bulk density) of sediments were routinely measured.

At the IFM-GEOMAR, Kiel/Germany, oxygen-isotope stratigraphy of Core MD01-2415 was established, including astronomical tuning. Radiocarbon AMS ^{14}C datings are available for some selected samples from the upper part of the sedimentary section. To study the biogenic sedimentation, TOC, biogenic opal, and $CaCO_3$ were determined and paleoproductivity estimated. Some paleoceanographic reconstructions were published in (Nürnberg and Tiedemann, 2004). A.G. Matul (2005) reported preliminary results of the paleoecological study of radiolarians.

One of the main goals of the present research was to study the history of terrigenous sedimentation, i.e., changes of provenance and transport processes of the terrigenous material. Methods are listed in (Levitan et al., 2007).

Results

Lithostratigraphic Characteristics

Figure 12.5 shows the lithostratigraphic characteristics of Core MD01-2415, including grain-size distribution, magnetic susceptibility and color reflectance (b*). The latter parameter correlates with the content of biogenic opal, i.e., the higher the opal content, the higher the index b*. Magnetic susceptibility depends on the content of ferrimagnetics and shows an inverse correlation with the content of biogenic opal in the studied sediment section.

In general, the sediment section has a rather simple structure, and two lithostratigraphic units were identified. The upper unit (called "Okhotomorsk sequence") represents the last ~ 415-ka of geological history. The lower unit ("Vanin sequence") continues downward at least to the bottom of the core.

Unit I, the Okhotomorsk sequence, is a product of cyclic and directional sedimentation. The cyclicity is manifested as alternation of cyclites (diatom sediments at the base and detrital-clayey and clayey sediments at the top). The base of Unit I at 23.4 m was identified by the first occurrence of a diatom ooze accumulated during MIS 11. Thus, boundary between the two units correlates approximately with the Early/Middle Neopleistocene. Light brown (in the upper interval) and olive green diatom ooze (clay fraction <0.005 mm up to 80–90%) has cheese-like and bioturbation structure. The biogenic opal is mainly related to frustules of planktonic diatoms; radiolarians, silicoflagellates, and siliceous sponge spicules only occur in very minor amounts. The opal content gradually decreases from 60–65% in the upper layer to 40% in the lowermost layer (Nürnberg and Tiedemann, 2004). The grain-size distribution is shown in Fig. 12.6a. According to (Nürnberg and Tiedemann, 2004), the diatom ooze contains 8–20% $CaCO_3$ (mainly foraminifers) and 1–1.4% TOC. Color reflectance decreases from 18 u. at the top of the sequence to 10 u. at the bottom, parallel to an increase of magnetic susceptibility from 0 to 200×10^{-5} SI (Fig. 12.5). The diatom ooze (as well as the other lithologies of the sequence) contains sporadic IRD presented by rock debris of 2–4 mm to 2–4 cm in size. Sandy admixtures were found at some places. Diatom ooze occurs in the following intervals of the unit: 0–2.2 m; 7.5–9.0 m; 17.7–18.8 m, and 22.1–23.4 m. These intervals correspond to MIS 1.1, 5.5, 9.3, and 11, i.e., to Terminations and

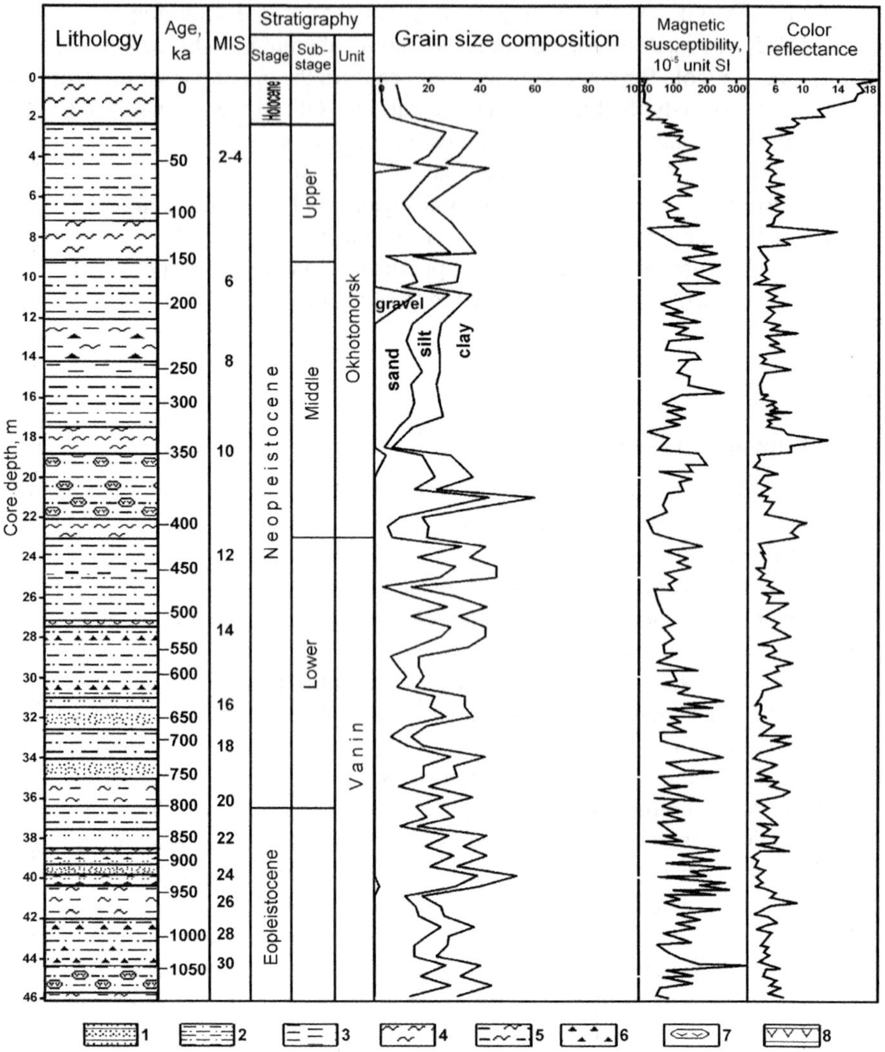

Fig. 12.5 Lithology and stratigraphy of sediment core MD01-2415 (Levitan et al., 2007b). 1 – sand; 2 – silty clay; 3 – clay; 4 – diatom ooze; 5 – diatom clay; 6 – IRD; 7 – volcanic ash lenses; 8 – volcanic ash layers

subsequent Interglaciations. A minor diatom enrichment was recorded in interlayers between 11.5 and 14.0 m, i.e., during Interstadials 7.3 and 7.5 (Fig. 12.6b). The sedimentation rate of diatom ooze in Unit I is generally about 4–5 cm/ky. In the upper Holocene layer of diatom sediments, however, sedimentation rates reached very high values of up to 22 cm/ky. Taking the core recovery technique into consideration, this value is probably exaggerated 2–2.5 times (Nürnberg and Tiedemann, 2004).

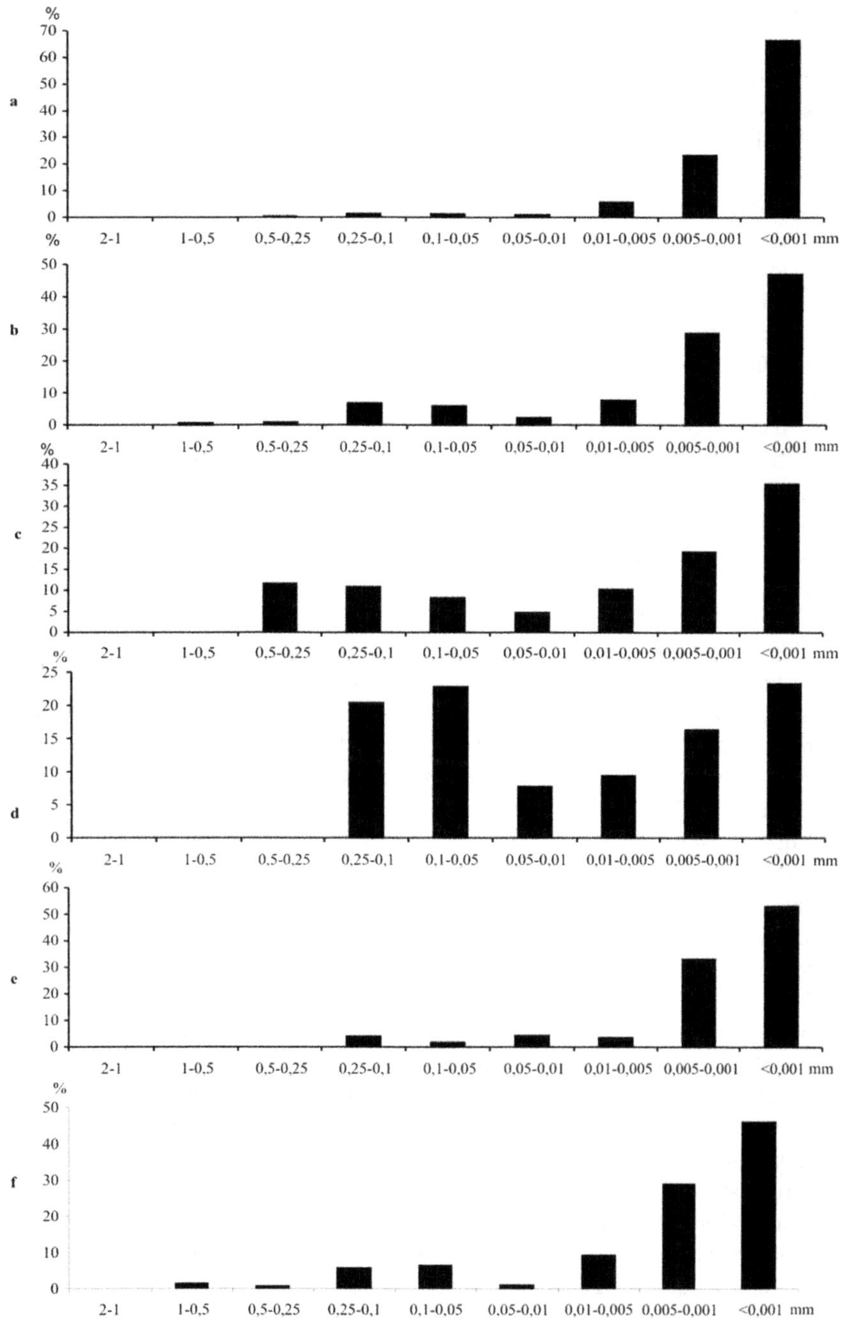

Fig. 12.6 Histograms of main lithologies and grain-size distribution (Levitan et al., 2007b). *a* – diatom ooze, *b* – diatom clay, *c* – silty clay, *d* – silty-clayey sand, *e* – vitric volcanic ash, *f* – host sediment for lenses of volcanic ash

The dominantly detrital-clayey and clayey sediments are represented by olive gray, yellowish green, and gray muds, silty-sandy clays with an admixture of gravel at some places (Fig. 12.6c), sandy clays (Fig. 12.6d), and clays with sand and gravel. The section commonly includes relatively fine-grained sediment varieties with a clay fraction of up to 70% of the bulk sediment. In general, bioturbation is either absent or vague. IRD is sporadically scattered throughout the section. Magnetic susceptibility ranges from 100 to 200 $\times$ 10^{-5} SI and color reflectance from 5 to 6 u. The average content of biogenic components is as follows: opal – 4–5%, $CaCO_3$ – 0–2%, and TOC – 0.3–1.2% (Nürnberg and Tiedemann, 2004).

In addition to the terrigenous varieties described above, in some intervals of the sequence fine (1–2 cm thick) layers of pink and light gray vitric volcanic ash (Fig. 12.6e) as well as black sandy vitroclastic and crystalloclastic volcanic ash were found. The latter ash varieties often occur in lenses of 2–3 cm in size, distinctly increasing in abundance in the silty clays with sandy admixture at 19–22 m core depth (Fig. 12.6f). This depth interval corresponds to approximately 350–395 ky. In the upper part of the unit, magnetic susceptibility is as high as 200 $\times$ 10^{-5} SI, whereas color reflection is minimal. Accumulation rate of abiogenic sediments is usually two or three times lower than that of diatom ooze.

Unit I (the Okhotomorsk sequence) includes at least four rhythmites (MIS 11–10, MIS 9–8, MIS 7–6, and MIS 5–2). Each rhythmite corresponds to a glacial-interglacial climate cycle of approximately 100 ky in duration; i.e., the rhythmites follow the periodicity of eccentricity variation (Bol'shakov, 2003). Furthermore, spectral analysis of the color reflectance also revealed distinct periodicities with a period of 41 ky (i.e., the obliquity cycle) and a period of 22 ky (i.e., the precession cycle) (Nürnberg and Tiedemann, 2004). The diatom ooze layer on top of the section belongs to the Holocene, i.e., the beginning of Cycle V. A distinct trend in sedimentation is obvious in the index b* of color reflectance, caused by the upward increase of biogenic opal content in the diatom ooze interlayers. As an example, regional processes especially related to subaerial explosive volcanism on Kamchatka or the Kuril Islands, are represented by volcanic ash layers and lenses, and by intervals with abundant lenses of sandy ash in the lower sedimentary section (MIS 10).

The chemical composition (based on 28 elements) supports the specific character of each of the lithotype (Table 12.2). For example, diatom ooze is enriched in Na and Sr. The remaining elements are intensely diluted by the biogenic opal and subordinate calcium carbonate. In general, diatom clays have a transitional composition between diatom ooze and detrital-clayey sediments. Because only one analysis of the vitroclastic volcanic ash is available, we cannot make any statistically significant conclusions about its composition. Nevertheless, it is noteworthy to mention that this ash is enriched in REE, Na, Sc, Fe, Br, and Hf. Concentrations of Rb, Cs, Ba, U, Ta, and Zn are significantly lower. It is clear that compositional differences between lithotypes of sandy and silty clays are primarily related to different grain size. Therefore, elements of the sand fraction (e.g., Cs, Ba, Cr, Th, U, Hf, and Zr) are concentrated in sandy clays, whereas elements of the clay fraction (Eu, Na, Rb, Sr, Sc, Fe, Co, As, Sb, and Br) are concentrated in silty clays (Table 12.2). REE spectra

Table 12.2 Distribution of chemical elements in the major lithotypes of sediment core MD01–2415

Lithotype	La	Ce	Nd	Sm	Eu	Tb	Yb	Lu	Na	Rb	Cs	Ca	Sr	Ba
Diatom ooze (n=14)	0.24–24.5 / 12.16	1.82–44.1 / 23.57	3.38–17.8 / 10.68	1.21–4.47 / 2.82	0.012–1.81 / 0.64	0.16–0.9 / 0.52	0.27–3.18 / 1.42	0.038–0.53 / 0.24	2.84–5.11 / 3.62	10.7–122.6 / 47.47	0.6–7.82 / 2.62	0.03–4.34 / 2.13	65–4445 / 755	35–2340 / 675
Diatom clay (n=2)	15.3–15.9 / 15.6	27.8–31 / 29.4	11.1–13.6 / 12.35	2.72–3.4 / 3.06	0.33–0.99 / 0.66	0.56–0.63 / 0.6	1.5–1.73 / 1.62	0.23–0.29 / 0.26	3.5–3.59 / 3.55	35.7	0.54–2.39 / 1.47	2.36–3.13 / 2.75	250–910 / 580	630–805 / 718
Volcanic ash (n=1)	20.1	40.1	19.9	5.09	0.59	0.91	2.26	0.38	4.86	7.28	1.28	2.59	480	495
Tuffite (n=5)	15.2–23 / 18.24	29.0–47.0 / 34.9	12.1–22.7 / 15.0	3.07–6.1 / 3.82	0.3–1.25 / 0.77	0.49–1.05 / 0.68	0.85–2.43 / 1.60	0.13–0.38 / 0.26	2.75–4.44 / 3.41	28.3–99.9 / 62.3	0.24–11.6 / 5.74	0.26–1.86 / 1.16	60–675 / 383	45–440 / 300
Sandy clay (n=27)	10.6–25.0 / 18.3	20.3–45.1 / 34.3	9.08–19 / 14.7	2.41–4.61 / 3.73	0.058–2.29 / 0.77	0.50–0.90 / 0.68	0.63–3.73 / 1.82	0.089–0.63 / 0.30	2.18–3.88 / 2.96	6.01–113.2 / 52.3	0.48–12.6 / 4.44	0.22–5.58 / 1.92	16–2405 / 436	15–2520 / 721
Silty clay (n=27)	7.52–29.8 / 17.4	16.8–54.2 / 33.7	9.52–22.5 / 15.4	2.89–5.7 / 3.96	0.038–2.37 / 0.93	0.41–1.05 / 0.70	0.30–3.80 / 1.78	0.037–0.64 / 0.29	1.84–4.97 / 3.26	10.7–284.7 / 60.5	0.32–10.3 / 3.60	0.21–4.48 / 2.09	49–2470 / 569	24–1425 / 622

Lithotype	Sc	Cr	Fe	Co	Ni	Se	As	Sb	Th	U	Br	Hf	Ta	Zr
Diatom ooze (n=14)	4.7–19 / 12.9	5.15–95.4 / 31.9	1.02–4.58 / 3.18	2.69–37 / 20.57	4.93–230 / 88.5	0.88–13.1 / 3.1	2.3–15.4 / 8.49	0.2–3.02 / 1.46	0.48–6.71 / 3.15	0.74–12.1 / 3.66	0.19–70.6 / 16.8	0.31–5.14 / 1.74	0.043–0.95 / 0.44	6–75 / 30
Diatom clay (n=2)	17.4–17.5 / 17.45	37.2–42.2 / 39.7	4.12–4.14 / 4.13	25.1–41 / 33.05	–	1.69–2.52 / 2.11	3.3	5.58–6.99 / 6.29	4.05–4.9 / 4.48	0.3–3.0 / 1.69	0.96–4.3 / 2.63	2.56–3.8 / 3.18	–	25–38 / 36.5
Volcanic ash (n=1)	18.1	49.5	4.54	24.3	–	3.14	3.89	2.89	5.39	1.09	47.9	3.81	0.27	16
Tuffite (n=5)	15.5–22.2 / 17.4	37.0–76.3 / 54.1	4.02–5.77 / 4.57	19.2–34.5 / 25.9	44.7–89.1 / 66.9	1.12–3.12 / 2.20	3.84–8.3 / 6043	0.68–3.18 / 2.41	3.75–8.03 / 6.08	2.85–5.61 / 3.87	1.31–61.5 / 15.7	0.85–3.92 / 2.64	0.035–0.73 / 0.42	11–40 / 25
Sandy clay (n=27)	7.99–20.5 / 14.4	4.18–131 / 54.9	2.29–5.65 / 4.07	6.63–36.3 / 23.5	21.0–600 / 191.1	0.48–24.8 / 3.5	1.12–44.4 / 8.51	0.076–2.85 / 1.45	3.07–11.3 / 5.6	0.14–18.1 / 3.81	1.48–41.6 / 14.9	0.35–6.94 / 3.04	0.26–2.19 / 0.75	5–330 / 47
Silty clay (n=27)	13.1–23.3 / 17.6	37.3–93.1 / 52.9	2.96–6.04 / 4.35	6.51–48.5 / 28.3	57.3–440 / 199	0.44–11.1 / 3.31	0.17–34.9 / 10.1	0.17–6.56 / 2.23	2.29–11.2 / 5.27	0.14–10.5 / 3.59	0.60–81.9 / 26.08	0.18–5.84 / 2.61	0.16–1.96 / 0.75	12–105 / 39

Note: Nominator – range of concentration, denominator – average values; Na, Ca, Fe are given in %, the rest elements – in ppm; n – number of samples; blank – no data

and its distribution pattern as well as data on the composition and distribution of clay minerals will be discussed later after presenting data from the entire sequence of Core MD01-2415.

Unit II (the Vanin sequence) is composed of lithotypes similar to those of Unit I (the Okhotomorsk sequence), except for the absence of diatom ooze in Unit II (Fig. 12.5). Instead, diatom clays have been accumulating during some warm periods (e.g., during MIS 19, 27, and 31). In the diatom clays of this sequence, the color reflectance does not exceed 6–8 u. and biogenic opal content does not exceed 20%. The sediments are enriched in TOC of up to 1.2% and commonly depleted in $CaCO_3$ (Nürnberg and Tiedemann, 2004). The role of diatom-rich sediments is significantly lower in the Vanin sequence relative to the Okhotomorsk sequence (Fig. 12.5). The radiolarian fauna in the Vanin sequence, however, supports a more intensive water exchange with the Pacific Ocean at that time (A.G. Matul, private communication, 2005).

The variations in terrigenous lithotypes, which prevail in the sections, are even more expressive. The Vanin sequence is also dominated by coarse-grained sediments (silty-sandy clays) that mainly occur in the lower part of the sedimentary section (i.e., 31–39 m core depth). This is reflected in the grain-size distribution and magnetic susceptibility of the sediments (Fig. 12.5). The content of abiogenic material is 80–90% at some places. The concentration of silty-sandy material may become as much as 50%. Magnetic susceptibility and color reflectance vary from 250 to 300×10^{-5} SI and from 2 to 6 u., respectively. Some interlayers, in particular at intervals representing MIS 30–28 and MIS 16–14, are strongly enriched in coarse-grained IRD (with diameters of up to 3–5 cm) that is well rounded.

As in the Okhotomorsk sequence, the Vanin sequence also contains rare thin layers of vitroclastic volcanic ash and layers of black sandy pyroclastic material, occurring in the lower part of the sedimentary section (interval 44.1–46.1 m, representing MIS 30). The sedimentation rate of all lithotypes of this sequence is slightly lower (up to 2∼3 cm/ky) than that determined for the Okhotomorsk sequence.

The spectral analysis of the continuous series of color reflectance shows that the Vanin sequence is also dominated by cycles related to eccentricity (Nürnberg and Tiedemann, 2004). However, the 41-ky-cycle plays a more important role in this sequence (particularly, in its lower part) than in the Okhotomorsk sequence. Cycles with a period of 22 ky were recorded as well.

Chemical Composition

The differences in grain size and biogenic content of the two units, i.e., the Okhotomorsk and Vanin sequences, are also clearly reflected in their chemical composition (Table 12.3). For example, relative to Okhotomorsk sequence, the Vanin sequence characterized by more coarse-grained terrigenous and less biogenic material (see above), is depleted in nearly all measured elements, except Eu, Na, Ca, Sb, and Cs.

The results of the correlation analysis allowed to distinguish five major associations of elements (except REE): (1) Fe, Sc, Cr, Co, Th, and Hf (associated with the sand and silt fraction); (2) Ba, Br, and U (associated with the clay fraction); (3) Rb,

Table 12.3 Average concentration of elements in the Okhotomorsk (I) and Vanin (II) sequences

Sequence	La	Ce	Nd	Sm	Eu	Tb	Yb	Lu	Na	Rb	Cs	Ca	Sr	Ba
I	15.3	29.1	12.7	3.24	0.8	0.58	1.49	0.24	3.45	51.9	4.14	2.18	365	455
II	17.8	34.3	15.5	3.99	0.78	0.72	1.89	0.31	3.1	53.9	3.5	1.87	679	806
Sequence	Sc	Cr	Fe	Co	Ni	Se	As	Sb	Th	U	Br	Hf	Ta	Zr
I	15.4	43.2	3.89	23.0	111.6	2.55	7.29	2.24	4.5	3.02	11.26	2.33	0.55	38
II	17	54.5	4.14	26.4	212.9	3.75	9.77	1.65	5.37	4.02	26.6	2.88	0.73	40

Note: Na, Ca, Fe are given in %, the rest elements – in ppm

Sr, Sb, Se, Co, and Zn; 4) Na and As (associated with the clay fraction) or Na and K (associated with the sand and silt fraction); and (5) Ca, Ta, and Cs (associated with the gravel fraction).

The varimax factor loading for several elements confirmed significantly the validity of associations listed above. For example, VF1 is typical of significant positive loads for Sc, Cr, Fe, Co, Th, and partly Hf; VF2 is typical of Ca and partly Na; and VF3 is typical of Na and Sb. The available data indicate that VF1 is related to the sand and silt fractions, VF2 to the biogenic material and partly to elements adsorbed on this material, and VF3 to the dissolved forms of mobile elements, particularly As adsorbed on clay particles.

The content of biogenic components (primarily diatoms) is rather high. Therefore, concentrations of some elements were recalculated to carbonate-opal-free matter (cofm) (for background and method of approach see (Nürnberg and Tiedemann, 2004). Results presented in Fig. 12.7 show a very curious regularity: the distribution of the majority of elements (without considering the dilution effect of biogenic material) is controlled by the distribution of abiogenic material. Diatom ooze contains terrigenous material derived from specific provinces (see, for example, the distribution of Na, Sr and Sc). This terrigenous material drastically differs from the material precipitating from sea ice or under sea-ice conditions during cold (glacial) periods. The Ba distribution is presumably related to variations in paleoproductivity that increased during times of accumulation of diatom oozes (Nürnberg and Tiedemann, 2004).

The Na concentration in the sediments at 32–36 m core depth is related to the enrichment of sand-sized plagioclases. Besides its correlation to the $CaCO_3$ content, Sr is concentrated in lenses of volcanic ash. Data presented in Fig. 12.7 not only confirm the lithostratigraphic subdivision of the studied core, but also indicate a specific composition of sediments at 44–46 m core depth corresponding to the period of 1050–1100 ka, particularly for the distribution of Ba, Sr, and Ni. This fact may support the enriched abundance of feldspars in the pyroclastic material.

According to (Maslov et al., 2004), an increase of the Th/Sc ratio suggests an increase in the total acidity of rocks in the source provinces. With respect to the Th/Sc ratio distribution, the core can be divided into three distinct intervals: (1) the interval from the core bottom up to 28 m core depth is characterized by rather

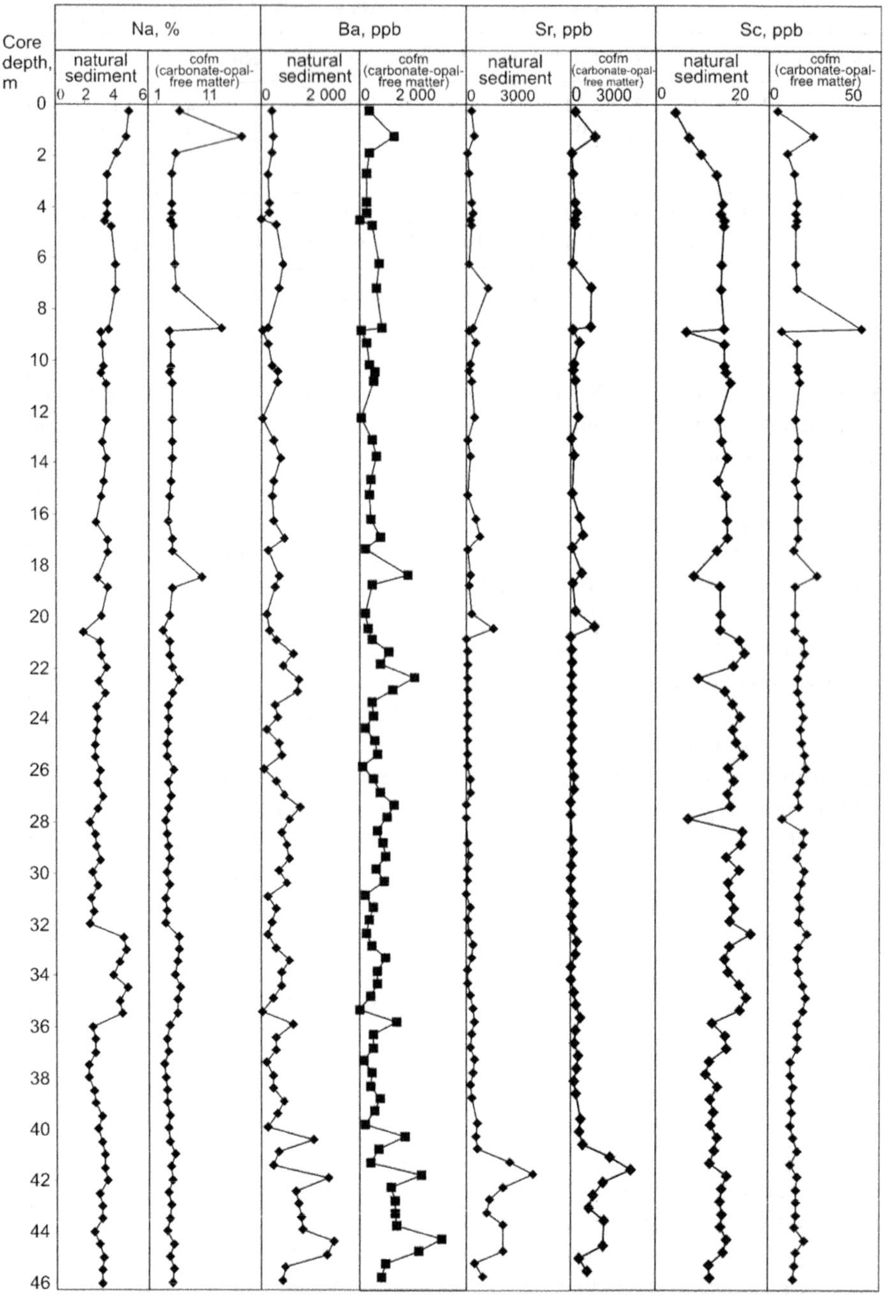

Fig. 12.7 Distribution of Na, Ba, Sr, and Sc in sediment core MD01-2415 (cofm – recalculated for carbonate-opal-free matter) (Levitan et al., 2007b)

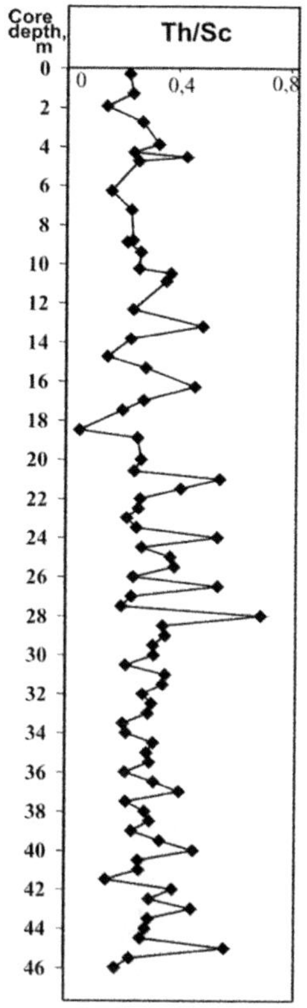

Fig. 12.8 Distribution of Th/Sc ratios in sediment core MD01-2415 (Levitan et al., 2007b)

stable Th/Sc values (~0.35-0.38); (2) the interval from 28 to 21 m core depth is also characterized by stable and moderate Th/Sc values but with a wider concentrations range (ca. 0.30–0.35, with a maximum value of 0.76); and (3) the interval from 21 m to the top of the section shows a distinct upward decrease in the Th/Sc ratio (Fig. 12.8). Thus, the Okhotsk volcanic belt was a more important source province during accumulation of the Vanin sequence compared to the Okhotomorsk sequence.

Rare Earth Elements

Rare earth elements (REE) were analyzed in 79 samples of bottom sediments. We could only determine concentrations of La, Ce, Nd, Sm, Eu, Tb, Yb, and Lu because of the specific character of the INAA method. Concentrations of the remaining REE (Pr, Gd, Dy, Ho, Er, and Tm) were deduced by calculation method.

The REE composition of the sediments is surprisingly uniform and similar to the average composition of Pacific terrigenous sediments beyond the iceberg zone reported in (Gurvich et al., 1980) (Table 12.3). An enrichment in the LREE, a distinct positive Ce anomaly, and a depletion in the MREE and HREE are clearly shown (Fig. 12.9a). Some variations of the absolute concentration of REE, particularly of La and Ce, are related to the dilution effect by carbonate and siliceous materials. It is worth to mention that the distinct negative Ce anomaly, which is typical

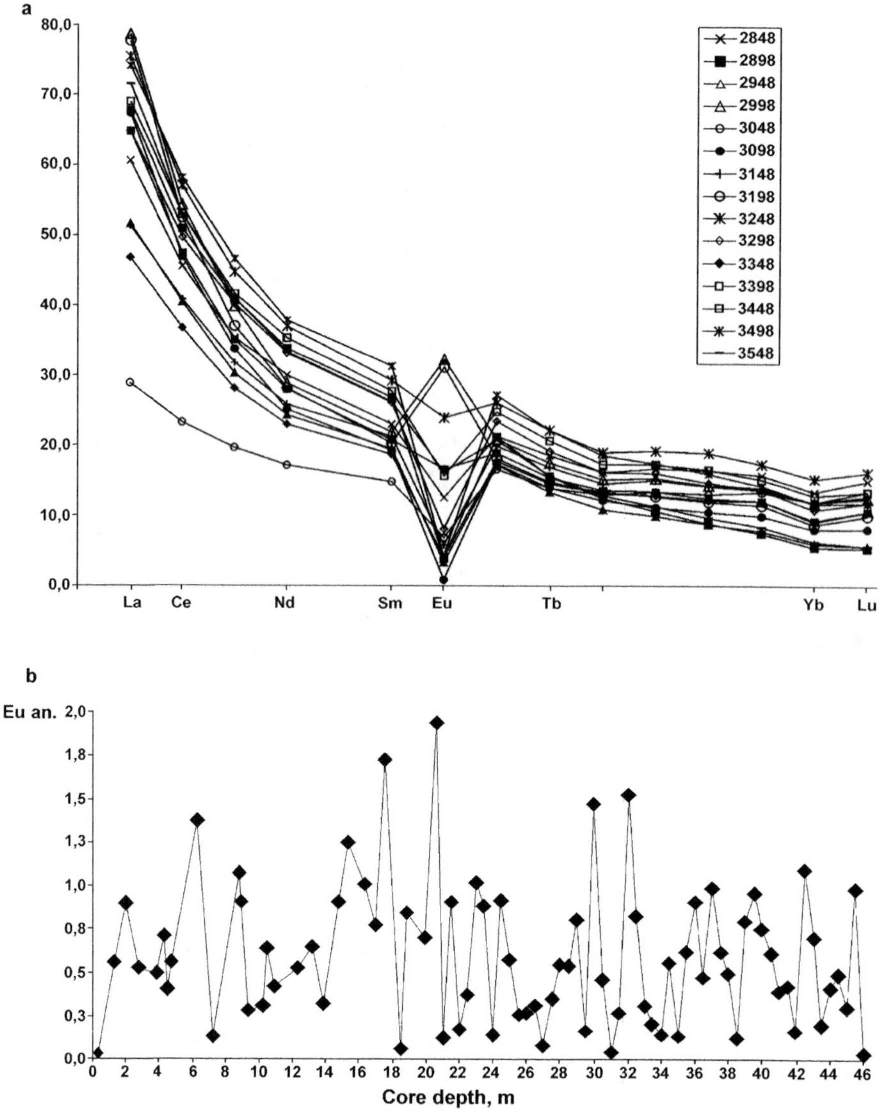

Fig. 12.9 REE spectra of sediments (chondrite normalized) (*a*), and distribution of Eu anomaly in sediment core MD01-2415 (*b*) (Levitan et al., 2007b). Eu anomaly for all samples was calculated according to formula "(Eu sample/Eu chondrite)/(Sm sample/Sm chondrite)". Those REE are shown which were determined by INAA method; other element concentrations are calculated. Numbers in Fig. 12.9*a:* sampling core depth (cm below sea floor)

of diatoms (Dubinin, 2004), was not manifested even in the Holocene diatom ooze with the highest content of biogenic opal because the REE concentration in the terrigenous matrix is much higher than that in diatom frustules.

In order to identify the sources of the terrigenous or volcanogenic-terrigenous material, we used a chondrite-normalization of REE compositions (Fig. 12.9a) taking into consideration the predominantly magmatic character of rock complexes in Kamchatka, the Kuril Islands, and the Okhotsk volcanic belt. The normalized spectra reflect a significant enrichment in the LREE and a moderate enrichment in the HREE. Furthermore, the sediments usually show a negative Eu anomaly (i.e., <1.0), except for seven samples with a positive Eu anomaly (>1.0).

Under the dominance of physical weathering in drainage basins, insignificant concentrations of soluble REE species in the shelf water, and a lack of REE sorption from seawater in the shelf zone, the REE composition in the bottom sediments is probably related to the mixing of components from various terrigenous and biogenic sources (Balashov, 1976; Gurvich et al., 1980). This assumption is also supported by recent data on bottom sediments from the continental-margin basins (Gaiero et al., 2004; Plank and Langmuir, 1998).

From Fig. 12.9b it is obvious that – based on the Eu anomaly – the sedimentary section can be divided into two parts. The lower part (>21 m core depth) is dominated by small amplitudes of Eu anomaly variation (average 0.3–0.5), whereas the upper part (<21 m core depth) shows a general decrease from >1.0 (maximum 1.85) at the base to nearly 0 at the top. Typical values range from 0.4 to 1.0.

In order to interpret the Eu anomaly data in the bottom sediments, we used reference data of REE compositions in various rock of eastern and western Kamchatka, the Kuril Islands, and the northern Okhotsk as well as other regions (Balashov, 1976; Gurvich et al., 1980; Ivanov, 1990; Volynets, 1994) (Table 12.4).

Based on the data listed in Table 12.4 as well as Figs. 12.5 and 12.9b, minimum Eu anomaly values are related to the presence of diatoms, whereas maximum values (>1.5) are related to radiolarians. Furthermore, the Vanin sequence contains

Table 12.4 Average value of Eu anomaly in different geologic objects (composed by M.A. Levitan)

Geological object	Eu anomaly
Diatom-radiolarian ooze of the Indian Ocean	1.44
Siliceous diatom ooze of the Indian Ocean	0.35
Terrigenous mud of the Indian Ocean	0.43
Terrigenous mud of the Pacific Ocean	0.62
Platform clays	0.50
Earth's crust	0.82
Granites ($SiO_2>70\%$)	0.35
Granitoids (SiO_2 60–70%)	0.62
Magmatic sequences of the eastern Kamchatka	0.82
Magmatic sequences of the western Kamchatka	0.61
Island-arc tholeits	0.93
Island-arc andesites	1.10

not only diatoms and sedimentary material transported from Kamchatka and partly the Kuril island arc, but also a significant portion of granitic material delivered from the Okhotsk volcanic belt. The relatively coarse-grained composition of the sediments in the study region excludes transportation by wind (Fig. 12.5). Based on the Eu anomaly data, the Okhotomorsk sequence is most likely dominated by material transported from western Kamchatka. The latter material was mixed with diatoms and other substances delivered from eastern Kamchatka, the Kuril Islands, and the Okhotsk volcanic belt. The age of sediments shows an inverse correlation with the share of diatoms. The atmospheric circulation was very intense during glaciations (Wolf, 2006) and abiogenic sediments in the studied sequence were generally characterized by very fine-grained composition. Therefore, we cannot exclude a partial eolian supply of clay material from Eurasia.

Clay Mineralogy

Based on X-Ray diffraction analysis, the following groups of layered silicates were identified in the studied sediments: illite, expanding minerals, chlorite, and kaolinite. The sedimentary section is mainly composed of expanding minerals and illite that may reach up to approximately 80–90% of the subcolloidal crystalline fraction. The group of expanding minerals includes smectite and irregular mixed-layer smectite-illite varieties dominated by the smectite-type interlayers (60–90%). Illite mainly presented by dioctahedral Fe-Al varieties ($b = 8.95$–9.00 Å), is commonly hydrated and its content is higher in sediments of the finest fraction with subcolloidal fraction not exceeding 50%. Chlorite commonly occurs as trioctahedral Fe-Mg variety.

In the diatom ooze developed in the uppermost, ten-cm-thick part of the sedimentary section, the subcolloidal fraction is composed of virtually X-ray amorphous material with a large halo induced by opal-A. In the underlying sediments, clay-mineral data allow a subdivision into two sequences. The Okhotomorsk sequence (Unit I; see above) demonstrates an upward increase in the smectite/illite ratio, particularly above 21 m core depth. The ratio reaches maximum values in the intervals with pyroclastic lenses (Fig. 12.10). This trend is also manifested in the age pattern of both diatom ooze and terrigenous-clastic sediments, although one can see certain specific features at some places. For example, at 274, 877 and 1088 cm core depth slight enrichments in homogeneous smectite dominated by monovalent Na-cation, were determined. Intervals enriched in mixed-layer phases (628, 938, and 1023 cm core depth) are dominated by bivalent cations (Ca, Mg) in the interlayers.

The Vanin sequence (Unit II), on the other hand, is characterized by an opposite trend, i.e., an upward decrease in the smectite/illite ratio (Fig. 12.10). Fluctuations of the smectite/illite ratio are more distinct than in the overlaying Okhotomorsk sequence. Furthermore, the smectite/illite ratio is sufficiently high in the pyroclastic sequence (Fig. 12.5). Upward the section up to the Eopleistocene/Neopleistocene boundary near 0.8 Ma, the ratio increases and reaches its maximum during MIS 20. After a significant decrease upward the section, a clear peak of the smectite/illite ratio was observed at 0.65 Ma. Then, the ratio again decreases up to the top of the

Fig. 12.10 Distribution of smectite/illite ratios in sediment core MD01-2415 (Levitan et al., 2007b)

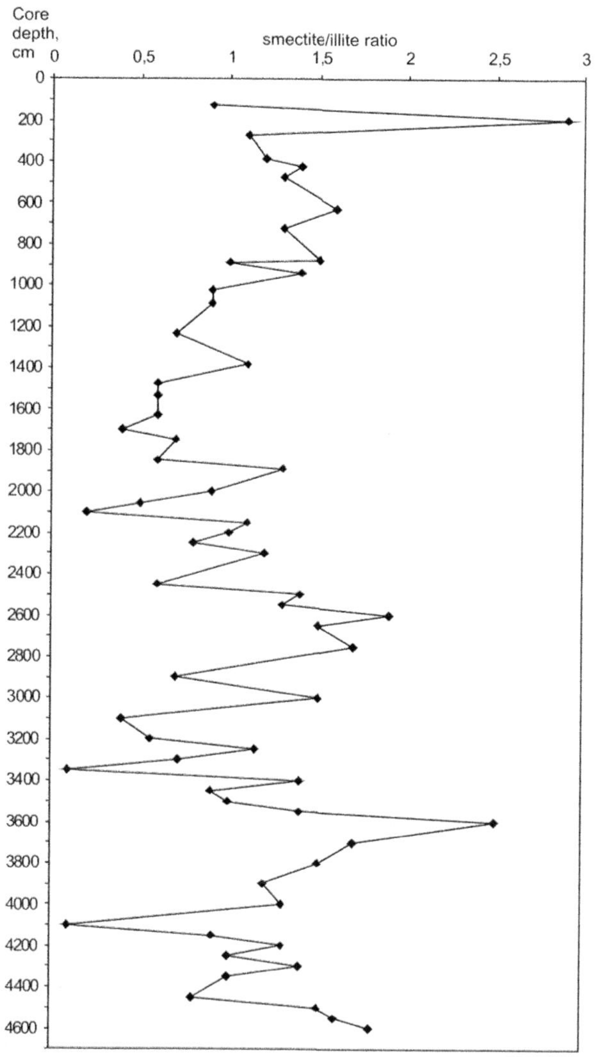

sequence. Based on our data, the smectites can be divided into two types. The type-1 smectite represents a product of transformation processes occurred in lenses of coarse-grained volcanic ash. The type-2 smectite is related to weathering and soil products of continental crust. Samples with a high content of expanding minerals sometimes found in the coarse-grained sediments, are dominated by uniform smectite structures (mixed-layer phases are subordinate). The concentration of illite is anomalously high (55–88%) in some samples (e.g., at depths of 3000–3200, 3348, and 4098 cm). Moreover, illite has a more perfect structure, and Fe concentration in octahedra is high. In general, the illite content shows a weak positive correlation with the chlorite content. Correlation with kaolinite is less evident.

Thus, in summary, our lithostratigraphy, chemistry, and mineralogy data from the whole sedimentary section support its subdivision into the Okhotomorsk and Vanin sequences.

A comparison of the records of the Eu anomaly and smectite/illite ratio (Figs. 12.9b and 12.10) revealed an interesting trend. The detailed analysis allowed to separate two zones. Zone 1 is characterized by synchronous variations of both records (i.e., a positive correlation between the Eu anomaly and the smectite/illite ratio), whereas zone 2 is marked by an opposite trend (i.e., a negative correlation between both parameters). Periods with positive and negative correlations can be divided into four stages: (1) 1.1–0.96 Ma; (2) 0.96–0.52 Ma; (3) 0.52–0.35 Ma; and (4) 0.35–0 Ma. This subdivision virtually matches the classification proposed by (Mudelsee and Schulz, 1997), according to which the Earth's climate history was marked by a Middle Pleistocene transitional period ("Mid-Pleistocene climate shift" between 0.95 and 0.55 Ma) from an epoch dominated by the 41-ky-cyclicity of obliquity (until 0.95 Ma B.P.) to an epoch dominated by the 100-ky-cyclicity of eccentricity (after 0.55 Ma B.P.). The termination of the Middle Brunhes climate transition (0.35 Ma) is also expressed in our record.

Discussion

A stratigraphic correlation of the Vanin sequence with records from the surrounding land is hampered, because the latter is poorly studied. The abundance of coarse-grained sediments that probably derived from the Okhotsk volcanic belt supports a rather high erosion rate in this source province.

The gravelly-sandy sequence accumulated between 0.92 and 0.84 Ma might be correlated with the Enemtem conglomerates and sandstones of western Kamchatka (Svitoch, 2003). A. Svitoch divides the recent geological history of the Kuril Islands into the Eopleistocene-Early Neopleistocene and Middle-Late Neopleistocene stages. The first stage is characterized by the general subsidence and extensive effusion of basaltic-andesite lava. This may be reflected in the lenses of volcanic ash and layers described above. Volcanic eruptions were particularly dramatic 1.1–1.05 Ma ago. Despite the development of glaciations, glaciers did not exist on the Kuril Islands. The present mountain topography of the island ridge has been developed during the period of tectonic uplift and volcanic eruptions of the second stage (Svitoch, 2003). The major mountain glaciations took place in the Middle and, particularly, Late Neopleistocene.

Thus, the stratigraphic boundary between the major stages of the Quaternary geological evolution of the Kuril Islands coincides with the Okhotomorsk/Vanin boundary. The composition of clay minerals suggests that the Vanin interval was characterized by a southward shift of the gyre mentioned above in accordance with the general trend of climate variations in the North Pacific toward more extreme glacial conditions (Jansen et al., 1986). It is worth to mention that the diatom data also indicate a southward shift of the Subpolar Front of the Sea of Okhotsk at approximately 1 Ma (Sancetta and Silvestri, 1986). The period of 1–0.8 Ma was marked by "significant fluctuations from intensely stratificated cold water masses with contrasting seasonal patterns to less extreme (but still cold water) conditions" (Matul, 2005, p. 261). The southward shift of the Subpolar Front is also recorded in the North Pacific 0.8–0.7 Ma (Sancetta and Silvestri, 1986). At the same time, the gyre was presumably subjected to sporadic short-term SW-oriented shifts at 0.95, 0.7, 0.62, and 0.4 Ma.

According to (Matul, 2005), the continuous occurrence of radiolarians in the sedimentary section of Core MD01-2415 indicates that the Sea of Okhotsk has never been completely covered by ice in the study region. The Vanin sequence developed during period of active water exchange with the Pacific during relatively warm intervals. However, paleoproductivity and intensity of vertical circulation both were appreciably lower relative to the later period, probably due to lesser climatic fluctuations.

In general, paleoproductivity reconstructions based on the content of biogenic opal, $CaCO_3$, and TOC, as well as the excess of Ba and chlorines (Nürnberg and Tiedemann, 2004), suggest periods of distinct intensification during times of accumulation of the Okhotomorsk sequence, especially during interglacials and across terminations (Fig. 12.11). Major subaerial explosions occurred at 0.35–0.3 Ma. According to data provided by (Sancetta and Silvestri, 1986), the Middle Brunhes transition took place 0.425–0.3 Ma and is characterized by a distinct glacial cyclicity and more extreme glacial conditions in the entire North Pacific (Jansen et al., 1986).

New primary production during the Holocene and MIS 9 is estimated to be about 110–140 g C/m^2/yr (Nürnberg and Tiedemann, 2004), which approximately matches the present data. The coincidence of peaks of primary productivity and maxima of accumulation rate of terrigenous material (Fig. 12.11) indicate the crucial role of biotransport for the accumulation of abiogenic material at the seafloor. Furthermore, our lithological and geochemical data suggest that the major part of the time interval represented in the sedimentary record of Core MD01-2415 was marked by glaciomarine sedimentation with sea ice as the main transport agent.

The relatively coarse-grained clastic material could enter sea ice (or settle on its surface) in the area of pack ice near cliffs. Large particles of surface sediments could be entrained by anchored ice along sea coast. Cryozoles were primarily released and accumulated at the seafloor during thawing of ice during summer. Some mountain glaciers, however, probably served as outlets during intensive sea-level rises in the Okhotsk region in the second half of Middle Neopleistocene (Anan'ev et al., 1984; Svitoch, 2003), e.g., during MIS 7. This conclusion is supported by the abundance of coarse material (probably IRD) in the MIS 7 sediments.

Of course, we cannot rule out a significant influence of atmospheric circulation in the Siberian Maximum in the formation of sea ice in the Okhotsk Sea during cooling episodes (Nürnberg and Tiedemann, 2004). Geochemical and mineralogical data, however, indicate that Kamchatka, the Kuril Islands, and the Okhotsk belt rather than Central Siberia served as the major source provinces for the Sea of Okhotsk. The situation has changed during interglacials and terminations: active water exchange with the Pacific fostered not only the increase of paleoproductivity owing to upwelling and diatom ooze accumulation and a much higher role of

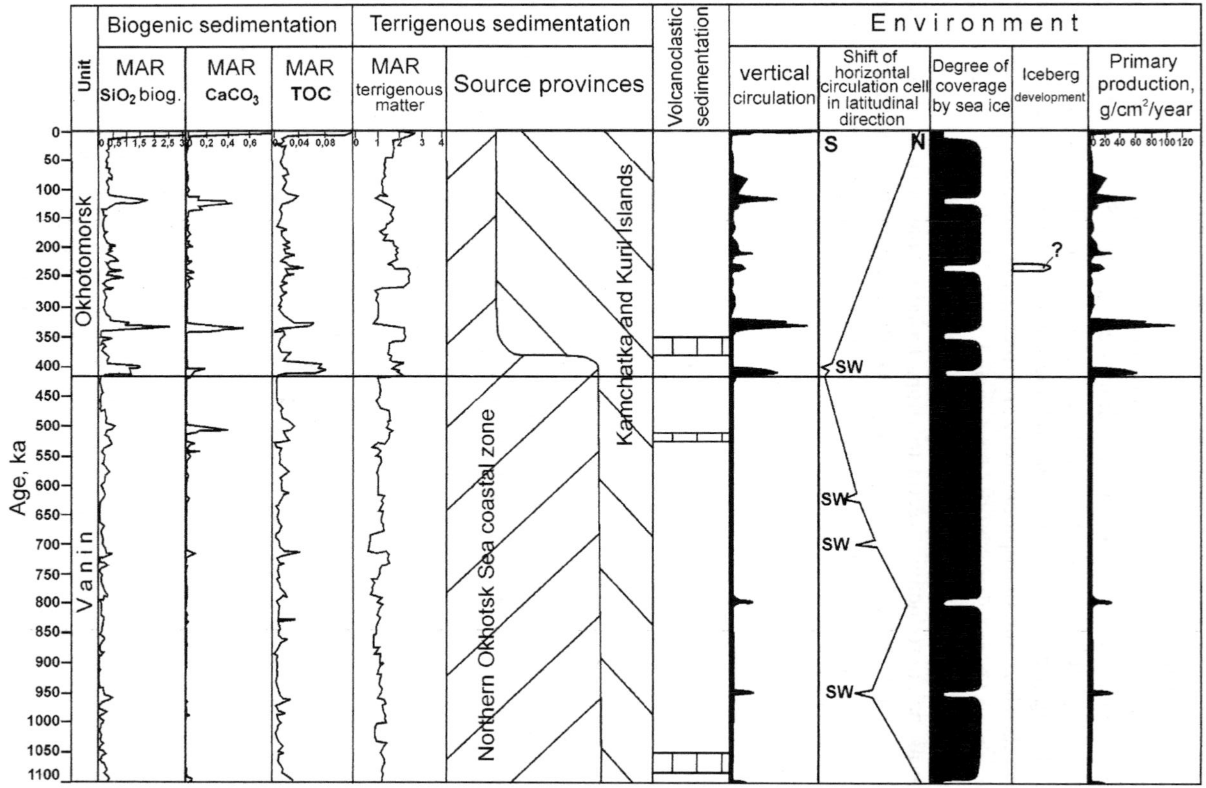

Fig. 12.11 Schematic summary compilation showing the sedimentation history of the Northern Sea of Okhotsk for the last 1.1 Ma (Levitan et al., 2007b). MAR – mass accumulation rates (g/cm^2/ky) (Nürnberg and Tiedemann, 2004), with change for terrigenous matter MAR. Paleoproductivity downcore up to level 130 ka is given according to (Nürnberg and Tiedemann, 2004), below – on our data

oceanic flora in diatom assemblage (Barash et al., 2003), but also the input of specific oceanic terrigenous suspension characterized by processes of geochemical differentiation of mature pelagic type. Clay mineral data indicate that the core of gyres has successively propagated northward during the last 400 ky, and the role of the Kuril Islands and western Kamchatka as source provinces has gradually increased.

Of special interest is the relationship between the changes in sedimentation processes described above and the global climate variations. Figure 12.12 illustrates recent results of the ice sheet drilling at the EPICA Dome C area of Antarctica (Brook et al., 2006). Figure 12.12a shows that variations in atmospheric temperature (based on oxygen isotopes) and concentrations of greenhouse gases (particularly CH_4) in Greenland and Antarctica demonstrate an amazing similarity during the last 90 ka. Figure 12.12b shows analogous similarity in the behavior of temperature (based on deuterium) and CO_2 in the Antarctic atmosphere over the last about 650 ka. The Middle Brunhes event (415 ka) is very well expressed in both plots. The earlier (>90 ka) stage of the Pleistocene could also be marked by global variations in the atmospheric circulation. According to (Liciecki and Raymo, 2005), the Middle Brunhes event is less prominent in the average global oxygen-isotope record of benthic foraminifers (Fig. 12.12b).

Thus, climate changes related to the integral impact of fluctuations of the Earth's insolation and orbital parameters (Bol'shakov, 2003) could have affected sedimentation in the Sea of Okhotsk mainly by fluctuations of atmospheric temperature and circulation. The temporal coincidence of the global Middle Brunhes event with the regional event in the evolution of the Kuril Island fostered notable changes in the provenance and the paleocirculation of surface waters in the Sea of Okhotsk.

According to (Shackleton, 1995), the period dominated by 41-ky fluctuations of obliquity has changed to the period dominated by 100-ky fluctuations of eccentricity approximately at 0.950 Ma. The transition period (final episode of the Middle Pleistocene Revolution) continued until 0.55–0.60 Ma (Mudelsee and Schulz, 1997). It is of interest that sediments accumulated during the transition period in the studied sedimentary section were enriched in sand fraction (Fig. 11.1).

Analysis of the material from the long sediment core MD01-2415 retrieved from the central northern Sea of Okhotsk allowed to divide the sedimentary section into the Okhotomorsk and Vanin sequences. They have accumulated during two major stages of geological evolution of the study region. The boundary between both sequences crossed the Early/Middle Neopleistocene (the so-called Middle Brunhes event). This boundary actually fits the boundary between the major evolution stages of the Kuril Islands (Svitoch, 2003).

The Vanin sequence is characterized by a combination of cyclicity and evolutionary trends with short-term episodes of explosive volcanism in the adjacent continental blocks (Kamchatka and the Kuril Islands). The cyclicity is expressed in glacial-interglacial rhythmites with the accumulation of diatom clays during warm (interglacial) intervals and the accumulation of sandy-silty clays during the cold (glacial) episodes. The trend is reflected in the upward decrease of the smectite/illite and Th/Sc ratios. We believe that the Okhotsk volcanic belt served as the main source province of volcanogenic-terrigenous material. This material (particularly

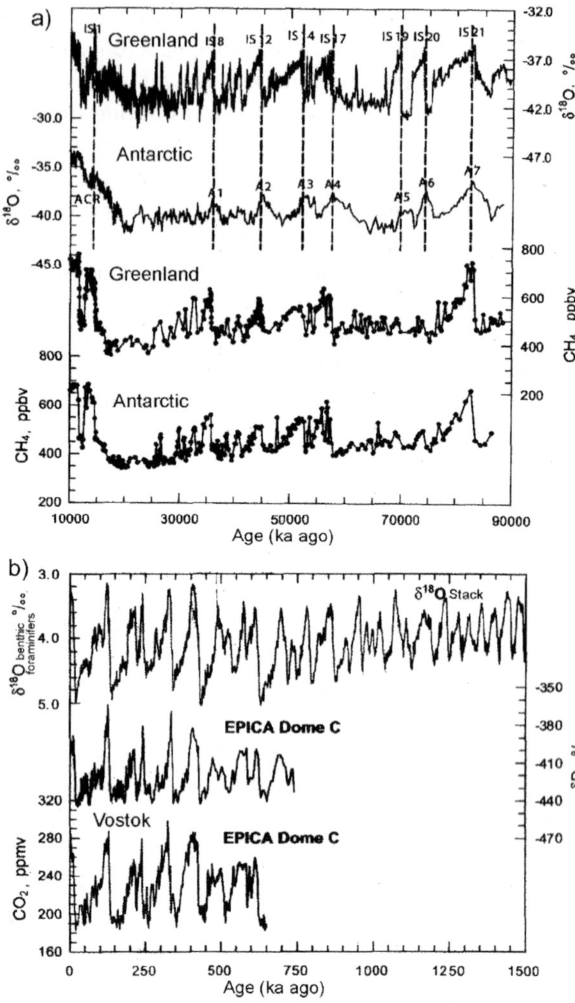

Fig. 12.12 Changes of oxygene-isotope composition and concentrations of greenhouse gases in ice cores of greenland and Antartic (Brook et al., 2006) for the last 90 ka (**a**) and for the last 750 ka (**b**) in comparison to the global benthic oxygen isotope stack from (Liciecki and Raymo, 2005). (**b**) (Levitan et al., 2007b)

its coarse-grained fraction) was primarily transported by sea ice. The gyre shifted in the southern direction with occasional south-eastern convergence (Fig. 12.11). Climatic distinctions between warm and cold intervals were insignificant.

Similar to the Vanin sequence and the Quaternary sedimentary section in general (Markov, 1986) the Okhotomorsk sequence is characterized by combination of cyclicity, evolutionary, and regional (in our case, explosive) effects. Among orbital cycles, cycles related to the 100-ky cyclicity of eccentricity play the major role (Nürnberg and Tiedemann, 2004). However, the Okhotomorsk sequence also demonstrates more contrasting distinctions between sediments of interglacial (diatom ooze) and glacials (mainly silty clays with coarse-grained admixture).

Increased paleoproductivity occurred because vertical water circulation enhanced; upcore it was an increase of Sm/Ill ratio and decrease in the Th/Sc ratio and Eu anomaly. It means the increasing role of Kuril Islands and Kamchatka as source provinces. As during the previous period, abiogenic matter was primarily transported by sea ice during cold intervals, although the role of icebergs cannot be ruled out for MIS 7 (Fig. 12.11). Warm intervals favored the intense water exchange with the Pacific Ocean. This environment intensified not only the horizontal circulation with the activation of the Kamchatka Current, but also the vertical circulation with formation of dichothermal waters and the intensification of primary productivity. In general, gyres successively shifted northward, probably due to changes in the circulation pattern after the uplift of the Kuril Islands.

In the second half of the Middle Neopleistocene, Late Neopleistocene, and Holocene, the volcanogenic-terrigenous material delivered from the Kuril Islands and southern area of western Kamchatka played a more important role in the sediment accumulation. Erosion of the Okhotsk volcanic belt during the second stage was less intense than during the first stage.

The Sea of Okhotsk is not a unique basin from the point of view of the strong influence of climatic (atmospheric) changes on sedimentation. Very similar results were obtained in the course of high-resolution paleoclimatic investigations of Quaternary bottom sediments of intercontinental lakes, such as Lake Baikal (Prokopenko et al., 2001) and marine continental margin basins, such as the South China Sea (Huang et al., 2005) and the Bering Sea (Wang and Chen, 2005). Marine geologists have long been aware that Quaternary fluctuations of temperature and atmospheric circulation play a crucial role in the sedimentation history of large upwelling zones.

We realize that many issues stated in the present paragraph have been insufficiently discussed, because many aspects of Quaternary geology of continental blocks surrounding the Sea of Okhotsk have been so far poorly studied. Nevertheless, we believe that our data from the (in terms of length, age, and continuity) unique sedimentary section of Core MD01-2415 recovered from the northern Sea of Okhotsk can stimulate further investigation of its regional features.

Part III
The Late Pleistocene Paleogeographic Events of Northern Eurasia and History of Sedimentation in the Subarctic Seas and the Arctic Ocean in Relation to the Northern Hemisphere Glaciation during the Last Climatic Cycle

The distinctive features of the Late Pleistocene and Holocene times are global and short-term climatic changes established by the modern methods. No special substantiation is required for the subglobal extent of the Eemian, or Mikulino-Kazantsev, Interglaciation, which has been recorded both in Eurasia and in North America and Greenland. In the eastern coast of Greenland marine and fluvial sediments were referred by means of biostratigraphic data to the Langelandselv Interglaciation correlative with the Eemian. In northern Canada, the interglacial deposits of the Collision Cape (the Banks Island), the Missinaibi Formation of the Hudson Bay, and the Kogalu Formation of the Brighton Cape (the Baffin Island) are correlated with the Eemian.

In spite of differences in natural environment, climatic conditions of all mentioned areas were close to or even warmer the present-time ones.

Chapter 13
Characteristic Features of the Mikulino Landscapes

Studies of paleolandscapes and paleoclimates of European Russia revealed some specific features of different regions.

In the central and northern regions, the forest landscapes dominated over the entire Mikulino Interglacial. Moreover, in the initial and terminal phases of the vegetation development (palynozones $M_{1,2,7,8}$ of V.P. Grichuk (1989)) forests grew all over European Russia. Birch and pine (M_1) forests were later replaced by fir ones (M_2). Pine-fir forests prevailed in the southern regions and pine-birch forests with *Betula nana* dominated in the north (Pleshivtzeva, 1972). At the end of the interglacial forests of the central regions demonstrated a backward succession of the same types: pine-fir and fir forests (M_7 is the upper maximum of fir distribution) were succeeded by pine-birch ones (M_8). The amount of *Betula nana* was increased in the north.

The optimum phases of the interglaciation (M_{3-6}) were characterized by essential differences in composition and structure of vegetation in the European Russia regions. In the central regions there was the consecutive-temporary culmination of the broad-leaved forests formed by oak, elm, linden, and hornbeam. A high percentage of pollen of filbert and alder-tree was recorded for M_4–M_5 (Grichuk, 1989).

In the northern regions (the Arkchangelsk and Leningrad regions and Karelia) boreal forests prevailed with participation of broad-leaved species, significant role of birch, and permanent presence of fir. The culmination of alder phase was fixed in spectra from all known sections. A shift of the alder phase to the north was shorter and less pronounced than to the south.

In the east, in the Dvina-Pechora and Kama floristic regions, the forests of the interglacial optimum were composed of alder, filbert, broad-leaved species, increased amount of pine and fir, and noticeably reduced participation of birch.

Thus, in contrast to the central regions where the broad-leaved forests were dominating, the northern regions were characterized by the southern taiga vegetation with permanent but variable participation of broad-leaved species. The zone of tundra was absent. In the western regions birch and fir were dominating forest-forming species, whereas in the eastern regions the role of pine and fir was considerably greater. Paleolandscape differences were also recorded in the southern regions (Spiridonova, 1991), along the sublatitudinal profile approximately at 52°N.

M.A. Levitan, Y.A. Lavrushin, *Sedimentation History in the Arctic Ocean and Subarctic Seas for the Last 130 kyr*, Lecture Notes in Earth Sciences 118,
DOI 10.1007/978-3-642-00288-5_13, © Springer-Verlag Berlin Heidelberg 2009

In spectra from the Doroshevichi outcrop at the Pripyat River pollen of alder and filbert amounts to 80% and 40% respectively. Participation of alder pollen does not exceed 30% in spectra from the Borkhov section at the Dnepr River because of very dry climate, whereas the filbert participation is still noticeable (to 23%) and confined to the maximal distribution of linden. In the Seim River basin of the Kursk area the role of linden and other wood species is decreased; the high participation of mesophilic grasses is recorded. This suggests a warming event and initial formation of steppe landscapes during the interglacial optimum in this region.

Thus, from the west to the east some aridization of climate and changes of forest landscapes at least into the forest-steppe ones has been documented.

To the south, in the region of Pavlovsk, forests of birch, pine, oak, and linden grew at the beginning of the interglaciation (Spiridonova, 1991). Grass-gramineous associations played a leading role in the flat interfluves (plakor). Balka-forests (M_2) included firs. Later (M_3) steppes with associations wormwood-gramineous vegetation and Chenopodiaceae were widely spread. The latter plants flourished in the Mikulino Interglacial optimum. Then the gramineous-wormwood associations appeared again were changed by grass vegetation.

The terminal phase was characterized by development of the broad-leaved–pine and broad-leaved forests of oak, less frequently of linden and elm; some role was played by gramineous–grass steppe associations.

Thus, the Mikulino optimum was marked by the important paleogeographic event. The hyperzonation of the southern taiga associations growing in large areas of the Russian Plain gave way to the latitudinal zonation of landscapes. Its main elements were the southern taiga landscapes in the northern regions, moderate forests at the Ilmen Lake and in the Volga-Oka interfluve, broad-leaved forests in the central regions, and steppe landscapes indicating semiarid environments in the south and the southeast. Such type of the latitudinal zonation was not a full analog of the modern zonation. It did not include zones of tundra, forest-tundra, desert, and semidesert. So, the latitudinal zonality of the Mikulino time can be regarded as the reduced type. Analysis of the general trend of the Mikulino paleolandscape alterations suggests a single optimum of the interglaciation without any additional or subordinate peaks of cooling and warming in the central and northern regions. In the southeastern regions a general pattern of landscape-climatic succession was more complex, as it was noted above.

Landscapes of the northern regions of European Russia demonstrated specific conservatism. The southern taiga landscapes did not undergo any essential changes during the interglacial optimum. Certain successive changes in broad-leaved forests took place in the central regions. Wide distribution of peat bogs and lacustrine sediments evidences of lake-boggy landscapes in these regions.

In the southern regions semiarid environments with characteristic steppe and forest–steppe landscapes and participation of broad–leaved species, especially in the river valleys, began to form in the interglacial optimum. Summarized available data allows conclusion that landscapes of southern European Russia acquired specific features under the influence of the Mediterranean air masses. The wide distribution of the boreal forests over the vast central and northern regions may indicate a decreasing effect of the Atlantic air masses and a growing influence of the Siberian anticyclone.

Finally, it is necessary to consider a debatable problem of thicknesses of the Mikulino deposits, duration of the interglaciation and, correspondingly, lifetime of the reconstructed landscapes. It should be reminded that the duration of the Eemian (Mikulino) Interglaciation was estimated as 20–25 or 35 ka by the U-Th method and as nearly 70 ka by the EPR method (Molodkov and Raukas, 1998a, 1998b; Kruk et al., 1998; Urban, 1998; Muhs, 1998). According to the estimations of 20 ka, the Eemian Interglaciation is commonly accepted to correspond to the isotopic stage 5e. However, the larger part of the Mikulino sequence in the central regions of European Russia, to which the "standard" palynological diagram with the corresponding palynological zones was applied, was recognized inside the lacustrine-boggy sediments up to 4–5 m thick. The peat is poorly consolidated. A question naturally arises about reasons of the low rate of peat accumulation, although the climatic conditions did not inhibit the increase in thickness. To answer the question, it is necessary to show what portion of the interglacial can be characterized by the Mikulino "standard" diagram. Figure 1.1 shows some oxygen–isotope data obtained for the Greenland ice sheet. It is clearly seen that the Eemian Interglaciation has more complex structure than one can imagine from the palynological spectra. However, it remains unclear, what exactly is reflected by numerous peaks of the Eemian. We do not know whether they are due to climatic fluctuations or redistribution of isotopes reflecting plastic flows of alternating melt and subsequently frozen waters. In other words, do many peaks reflect processes of dynamic metamorphism of ice? The problem is undoubtedly of significant interest. It is still impossible to solve it on base of available data.

Chapter 14
Results of Paleoclimate Studies

Application of physical and biostratigraphic methods enabled establishing short-term climatic fluctuations in the Late Pleistocene and Holocene in the Arctic latitudes. It is still impossible now to reconstruct their succession and scale. Nevertheless, some of them are of great interest, since they give an idea about the high climate dynamics stemmed from migration and influence of different air masses as well as the ocean level fluctuations. As noted above, the macrofauna records the short-term invasions of the Arctic waters into the paleoshelf areas immediately after and before the Boreal transgression. The invasions brought about climatic coolings. Similarly, invasions of the Atlantic water caused the short-term warmings.

For example, the recent investigations of the Arctic Ocean revealed several short-term warming episodes in the Middle Pleistocene and the Eo-Valday. During the last interglacial and the Holocene climatic optimum the temperature rise was so significant that materials obtained from the Lomonosov Ridge suggest existence of seasonal sea ice (Nowaczyk et al., 1992). In the Barents Sea no glaciomarine sediments were accumulated during the Holocene climatic optimum, and a zone of the winter sea ice was very limited. At the same time glacial cupolas may disappear in the Franz-Joseph Land (Lavrushin et al., 1990).

Dynamical climatic changes in Greenland and the Norwegian Sea were caused by the temporal-spatial distribution of surface waters. Penetration of the Atlantic waters brought about climatic conditions close to the interglaciation ones in the time intervals corresponding to MIS 5a and MIS 1. Short-term intrusions of warm Atlantic water were recorded for MIS 5d; MIS 5c; and MIS 3. Cold glacial and, partially, interstadial climatic situations correspond to MIS 5b; MIS 4; and MIS 2 (Henrich, 1998). Sometimes surface water temperatures fell by 8–10°C in the North Atlantic (Ruddiman et al., 1986).

The distribution of the warm Atlantic surface water masses caused sea ice melting of different intensity, as illustrated well by data from the Norwegian-Greenland basin. In the paleo-Gulf Stream zone significant amounts of melt waters occurred in the time spans corresponding to MIS 1; MIS 5a; and MIS 5e. The ice melting was substantially less significant at the same time spans in the north of the Greenland Sea and the Fram Strait (Baumann et al., 1995). These facts are indicative of regional variations of climatic changes in the Arctic regions.

M.A. Levitan, Y.A. Lavrushin, *Sedimentation History in the Arctic Ocean and Subarctic Seas for the Last 130 kyr*, Lecture Notes in Earth Sciences 118,
DOI 10.1007/978-3-642-00288-5_14, © Springer-Verlag Berlin Heidelberg 2009

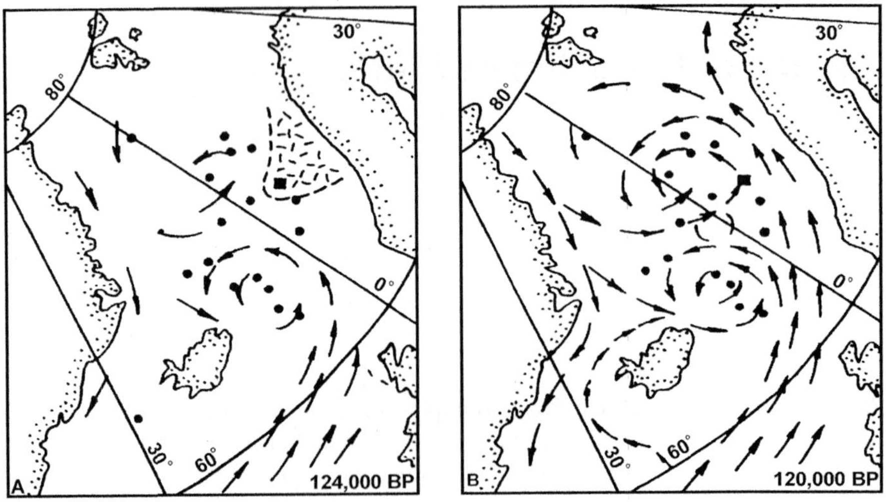

Fig. 14.1 Surface currents during the Eemian Interglaciation (Haake and Pflaumann, 1989). *a* – currents of MIS 5e beginning; *b* – currents of MIS 5e end. Area of distribution of pack sea ice in the Vøring Plateau area is hatched

Studies in the Fram Strait allowed dating the glacial events in Spitsbergen as 140–130, 112–110, 75–50 and 20–10 ka ago.

The paleoceanographic data (Fig. 14.1) on rapid changes in the oceanic currents at the Vøring Plateau, Norway (Haake and Pflaumann, 1989) are of interest. The authors reconstructed a system of oceanic currents on the base of thorough analysis of foraminiferal assemblages. So, the Gulf Stream which played the role of peculiar "stove" not only for Europe, but also for the Arctic latitudes, underwent serious changes during 4 ka. This caused considerable climatic changes at least in the Arctic shelf seas and decreased a role of floating sea ice in sedimentation (Chistyakova, 1997).

The oxygen–isotope studies of foraminifers from the Mikulino sediments of the Arkhangyelsk region suggest positive temperatures of bottom waters in this paleobasin (Lavrushin et al., 1990; Khoreva et al., 1991). The published paleogeographic reconstructions (Lavrushin and Spiridonova, 1995; Pleshivtzeva, 1972) are based on palynological data and do not deal with paleoclimatic indices.

In general, noticeable temperature rises occurred in the Eemian (Mikulino) Interglaciation in the high latitudes: in winter by 10–12°C, in summer by 6–8°C. According to I. I. Borzenkova, in the high latitudes the temperature difference between the Eemian and the present time reached 7.6–4.8°C in summer and 8.0–6.5°C in winter (Zubakov, 1986). At that time the climate was less continental.

The Late Pleistocene climatic variations are usually defined by the position of the Polar Front in the western Arctic regions. The works of the Scandinavian researchers in the northeastern part of the Norwegian Sea showed intensive and long shifts of the Polar Front during the last 50 ka. There was an open basin at MIS 3, the Polar

Front and a zone of mixture the Arctic and Polar water masses at the MIS 3–MIS 2 boundary, and an ice-free basin 20–15 ka ago.

Studies of the complicated post-Eemian history of the Scandinavian ice sheet were carried out by the international scientific team (Baumann et al., 1995), who developed a new concept of the ice sheet dynamics by analyzing moraines and sections of the bore holes drilled in the Norwegian Sea (IRD and $CaCO_3$ content). They distinguished two stages of the ice sheet growth in the Early Valday, which correspond to MIS 5d and MIS 5b and are separated by the Brerup Interstadial (MIS 5c). The end of the Early Valday is correlated with the Redestall Interstadial (MIS 5b). In the Middle Valday the ice sheet experienced a noticeable growth 63–54 ka ago (the growth began at MIS 4 and reached a maximum at early MIS 3) and 47–43 ka ago (MIS 3; the Lashamp–Olbi paleomagnetic event). The Olesunn Interstadial is fixed 38.5–32.5 ka ago. The beginning of the Late Valday glaciation is related to the terminal Middle Valday (Fig. 14.2).

According to the published map of I. I. Borzenkova (Zubakov, 1986), 18 ka ago the deviation of summer temperatures from the present-time ones was different in the western and eastern high latitude Arctic regions. The greatest deviation was recorded in the West Arctic, where it ranged from –22 to –18°C in the Canadian-Greenland sector. Temperature of the North Atlantic surface water was 6–10°C lower than the present-time one, and the sea ice cover was extended up to 50°N.

Detailed chronostratigraphic studies in the Severnaya Zemlya revealed that in the Late Pleistocene a considerable growth of ice sheet occurred 120–170, 70–80, 12–19 ka ago (Makeev et al., 1992).

The mentioned map shows that 18 ka ago summer temperatures in the eastern part of the Arctic deviated by 12–10°C from the present-time ones. Thus, in this region during the last glaciation the annual temperature was lower than the present one by approximately 10–12°C.

Thus, the Late Pleistocene climate of the Arctic regions was characterized by periodic development and disappearance of continental and sea ice cover as well as variable thickness of subsurface glaciation. The development of continental glaciation coincided with glacioeustatic regressions, and its degradation was synchronous with glacioeustatic transgressions.

Subsurface glaciations gave rise to permafrost deposits of considerable thickness. There are different manifestations of the glaciation. The most impressive are thick ice veins. Climatic conditions of their formation are interpreted in different ways: on the base of "emotions" and oxygen-isotope data. The former interpretation suggests that the thick syngenetic ice veins were formed at winter temperatures of –70 and even –100°C, but according to the oxygen-isotope data, winter temperatures were 5–15°C lower than the present-time ones (Vasilchuk et al., 1985). In the seaside lowlands of Yakutiya the warmings caused thermokarst processes (Lavrushin, 1963) while the coolings increased thickness of frozen ground.

The terminal Late Pleistocene (the Late Glacial) was characterized by the most intensive climatic fluctuations of short time, which reflected extreme instability of environments. Climatic changes of low taxonomic rank were recorded for the interval of 15–11 ka ago in the western part of the Arctic Ocean and the adjacent

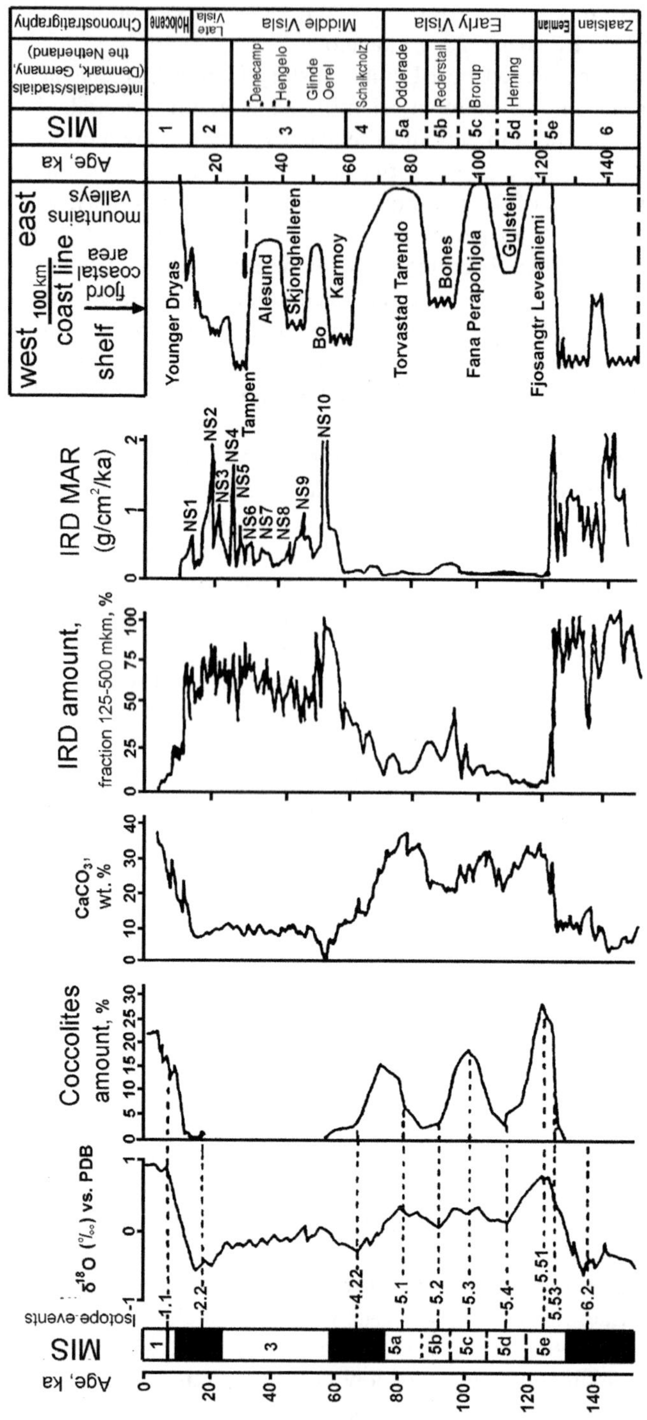

Fig. 14.2 Stadial-interstadial events of the Late Pleistocene in the West Scandinavia and their reflections in sediments of the adjacent shelf (cal. age scale) (Baumann et al., 1995)

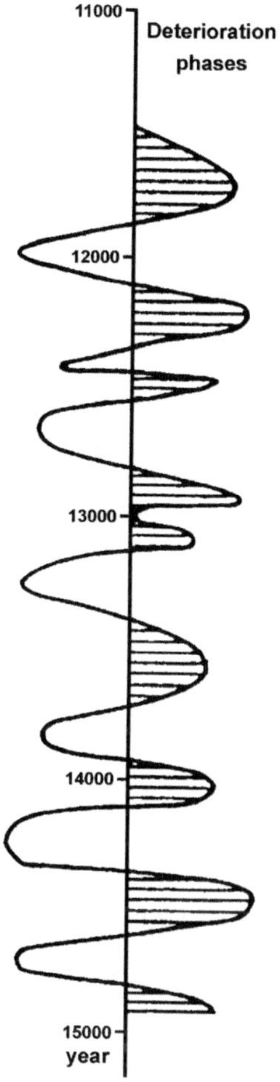

Fig. 14.3 Climate changes in the West Arctic between 15 000 and 11 000 years ago (by Yu.A. Lavrushin)

seas (Fig. 14.3): warming episodes occurred 14600, 14200, and 13200 years ago (Karpuz, 1992) and the lowest temperatures of surface water masses was fixed for the intervals of 12900–12800, 12500–12400, 12300–12000, 11900–11500 years ago (the terminal Bølling – Allerød and the Younger Dryas) (Karpuz and Jansen, 1992).

The complex studies of bottom sediments in the central part of the Arctic Ocean allowed reconstructing the paleoceanographic environments of the Late Quaternary deglaciation (Stein et al., 1994a). Analyses of $\delta^{18}O$, AMS ^{14}C of *N. pachyderma* tests as well as geochemical and sedimentological studies of the Gakkel and Lomonosov ridges and the Amundsen Basin found out considerable supply of melt waters 15700 and 7200 cal. yrs. BP. These data are well consistent with the temperature rises, degradation of continental ice covers and reduction of frozen ground areas in the periglacial regions in Northern Eurasia, Canada, and Alaska.

In the Younger Dryas the Arctic conditions existed in the Greenland-Norwegian Basin. In northern Norway the advances of glaciers were recorded for the Younger Dryas (10.4 and 11.2 ka ago) and the glacier shifts occurred during the Holocene deglaciation (10.2–10.1; 9.9; 9.6; and 9.3 ka ago) (Andersen et al., 1995). In Greenland sharply unstable climatic environments existed 14–10 ka ago.

The dynamic glacial events took place in the Franz-Joseph Land. Glaciers almost completely disappeared in the Allerød and revived in the Younger Dryas (Rubini′s stage) (Grossvald et al., 1973).

The Late Glacial climatic events in Eurasia can be described only in fragments. As showed by A.A. Velichko et al. (1994) and V.A. Klimanov (1994), in Taimyr the mean temperatures of July and January were close to the present-time ones during the Allerød phase and 2–3°C lower in the maximum of the Younger Dryas cooling; precipitation being less by 100–150 mm. In the Sverdrup Islands the Allerød period was distinguished by higher summer temperatures and lower winter temperatures. Periglacial flora prevailed there during the Younger Dryas. The seaside lowlands of Yakutiya were covered 14–15 ka ago by the Arctic tundra vegetation, i.e., bushy tundra and forest-tundra during the Allerød period and dominating tundra landscapes during the Younger Dryas. In the Kotelny Island there existed shrubby vegetation 12200–12500 years ago, while the long interval of 12.2–10 ka ago was characterized by more continental climate, increasing amount of xerophytes, and by disappearance of tree pollen from spectra. All available data on the Allerød–Younger Dryas climatic fluctuations were summarized in the cartographic form by V.A. Klimanov (1994).

The short-term climatic changes in the high latitudes were described in many publications. It is impossible to observe all of them herein. Therefore attention should be focused on reconstructed quantitative indicators of paleoclimatic conditions and on qualitative characteristics of inadequately studied events and regions.

One more point should be kept in mind when considering the Holocene climate variations. Unlike the climatic events of the time intervals described above, those of the Holocene lasted several hundred years (or less). Therefore taxonomically they can be referred to microclimatic events. Nevertheless, their effect may be of not only regional, but also a global scale.

The isotope-biostratigraphic studies in the North Atlantic revealed that short-term drops of surface water temperature occurred approximately 8200, 5800 and 1400 cal. yrs. BP. The importance of these data lies in the fact that the West Arctic has been experienced and experiences now a powerful influence of both Arctic and Atlantic air masses. Therefore these data can contribute to genetic interpretation of the climatic events.

A warming event took place in the Boreal. The Gulf Stream penetrated into the Bear Island area of the Barents Sea. The forest taiga vegetation prevailed on the Pechora Sea coast (Spiridonova and Lavrushin, 1997). Mean temperatures of July and January rose by 3–4°C and 2–3°C respectively in Taimyr. The warming was also recorded in the eastern areas excluding the Kotelny Island.

A lack of materials prevents subdivision of the Atlantic and other phases of the Holocene into climatic episodes. The most noticeable warming occurred about 5000 years ago (Svendsen et al., 1992). In Spitsbergen, plant remains from the lacustrine sediments suggest that mean temperatures of July in the interval of 3000–4000 years ago were 1–2°C higher than the present ones.

This interval in the Barents Sea can be described only in general terms. Here a warming brought about almost complete disappearance of sea floating ice even in wintertime (Lavrushin et al., 1990), as inferred from very small content of IRD in bottom sediments of this age (Fig. 14.4).

A complicated concept of the Holocene glacial events in the Franz-Joseph Land, which were caused by climatic changes, was developed by M.G. Grossvald et al. (1973). The reduction and even complete disappearance of glaciers was hypothesized for the interval of 8.5–2.5 ka ago. A cooling and formation of glacial cupolas (the Sedov Stage) are assumed to be about 2.5 ka ago. A short-term degradation of glaciers approximately 1000 years ago was succeeded by their advance (the Victoria Stage). A temperature rise up to +3°C was reconstructed for the area of

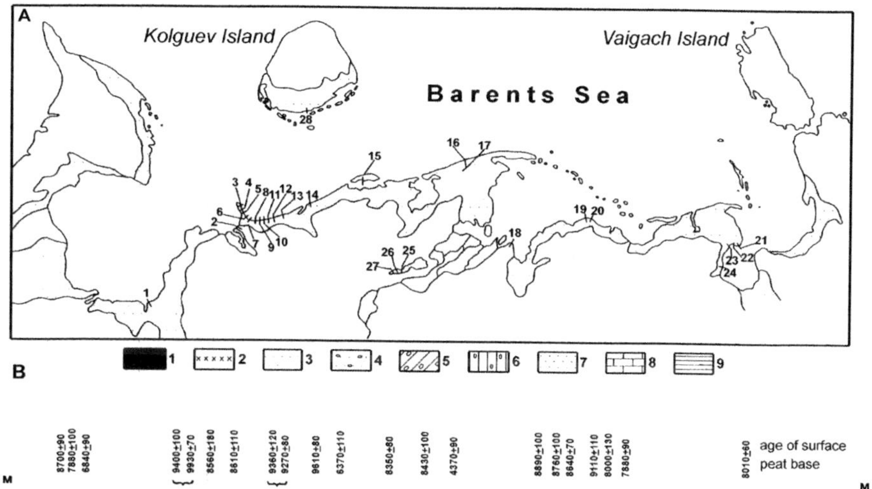

Fig. 14.4 Structure of coastal late glacial terraces (Lavrov and Potapenko, 2005). *a* – location of main sedimentary sections; *b* – terrace structure: 1 – peat, 2 – peat under moraine deposits, 3 – sand, 4 – sand with gravel and pebble, 5 – boulder clays, 6 – flow till, 7 – sand and silt ("kame" sediments), 8 – limestone, dolomite, 9 – clay, siltstone

Khaipudyrskaya Inlet and Taimyr. This warming was less pronounced in northeastern Asia, probably because of influence of the Arctic air masses but the climatic situation was far from stable (Velichko et al., 1994).

In the marine environments the Holocene optimum is usually related to intensification of the Atlantic waters advection (in particular, a rise of their volume and temperature). The optimum was pronounced 8.7–7.7 cal. kyrs. BP in the coastal waters of western and northern Spitsbergen (Salvigsen et al., 1992), 7.8–6.8 cal. kyrs. BP in Franz Victoria Trough (Duplessy et al., 2001), 8.0–7.3 cal. kyrs. BP in the Kara Sea (Stein et al., 2002b), 9.5–5.5 cal. kyrs. BP in the Chukchi Sea (a peak around 8.0). The penetration of the Atlantic waters into the Laptev Sea 10.7–9.2 cal. kyrs. BP (Polyakova et al., 2005) may cause a warming in the Preboreal. Ostracod records from the central area of the Arctic Ocean indicate that a layer of the intermediate Atlantic waters increased in thickness upward in the Middle Holocene (Cronin et al., 1995).

In the Early Subboreal the global cooling caused a sharp change in the climatic conditions of the regions studied. In the Barents Sea the hydrological picture became similar to the present-time one. As a result, the cold East Spitsbergen current appeared. The influence of the Atlantic air masses and, correspondingly, the Gulf Stream was brought to the minimum, whereas the effect of the Arctic air masses increased. Tundra landscapes spread widely along the Barents Sea coasts (Spiridonova and Lavrushin, 1997). The Early Subboreal cooling was pronounced in all Arctic regions.

Repeated climate variations during the Subboreal and Subatlantic are also recorded in the Arctic latitudes of Eurasia. Their quantitative parameters varied between different areas. A possible general trend is a climate warming between 4 and 3 ka ago, which was fixed by many authors.

Very few data were accumulated on the climatic changes in the last millennium. Of interest are results of studying the glacier shifts in the north of Spitsbergen. For the last 2500 years five short episodes of glacier advance have been established there (Fürrer and Stapfer, 1991; Fürrer, 1992). Comprehensive studies of bottom sediments of the Russkaya Gavan' (the fjord on the western coast of Northern Island of the Novaya Zemlya Archipelago, into which the outlet the Shokalsky Glacier falls), revealed two advance–retreat cycles of this glacier during the last millennium: 1400–1700 years ago and since 1700 until the present time (Zeeberg et al., 2003).

Chapter 15
Particularities of Sedimentation Processes Within the Continental Blocks and Marine Basins

Deglaciation Peculiarities

The fundamental differences in the processes of sedimentogenesis appear in the studied environments at the stage of glacial cover degradation. At this time the glacioeustatic transgression had a significant effect on the process of degradation in the limits of glacial shelves that actively acted on the destruction of the ice cover which waters penetrated through the cracked zones, formed bays with the ice shores, and thus contributed to the breakdown of the lateral part of the ice massif. Accumulation of glacioturbidites of different type and also the sediments of the slurry flows, which are connected with avalanche sedimentation of varve-like rhythmic-layered sediments, has occurred in the Late Glacial period (Chistyakova and Lavrushin, 2004).

Several factors acted on land to influence acceleration of the glacier degradation. The first one is the above noted surge development in a number of areas along the European North. The second one is the simultaneous destruction of glacial cover by waters of glacioeustatic transgression and ice-dammed lakes. The role of the latter was reduced to breakthrough and wash-out of ice weirs and mass catastrophic discharge of waters of ice-dammed reservoirs.

One more fact should be examined specially. The role of icebergs in the destruction of the limb zone of glacial cover has not been emphasized above. From our point of view, the glacial covers, which extended on the larger part of the Barents and Kara Seas due to their shallowness, were not floating, but moved directly along the glacial bed. Therefore the process of iceberg formation was insignificant, and, correspondingly, the role of iceberg glacial-marine sediments in the accumulation was negligible. Floating sea ice, which could seize fragmental material, for example, from the island coast, which contour the Barents Sea played a large role in this environment. The role of icebergs increased only under the conditions of the deep water fjords, similar to those, which are located on Spitsbergen and in Scandinavia.

A few words are worthy to mention about the ecological role of slurry flows. This role is reduced to two circumstances. First of all, because of the large inflow of melt glacial waters, the bays shaped near the edge of the ice sheet, and in spite of rising sea level, have been filled to the significant degree with fresh water

M.A. Levitan, Y.A. Lavrushin, *Sedimentation History in the Arctic Ocean and Subarctic Seas for the Last 130 kyr*, Lecture Notes in Earth Sciences 118,
DOI 10.1007/978-3-642-00288-5_15,

in the course of the Late Glacial transgression that prevented the penetration of marine biota. A huge amount of slurry material also did not contribute to entry of marine organisms because of the high turbidity, which hampered photosynthesis. As a result, a sufficiently short-term ecological hiatus has appeared. Analogous phenomena were observed in some fjords of Spitsbergen, where only some pioneer forms of polychaets were able to penetrate the waters, enriched by slurry material. Figure 15.1 demonstrates the schematic diagram of the sedimentation zone dynamics and zones of marine ecosystem of the trough glacier which terminated in one of the fjords of Western Spitsbergen Island in the transgressive (A) and regressive (B) glacial stages.

As a whole, it is necessary to emphasize its most important special features. The destruction of glacial cover occurred in the shallow glacial shelves in the course of the Late Glacial transgression at a high rate. Not only the surface, but also the extreme submarine ablation of glacial cover contributed to above process. As a

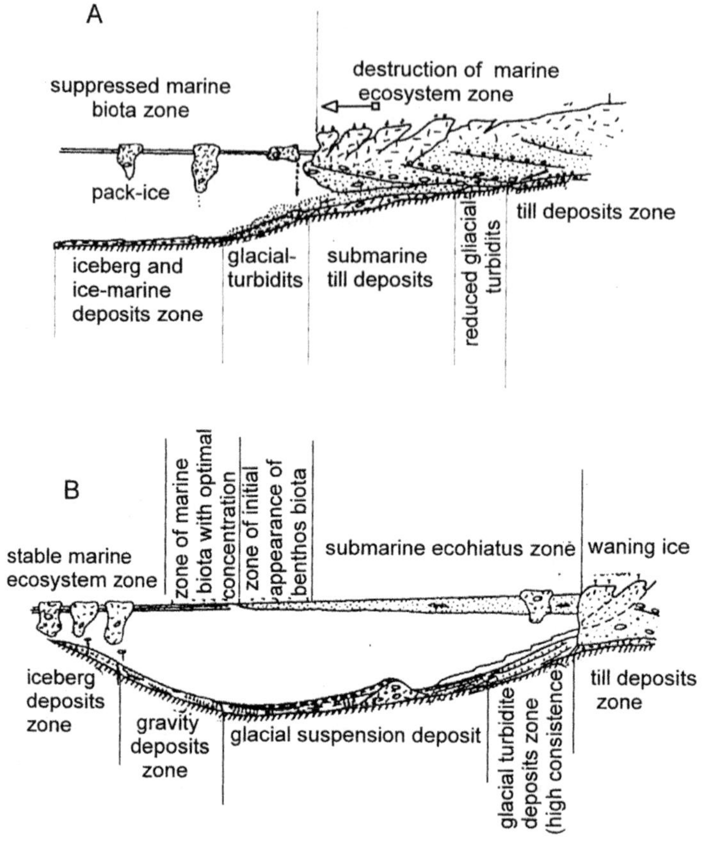

Fig. 15.1 Sedimentation zones and ecological conditions during glacial expansion (**A**) and waning (**B**) (fjord variant)

result, in the nascent Late Glacial basin the specific sedimentation processes has prevailed, consequently, the most important special feature of which was the predominance of glacioturbidites. Among the latter a significant role belonged to the slurry-turbid flows, which caused the avalanche-type of sedimentogenesis (Lavrushin, 2005).

This sedimentation process is connected with the accumulation of the rhythmically layered sediments with the straight gradation layering. The maximum thickness of sediments reaches 70 m (Chistyakova and Lavrushin, 2004). L. Polyak with co-authors (1995) established that their accumulation occurred into two short-term stages of the glaciomarine sedimentation. The first stage covered the interval of 12.7–12.2 ka (^{14}C age) – Bølling, and the second stage is 10.5–9.9 ka corresponding to the second half of the Younger Dryas–beginning of Preboreal, which in turn is equivalent to two extreme pulses of the glaciomarine sedimentation in the Norwegian Sea in the course of the Late Glacial transgression (Polyak et al., 1995).

In recent years the published materials of A.S. Lavrov and L.M. Potapenko (2005) allowed to reconstruct the more detailed dynamics of Late Glacial transgression. According to the data of the above researchers, along southeastern coast of the Barents Sea almost everywhere from the Czhöshskaya Guba to the Khaypudyrskaya Guba so called coastal terrace is extended for 800 km, its maximum altitude reaches 30 m; in the direction to the east it is reduced (Fig. 14.4). Presence of the interlayers of the unequal-age peat bogs in the offshore-marine sediments, which testify about the regressive short-term phases in the development of the Late Glacial transgression, is one of the important special features of the structure of this terrace. The analysis of the available numerous radiocarbon dates of these peat bogs (Lavrov and Potapenko, 2005), and also the published materials on the adjacent regions (Polyak et al., 1995) made it possible to develop the more detailed dynamics of changes in the level of the Barents Sea during the Late Glacial transgression (Fig. 15.2).

First of all, it was established that its maximum level was at about 11 ka, i.e., in the Allerød Period. More lately the intermittent-lowered trend of transgression has appeared which reached its minimum about 10.3 ka. This major regressive phase named by us as Varandey Event had an age of approximately 10.5–10.3 ka. This

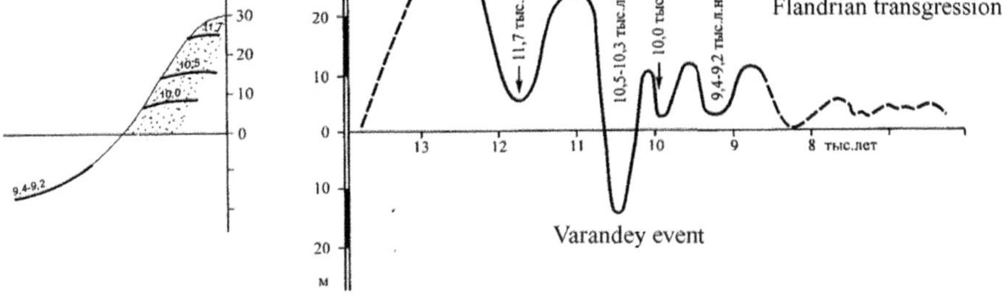

Fig. 15.2 Barents sea level changes during post-glacial time

event was established on the basis of the peat bog revealed on the sea bottom at the depth of approximately 15 m near the Varandey Settlement, which reflects the lowest position of the sea level. Furthermore, two additional short-term regressive phases were established, which age, judging by the datings of peat bogs, was about 10.0–9.7 and 9.3–9.0 ka. In general, the mentioned phases of decrease in the level of Late Glacial transgression is most likely to be connected with the short-term global cooling, which significantly slowed down the processes of glacial cover degradation. In addition, an older regressive phase is observed at the level of approximately 11.7 ka.

It is necessary to recall that the active destructive action of the glacier waters of Late Glacial transgression is one of the important special features of the degradation of those glacial cover parts, which were located in the shelves. These waters from the ice cracks along with a simultaneous rise of sea level formed the numerous molded edges, which deeply penetrated into the shelf glacial cover, which in turn contributed to the increase of its degradation (Grosswald, 1973). A similar process of destroying shelf ice is related to one of the types of the submarine ablation.

Finally, it is important to note one more extreme event of the Late Glaciation period. In this case, the discussion deals with the catastrophic descents of pondage by the last glaciation of the lake reservoirs, which arose in the river valleys. Similar ponds existed in the valleys of Pechora, Ob′, and Yenisei rivers, and their descent was the complementary factor, which contributed both to the destruction of glacial cover and to extension in the shelf of an enormous quantity of terrigenous material, which formed the significant fans. Thus, for example, there are the sandy islets of so called Gulyaevskie Koshki in the estuarine part of the Pechora River in the Pechora Sea that are the preserved fragment of one of the above fans.

Thus, let us bring some sums. The uniqueness of deglaciation process of the Arctic shelves was connected not only with the climate action of which in the Arctic was less intensive than in the middle latitudes, but also with the waters of the Late Glacial transgression. The rate of glacial cover destruction proved to be catastrophic; the changes in the relationship of dry land and sea proved to be similar by the rate of the described above process. The calculations show that the water area of the Barents Sea (including the Pechora River) reappeared approximately 2 ka ago. In this case, in this short time interval in the Arctic latitudes the shelf glacial covers were destroyed and the sea penetrated far into land – from the west to the east – from the break of continental slope for more than 1000 km. As it was mentioned before, a significant influence on the destruction of glacial cover also rendered the catastrophic descent of ice-dammed lake reservoirs, which, however, were of a local value. Finally, the internal dynamics of the Late Glacial transgression, complicated by the repeated short-term decreases and rises of sea level, was of significant interest.

Now let us briefly examine changes in the Late Glacial time in the relationship of dry land and sea in the Arctic region, which did not go through the mainland glaciation. For the most part, the greatest amount of data is related to the Laptev Sea, where international Russian-German expedition has been working for years.

Furthermore, according to the data of Russian researchers on Kotelny Island and Novosibirskie Islands, and adjacent from the east the territory of the Primorie Lowland, only the terrestrial sediments are found in the Late Pleistocene formations. The characteristic property of these sediments is their frozen state and content of different types of ice, part of which has a significant thickness. According to the data from boring in the bottom of the Laptev Sea the terrestrial sediments are laid under the thin cover of marine sediments.

The analysis of available material made it possible to establish that, judging by the existing dates, for the period of 800–1000 years the waters of the Arctic Ocean have penetrated into the limits of the continental land approximately for 500 km, forming the aquatory of the Laptev Sea. A similar catastrophic change in the relationship between land and sea was caused by three factors: by raising sea level at the end of the Late Glacial transgression, with the intensity of the processes of the abrasion of frozen rocks, which is characteristic up to the present as well, and also by propagating of the Gakkel Ridge, actually by tectonic processes. It is possible that in the beginning of Younger Dryas the processes of propagating became more intensive and, as a result of it, the rift-like depression was formed, where the penetration of transgression into the depth of land occurred as the ingressive bay, which contributed to the increase in the intensity of the abrasion processes of the permafrost rocks. The next impulse of propagating possibly took place in the late Holocene.

Facies Variability during Glaciations, Deglaciations, Interglacials

Facies types of the maximum development of glaciations, deglaciations and interglacials in the continent-ocean profile strikingly differ from each other.

In fact, it is possible to assume a maximum increase in the level of ocean, the noticeable sea transgressions and the essential displacement of coastline into the continents during interglacial periods (Fig. 15.3). In this case, facies differences in sedimentation basins were relatively insignificant and were reduced, in essence, to the traditional dependence of marine sedimentation processes from the distance on the source of sedimentary material, by bathymetry (relief) and with the degree of sea ice development.

In the periods of the maximum glacial development, on the contrary, the lowest position of sea level, the regression, the displacement of coastline sometimes up to the break of the present shelves were typical. In the limits of the non-glacial shelves, alluvial plains with the appropriate drainage network and alluvium were actively developed. In the regions of the continental glacier development the complexes of continental-glacial sediments were formed, changed on the facies profile by the complex of the terrestrial periglacial sediments, which pass into the sediments of marine periglacial (sometimes shelf and more frequently sediments of continental slope and rise). Deep-water sediments in the collection of genetic types noticeably differ from the sediments of the continental margins. In this case sedimentation rates

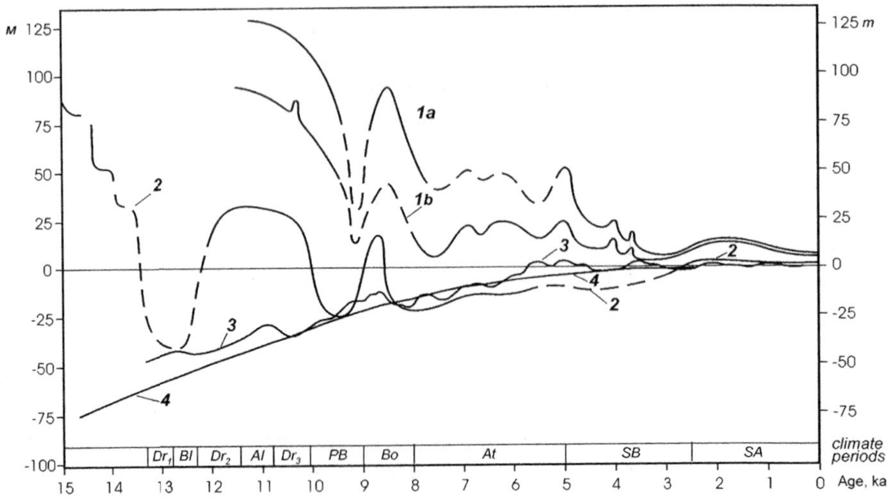

Fig. 15.3 Movements of the Barents and White Seas coastal lines (Lavrov and Potapenko, 2005). 1 – Kola Peninsula (Koshechkin, 1975) – highest (1*a*) and lowest (1*b*) position; 2 – Malozemel'skaya and Bol'shezemel'skaya tundra (Arslanov et al., 2005); 3, 4 – sea level: 3 – according to (Fairbanks, 1989), 4 – according to (Shepard, 1973)

noticeably grew. The marine ice sheets often merged with the continental glacial covers. In those regions, where the continental glaciers were absent, in the studied epochs the permafrost rocks reached a maximum development, while in the adjacent deep-water regions sedimentation rates were reduced.

During deglaciation sea level rose, rapid transgression occurred coastline was displaced in the direction of the land. Correspondingly, the drainage network existed in the non-glacial shelves was reduced. Melting water in the form of high-density flows carried the enormous masses of terrigenous material into sedimentation basins.

Geological History of the Arctic Ocean Sea Ice during the Last 60 ka

At present the ideas about the thermodynamic stability of the sea ice cover of the Arctic Ocean are widely spread among the significant group of researchers. Moreover, in some works ice cover existence is supposed for a period of several million years. However, important changes in the structure of water column, in the marine biota and in the ice cover of ocean are noted in the last decade in connection with the global warming of climate. With respect to the latter, as was shown by space survey, recently the enormous polynyas began to appear in the Arctic Ocean up to the North Pole, which reflects the sudden event of the partial, but rapid destruction of the sea ice cover of the Arctic Ocean.

The analysis of the existing publications, in which this problem is discussed, showed that in the geological history of the Arctic Ocean sea ice cover the episodes of the extreme decrease of ice-covered areas have repeatedly appeared, having a larger scale than those arisen at present. The main indicators of this process were the layers of bottom sediments enriched by IRD (Heinrich layers in the North Atlantic (Heinrich, 1988), and same type layers in the Nordic Seas – HA (Dahlgren and Vorren, 2003), and in the Arctic Ocean – AL (Darby et al., 2002)), and the so called highly productive layers with the Atlantic species of planktonic foraminifers. Highly productive (HP) layers fix the repeated (in composition of planktonic species) processes of the Atlantic Water advection into the high-latitude regions of the Atlantic Ocean, which contributed to an increase in the intensity of melting processes of significant sea ice volumes and icebergs, near-surface reduction of salinity and, correspondingly, the vertical stratification of water column. It seems that duration of individual HP layer formation does not exceed several ten or hundred years.

The ideas about the sudden processes of the Atlantic Water advection and the rapid climatic changes in the polar part of the Atlantic Ocean during MIS 2–4 are in detail presented in (Dokken and Hald, 1996). In the course of the advections of surface Atlantic Water into the high-latitude Arctic region, between the parts of the ocean, which were covered with sea ice and relatively warm ice-free, the significant temperature gradient could be generated, which contributed to strengthening gale activity and to growth of the circulation intensity of surface water mass. In the final this also contributed to the destruction of sea ice.

As far as the mentioned Heinrich events are concerned, in the studied sediment cores of Late Pleistocene in the northeastern part of the Atlantic Ocean six intervals of the increased IRD content (fraction 180–3000 mm) were established. Planktonic foraminifers were studied, among which the following foraminifers characteristic of different water masses were identified: *Neogloboqudrina pachyderma* (sin.) – surface cold waters; surface warm waters: *G. truncatulinoides*, *G. scitula*, *G. ruber* and *G. hirsute*; surface moderate (intermediate) waters: *G. inflata*. As a whole, the Heinrich events (with duration ca. 1–2 ka) reflect highly dynamic variations of the Earth climate, at least in the Arctic and Subarctic regions as well as glaciodynamic processes in the continental ice sheets, not always caused by climate. Nevertheless, these events are connected with the emergence of the large volumes of sea ice in the Arctic and Subarctic regions of the Atlantic. This was caused by strengthening of cold East Greenland Current, and, correspondingly, it is possible to assume the advection of Polar Water into the temperate latitudes.

The duration of the advections of Atlantic and Polar Waters was different according to the published geochronological materials. But what is very important, these advections, according to obtained data, coincided, with thermochrons and kriochrons of the Middle Valday of the Russian Plain (Table 15.1).

In conclusion let us bring the brief sums:

1. The Atlantic-Arctic oceanic-atmospheric climate machine determined the basic trends of the development of the natural situation in the boreal (moderate) belt of the adjacent land in the recent geological past.

Table 15.1 Geochronological correlations of high productive events, Heinrich events (Bonds et al. 1993; Shaw, 1995, Dokken and Hald, 1999; Hald, 2001) and the Middle Valday Interstadials of the Russian Plain (by Yu.A. Lavrushin)

High productive events of the Atlantic Ocean (HP), age (kyr B.P.)	Heinrich events, age (kyr B.P.)	The Middle Valday interstadials of the Russian Plain, age (kyr B.P.)
	Hc (X-XII centuries)	
	Ha (5,0)	
	H0 (10,0)	
	H1 (13,5 - 14,5)	
HP1 (14,5 - 19,5)		
	H2 (19,4 - 21,6)	Gmelin (19 - 21)
HP2 (22,5 - 29,5)	H3 (26,6 - 29,1)	Danube-Bryansk (24 - 29)
HP3 (34,5 - 37,5)	H4 (33 - 39,1)	Kashin (34 - 38)
HP4 (ca. 42,5 - 47,5)	H5 (ca. 43 - 50)	Grazhdansky prospect (40 - 44) [Glinde, the Netherlands] (40 -50)
HP5 (ca. 49,5 - 51,5)		
HP6 (ca. 54 - 60)		[Oerel, the Netherlands] (55 - 58)
	H6 (ca. 60 - 66)	

Age, kyr: 0, 10, 20, 30, 40, 50, 60, 70

2. The most important processes, which had an effect on the directions of the environment conversions, were advections of Atlantic Water (AAW) into the Arctic Ocean and the advections of the Polar Water (APW) into the Atlantic. Disturbance in the continuity of sea ice cover was connected with the penetration of the Atlantic Water, while on the land it was related to climate warming. The sufficiently clear correlation between the advections and the topographical-climatic events in the Russian Plain is geochronologically established.
3. There are two types of the events in AAW and APW:

    a) duration of several thousand years:
    b) duration of several ten or hundreds years.

4. The first type is related to the topographical-climatic conversions; the second type is connected with the weather and insignificant climatic events.
5. Discontinuity of AAW, their different duration, and also topographical reconstructions in the temperate latitudes of land give grounds to assume the changing value of the related heat flux brought into the high-latitude regions. Furthermore, the routine of AAW manifestation makes it possible to speak about the high dynamics of the West Spitsbergen Current. The curtailment of AAW events testifies about the discontinuity of the geological history of this current. The different AAW types were are connected with the destruction of the sea ice cover in the

Arctic Ocean and penetration of the Atlantic Water into the seas of the eastern margins of Eurasia.

APW have contributed to extension of the large volumes of sea ice from the Arctic Basin. The subsequent advections are confidently geochronologically correlated with the climate cooling on the territory of the Russian Plain.

AAW and APW controlled the migration of the Polar Front.

6. The main proxies of indicated paleoceanographic processes are increases of IRD amount (>150 mkm) at the initial stage and planktonic foraminifer content.

Intercoupling of Atmo-, Hydro-, Cryo-, Bio-, and Lithospheres

There is the most common assumption that the history of sedimentation in enormous examined region for the last 130 ka was mainly determined by climatic changes. The latter, in turn, produced changes in the interacted covers: atmo -, hydro- , cryo- and to sedimentosphere. In the inner and marginal sedimentation basins, changes in the temperature and circulation of atmosphere played probably the leading part. In the studied high-latitude regions of Northern Hemisphere it is difficult to overestimate the role of glaciation history (continental, mountain, marine): its emergency, development and curtailment. The alternation of interglacial and glacial periods composes the essence of natural process in the Quaternary period in the high latitudes. We attempted to illustrate this on the base of the last 130 ka.

We have mentioned above that in the shallow shelves of Eurasia the possibilities of studying the history of sedimentation in the times, which preceded the second isotope-oxygen stage, were limited by the natural and technical reasons. However, revealed by the authors and their associates the special features of changes of the largest rivers discharge at the end of the Pleistocene-Holocene in the shelf of North Asia indicate the high degree of similarity that can be explained only by common for the enormous regions climatic changes. Accordingly, the attention should be payed to the qualitative similarity of the histogram showing the beginning of an increase in the largest peat swamps of West Siberia and the typical form of the graph of the river drainage change for the Ob and Yenisei Rivers.

At the same time the history of Holocene drainage for the rivers of the European North of Russia and for the rivers of the north of Siberia is different: in the first case (according to present rare data) its uneven growth is observed (Sidorchuk et al., 1999), and in the second – the uneven weakening. Probably, this difference is caused by reverse trends in Holocene history of the nourishment of rivers in both regions: as a whole, in the course of time the role of atmospheric precipitations has been growing in comparison with the melting of permafrost rocks; however, in Siberia the role of the latter process in first half of Holocene was considerably higher than in the European part of the country because of incomparably larger volume of permafrost in the drainage basins of the Siberian rivers, and the humidity of the atmosphere was always higher in the north of European part.

We should focus attention on the revealed connections between the volumes of ice in Northern Hemisphere during the last climate cycle and sedimentation rates in the seas of the Subarctic and the Arctic Ocean (Levitan and Stein, 2005, 2008). Also, the relationships between the terrigenous and biogenic components in the marine and oceanic cross-sections of Upper Pleistocene-Holocene depending on climate changes are of great interest.

Thus, sedimentation rates are directly proportional to the volume of continental ice sheets of the Northern Hemisphere (for the identical temporary cross-sections) in the Nordic Seas and in the glacial continental margins (slopes and rises) of the Arctic Ocean. Actually, this circumstance attests to the fact that the precisely continental glaciers served (due to the bulldozer effect, the activity of icebergs and under-ice flows) as the basic suppliers of terrigenous material into the final basins of sedimentation and it is the main reason for circumcontinental zonality in the area of sea ice sedimentation.

In the deep-water bed of the Arctic Ocean (in its underwater ridges of different genesis and in the deep-water basins) and the seas of the eastern sub-Arctic the quite different regularities acted: sedimentation rates were inversely proportional to the volume of ice in the Northern Hemisphere (Levitan and Stein, 2007). Most likely, the volumes of sea ice grew in the epochs of glaciations and stadials and decreased in interglacial periods, terminations and interstadials. Accordingly, the rate of sea ice melting was inversely proportional to its volume. Thus, terrigenous sedimentation under these conditions was mainly caused by accumulation of the former cryosoles (here we do not examine the problem of terrigenous turbidites, so numerous along the continental margins and in the deep-water basins of the Arctic Ocean).

Consequently, one and the same reason – large climate changes, which were caused by changes in the insolation and also by Earth orbital and suborbital cycles, led to strengthening of terrigenous sedimentation in the cold epochs in those basins of the Northern Hemisphere, where prevailed continental (or shelf) ice sheets, also, to its weakening in those basins, where pack ices played the determining role.

In this case for all studied types of marine and oceanic sedimentation basins practically one and the same regularity is observed: the sediments of warmer epochs to one degree or another are always enriched by the biogenic remains (carbonate, in essence, in the western Subarctic and the Arctic Ocean and siliceous, as a rule, in the eastern Subarctic and in the Chukchi Sea). Most frequently this enrichment leads to a change in the component composition of sediments, but sometimes reflects only in a change in composition of micropaleontological remains (for example, in the west of the Nordic Seas, located under great influence of cold East Greenland Current).

The mechanisms of such changes in the lithological composition can be different depending on the changes in the relationship of the fluxes of the biogenic and abiogenic matter: here are the fluctuations of paleoproductivity, and the supply of terrigenous sedimentary matter from the land or from the sea ice and the icebergs, and, to some extent the degree of the dissolution of biogenic material should be taken into consideration. The fluctuations of paleoproductivity are connected, first of all, with the changes in amount of the nutrients supplied in photosynthesis zone. Basic mechanisms are the following: water exchange with the surrounding ocean basins and, first of all, the advection of Atlantic and Pacific Waters into the Arctic Ocean; the intensification of thermohaline circulation expressed both in the horizontal and vertical directions; the melting of sea ice and icebergs (development of the edge areas of the increased primary production) and the delivery of biogens by rivers.

As it was shown above, the action of all listed mechanisms grew in the warm epochs (termination, interglacial period, to a certain degree of interstadials) and weakened in the periods of glaciations. Concerning the actively investigated in recent years mechanism of nutrient delivery (first of all, Fe) to the surface of sedimentation basin by eolian way, it should be noted that atmosphere circulation sharply grew in the periods of glaciation and climate cooling. Thus, in the majority of the investigated regions the supply of nutrients by waterway played more important role compared with eolian supply. In any case in the studied region the discussion deals with the climatically caused changes in all these mechanisms, and future studies should be concentrated on the establishment and the quantitative assessment of its contribution, including with the aid of the mathematical modelling.

It is interesting that data on the Southern Ocean (still needed a statistical processing) testify about the manifestation of the already noted regularities and the high latitudes of Southern Hemisphere. For example, the content of biogenic (diatomaceous) opal is increased in the sediments of interglacial and interstadial periods and is reduced in the sediments, which were accumulated during the glaciations in the Ross Sea. In this case sedimentation rates grew in the cold epochs and were reduced in the warm epochs (Pistolato et al., 2006). Practically the same regularities were revealed during the study of mass accumulation rates of terrigenous material for the Scotia and Weddell Seas: its noticeable growth in cold epochs and decrease in the warm ones (Diekmann et al., 2003).

It is interesting to note that atmosphere circulation was strongly activated in the periods of maximum glaciations, and thermohaline circulation, on the contrary, weakened up to the almost complete curtailment during the especially powerful surges. The interglacial periods are characterized by opposite trends. Biosphere reacted to all these changes through a regular change of organism assemblages, which dwell under varying conditions according to the temperature, salinities, primary productions (nourishment), illumination (as the functions of the development of sea ice and turbidity). As a consequence, not only the relationship of biogenic and abiogenic components in bottom sediments, but also the intensity of biofiltration, biotransport and bioaccumulation were changed.

XXX

Thus, as a result of the undertaken study, it was possible to make the specific contribution to the solution of some problems connected with the influence of climatic changes in the high latitudes of Northern Hemisphere on the history of sedimentation of the last 130 ka (in some regions – within the larger period of time, in some – for shorter one). However, many of issues under discussion in connection with the uneven and often insufficient research works of a number of regions both on land and in the sea remain thus far unsolved. We hope that our book will help researchers to determine more clearly these unsolved problems and to make a contribution to their study.

References

Aagaard, K., Carmack, E.C. The role of sea ice and other fresh water in the Arctic circulation. J. Geophys. Res., 1989, Vol. 94 (C 10), pp. 14485–14498.

Aldahan, A.A., Shi Ning, P., Possnert, G. et al. ^{10}Be records from sediments of the Arctic Ocean covering the past 350 ka. Mar. Geol., 1997, Vol. 144, pp. 147–162.

Alekseeva, T.N., Svalnov, V.N. To the method of grain size analysis of fine-grained sediments. Oceanology, 2000, Vol. 40, N 2, pp. 304–312 (in Russian).

Anan'ev, G.S., Smirnova, T.I., Kulikov, O.A. et al. Quaternary history of the Northern Okhotsk region. In: Geographic studies of the Quaternary. Moscow: MSU, 1984, pp. 97–120 (in Russian).

Andersen, E.S., Ostmo, S.R., Forsberg, C.F., Lehman, S.J. Late- and post-glacial depositional environments in the Norwegian Trench, northern North Sea. Boreas., 1995, Vol. 24, pp. 47–64.

Anderson, P.M., Kotov, A.N., Lozhkin, A.V., Trumpe, M.A. Palynological and radiocarbon data from late Quaternary deposits of Chukotka. In: Late Quaternary Vegetation and Climate of Siberia and the Russian Far East (palynological and radiocarbon database). Ed. K.V. Simakov. Magadan: North East Science Center RAS, 2002, pp. 35–79.

Andreeva, I.A., Tarasov, G.A., Kukina, N.A., Krupskaya, V.V. Granulometric composition of Upper Quaternary sediments in the St. Anna Trough. Ber. Polarforsch., 1999, N 342, pp. 205–214.

Andreicheva, L.N. Pleistocene of the European north-east. Ekaterinburg: UrB RAS, 2002, 320p. (in Russian).

Andrew, J.A., Kravitz, J. Sediment distribution in deep areas of the Northern Kara Sea. In: Marine Geology and Oceanography of the Arctic Seas. Ed. Y. Herman. Berlin: Springer, 1974, pp. 231–256.

Andrews, J.T., Dunhill, G. Early to mid-Holocene Atlantic water influx and deglacial meltwater events, Beaufort Sea slope, Arctic Ocean. Quatern. Res., 2004, Vol. 61, pp. 14–21.

Andrews, J.T., Jennings, A.E., Cooper, T. et al. Late Quaternary sedimentation along fjord to shelf (trough) transect, East Greenland (68ºN). Geol. Soc. Spec. Publ., 1996, Vol. 197, N 111, pp. 153–166.

Andrews, J.T., Principato, S.M. Grain-size characteristics and provenance of ice-proximal glacial marine sediments. Geol. Soc. London Spec. Publ., 2002, Vol. 203, pp. 305–324.

Anthropogen of Taymyr (Eds N.V. Kind and B.N. Leonov). Moscow: Nauka, 1982, 182p. (in Russian).

Antonow, M., Goldschmidt, P.M., Erlenkeuser, H. The climate-sensitive Vesterisbanken area (central Greenland Sea): Depositional environment and paleoceanography during the past 250.000 years. Crzybowski Foundation Spec. Publ., 1997, N 5, pp. 101–118.

Aota, M., Uematsu, E. The study on the polar ocean and the Sea of Okhotsk. J. Geogr. 1989, Vol. 98 (5), pp. 70–82.

Arkhipov, S.A., Isayeva, L.L., Bespaly, V.G., Glushkova, O.A. Glaciation of Siberia and north-east USSR. Quatern. Sci. Rev., 1986, Vol. 5, pp. 463–473.

Arkhipov, S.A. Chronology of geological events of the West Siberia Late Pleistocene. Geol. Geophys., 1997, Vol. 38 (12), pp. 1863–1884 (in Russian).

Arkhipov, S.A. Main geological events of the Late Pleistocene (West Siberia). Geol. Geophys., 2000, Vol. 41 (6), pp. 792–799 (in Russian).

Arkhipov, S.A., Levchuk, L.K., Shelkoplyas, V.N. Marine Quaternary deposits of lower part of Ob River. In: Geochronology of the Quaternary. Moscow: Nauka, 1992, pp. 90–102 (in Russian).

Arslanov, Kh.A., Maximov, F.E., Kuznetsov, V.Yu. et al. Uranium-thorium age and paleobotanic characteristics of interglacial peat layer in the reference cross-section Rodionovo. In: Quarter-2005. Syktyvkar: Geoprint, 2005, pp. 21–23 (in Russian).

Artem'yev, V.E. Geochemistry of organic matter in the river–sea system. Moscow: Nauka, 1993, 204p. (in Russian).

Axenov, A.A., Dunaev, N.N., Ionin, A.S. et al. Arctic shelf of Eurasia in the Late Quarternary. Moscow: Nauka (Science), 1987, 278p. (in Russian).

Backman, J., Jakobsson, M., Løvlie, R. et al. Is the central Arctic Ocean a sediment starved basin?. Quatern. Sci. Rev. 2004, Vol. 23, pp. 1435–1554.

Balashov, Yu.A. Geochemistry of rare earth elements. Moscow: Nauka, 1976, 268p. (in Russian).

Barash, M.S., Matul, A.G., Khusid, T.A. et al. Microfauna in sediments of the eastern Sea of Okhotsk and Late Quaternary oceanology. In: Geology of the seas and oceans. V. I. M.: GEOS, 2003, pp. 194–195 (in Russian).

Barash, M.S., Os'kina, N.S. Distribution of tests *Globigerina pachyderma* (Ehr.) in the ocean sediments depending from surface sea water temperature. In: Marine Micropaleontology. Moscow: Nauka, 1978, pp. 196–205 (in Russian).

Bassinot, F., Baltzer, A., Chen, M.-T. et al. WEPAMA cruise MD 122/IMAGES VII. Plousane: Institut Polaire Français, 2003, 453p.

Bauch, H., Pavlidis, Yu.A., Matishov, G.G. et al. Pechora Sea coastal zone during Late Pleistocene and Holocene period (results in frame INTAS project N 1489–99). Abstracts of Papers, 5-th Workshop on LOIRA, Moscow: IORAS, 2002, pp. 14–17.

Bauch, H., Polyakova, Ye.I. Late Holocene variations in Arctic shelf hydrology and sea-ice regime: Evidence from north of the Lena Delta. Int. J. Earth Sci., 2000, Vol. 89, pp. 569–577.

Bauch, H.A. Paleoceanography of the North Atlantic Ocean (68–76º N) during the past 450 ky deduced from planktic foraminiferal assemblages and stable isotopes. Crzybowski Foundation Spec. Publ., 1997, N 5, pp. 83–100.

Bauch, H.A., Kassens, H., Erlenkeuser, H. et al. Depositional environment of the Laptev Sea (Arctic Siberia) during the Holocene. Boreas., 1999, Vol. 28 (1), pp. 194–204.

Bauch, H.A., Mueller-Lupp, T., Taldenkova, E. et al. Chronology of the Holocene transgression at the North Siberian margin. Glob. Planet. Change., 2001, Vol. 31, pp. 125–139.

Baumann, K.-H., Lackschewitz, K.S., Mangerud, J. et al. Reflection of Scandinavian ice sheet fluctuations in Norwegian Sea sediment during the past 150,000 years. Quatern. Res., 1995, Vol. 43, pp. 185–197.

Behre, K.E. Biostratigraphy of the last glacial period in Europe. Quatern. Sci. Rev., 1989, Vol. 8, pp. 25–44.

Behre, K.E., Plicht, J. Towards an absolute chronology for the last glacial period in Europe: Radiocarbon dates from Oerel, northern Germany. Veget. Hist. Archaeobot., 1992, Vol. 1, pp. 111–117.

Behrends, M. Rekonstruktion von Meereisdrift und terrigenem Sedimenteintrag im Spätquartär: Schwermineralassoziationen in Sedimenten des Laptev-See-Kontinentalrandes und des zentralen Arktischen Ozean. Ber. Polarforsch., 1999, N 310, 167s.

Behrends, M., Hoops, E., Peregovich, B. Distribution pattern of heavy minerals in Siberian rivers, the Laptev Sea and the eastern Arctic Ocean: An approach to identify sources, transport and pathways of terrigenous matter. In: Land-ocean systems in the Siberian Arctic: Dynamics and history. Berlin: Springer-Verlag, 1999, pp. 265–286.

Belov, N.A., Lapina, N.N. Bottom sediments of the Arctic Ocean. Leningrad: Hydrometeoizdat, 1961, 214p. (in Russian).

Berner, H. Mechanismen der Sedimenbildung in der Framstrasse, im Arktischen Ozean und in der Norwegischen See. Ph. D. Thesis. Bremen University, 1991, 167s.

Bezrodnykh, Yu.P., Nazarov, B.V. Structure, composition peculiarities, and accumulation history of Pliocene–Pleistocene sediment sequence of the Central Chukotka. In: Shelf Structure of the USSR Seas as Base for Evaluation of Engineer-Geological Conditions. Riga: VNIIMorgeo, 1984, pp. 56–63 (in Russian).

Birgel, D., Hass, H.C. Oceanic and atmospheric variations during the last deglaciation in the Fram Strait (Arctic Ocean): A coupled high-resolution organic-geochemical and sedimentological study. Quatern. Sci. Rev., 2004, Vol. 23, pp. 29–47.

Birgel, D., Stein, R. Northern Fram Strait and Yermak Plateau: Distribution, variability and burial of organic carbon and paleoenvironmental implications. In: The organic carbon cycle in the Arctic Ocean. Eds. R. Stein and R.W. Macdonald. Toronto: Springer-Verlag, 2004, pp. 279–294.

Biscaye, P.E. Mineralogy and sedimentation of recent deep-sea clay in the Atlantic Ocean and adjacent seas and oceans. Geol. Soc. Am. Bull., 1965, Vol. 76, pp. 803–832.

Bischof, J. Dropstones in the Norwegian-Greenland Sea: Indications of Late Quaternary circulation patterns? In: Geological History of the Polar Oceans: Arctic Versus Antarctic. Eds. U. Bleil and J. Thiede. NATO ASI Ser. C., 1990, Vol. 308, pp. 499–518.

Bischof, J. Ice drift, ocean and climate change. Berlin: Springer, 2000, 214p.

Bischof, J.F., Clark, D.L., Vincent, J.-C. Origin of ice-rafted debris: Pleistocene paleoceanography in the western Arctic Ocean. Paleoceanography, 1996, Vol. 11, pp. 743–756.

Biske, G.S. Quaternary sediments and geomorphology of Karelia. Petrozavodsk, 1959, 307p. (in Russian).

Bobylev, L.P., Kondratyev, K.Y., Johannessen, O.M. Arctic Environment variability in the Context of Global Change. Chichester: Springer Praxis Publishing, 2003, 428p.

Bogdanov, Yu.A., Murdmaa, I.O., Gurvich, E.G. et al. Study of the structure of upper part of the Barents Sea sediment sequence for description of sedimentation history and paleoceanographic speculations. In: Experience of system oceanological researchs in the Arctic. Moscow: Nauka, 2001, pp. 598–615 (in Russian).

Bol'shakov, V.A. New conception of the orbital theory of paleoclimate. Moscow: MSU, 2003, 256p. (in Russian).

Bond, G., Heinrich, H., Broecker, W. et al. Evidence for massive discharges of icebergs into the North Atlantic Ocean during the last glacial period. Nature, 1992, Vol. 360, pp. 245–249.

Boucsein, B. Organic carbon in Late Quaternary sediments: Responses to paleoenvironmental changes in the Laptev and Kara Seas. Ber. Polarforsch., 2000, N 365, 133p.

Bourtman, M.V., Levitan, M.A. New data on heavy mineral distribution in Southern Kara Sea bottom sediments. Ber. Polarforsch., 2002, N 419, pp. 87–100.

Broecker, W.S. The great ocean conveyor. Oceanography, 1991, Vol. 4, pp. 79–89.

Brook, E.J., Wolff, E., Dahl-Jensen, D. et al. The future of ice coring: International Partnerships in Ice Core Sciences (IPICS). PAGES News, 2006, Vol. 14 (1), pp. 6–10.

Burenkov, V.L., Vasil'kov, A.V. On the fresh waters influence at spatial distribution of the Kara Sea hydrological characteristics. Oceanology, 1994, N 5, pp. 652–661 (in Russian).

Calvo, E., Grimalt, J., Jansen, E. High resolution $U_{37}{}^{k}$ sea surface temperature reconstruction in the Norwegian Sea during the Holocene. Quatern. Sci. Rev., 2002, Vol. 21, pp.1385–1394.

Cheremisinova, E.A. Marine diatom flora of interglaciation sediments in valleys of Mga and Vytegra Rivers, and in the Ladoga Lake Basin. Abstr. Ph. D. thesis. Leningead: LSU, 1952, 25p. (in Russian).

Chernyavsky, V.I., Zhigalov, I.A., Matveev, V.I. Oceanological fundamentals of high biological productivity zones formation. In: Seas Hydrometeorology and Hydrochemistry. Vol. 2, N 9. The Sea of Okhotsk. St. Petersburg: Hydrometeoizdat, 1993, pp. 157–160 (in Russian).

Chistyakova, I.A. Sedimentation and history of Quaternary shallow-water glacial shelf (the Pechora Sea as an example). Abstr. Ph. D. thesis. Moscow: GIN RAS, 1997, 24p. (in Russian).

Chistyakova, I.A., Lavrushin, Yu.A. Suspensites of the last Late Glacial period on the territory of Russian Plain and adjacent shelves: Types, composition and sedimentogenesis peculiarities. Bulletin Comission Quaternary Period Research. N 65, Moscow: GEOS, 2004, pp. 36–43 (in Russian).

Clark, D.L., Arctic Ocean ice cover and its Late Cenozoic history. Geol. Soc. Amer. Bull., 1971, Vol. 82, pp. 3313–3324.

Clark, D.L., Magnetic reversals and sedimentation rates in the Arctic Basin. Geol. Soc. Amer. Bull., 1970, Vol. 81, pp. 3129–3134.

Clark, D.L., Whitman, R.R., Morgan, K.A. and Mackay, S.D. Stratigraphy and glacio-marine sediments of the Amerasian, Basin, central Arctic Ocean. Geol. Soc. Amer. Spec. Pap., 1980, Vol. 181, 57p.

Coachman, L.K., Aagaard, K. Physical oceanography of Arctic and subarctic seas. In: Marine Geology and Oceanography of the Arctic Seas. Ed. Y. Herman. New York: Springer, 1974, pp. 1–72.

Corner, G.D., Yevzerov, V.Y., Kolka, V.V., Møller, J.J. Isolation basin stratigraphy and Holocene relative sea-level change at the Norwegian-Russian border north of Nikel, northwest Russia. Boreas., 1999, Vol. 28 (1), pp. 146–166.

Cronin, T.M., Holtz, T.R., Stein, R. et al. Late Quaternary paleoceanography of the Eurasian Basin, Arctic Ocean. Paleoceanography, 1995, Vol. 10, pp. 259–281.

Dahlgren, K.I.T., Vorren, T.O. Sedimentary environment and glacial history during the last 40 ka of the Vøring continental margin, mid-Norway. Mar. Geol. 2003, Vol. 193, pp. 93–127.

Dai, M.-H., Martin, J.-M. First data on trace metal level and behavior in two major Arctic river-estuarine systems (Ob and Yenisei) and in the adjacent Kara Sea, Russia. Earth Planet. Sci. Lett., 1995, Vol. 131, pp. 127–141.

Darby, D.A., Bischof, J.F., Jones, G.A. Radiocarbon chronology of depositional regimes in the western Arctic Ocean. Deep-Sea Res. II., 1997, Vol. 44 (8), pp. 1745–1757.

Darby, D.A., Bischof, J.F., Spielhagen, R.F. et al. Arctic ice export events and their potential impact on global climate during the late Pleistocene. Paleoceanography, 2002, Vol. 17, pp. 15–17.

Darby, D.A., Naidu, A.S., Mowatt, T.C., Jones, G.A. Sediment composition and sedimentary processes in the Arctic Ocean. In: The Arctic Seas: Climatology, Oceanography, Geology, and Biology. Ed. Y. Herman. New York: VanNostrand Reinhold, 1989, pp. 657–720.

de Vernal, A., Hillaire-Marcel, C., Darby, D.A. Variability of sea ice cover in the Chukchi Sea (western Arctic Ocean) during the Holocene. Paleoceanography, 2005, Vol. 20, PA4018, doi:10.1029/2005PA001157.

Demina, L.L., Levitan, M.A., Politova, N.V. About speciation forms of several heavy metals in the bottom sediments of Ob and Yenisei estuaries. Geochemistry, 2006, N 2, pp. 212–226 (in Russian).

Dethleff, D., Rachold, V., Tintelnot, M., Antonow, M. Sea-ice transport of riverine particles from the Laptev Sea to Fram Strait based on clay mineral studies. Int. J. Earth Sci., 2000, Vol. 89, pp. 496–502.

Devyatova, E.I. Environment of the Late Pleistocene and its influence at people distribution in Severnaya Dvina Basin and Karelia. Petrozavodsk, 1982, 155p. (in Russian).

Diekmann, B., Fütterer, D.K., Grobe, H. et al. Terrigenous sediment supply in the polar to temperate South Atlantic: Land-ocean links of environmental changes during the Late Quaternary. In: The South Atlantic in the Late Quaternary: Reconstruction of Material Budgets and Current Systems. Berlin: Springer-Verlag, 2003, pp. 375–399.

Dittmers, K. Late Weichselian to Holocene sedimentation in the inner Kara Sea: qualification and quantification of processes. Ber. Polarforsch., 2006, N 523, 171p.

Dittmers, K., Niessen, F., Stein, R. Holocene sediment budget and sedimentary history of the Ob and Yenisei estuaries. In: Siberian Rivers Run-off in the Kara Sea. Characterisation, Quantification Variability and Environmental Significance. Eds. R Stein et al. Amsterdam: Elsevier, 2003, pp. 457–488.

Dittmers, K., Niessen, F., Stein, R. Late Weichselian fluvial evolution on the Southern Kara Sea shelf, North Siberia, Global Planet, Change, 2008, Vol. 60, pp. 327–350.

Dittmers, K., Schoster, F. Ancient riverine channels at the Kara Sea floor. Ber. Polarforsch., 2004, N 479, pp. 43–59.

Dittmers, K., Stein, R., Steinke, T. Core logging: Magnetic susceptibility. Ber. Polarforsch., 2001, N 393, pp. 89–91.

Dobrovolsky, A.D., Zalogin, B.S. The USSR Seas. Moscow: MSU, 1982, 192p. (in Russian).

Dokken, T. Last interglacial/glacial cycle on the Svalbard/Barents sea margin. Ph.D. thesis. University of Tromsø. 1995, 127p.

Dokken, T.M., Jansen, E. Rapid changes in the mechanism of ocean convection during the last glacial period. Nature, 1999, Vol. 401, pp. 458–461.

Dowdeswell, J.A., Whittington, R.J., Jennings, A.F. et al. An origin for laminated glaciomarine sediments through sea-ice build-up and suppressed iceberg rafting. Sedimentology, 2000, Vol. 47, pp. 557–576.

Dubinin, A.V. Geochemistry of rare earth elements in the ocean. Abstr. Dr. Sci. thesis. Moscow: GEOKHI, 2004, 54p. (in Russian).

Duplessy, J.-C., Cortijo, E., Ivanova, E. et al. Paleoceanography of the Barents Sea during the Holocene. Paleoceanography, 2005, Vol. 20, PA4004, doi: 10.1029/2004PA001116.

Duplessy, J.-C., Ivanova, E., Murdmaa, I. et al. Holocene paleoceanography of the northern Barents Sea and variations of the northward heat transport by the Atlantic Ocean. Boreas., 2001, Vol. 30, pp. 2–16.

Duryagina, D.A., Konovalenko, L.A. Palinology of Pleistocene depositis of the north-eastern European part of Russia. St. Petersburg: Nauka, 1993, 124p. (in Russian).

Eicken, H. The role of Arctic ice in transporting and cycling of terrestrial organic matter. In: The Organic Carbon Cycle in the Arctic Ocean. Eds. R. Stein and R. McDonald. Berlin: Springer, 2004, pp. 45–52.

Eisenhauer, A., Spielhagen, R.F., Frank, M. et al. ^{10}Be records of sediment cores from high northern latitudes: Implications for environmental and climatic changes. Earth Planet. Sci. Lett. 1994, Vol. 124, pp. 171–184.

Elverhøi, A., Andersen, E.S., Dokken, T. et al. The growth and decay of the Late Weichselian ice sheet in western Svalbard and adjacent areas based on provenance studies of marine sediments. Quatern. Res. 1995, Vol. 44, pp. 303–316.

Elverhøi, A., Dowdeswell, J., Funders, S. et al. Glacial and oceanic history of the polar North Atlantic margins: An overview. Quartern. Sci. Rev. 1998, Vol. 17, pp. 1–10.

Epstein, O.G. Outcrop Vas'tyansky Kon' in the Lower Pechora River – cross-section of the powerfull end-moraine in the active marginal zone of Novaya Zemlya Ice Sheet. Bulletin Comission Quaternary Period Research. 1990, N 59, pp. 14–28 (in Russian).

Epstein, O.G., Chistyakova, I.A. The Pechora Sea shelf in the Late Valday-Holocene: Main sedimentological and paleogeographic events. Bulletin Comission Quaternary Period Research. 2005, N 66, pp. 107–123 (in Russian).

Evans, J., Dowdeswell, J.A., Grobe, H. et al. Late Quaternary sedimentation in Kejser Joseph Fjord and the continental margin of East Greenland. Geol. Soc. Lond. Spec. Publ., 2002, Vol. 203, pp. 149–179.

Explanatory note to tectonic map of the Barents and Kara Seas. Moscow: ILMIS RAS, 1998, 178p. (in Russian).

Fahl, K., Stein, R., Gaye-Haake, B. et al. Biomarkers in surface sediments from the Ob and Yenisei estuaries and the southern Kara Sea: Evidence for particulate organic carbon sources, pathways, and degradation. In: Siberian River Run-off in the Kara Sea: Characterization, Quantification, Variability, and Environmental Significance. Proc. Mar. Sci. Vol. 6. Amsterdam: Elsevier, 2003, pp. 329–348.

Fairbanks, R.G. A 17,000-year glacio-eustatic sea level record: Influence of glacial melting rates on the Younger Dryas event and deep ocean circulation. Nature, 1989, Vol. 342, pp. 637–642.

Flower, B.P. Late Quaternary stable isotopic stratigraphy of Hole 910A, Yermak Plateau, Arctic Ocean: Relations with Svalbard/Barents Sea ice sheet history. Proc. ODP. Sci. Res., 1996, Vol. 151, pp. 445–454.

Forman, S.L., Lubinski, D.L., Zeeberg, J.J. et al. Postglacial emergence and Late Quaternary glaciation on northern Novaya Zemlya, Arctic Russia. Boreas., 1999, Vol. 28 (1), pp. 133–145.

Frolov, V.T. Genetic typification of marine sediments. Moscow: Nauka, 1984, 221p. (in Russian).

Frolov, V.T. Lithology. Book 3. Moscow: MSU, 1995, 352p. (in Russian).

Funder S., Hjort C., Landvik J.Y. et al. History of stable ice margin – East Greenland during the middle and upper Pleistocene. Quatern. Sci. Rev., 1998, Vol. 17, pp. 77–123.

Fürrer G. Zur gletschergeschite des Liefdefjords. Stuttgarter geographische Studien. 1992, Bd. 117. S. 267–278.

Fürrer, G., Stapfer, A. Zur Nacheiszeitlichen Gletschergesche des Liefdefjords (Spitzbergen). Geographica Helvetica, 1991, N 4. S. 147–155.

Gaiero, D.M., Depetris, P.J., Probst, J.-L. et al. The signature of river- and wind-borne materials exported from Patagonia to the southern latitudes: A view from REEs and implications for paleoclimatic interpretations. Earth Planet. Sci. Lett., 2004, Vol. 219, pp. 357–376.

Gard, G. Late Quaternary coccoliths at the North Pole: Evidence of ice-free conditions and rapid sedimentation in the Central Arctic Ocean. Geology, 1993, Vol. 21, pp. 227–230.

Gard, G., Backman, J. Synthesis of Arctic and Sub-Arctic coccolith biochronology and history of North Atlantic drift water influx during the last 500.000 years. In: Geological History of the Polar Oceans: Arctic versus Antarctic. Eds. U. Bleil and J. Thiede. Dordrecht: Kluwer, 1990, pp. 417–445.

Garetsky, R.G., Aizberg, R.E., Karabanov, A.K. et al. Latest tectonics and geodynamics of the Central Europe. Geotectonics, 1999, N 5, pp. 3–14 (in Russian).

Gataullin, V., Mangerud, J., Svendsen, J.I. The extent of the Late Weichselian ice sheet in the southeastern Barents Sea. Glob. Planet. Change., 2001, Vol. 31 (1–4), pp. 453–474.

Gataullin, V., Polyak, L. Morainic ridge complex, eastern Barents Sea. In: Glaciated Continental Margins: An Atlas of Acoustic Images. London: Chapman and Hall, 1977, pp. 82–83.

Gataullin, V., Polyak, L., Epstein, O., Romanyuk, B. Glacigenic deposits of the Central Deep: A key to the Late Quaternary evolution of the eastern Barents Sea. Boreas, 1993, Vol. 22, pp. 47–58.

Gataullin, V.N., Polyak, L.V. About presence of glacial deposits in the Central Deep of the Barents Sea. Repts. Acad. Sci., 1992, Vol. 314 (6), pp. 1463–1467 (in Russian).

Geological structure of the USSR and regularities of mineral deposits distribution. Vol. 9. The Soviet Arctic Seas. Leningrad: Nedra, 1984, 280p. (in Russian).

Geology of Quaternary deposits of the north-western European part of the USSR. Leningrad: Nedra, 1967, 342p. (in Russian).

Geomorphology and Quaternary deposits of the north-western European part of the USSR. Leningrad: Nauka, 1969, 253p. (in Russian).

Gol'bert, A.V., Gudina, V.I., Zudin, A.N. et al. Vast'yansky Kon' – reference cross-section of the northern Pechora region. In: Lithology and Environment of Quaternary Deposits Formation at the North Eurasia. Novosibirsk: Nauka, 1973, pp. 137–210 (in Russian).

Goldschmidt, P.M. The ice-rafting history of coarse-grained sediments in the Norwegian-Greenland sea for the last two glacial/interglacial cycles. Crzybowski Foundation Spec. Publ., 1997, N 5, pp. 119–139.

Gorbarenko, S. Stable isotope and lithologic evidence of Late-Glacial and Holocene oceanography of the northwestern Pacific and its marginal seas. Quatern. Res., 1996, Vol. 46, pp. 230–250.

Gorbarenko, S.A. Paleoceanography of the Far East Seas and the north-western Pacific in the Late Pleistocene and Holocene. Abstr. Dr. Sci. thesis. Moscow: IO RAS. 2004, 44p. (in Russian).

Gordeev, V.V., Martin, J.M., Sidorov, I.S., Sidorova, M.V. A reassessment of the Eurasian river input of water, sediment, major elements, and nutrients to the Arctic Ocean. Am. J. Sci., 1996, Vol. 296, pp. 664–691.

Gordeev, V.V., Rachold, V. River input. In: The Organic Carbon Cycle in the Arctic Ocean. Eds. R. Stein and R. Macdonald. Berlin: Springer, 2004, pp. 33–40.

Gordienko, P.A., Laktionov, A.F. Circulation and physics of the Arctic Basin waters. Annals of the International Geophysical Year. Oceanography, 1969, Vol. 46, pp. 94–112.

Gorshkova, T.N. The Kara Sea sediments. Repts All-Union Hydrobiol. Soc., 1957, Vol. 8. pp. 66–72 (in Russian).

Grantz, A., Phillips, R.L., Jones, G.A. Holocene pelagic and turbidite sedimentation rates in the Amerasia Basin, Arctic Ocean from radiocarbon age-depth profiles in cores. GeoResearch Forum., 1999, Vol. 5. pp. 209–222.

Grichuk, V.P. Ancient floras as paleontological base for stratigraphy of Quaternary deposits. In: Relief and Stratigraphy of the North-Eastern Russian Plain Quaternary Deposits. Moscow: Acad. Sci. USSR, 1961, pp. 25–71 (in Russian).

Grichuk, V.P. History of flora and vegetation of the Russian Plain in the Pleistocene. Moscow: Nauka, 1989. 182p. (in Russian).

Grichuk, V.P. Vegetation of the Russian plain in the early-middle Quaternary. Moscow: Acad. Sci. USSR, 1950. 202p. (in Russian).

Grigoriev, M.N., Rachold, V., Hubberten, H.-W., Schirmeister, L. Organic carbon input to the Arctic Seas through coastal erosion. In: The Organic Carbon Cycle in the Arctic Ocean. Eds. R. Stein and R. Macdonald. Berlin: Springer, 2004, pp. 41–45.

Gritzenko, I.I., Krapivner, R.B., Bondarev, V.N., Onishchenko, S.V. Geology and paleogeography of the Upper Cenozoic in the Barents Sea region. In: Problems of the Arctic Seas Cenozoic Paleoecology and Paleogeography. Apatity: KSC RAS, 1989, pp. 143– 149 (in Russian).

Grosswald, M.G. Late-Weichselian ice sheets in Arctic and Pacific Siberia. Quatern. Intern., 1998, Vol. 45/46, pp. 3–18.

Grosswald, M.G., Krenke, A.N., Vinogradov, O.N. et al. Glaciation of Franz-Iosef Land. Moscow: Nauka, 1973, 347p. (in Russian).

Grousset, F.E., Pujol, C., Labeyrie, L. et al. Were the North Atlantic Heinrich events triggered by the behavior of the European ice sheets? Geology, 2000, Vol. 28, pp. 123–126.

Gudina, V.I. Foraminifers, stratigraphy, and paleogeography of the northern USSR marine Pleistocene. Novosibirsk: Nauka, 1976, 125p. (in Russian).

Gudina, V.I., Yevserov, V.Ya. Stratigraphy and foraminifers of the Kola Peninsula Upper Pleistocene. Novosibirsk: Nauka, 1973, pp.7–63 (in Russian).

Gurevich, V.I. Recent sedimentogenesis and environment of the Arctic shelf of Western Eurasia. Oslo: Norsk Polarinstitut, 1995, 92p.

Gurvich, E.G., Isaeva, A.B., Demina, L.V. et al. Chemical composition of bottom sediments in the Kara Sea and Ob and Yenisei estuaries. Oceanology, 1994, N 5, pp. 766–775 (in Russian).

Gurvich, E.G., Lukashin, V.N., Lisitsin A.P., Kurinov, A.D. Rare earth elements and ittrium. In: Geochemistry of Hydrolyzate Elements. Moscow: Nauka, 1980, pp. 71–116 (in Russian).

Haake, F.W., Pflaumann, U. Late Pleistocene foraminiferal stratigraphy on the Vøring Plateau, Norwegian Sea. Boreas., 1989, Vol. 18 (4), pp. 343–356.

Hald, M. Climate change and paleoceanography. In: The Northern North Atlantic: A Changing Environment. Berlin: Springer, 2001, pp. 281–290.

Hald, M., Aspeli, R. Rapid shifts of the northern Norwegian Sea during the last deglaciation and the Holocene. Boreas., 1997, Vol. 26, pp. 15–28.

Hald, M., Dokken, T., Hagen, S. Paleoceanography of the European Arctic margin during the last deglaciation. In: Late Quaternary Paleoceanography of the North Atlantic margins. Geol. Soc. Spec. Publ., 1996. Boulder 111, pp. 275–287.

Hald, M., Husum, K., Vorren, T.O., et al. Holocene climate in the subarctic fjord Malangen, northern Norway: A multi-proxy study. Boreas., 2003, Vol. 32, pp. 543–559.

Hald, M., Kolstad, V., Polyak, L. et al. Late-glacial and Holocene paleoceanography and sedimentary environments in the St. Anna Trough, Eurasian Arctic Ocean. Palaeogeogr. Palaeoclimatol. Palaeoecol., 1999, Vol. 146, pp. 229–249.

Hald, M., Steinsund, P.I. Benthic foraminifera and carbonate dissolution in surface sediments of the Barents and Kara Seas. Ber. Polarforsch., 1996, N 212, pp. 285–307.

Hass, H.C. A method to reduce the influence of ice-rafted debris on grain-size record from the northern Fram Strait, Arctic Ocean. Pol. Res., 2002, Vol. 21, pp. 305–314.

Hass, H.C., Birgel, D., Didie, C. et al. Marine Geology. Ber. Polarforsch., 2000, N 368, pp. 27–56.

Hebbeln, D., Berner, H. Surface sediment distribution in the Fram Strait. Deep-Research I., 1993, Vol. 40 (9), pp. 1731–1745.

Hebbeln, D., Henrich, R., Baumann, K.-H. Paleoceanography of the last interglacial/glacial cycle in the polar North Atlantic. Quatern. Sci. Rev., 1998, Vol. 17, pp. 125–153.

Hebbeln, D., Wefer, G. Late Quaternary paleoceanography in the Fram Strait. Paleoceanography, 1997, Vol. 12, pp. 65–78.

Heinrich, H. Origin and consequences of cyclic ice Rafting in the Northeast Atlantic Ocean during the past 130,000 years. Quat. Res, 1988, Vol. 29, pp. 142–152.

Henrich, R. Dynamics of Atlantic water advection to the Norwegian-Greenland Sea – time-slice record of carbonate distribution in the last 300 ka. Mar. Geol., 1998, Vol. 145, pp. 95–131.

Henrich, R. Glacial/interglacial cycles in the Norwegian Sea: Sedimentology, paleoceanography and evolution of Late Pliocene to Quaternary Northern Hemisphere climate. Proc. ODP. Sci. Res., 1989, Vol. 104, pp. 267–291.

Henrich, R., Kassens, H., Vogelsang, E., Thiede, J. Sedimentary facies of glacial-interglacial cycles in the Norwegian Sea during the last 350 ka. Mar. Geol., 1989, Vol. 86, pp. 283–319.

Hevrøy, K., Lavik, G., Jansen, E. Quaternary paleoceanography and paleoclimatology of the Fram Strait/Yermak Plateau region: Evidence from sites 909 and 912. Proc. ODP. Sci. Res., 1996, Vol. 151, pp. 469–482.

Huang, B., Jian, Z., Wang, P. Paleoceanographic evolution recorded in the northern South China Sea since 4 Ma. Sci. in China. Ser. D. Earth Sci., 2005, Vol. 48 (12), pp. 2166–2173.

Ivanov, B.V. Types of andesitic volcanism of the Pacific mobile belt. Moscow: Nauka, 1990, 213p. (in Russian).

Ivanova, E.V. Benthic foraminifera in sediments from the southern Kara Sea: Preliminary results. Ber. Polarforsch., 2001, N 393, pp. 141–150.

Ivanova, E.V. Benthic foraminiferal assemblages of Kara Sea: Response to salinity variation during the Holocene. Progr. and Abstr. 4th Intern. Workshop SIRRO. Moscow: GEOKHI, 2003, pp. 17.

Ivanova, E.V. Holocene foraminiferal assemblages of Barents and Kara Seas: A response to paleoceanographic changes. Progr. and Abstracts EMMM'2002. Vienna, 2002, pp. 106–107.

Ivanova, E.V., Murdmaa, I.O., Duplessy, J.-C., Paterne, M. Late Weichselian to Holocene Paleo-environments of the Barents Sea. Glob. Planet. Change., 2002, Vol. 34 (3–4), pp. 209–218.

Jakobsson, M., Backman, J., Murray, A., Løvlie, R. Optically stimulated luminescence dating supports central Arctic Ocean cm scale sedimentation rates. Geochem., Geophys., Geosys., 2003, Vol. 4 (2). 1016 doi:10.1029/2002GC000423, pp. 1–11.

Jakobsson, M., Grantz, A., Kristoffersen, Y., Macnab, R. Physiography and bathymetry of the Arctic Ocean. In: The Organic Carbon Cycle in the Arctic Ocean. Eds. R. Stein and R. Macdonald. Berlin: Springer, 2004, pp. 1–5.

Jakobsson, M., Løvlie, R., Al-Hanbali, H. et al. Manganese and color cycles in Arctic ocean sediments constrain Pleistocene chronology. Geology, 2000, Vol. 28 (1), pp. 23–26.

Jansen, J.H.F., Kuijpers, A., Troelstra, S.R. Mid-Bruhnes climatic event: Long-term changes in global atmosphere and ocean circulation. Science, 1986, Vol. 232, pp. 619–622.

Johannessen, O.M. Brief overview of the physical oceanography. In: The Nordic Seas. Ed. B.G. Hurdle. Heidelberg: Springer, 1986, pp. 103–127.

John, B.S., Sugden, D.E. Coastal geomorphology of high latitudes. Progress in Geography. 1975, Vol. 7, pp. 53–132.

Johnsen, S.J., Clausen, H.B., Dansgaard, W. et al. The Eemian stable isotope record along the GRIP ice core and its interpretation. Quatern. Res., 1995, Vol. 43, pp. 117–124.

Johnson, G.L. Morphology and plate tectonics: The modern polar oceans. In: Geological history of the polar oceans: Arctic versus Antarctic. Eds. U. Bleil and J. Thiede. Dordrecht: Kluwer Academic Publishers, 1990, pp. 11–28.

Jünger, B. Deep water renewal in the Greenland Sea during the past 340.000 years. Geomar Report. 1994, Vol. 35, pp. 35–73.

Karpuz, N.K. A high-resolution diatom record of the last glaciation from the SE Norwegian Sea: Documentation of rapid climatic changes. In: Abstr. Fourth Intern. Conf. Paleoceanography. Kiel. 1992, pp. 160.

Karpuz, N.K., Jansen, E. Paleoceanographic development of the Greenland, Iceland and Norwegian Seas through the last 15 ka: The diatom and $\delta^{18}O$ evidence. In: Abstr. Fourth Intern. Conf. Paleoceanography. Kiel. 1992, pp. 160.

Kassens, H. Verfestigte Sedimentlagen und seismische Reflektoren: Frühdiagenese und Palä-Ozeanographie in der Norwegischen See. Ber. Sonderforschungsbereich. Univ. Kiel. 1990. Bd. 24. S. 1–117.

Khabakov, A.V. About indicies of gravel roundness. Sov. Geol., 1946. (10), pp. 98–99 (in Russian).

Khasankaev, V.B. Study of bottom stone matter as the source of information about composition of sea floor outcrops in the south-eastern Barents Sea. Lithol. Miner. Resour. 1978. N 3, pp. 118–120 (in Russian).

Khoreva, I.M., Vasil'ev, V.P., Lavrushin, Yu.A. Paleoecology of the Mikulino Sea on the North of the USSR European part. In: Geological and paleoecological Quaternary environments. Moscow: Nauka, 1991, pp. 57–67 (in Russian).

Khotinsky, N.A. Holocene of the Northern Eurasia. Moscow: Nauka, 1977, 198p. (in Russian).

Khusid, T.A., Korsun, S.A. Modern benthic foraminiferal assemblages in the Kara Sea. Ber. Polarforsch. 1996. N 212, pp. 308–314.

Kleiber, H.P., Niessen, F., Weiel, D. The Late Quaternary evolution of the western Laptev Sea continental margin, Arctic Siberia – implications from sub-bottom profiling. Glob. Planet. Change., 2001, Vol. 31, pp. 105–124.

Klimanov, V.A. Climate of the North Eurasia in late glacial time. In: Short-Period and Abrupt Environment-Climatic Changes in the Last 15 000 yrs. Moscow: IG RAS, 1994, pp. 61–93 (in Russian).

Klitgaard-Kristensen, D., Sejrup, H.P., Hafliclason, H. The last 18 kyr fluctuations in Norwegian Sea surface conditions and implications for the magnitude of climatic change: Evidence from the North Sea. Paleoceanography, 2001, Vol. 16 (5), pp. 455–467.

Knies, J., Matthiesen, J., Mackensen, A. et al. Effects of Arctic freshwater forcing on thermohaline circulation during the Pleistocene. Geol. Soc. Amer. Bull., 2007, Vol. 35, pp. 1075–1078.

Knies, J., Nowaczyk, N., Müller, C. et al. A multiproxy approach to reconstruct the environmental changes along the Eurasian continental margin over the last 150 000 years. Mar. Geol., 2000, Vol. 163, pp. 317–344.

Koç, N., Jansen, E., Hald, M., Labeyrie, L. Late glacial-Holocene sea surface temperatures and gradients between the North Atlantic and the Norwegian Sea: Implications for the Nordic heat pump. Geol. Soc. Spec. Publ., 1996, Vol. 111, pp. 177–185.

Kodina, L.A., Tokarev, V.G., Sukhoruk, V.I. et al. Preliminary results of geochemical investigations of sediment cores along the Yenisei profile (acoustic units I-II). Ber. Polarforsch., 2001, N 393, pp. 179–184.

Kodina, L.A., Tokarev, V.G., Vlasova, L.N., Korobeinik, G.S. Contribution of biogenic methane to ikaite formation in the Kara Sea: Evidence from the stable carbon isotope geochemistry. In: Siberian River Run-Off in the Kara Sea. Characterisation, Quantification, Variability and Environmental Significance. Amsterdam: Elsevier, 2003, pp. 349–375.

Köhler, S.E.I. Spätquartäre paläo-ozeanographische Entwicklung des Nordpolarmeeres und Europäischen Nordmeeres anhand von Sauerstoff- und Kohlenstoffisotopenverhältnissen der planktischen Foraminifere *Neogloboquadrina pachyderma* (sin.). Geomar Report. 1992, Vol. 13, 104p.

Kolesov, G.M. Determination of elements: Neutron-activation analysis in geochemistry and cosmochemistry. J. Analyt. Chem., 1994, Vol. 49 (1), pp. 55–66 (in Russian).

Kolstad, V. Paleomiljøendringer I St. Anna-renna fra siste istid og til I dag, belyst ved litolog foraminiferstratigrafi. Ph. D. Thesis. Univ. Tromsø, 1996, 131p.

Kordikov, A.A. The Kara Sea sediments. Repts NIIGA., 1953, Vol. 56, 142p. (in Russian).

Koshechkin, B.I. Movements of coastal line of the Barents and White Seas in the late- and postglacial time. News Acad. Sci. USSR. Ser. Geogr., 1975, N 4, pp. 91–100 (in Russian).

Kosheleva, V.A., Yashin, D.S. Bottom sediments of the Arctic Seas. St. Petersburg: VNIIO, 1999, 286p. (in Russian).

Kraus, M., Matthiessen, J., Stein, R. A Holocene marine Pollen Record from the northern yenisei Estuary (Southern Kara Sea, Siberia). Proceedings in Marine Science, 2003, Vol. 6, Elsevier, pp. 435–457.

Kremenetski, K.V., Velichko, A.A., Borisova, O.K. et al. Peatlands of the Western Siberian lowlands: Current knowledge on zonation, carbon content and late Quaternary history. Quatern. Sci. Rev., 2003, Vol. 22, pp. 703–723.

Kruk, R.W., Beets, Ch., Elliott, T. et al. U-series radiometric dating of mollusks from Eemian deposits in the Amsterdam Basin. In: The Eemian: Local sequences, global perspectives. Abstr. SEQS Symposium. Kerkrade, 1998, pp. 42.

Krupskaya, V.V., Levitan, M.A. Sediment types of surface sediments. Ber. Polarforsch, 2003, N 450, pp. 35–40.

Krylov, A.A. Lithology of recent sediments of the north-eastern Kara Sea. Abstr. Ph. D. Thesis. St. Petersburg SU, 2000, 19p. (in Russian).

Kubisch, M. Die Eisdrift im Arktischen Ozean ährend der letzten 250.000 Jahre. Geomar Report, 1992, Vol. 16, 100p.

Kuhlemann, J., Lange, H., Paetsch, H. Implications of connection between clay mineral variations and coarse grained debris and lithology in the central Norwegian-Greenland Sea. Mar. Geol., 1993, Vol. 114, pp. 1–11.

Kuijpers, A., Troelstra, S.R., Prins, M.A. et al. Late Quaternary sedimentary processes and ocean circulation changes at the Southeast Greenland margin. Mar. Geol., 2003, Vol. 195, pp. 109–129.

Kukina, N.A., Levitan, M.A., Tarasov, G.A. Distribution of light minerals in surface sediments of the St. Anna Trough. Ber. Polarforsch., 1999, N 342, pp. 134–138.

Kukla, G., McManus, J.F., Roussen, D.-D., Chuine, I. How long and how stable was the last interglacial?. Quatern. Sci. Rev., 1997, Vol. 16, pp. 605–612.

Kulikov, N.N. Sedimentation in the Kara Sea. In: Recent sediments of the seas and oceans. Moscow: Nauka, 1961, pp. 437–447 (in Russian).

Kurnosov, V.B., Murdmaa, I.O. Clay minerals in the recent sediments of the Sea of Okhotsk. Oceanology, 1978, N 4, pp. 671–680 (in Russian).

Kuznetsov, V.Yu., Arslanov, Kh.A., Kozlov, V.B. et al. Absolute age of buried peat from strathotypic cross-section "Mikulino" and parastrathotypic cross-section "Niznyaya Boyarshina" according to U-Th dating. Repts. 3 All-Russian Meeting on Quaternary period. Vol. 1. Smolensk. 2002, pp. 135–136 (in Russian).

Laberg, J.S., Vorren, T.O., Mienert, J. et al. Late Quaternary paleoenvironment and chronology in the Trænadjupet Slide area offshore Norway. Mar. Geol., 2002, Vol. 188, pp. 35–60.

Lackschewitz, K.S. Sedimentationsprozesse am Aktiven Mittelozeanischen Kolbeinsey Rücken (Nördlich von Island). Geomar Report, 1991, Vol. 9, pp. 16–116.

Landvik, J.Y., Bondevik, S., Elverhøi, A. et al. The last glacial maximum of Svalbard and the Barents Sea area: Ice sheet extent and configuration. Quatern. Sci. Rev., 1998, Vol. 17, pp. 43–75.

Larsen, E., Funder, S., Thiede, J. Late Quaternary history of northern Russia and adjacent shelves – a synopsis. Boreas., 1999, Vol. 28, pp. 6–11.

Lavrov, A.S., Potapenko, L.M. Neopleistocene of the north-eastern Russian Plain. Moscow: Aerogeologiya, 2005, 222p. (in Russian).

Lavrova, M.A. About marine interglaciation transgressions in the Pechora region. Sci. Repts LSU. Ser. Geogr., 1949b, Vol. 124 (6), pp. 28–49 (in Russian).

Lavrova, M.A. Quaternary geology of the Kola Peninsula. Moscow-Leningrad: Acad. Sci. USSR, 1960, 233p. (in Russian).

Lavrova, M.A. To the question about marine interglaciation transgressions in the Pechora region. Sci. Repts LSU. Ser. Geol., 1949a, N 6, pp. 17–29 (in Russian).

Lavrova, M.A., Troitsky, S.A. Interglaciation transgressions of the northern regions of Europe and Siberia. In: Chronology and climates of the Quaternary. Moscow: Nauka, 1960, pp. 124–136 (in Russian).

Lavrushin, Yu.A., Spiridonova, E.A. The Late Pleistocene events in the North of European Russia: Geology and paleoenvironments. In: Climate and Environmental Changes of the East Europe During Holocene and Late-Middle Pleistocene. Moscow: IG RAS, 1995, pp. 63–68.

Lavrushin, Yu.A. Alluvium of plain rivers of Subarctic belt and periglacial areas of continental glaciations. GIN Repts. Vol. 87. Moscow: Acad. Sci. USSR, 1963, 265p. (in Russian).

Lavrushin, Yu.A. Fundamentals of conception of Quaternary subaeral and shelf sedimentation of different density flows. In: Recent Problems of Geology. Moscow: Nauka, 2004, pp. 606–622 (in Russian).

Lavrushin, Yu.A., Alekseev, M.N. Arctic regions. In: Climate and Environmental Changes for the Last 65 Ma. Moscow: GEOS, 1999, pp. 21–42 (in Russian).

Lavrushin, Yu.A., Alekseev, V.M., Khasankaev, V.B. et al. To paleomarinology of the Holocene climate optimum in the north-western part of the Barents Sea. News Acad. Sci. Estonia. Geol., 1990, Vol. 39 (2), pp. 76–82 (in Russian).

Lavrushin, Yu.A., Epstein, O.G. Geological events of the Pleistocene at the north of the East Europe and in the southern Barents Sea. Bull. Comission Quatern. Res., N 64. Moscow: GEOS, 2001, pp. 35–60 (in Russian).

Lavrushin, Yu.A., Spiridonova, E.A., Kholmovoy, G.V. et al. Calendar-event scale of the Late Neopleistocene. Repts 3 All-Russian Meeting on Quaternary Period. Vol. 1. Smolensk, 2002, pp. 143–145 (in Russian).

Levchuk, L.K. Biostratigraphy of Upper Pleistocene of the northern Siberia according to foraminifers. Novosibirsk: Nauka, 1984, 128p. (in Russian).

Levitan, M. Facies variability of surface sediments along the Yenisei transect based on grain-size composition, heavy and light mineral data. Ber. Polarforsch., 2001, N 393, pp. 92–106.

Levitan, M., Stein, R. Influence of the Northern Hemisphere glaciation at sedimentation rates and composition of the Arctic Ocean and SubArctic Seas bottom sediments during the last climate cycle. Abstr. Intern. School Mar. Geol., Moscow: IORAN, 2005, Vol. 1, pp. 23–24.

Levitan, M.A. Composition of fraction >63 mkm of surface sediments from Ob, Taz, and Yenisei rivers and the southern Kara Sea. Ber. Polarforsch., 2002, N 419, pp. 80–86.

Levitan, M.A., Andreeva, I.A., Bourtman, M.V., Smirnova, L.S. Heavy minerals in Upper Quaternary sediments of the northern and eastern Kara Sea. Ber. Polarforsch., 1999b, N 342, pp. 214–228.

Levitan, M.A., Dekov, V.M., Gorbunova, Z.N. et al. The Kara Sea: A reflection of modern environment in grain size, mineralogy, and chemical composition of the surface layer of bottom sediments. Ber. Polarforsch., 1996, N 212, pp. 58–80.

Levitan, M.A., Ivanov, G.I., Bourtman, M.V. et al. Provenance of the Kara Sea surface sediments based on heavy mineral data. Ber. Polarforsch., 1999a, N 342, pp. 160–171.

Levitan, M.A., Kolesov, G., Chudetsky, M. Chemical characteristics of main lithofacies based on instrumental neutron-activation analysis data. Ber. Polarforsch., 2002, N 419, pp. 101–111.

Levitan, M.A., Krupskaya, V.V. Composition of fraction >125 mkm from surface sediments (on BP01 and BP02 data). Ber. Polarforsch., 2003, N 450, pp. 27–34.

Levitan, M.A., Krupskaya, V.V., Frolova, E.A., Vlasova, L.N. First results of the sediment studies. Ber. Polarforsch., 2004, N 479, pp. 44–54.

Levitan, M.A., Kuptsov, V.M., Romankevich, E.A., Kondratenko A.V. Some indication for late Quaternary Pechora River discharge: Results of vibrocorer studies in the southeastern Pechora Sea. Int. J. Earth Sci., 2000, Vol. 89, pp. 533–540.

Levitan, M.A. Super-long sediment cores. Nature, 2003, N 8, pp. 84–85 (in Russian).

Levitan, M.A., Arnold, M., Bourtman, M.V. et al. To the history of the Holocene sedimentation in the eastern Kara Sea. Oceanology, 2000a, N 4, pp. 614–620 (in Russian).
Levitan, M.A., Belyaev, N.A., Bourtman, M.V. et al. History of the Holocene sedimentation in the South Novaya Zemlya Trough. Lithol. Miner. Resour., 2003a, N 6, pp. 660–672 (in Russian).
Levitan, M.A., Bourtman, M.V., Dara, O.M. Lithology of the Upper Quaternary sediments. In: Pechora Sea. Experience of System Researchs. Moscow: More, 2003b, pp. 255–276 (in Russian).
Levitan, M.A., Bourtman, M.V., Demina, L.L. et al. Facies variability of the surface-layer sediments of Ob-Yenisei Shoal, and Ob and Yenisei estuaries. Lithol. Miner. Resour., 2005a, N 5, pp. 472–484 (in Russian).
Levitan, M.A., Bourtman, M.V., Demina, L.L. et al. History of the Holocene sedimentation in the southern Kara Sea. Lithol. Miner. Resour., 2004. N 6, pp. 651–666 (in Russian).
Levitan, M.A., Bourtman, M.V., Gorbunova, Z.N., Gurvich, E.G. Quartz and feldspars in surface layer of the Kara Sea bottom sediments. Lithol. Miner. Resour.. 1998. N 2, pp. 115–125 (in Russian).
Levitan, M.A., Byakov, A.F., Dmitrievsky, A.N. First finding of gas pockmark in the Russian Arctic shelf. Repts Russian Acad. Sci., 1999, Vol. 368, (3), pp. 364–367 (in Russian).
Levitan, M.A., Khusid, T.A., Kuptsov, V.M. et al. Types of Upper Quaternary cross-sections of the Kara Sea. Oceanology, 1994, N 5, pp. 776–788 (in Russian).
Levitan, M.A., Kukina, N.A. Mineral composition of the light fraction from Upper Quaternary sediments of Saint Anna Trough and its paleoceanographic interpretation. Lithol. Miner. Resour., 2002, N 3, pp. 306–315 (in Russian).
Levitan, M.A., Lavrushin, Yu.A., Stein, R. Outlines of sedimentation history of the Arctic Ocean and Subarctic Seas for the last 130 ka. Moscow: GEOS, 2007a. 404p. (in Russian).
Levitan, M.A., Luksha, V.L., Tolmacheva, A.V. Sedimentation history in the northern Okhotsk Sea for the last 1.1 Ma. Lithol. Miner. Resour., 2007b, N 3, pp. 227–246 (in Russian).
Levitan, M.A., Mityaev, M.M., Ivanov, V.V. Facies variability of Yermak Plateau Holocene sediments based on fraction >0.063 mm study. Lithol. Miner. Resour., 2000b, N 3, pp. 235–245 (in Russian).
Levitan, M.A., Murdmaa, I.O., Ivanova, E.V. et al. Sedimentation history of Vøring Plateau (the Norwegian Sea) for the last 25 kyr.. Lithol. Miner. Resour., 2005b, N 6, pp. 563–580 (in Russian).
Levitan, M.A., Musatov, E.E., Bourtman, M.V. Sedimentation history of Yermak Plateau during the last 190 kyr. Communication 1. Lithology and mineralogy of Middle Pleistocene-Holocene sediments. Lithol. Miner. Resour., 2002a, N 6, pp. 563–576 (in Russian).
Levitan, M.A., Musatov, E.E., Bourtman, M.V. Sedimentation history of Yermak Plateau during the last 190 kyr. Communication 2. Paleoceanographic interpretation. Lithol. Miner. Resour., 2002b, N 6, pp. 577–588 (in Russian).
Levitan, M.A., Nürnberg, D., Stein, R. et al. About role of cryosoles in accumulation of the recent Arctic Ocean sediments. Repts Russian Acad. Sci., 1995b, Vol. 344 (4), pp. 506–509 (in Russian).
Levitan, M.A., Roshchina, I.A., Tolmacheva, A.V. Holocene history of Ob riverine discharge on lithological and geochemical data. Geochemistry, 2007c, N 6, pp. 654–669 (in Russian).
Levitan, M.A., Stein, R. History of sedimentation rates in the Arctic Ocean during the last 130 ka.. Bulletin Comission Quaternary Period Research, 2007, N 67, pp. 33–43 (in Russian).
Levitan, M.A., Stein, R. History of sedimentation rates in the glacial zone of sedimentation during the last 130 ka. Lithol. Miner. Resour., 2008, N 1, pp. 74–86 (in Russian).
Levitan, M.A., Tarasov, G.A., Bourtman, M.V., Kukina, N.A. Mineral composition of surface-layer sediments of Saint Anna Trough. Oceanology, 1999, N 6, pp. 903–911 (in Russian).
Levitan, M.A., Wahsner, M., Nürnberg, D., Shelekhova, E.S. Average composition of clay mineral assemblages in the surface-layer sediments of the Arctic Ocean. Repts Russian Acad. Sci., 1995a, Vol. 344 (3), pp. 364–366 (in Russian).
Liciecki, L.E., Raymo, M.E. A Pliocene-Pleistocene stack of 57 globally distributed benthic $\delta^{18}O$ records. Paleoceanography, 2005, Vol. 20. Doi: 10.1029/

Lightfoot, P.C., Naldrett, A.J., Gorbachev, N.S. et al. Geochemistry of the Siberian Trap of the Norilsk area, USSR, with implication for the relative contributions of crust and mantle to flood basalt magmatism. Contrib. Mineral. Petrol., 1990, Vol. 104, pp. 631–644.

Lisitsin, A.P. Sea-ice and iceberg sedimentation in the Ocean: Recent and past. Berlin: Springer, 2002, 563p.

Lisitsin, A.P. Marginal filter of oceans. Oceanology, 1994, N 5, pp. 735–747 (in Russian).

Liyvrand, E.D. Age and correlation of deposits recovered in cross-sections Pas'va and Koleshki. In: Pleistocene Geology of the North-Western USSR. Apatity: KSC USSR Acad. Sci., 1981, pp. 72–86 (in Russian).

Lloyd, J.M., Kroon, D., Boulton, G.S. et al. Ice rafting history from the Spitsbergen ice cap over the last 200 kyr. Mar. Geol., 1996, Vol. 131, pp. 103–121.

Loseva, E.I. Middle Valday sea-lake at the west of Bol'shezemelskaya tundra. Bulletin Comission Quaternary Period Research, 1978, N 48, pp. 735–747 (in Russian).

Loutre, M.F., Berger, A. Marine isotope stage 11 as an analogue for the present interglacial. Glob. Planet. Change., 2003, Vol. 36, pp. 209–217.

Lubinski, D.J., Korsun, S., Polyak, L. et al. The last deglaciation of the Franz Victoria Trough, northern Barents sea. Boreas., 1996, Vol. 25, pp. 89–100.

Lubinski, D.J., Polyak, L., Forman, S.L. Deciphering the Latest Pleistocene and Holocene inflows of freshwater and Atlantic water to the deep northern Barents and Kara seas: Foraminifera and stable isotopes. Quatern. Sci. Rev., 2001, Vol. 20, pp. 1851–1879.

MacDonald, G.M., Velichko, A.A., Kremenetski, C.V. et al. Holocene treeline history and climate change across Northern Eurasia. Quatern. Res., 2000, Vol. 53, pp. 302–311.

Macdonald, R.W., Harner, T., Fyfe, J. et al. AMAP assessment 2002: The influence of global change on contaminant pathways to, within and from the Arctic. Oslo, AMAP, 2003, 65p.

Macdonald, R.W., Sakshaug, E., Stein, R. The Arctic Ocean: Modern status and recent climate change. In: The organic carbon cycle in the Arctic Ocean. Eds. R. Stein and R.W. Macdonald. Toronto: Springer-Verlag, 2004, pp. 6–20.

Magnus, S. Benthic Foraminifera in the Boreas Basin, Greenland Sea: Distribution and paleoceanographic reconstructions of the last 450 000 yeares. Ber. Polarforsch., 2000, N 373, pp. 1–137.

Makeev, V.M., Bol'shiyanov, D.Yu., Makakhovsky, D.B. et al. Stratigraphy and geochronology of the Severnaya Zemlya Pleistocene deposits. In: Geochronology of Quaternary Period. Moscow: Nauka, 1992, pp. 132–137 (in Russian).

Mangerud, J. Correlation of the Eemian and the Weichselian with deep sea oxygen isotope stratigraphy. Quatern. Intern., 1989, Vol. 3/4, pp. 1–4.

Mangerud, J., Svendsen, J.I., Astakhov, V.I. Age and extent of the Barents and Kara ice sheets in Northern Russia. Boreas., 1999, Vol. 28, pp. 46–80.

Map of Quaternary deposits of European part of the USSR and adjacent territories. Scale 1:1 500 000. Leningrad: VSEGEI, 1971 (in Russian).

Marina, M.M., Levitan, M.A., Lyutzarev, S.V. Accumulation history of organic carbon in the Holocene sediments of the north-western Bering Sea shelf. Oceanology, 1985, N 4, pp. 602–609 (in Russian).

Markov, K.K. Selected works. Paleogeography and the most recent deposits. Moscow: Nauka, 1986, 280p. (in Russian).

Martinson, D.G., Pisias, N.G., Hays, J.D. et al. Age dating and the orbital theory of the ice ages: Development of high-resolution 0 to 300 000 years chronostratigraphy. Quatern. Res., 1987, Vol. 27, pp. 1–29.

Maslov, A.V., Krupenin, M.T., Ronkin, Yu.L. et al. Fine alumosiliciclastic sediments of the stratho-typic cross-section of the South Ural Middle Rhyphean: Peculiarities of formation, composition and evolution of source provinces. Lithol. Miner. Resour., 2004, N 4, pp. 414–441 (in Russian).

Matishov, G.G. Ocean floor during the glacial period. Leningrad: Nauka, 1984, 176p. (in Russian).

Matishov, G.G., Matishov, D.G., Shchipa, E., Rissanen, K. Radionuclides in ecosystem of the Barents and Kara Seas region. Apatity: KSC RAS, 1994. 238p. (in Russian).

Matthiessen, J. Distribution of palynomorphs in surface sediments from the Ob and Yenisei estuaries (Kara Sea, Arctic Ocean). Ber. Polarforsch., 1999, N 300, pp. 222–235.

Matthiessen, J., Baumann, A. Dinoflagellate cyst records from the East Greenland continental margin during the last 15.000 years: Implications for paleoceanographic reconstructions. Grzybowski Foundation Special Publication, 1997, N 5, pp. 101–118.

Matthiessen, J., Baumann, K.-H., Schröder-Ritzrau, A. et al. Distribution of calcareous, siliceous and organic-walled planktic microfossils in surface sediments of the Nordic Seas and their relation to surface-water masses. In: The Northern North Atlantic: A Changing Environment. Berlin: Springer, 2001, pp. 105–128.

Matthiessen, J., Stepanets, O.V. Scientific cruise report of the Kara Sea expedition of RV ≪Academik Boris Petrov≫ in 1997. Ber. Polarforsch., 1998, N. 266, 102p.

Matul, A.G. Large-scale environmental changes in the North Pacific during the last one million years (on micropaleontological data). Materials of IV All-Russian Meeting on Quaternary period research. Syktyvkar. 2005, pp. 261–262 (in Russian).

McManus, J.F., Major, C.O., Flower, B.P., Fronval, T. Variability in sea-surface conditions in the North Atlantic – Arctic gateways during the last 140.000 years. Proc. ODP. Sci. Res., 1996, Vol. 151, pp. 437–444.

Mel'nikov, V.P., Spesivtzev, V.I. Engineer-geological and geocryological conditions of the Barents and Kara Seas shelves. Novosibirsk: Nauka, 1995, 195p. (in Russian).

Merkt, J. Varve chronology and palinology of the Late Glacial in north-west Germany from lacustrine sediments of Hämelsee in Lower Saxony. Quatern. Intern., 1999, Vol. 61, pp. 41–59.

Midttun, L. Formation of dense bottom water in the Barents Sea. Deep-Sea Res., 1985, Vol. 32, pp. 1233–1241.

Modern and Late Quaternary depositional environment of the St. Anna Trough area, northean Kara sea. Ber. Polarforsch., 1999, N 342, 245p.

Molodkov, A.N., Raukas, A.V., Makeev, V.M., Baranovskaya, O.F. To EPR-chronostratigraphy of the North Asia marine deposits and their correlation with Pleistocene events. In: Geochronology of Quaternary Period. Moscow: Nauka, 1992, pp. 41–46 (in Russian).

Molodkov, A., Raucas, A. ESR age of the Late Pleistocene transgressions in the eastern part of the White Sea coast. Geologija, 1998a, N 25, pp. 62–69.

Molodkov, A., Raukas, A. The Eemian in the south-eastern periphery of the Fennoscandian shield. In: The Eemian: Local Sequences, Global Perspectives. Abstr. SEQS Symp. Kerkrade, 1998b, pp. 58.

Moretzky, V.N. Distribution and dynamics of freshened waters in the Kara Sea. Repts AANII, 1985, Vol. 389, pp. 33–39 (in Russian).

Mudelsee, M., Schulz, M. The Mid-Pleistocene climate transition: Onset of 100 ka cycle lags ice-volume build-up by 280 ka. Earth Planet. Sci. Lett., 1997, Vol. 151, pp. 117–123.

Muhs, D.R. The timing and climate of the last interglacial in North America. In: The Eemian: Local Sequences, Global Perspectives. Abstr. SEQS Symp. Kerkrade, 1998, pp. 59.

Müller, C. Rekonstruktion der Paläo-Umweltbedingungen am Laptev-See-Kontinentalalrand während der beiden letzten Glazial/Interglazial-Zyclen anhand sedimentologischer und mineralogischer Untersuchungen. Ber. Polarforsch., 1999, N 328. 146s.

Müller, C., Stein, R. Grain-size distribution and clay-mineral composition in surface sediments and suspended matter of the Ob and Yenisei rivers. Ber. Polarforsch., 1999, N 300, pp. 179–187.

Murdmaa, I., Ivanova, E., Levitan, M. et al. Facies System of the Central and Easten Barents Sea since the last Glaciation to Recent. Mar. Geol., 2006, Vol. 230, pp. 275–303.

Murdmaa, I.O. Ocean facies. Moscow: Nauka, 1987, 304p. (in Russian).

Murdmaa, I.O., Ivanova, E.V. Postglacial sedimentation history in the Barents Sea shelf depressions. Lithol. Miner. Resour., 1999, N 6, pp. 142–154 (in Russian).

Musatov, E.E. Development of the Barents-Kara Sea shelves bottom relief in Cenozoic. Geomorphology, 1989, N 3, pp. 76–84 (in Russian).

Nam, S.I. Late Quaternary glacial history and paleoceanographic reconstructions along the East Greenland continental margin: Evidence from high-resolution records of stable isotopes and ice-rafted debris. Ber. Polarforsch., 1997, N 241, 154p.

Nam, S.I., Stein, R. Late Quaternary variations in sediment accumulation rates and their paleoenvironmental implications: A case study from the East Greenland continental margin. GeoResearch Forum, 1999, Vol. 5, pp. 223–240.

Nam, S.I., Stein, R., Grobe, H., Hubberten, H. Late Quaternary glacial-interglacial changes in sediment composition at the East Greenland continental margin and their implications. Mar. Geol., 1995, Vol. 122, pp. 243–262.

Naugler, F.P. Recent sediments of the East Siberian Sea. M.S. Thesis. Univ. Washington, Seattle, 1967, 71p.

Nees, S. High-resolution benthic foraminiferal records of the last glacial termination in the northern North Atlantic. Grzybowski Foundation Special Publication, 1997, N 5, pp. 167–197.

Neishtadt, M.I. History of forests and paleogeography of the USSR in the Holocene. Moscow: Acad. Sci. USSR, 1957, 403p. (in Russian).

Niessen, F., Dittmers, K. GeoChirp and ELAC sediment echograph profiling. Ber. Polarforsch., 2002, N 419, pp. 64–73.

Niessen, F., Kleiber, H.-P. Marine sediment echosounding using PARASOUND. Ber. Polarforsch., 1997, N 255, pp. 99–105.

Nørgaard-Pedersen, N. Late Quaternary Arctic Ocean sediment records: Surface ocean conditions and provenance of ice-rafted debris. Geomar Reports, 1997, N 65, 107p.

Nørgaard-Pedersen, N., Spielhagen, R.F., Erlenkeuser, H. et al. Arctic Ocean during the Last Glacial maximum: Atlantic and Polar domains of surface water mass distribution and ice cover. Paleoceanography, 2003, Vol. 18 (3). 1063, doi: 10.1029/2002PA000781.

Nørgaard-Pedersen, N., Spielhagen, R.F., Thiede, J., Kassens, H. Central Arctic surface ocean environment during the past 80.000 years. Paleoceanography, 1998, Vol. 13, pp. 193–204.

Notholt, H. Die Auswirkung der ≪NorthEastWater≫-Polynya auf die Sedimentation vor NO-Grönland und Untersuchungen zur Paläo-Ozeanographie seit dem Mittelweichsel. Ber. Polarforsch, 1998, N 275, 183p.

Nowaczyk, N.R., Baumann, M. Combined high-resolution magnetostratigraphy and nannofossil biostratigraphy for late Quaternary Arctic Ocean sediments. Deep-Sea Res., 1992, Vol. 39, Suppl. 2, pp. 567–601.

Nowaczyk, N.R., Frederichs, T.W., Eisenhauer, A., Gard, G. Magnetostratigraphic data from late Quaternary sediments from the Yermak Plateau, Arctic Ocean: Evidence for four geomagnetic polarity events within the last 170 ka of the Brunhes Chron. Geophys. J. Int., 1994, Vol. 117, pp. 453–471.

Nürnberg, D. Biogenic barium and opal in shallow Eurasian shelf sediments in relation to the pelagic Arctic Ocean environment. Ber. Polarforsch., 1996, N 212, pp. 96–118.

Nürnberg, D., Levitan, M., Pavlidis, Yu., Shelekhova, E. Distribution of clay minerals in surface sediments from the eastern Barents and south-western Kara seas. Geolog. Rundschau., 1995, Vol. 84, pp. 665–682.

Nürnberg, D., Tiedemann, R. Environmental change in the Sea of Okhotsk during the last 1,1 million years. Paleoceanography, 2004, Vol. 19, PA4011, doi: 10.1029/2004PA001023.

Nürnberg, D., Wollenburg, I., Dethleff, D. et al. Sediments in Arctic sea ice: Implications for entrainment, transport and release. Mar. Geol., 1994, Vol. 119, pp. 185–214.

Ó Cofaigh, C., Taylor, J., Dowdeswell, J.A. et al. Sediment reworking on high-latitude continental margins and its implications for paleoceanographic studies: Insights from the Norwegian-Greenland Sea. In: Glacier-Influenced Sedimentation on High-Latitude Continental Margins. Geol. Soc. London. Spec. Publ. 2002, Vol. 203, pp. 325–348.

Ohtani, K. To confirm again the characteristics of the Oyashio water. Bull. Hokkaido Nation. Fish. Res. Inst., 1991, Vol. 55, pp. 1–24.

Okuneva, O.G. Development methods and proves of proposed stratigraphical schemes for Upper Pleistocene deposits of the Pechora Depression shelf. Riga: NIIMorgeo, 1991, 68p. (in Russian).

Paetsch, H., Botz, R., Scholten, J.C., Stoffers, P. Accumulation rates of surface sediments in the Norwegian-Greenland Sea. Mar. Geol., 1992, Vol. 104, pp. 9–30.

Pavlidis, Yu.A. Sedimentation environment in the Chukchi Sea and facies-sedimentation zones of its shelf. In: Problems of Shelf Geomorphology, Lithology, and Lithodynamics. Moscow: Nauka, 1982, pp. 47–76 (in Russian).

Pavlidis, Yu.A., Dunaev, N.N., Shcherbakov, F.A. Actual problems of the Barents Sea Quaternary geology. In: Recent Processes of the World Ocean Shelves Sedimentation. Moscow: Nauka, 1990, pp. 76–93 (in Russian).

Peinert, R., Antia, A., Bauerfeind, E. et al. Particle flux variability in the Polar and Atlantic biogeochemical provinces of the Nordic Seas. In: The Northern North Atlantic: A Changing Environment. Berlin: Springer, 2001, pp. 53–68.

Peregovich, B. Die postglaziale Sedimentationsgeschichte der Laptewsee: schwermineralogische und sedimentologische Untersuchungen. Ber. Polarforsch., 1999, N 316, 88s.

Petelin, V.P. Grain size analysis of marine bottom sediments. Moscow: Nauka, 1967. 128p. (in Russian).

Petrov, O.M. Stratigraphy and fauna of marine mollusks from Chukotka Peninsula Quaternary deposits. Repts GIN Acad. Sci. USSR., Vol. 155, Moscow: Nauka, 1966, 251p. (in Russian).

Pfirman, S., Wollenburg, I. J., Lange, M.A. Lithogenic sediment on Arctic pack ice: Potential aeolian flux and contributions to deep sea sediments. In: Paleoclimatology and Paleometeorology. Eds. M. Sarnthein and M. Leinen. Dordrecht: Kluwer Acad. Publ., 1989, pp. 463–493.

Phillips, R.L., Grantz, A. Regional variations in provenance and abundance of ice-rafted clasts in Arctic Ocean sediments: Implications for the configuration of late Quaternary oceanic and atmospheric circulation in the Arctic. Mar. Geol., 2001, Vol. 172, pp. 91–115.

Pistolato, M., Quaia, T., Marinoni, L. et al. Grain size, mineralogy and geochemistry in Late Quaternary sediments from the western Ross Sea outer slope as proxies for climate changes. In: Antarctica. Contributions to Global Earth Sciences. Berlin, Heidelberg: Springer, 2006, pp. 423–432.

Plank, T., Langmuir, C.H. The chemical composition of subducting sediment and its consequences for the crust and mantle. Chem. Geol., 1998, Vol. 145, pp. 325–394.

Pleshivtzeva, E.S. Palinological characteristics of the boreal transgression sediments reference cross-section in the north-west Arkhangelsk District (the North Dvina Depression area). In: Pleistocene palinology. Moscow: Nauka, 1972, pp. 93–104 (in Russian).

Pogodina, I.A. Micropaleontological research of the Barents Sea Upper Pleistocene deposits. In: Foraminifers of the Barents Sea (Hydrobiology and Quaternary Paleoecology). Apatity: KSC RAS, 1994, pp. 71–83 (in Russian).

Polyak, L., Forman, S.L., Herlihy, F.A. et al. Late Weichselian deglacial history of the Svyataya (Saint) Anna Trough, northern Kara Sea, Arctic Russia. Mar. Geol., 1997, Vol. 143, pp. 169–188.

Polyak, L., Gataullin, V., Okuneva, O., Stelle, V. New constraints on the limits of the Barents-Kara ice sheet during the Last Glacial Maximum based on borehole stratigraphy from the Pechora Sea. Geology, 2000a, Vol. 28, pp. 611–614.

Polyak, L., Lehman, S.J., Gataullin, V., Jull, A.J.T. Two-step deglaciation of the southeastern Barents Sea. Geology, 1995, Vol. 23, pp. 567–571.

Polyak, L., Levitan, M., Gataullin, V. et al. The impact of glaciation, river-discharge and sea-level change on Late Quaternary environments in the southwestern Kara Sea. Int. J. Earth Sci., 2000b, Vol. 89, pp. 550–562.

Polyak, L., Levitan, M., Khusid, T. et al. Variations in the influence of riverine discharge on the Kara Sea during the last deglaciation and the Holocene. Glob. Planet. Change., 2002, Vol. 32, pp. 291–309.

Polyak, L., Mikhailov, V. 1996. Post-glacial environments of the southeastern Barents Sea: Foraminiferal evidence. In: Late Quaternary Paleoceanography of the North Atlantic Margins. Geol. Soc. Spec. Publ., 1996, Vol. 111, pp. 323–337.

Polyak, L., Murdmaa, I., Ivanova, E. A high-resolution 800-year glaciomarine record from Russkaya Gavan, Novaja Zemlya fiord, eastern Barents Sea. The Holocene, 2004, Vol. 14, (4), pp. 628–634.

Polyak, L., Solheim, A. Late- and post-glacial environments in the northern Barents Sea west of Franz-Josef Land. Polar Res., 1994, Vol. 13, pp. 197–207.

Polyakova, Ye. Diatom assemblages in surface sediments of the Kara Sea (Siberian Arctic) and their relationship to oceanological conditions. In: Siberian River Run-Off in the Kara Sea: Characterization, Quantification, Variability, and Environmental Significance. Proc. in Mar. Sci. Amsterdam: Elsevier, 2003, Vol. 6, pp. 375–400.

Polyakova, Ye., Stein, R. Holocene variations in the Eurasian Arctic shelf surface water salinity and sea ice regime: Evidences from diatom assemblages of the Siberian Arctic shelf, Kara Sea. In: Program and Abstracts EMMM'2002, Vienna, 2002, pp. 164–165.

Polyakova, Ye.I., Bauch, H.A., Kassens, H. Ice-hydrological regime changes in the Late Holocene Laptev Sea. Earth Sci. Rep., 2000, Vol. 370, (5), pp. 686–688.

Polyakova, Ye.I., Bauch, H.A., Klyuvitkina, T.S. Early to middle Holocene changes in Laptev Sea water masses deduced from diatom and aquatic palynomorph assemblages. Glob. Planet. Change., 2005, Vol. 48, pp. 208–222.

Polyakova, Ye.I. Arctic Seas of Eurasia in the Late Cenozoic. Moscow: Nauchny mir, 1997, 146p. (in Russian).

Poore, R.Z., Osterman, L., Curry, W.B., Phillips, R.L. Late Pleistocene and Holocene meltwater events in the western Arctic Ocean. Geology, 1999, Vol. 27, pp. 759–762.

Popov, A.I. Pleistocene deposits in the lower part of Pechora River. In: Cenozoic Cover of Bolshezemel'skaya Tundra. Moscow: MSU, 1963, pp. 24–49 (in Russian).

Prokopenko, A.A., Karabanov, E.B., Williams, D.F. et al. Biogenic silica record of the Lake Baikal response to climatic forcing during the Brunhes. Quatern. Res., 2001, Vol. 55, pp. 123–132.

Rachor, E. Scientific cruise report of the Arctic expedition ARK-XI/1 of RV ≪Polarstern≫ in 1995. Ber. Polarforsch., 1997, N 226, 104p.

Rasmussen, T.L., Thomsen, E., Kuijper, A., Wastegard, S. Late warming and early cooling of the sea surface in the Nordic Seas during MIS 5e (Eemian interglacial). Quatern. Sci. Rev., 2003, Vol. 22, pp. 809–821.

Romankevich, E.A., Korneeva, G.A., Shevchenko, V.P. et al. Suspended organic matter in the Barents Sea. Oceanology, 2000, Vol. 40 (2), pp. 208–216 (in Russian).

Romankevich, E.A., Vetrov, A.A. Carbon cycle in the Russian Arctic Seas. Moscow: Nauka, 2001, 302p. (in Russian).

Romanovsky, S.I. Sedimentological fundamentals of lithology. Leningrad: Nedra, 1977, 405p. (in Russian).

Ruddiman, W.F., Shackleton, N.J., McIntyre, A. North Atlantic sea-surface temperatures for the last 1,1 million years. In: North Atlantic paleoceanography. Eds. C.P. Summerhayes and N.J. Shackleton. Geol. Soc. Publ., 1986, pp. 155–173.

Rudels, B. On the mass balance of the Polar Ocean with special emphasis on the Fram Strait. Norsk Polarinst. Skr., 1987, Vol. 199, pp. 1–53.

Rudels, B., Jones, E.P., Anderson, L.G., Kattner, G. On the intermediate depth waters of the Arctic Ocean. In: The Polar Oceans and Their Role in Shaping the Global Environment. Geophys. Monogr. Ser., Vol. 85. AGU, Washington D.C., 1994, pp. 33–46.

Rudels, B., Meyer, R., Fahrbach, E. et al. Water mass distribution in Fram Strait and over the Yermak Plateau in summer 1997. Ann. Geophys., 2000, Vol. 18, pp. 687–705.

Saidova, Kh. M. Stratigraphy and paleogeography of the Chukchi Sea and Bering Strait Holocene according to foraminifers. In: Problems of Shelf Geomorphology, Lithology, and Lithodynamics. Moscow: Nauka, 1982, pp. 92–115 (in Russian).

Sakshaug, E. Primary and secondary production in the Arctic Seas. In: The Arctic Ocean Organic Carbon Cycle: Present and Past. Eds. R. Stein and R.W. Macdonald. Berlin: Springer Verlag, 2004, pp. 57–82.

Salvigsen, O., Forman, S.I., Miller, G.H. Thermophilous molluscs on Svalbard during the Holocene and their paleoclimatic implications. Pol. Res., 1992, Vol. 11, pp. 1–10.

Sancetta, C., Silvestri, S.M. Diatom stratigraphy of the Late Pleistocene (Bruhnes) Subarctic Pacific. Mar. Micropaleontol., 1986, Vol. 9, pp. 263–274.

Sarnthein, M., Jansen, E., Weinelt, M. et al. Variations in Atlantic surface ocean paleoceanography, 50–80°N: A time-slice record of the last 30.000 years. Paleoceanography, 1995, Vol. 10, pp. 1063–1094.

Sarnthein, M., Pflaumann, U., Weinelt, M. Past extent of sea ice in the northern North Atlantic inferred from foraminiferal paleotemperature estimates. Paleoceanography, 2003, Vol. 18, (2), doi: 1047 (25.1–25.8).

Sarnthein, M., Stattegger, K., Dreger, D. et al. Fundamental modes and abrupt changes in North Atlantic circulation and climate over the last 60 ky – concepts, reconstruction and numerical modeling. In: The Northern North Atlantic: A changing environment. Berlin: Springer, 2001, pp. 365–410.

Schäfer, P., Thiede, J., Gerlach, S. et al. The environment of the northern North Atlantic Ocean: Modern depositional processes and their historical documentation. In: The Northern North Atlantic: A Changing Environment. Berlin: Springer-Verlag, 2001, pp. 1–18.

Schneider, D.A., Backman, J., Curry, W.B., Possert G. Paleomagnetic constraints on sedimentation rates in the Eastern Arctic Ocean. Quatern. Res., 1996, Vol. 46, pp. 61–71.

Schoster, F., Levitan, M. Scientific cruise report of the joint Russian–German Kara Sea expedition in 2002 with RV ≪Academik Boris Petrov≫. Ber. Polarforsch., 2003, N 450, 109p.

Schoster, F., Levitan, M. Scientific cruise report of the Kara Sea expedition with RV ≪Academik Boris Petrov≫ in 2003 within the frames of the Russian–German project ≪SIRRO≫ and the Russian–Norwegian project ≪MAREAS≫. Ber. Polarforsch., 2004, N 479, 147p.

Schoster, F., Stein, R. Major and minor elements in surface sediments of Ob and Yenisei estuaries and the adjacent Kara Sea. Ber. Polarforsch., 1999, N 300, pp. 196–207.

Schubert, C.J., Stein, R., Calvert, S.E. Tracking nutrient and productivity variations over the last deglaciation in the Arctic Ocean. Paleoceanography, 2001, Vol. 16 (2), pp. 199–211.

Scourse, J.D., Hall, I.R., McCave, I.N. et al. The origin of Heinrich layers: Evidence from H2 for European precursor events. Earth Planet. Sci. Lett., 2000, Vol. 182, (2), pp. 187–195.

Sedykh, E.M., Starshinova, N.P., Bannykh, L.N. et al. Identification of heavy metals and their forms in the water and sediments of water reservoirs by means of AES-ICP and ETAAS methods. J. Analyt. Chem., 2000, Vol. 55 (4), pp. 385–391 (in Russian).

Serebryanny, L., Malyasova, E. The Quaternary vegetation and landscape evolution of Novaja Zemlja in the light of palynological records. Quatern. Intern., 1998, Vol. 45/46, pp. 59–70.

Shackleton, N. New data on the evolution of Pliocene climatic stability. In: Paleoclimate and evolution with emphasis on human origin. London: Yale University Press, 1995, pp. 242–248.

Shackleton, N.J., Chapman, M., Sánchez-Coñi, M.F. et al. The classic Marine Isotope Substage 5e. Quatern. Res., 2002, Vol. 58, pp. 14–16.

Shelekhova, E.S. Regularities of clay mineral distribution in the Barents and Kara Seas surface-layer sediments. Abst. Ph. D. thesis. Moscow: IO RAS, 1998, 29p. (in Russian).

Shepard, F.P. Submarine geology. New York: Harper and Pow, 1973, 215p.

Shevchenko, V.P., Lisitsin, A.P. Aeolian input. In: The Organic Carbon Cycle in the Arctic Ocean. Eds. R. Stein and R.F. Macdonald. Berlin: Springer, 2004, pp. 53–54.

Shevchenko, V.P. Aerosols: Influence on sedimentation and environment in the Arctic. Abstr. Ph. D. thesis. Moscow: IO RAS, 2000, 32p. (in Russian).

Sidorchuk, A.Y., Borisova, O.K., Kovalukh, N.N. et al. Paleohydrology of Low Vychegda in late-glacial and Holocene. Herald MSU. Ser Geogr. 1999, N 5, pp. 35–41 (in Russian).

Siegel, F.R., Galasso, J.J., Kravitz, J.H. Geochemistry of thirteen Voronin Trough cores, Kara Sea, European Arctic: Hg and As contaminants at 1965 timeline. Appl. Geochem., 2001b, Vol. 16, pp. 19–34.

Siegel, F.R., Kravitz, J.H., Galasso, J.J. Arsenic and mercury contamination in 31 cores taken in 1965, St. Anna Trough, Kara Sea, Arctic Ocean. Environ. Geol., 2001a, Vol. 40, pp. 528–542.

Simstich, J., Stanovoy, V., Bauch, D. et al. Holocene variability of bottom water hydrography on the Kara Sea shelf depicted in multiple single-valve analyses of stable isotopes in ostracods. In: Progr. and Abstr. 4th Intern. Workshop SIRRO. Moscow: GEOKHI, 2003, pp. 27.

Spielhagen, R.F. Die eisdrift in der Framstrasse während der letzten 200.000 jahre. Geomar Report, 1991, N 4, 133s.

Spielhagen, R.F., Baumann, K.-H.., Erlenkeuser, H. et al. Arctic Ocean deep-sea record of northern Eurasian ice sheet history. Quatern. Sci. Rev., 2004, Vol. 23, pp. 1455–1484.

Spiridonova, E.A. Evolution of the Don area vegetation cover in the Late Pleistocene-Holocene. Moscow: Nauka, 1991, 220p. (in Russian).

Spiridonova, E.A. Palinological characteristics of the Middle Valday Interstadial and its significance for reconstruction of flora and vegetation development history of the Russian Plain. Bulletin Comission Quatern Period Research, 1983, N 52, pp. 42–57 (in Russian).

Spiridonova, E.A., Lavrushin, Yu.A. Correlation of geological and ecological events in the Holocene of the arctic, boreal, and arid zones of the East Europe. In: Quaternary Geology and Paleogeography of Russia. Moscow: GEOS, 1997, pp. 151–170 (in Russian).

Stein, R. Lithostratigraphy of gravity corers and correlation with sediment echograph profiles (≪Academik Boris Petrov≫ Kara Sea expeditions 1999 and 2000). Ber. Polarforsch., 2001, N 393, pp. 120–140.

Stein, R., Boucsein, B., Fahl, K. et al. Accumulation of particulate organic carbon at the Eurasian continental margin during late Quaternary times: Controlling mechanisms and paleoenvironmental significance. Glob. Planet. Change., 2001, Vol. 31 (1–4), pp. 87–104.

Stein, R., Boucsein, B., Hefter, J. et al. Marine geology. Ber. Polarforsch., 2000, N 360, pp. 49–69.

Stein, R., Dittmers, K., Fahl, K. et al. Arctic (palaeo) river discharge and environmental change: Evidence from the Holocene Kara Sea sedimentary record. Quatern. Sci. Rev., 2004, Vol. 23, pp. 1485–1511.

Stein, R., Fahl, K. Holocene accumulation of organic carbon at the Laptev Sea continental margin (Arctic Ocean): Sources, pathways and sinks. Geo-Marine Lett., 2000, Vol. 20, pp. 27–36.

Stein, R., Fahl, K. The Laptev Sea: Distribution, Sources, Variability and Burial of Organic Carbon. R. Stein, R. McDonald (Eds). The organic carbon cycle in the Arctic Ocean. Berlin: Springer, 2004, pp. 213–236.

Stein, R., Grobe, H., Wahsner, M. Organic carbon, carbonate and clay mineral distributions in eastern central Arctic Ocean surface sediments. Mar. Geol., 1994b, Vol. 119, pp. 269–285.

Stein, R., Fahl, K. Scientific Cruise Report of the Arctic Expedition ARK-XIII/2 ≪Polarstern≫ in 1997. Ber. Polarforsch., 1997, N 255. 235p.

Stein, R., Fahl, K., Fütterer, D.K. et al. (Eds.) Siberian River Run-off in the Kara Sea. Characterisation, quantification, variability and environmental significance. Amsterdam: Elsevier, 2003b, 488pp.

Stein, R., Fahl, K., Nissen, F., Siebold, M. Late Quaternary Organic Carbon and Biomarker Records from the Laptev Sea Continental Margin (Arctic Ocean): Implication for organic carbon flux and composition, dynamics and history. In: Land-Ocean Systems in the Siberian Arctic: Dynamics and history. Berlin: Springer-Verlag, 1999, pp. 635–655.

Stein, R., Fahl, K., Dittmers, K. et al. Holocene siliciclastic and organic carbon fluxes in the Ob and yenisei estuaries and the adjacent inner Kara Sea: Quantification, variability, and paleoenvironmental implications. R. Stein, K. Fahl, D.K. Fütterer, E.M. Galimov and O.V. Stepanets (Eds). Siberian River Run-off in the Kara Sea. Characterisation, quantification, variability and environmental significance. Amsterdam: Elsevier, 2003a, pp. 401–434.

Stein, R., Ivanov, G.I., Levitan, M.A., Fahl, K. Surface-sediment composition and sedimentary processes in the central Arctic Ocean and along the Eurasian continental margin. Ber. Polarforsch., 1996, N 272, 324p.

Stein, R., Knies, J. Late Quaternary Organic Carbon Records in the St. Anna Trough (Kara Sea). Ber. Polarforsch., 1999, N 342, pp. 228–234.

Stein, R., Levitan, M. Lithological core description. Ber. Polarforsch., 2001, N 393, pp. 247–283.

Stein, R., Macdonald, R.. The organic carbon cycle in the Arctic Ocean. Berlin: Springer, 2004, 363p.

Stein, R., Nam, S.I., Grobe, H., Hubberten, H. Late Quaternary glacial history and short-term ice-rafted fluctuations along the East Greenland continental margin. Geol. Soc. Spec. Publ., 1996, N 111, pp. 135–151.

Stein, R., Nam, S.I., Schubert, C. et al. The last deglaciation event in the eastern central Arctic Ocean. Science, 1994a, Vol. 264, pp. 692–696.

Stein, R., Niessen, F., Dittmers, K. et al. Siberian river run-off and Late Quaternary glaciation in the southern Kara Sea, Arctic Ocean: Preliminary results. Polar Res., 2002, Vol. 21, pp. 315–322.

Stein, R., Schubert, C.J., Vogt, C. et al. Stable isotope stratigraphy, sedimentation rates and salinity changes in the latest Pleistocene and Holocene eastern central Arctic Ocean. Mar. Geol., 1994, Vol. 119, pp. 250–268.

Stein, R., Stepanets, O. Scientific cruise report of the joint Russian–German expedition of RV ≪Academik Boris Petrov≫ in 1999. Ber. Polarforsch., 2000, N 360, 141p.

Stein, R., Stepanets, O. Scientific cruise report of the Kara Sea expedition 2001 of RV ≪Academik Boris Petrov≫: The German–Russian project on Siberian river run-off (SIRRO) and the EU project ≪ESTABLISH≫. Ber. Polarforsch, 2002, N 419, 278p.

Stein, R., Stepanets, O. The German–Russian project on Siberian river run-off (SIRRO): Scientific cruise report of the Kara Sea expedition ≪SIRRO 2000≫ of RV ≪Academik Boris Petrov≫ and first results. Ber. Polarforsch, 2001, N 393, 287p.

Steinke, T. Rekonstruktion spätquartärer Paläo-Umweltbedingungen in der Kara See anhand sedimentologischer und mineralogischer Untersuchungen. Master Thesis. Bremen University, 2002, 98p.

Stepanets, O.V., Borisov, A.P., Komarevsky, V.M. et al. Research of ecological conditions in the Ob and Yenisei estuaries and adjacent Kara Sea shelf. Ber. Polarforsch., 2000, N 360, pp. 84–91.

Stepanets, O.V., Borisov, A.P., Ligaev, A.N. et al. Study of anthropogenic pollution in the Kara Sea and adjacent estuaries of Yenisei and Ob in 2002. Ber. Polarforsch., 2003, N 450, pp. 72–84.

Stepanets, O.V., Borisov, A.P., Solov'eva, G.Yu. Distribution of anthropogenic radionuclides in the estuaries of Ob and Yenisei rivers and adjacent Kara Sea. Ber. Polarforsch., 1999, N 300, pp. 132–140.

Struck, U. Paleoecology of benthic foraminifera in the Norwegian-Greenland Sea during the past 500 ka. Grzybowski Foundation Spec. Publ., 1997, N 5, pp. 51–82.

Stuiver, M., Braziunas, T.F. Sun, ocean, climate and atmospheric $^{14}CO_2$; an evaluation of causal and spectral relationships. The Holocene, 1993, Vol. 3, pp. 289–305.

Stuiver, M., Reimer, P.J. Extended ^{14}C data base and revised CALIB 3,0 ^{14}C age calibration program. Radiocarbon. 1993, Vol. 35 (215–230), pp. 880–935.

Stuiver, M., Reimer, P.J., Bard, E. et al. INTCAL 98 radiocarbon age calibration, 24,000-0 cal BP. Radiocarbon, 1998, Vol. 40, pp. 1041–1083.

Subba Rao, D.V., Platt, T. Primary Production of Arctic Waters. Pol. Biol., 1984, Vol. 3, pp. 191–201.

Svendsen, J.I., Alexanderson, H., Astakhov, V.I. et al. Late Quaternary ice sheet history of Northern Eurasia. Quatern. Sci. Rev., 2004, doi: 10.1016.

Svendsen, J.I., Astakhov, V.I., Bolshiyanov, D.Yu. et al. Maximum extent of the Eurasian ice sheets in the Barents and Kara Sea region during the Weichselian. Boreas., 1999, Vol. 28 (1), pp. 234–242.

Svendsen, J.I., Mangerud, J., Elverhøi, A.A. et al. The late Weichselian glacial maximum in western Spitsbergen inferred from offshore sediment cores. Mar. Geol., 1992, Vol. 104, pp. 1–17.

Svindland, K.T., Vorren, T.O. Late Cenozoic sedimentary environments in the Amundsen Basin, Arctic Ocean. Mar. Geol., 2002, Vol. 186, pp. 541–555.

Svitoch, A.A. Marine Pleistocene of the Russia coastal zones. Moscow: GEOS, 2003, 362p. (in Russian).

Taldenkova, E., Bauch, H.A., Stepanova, A. et al. Last glacial environmental evolution of the Laptev Sea shelf as reflected in molluscan, ostracodal, and foraminiferal faunas. Glob. Planet. Change. 2005, Vol. 48, pp. 223–251.

Taldenkova, E., Stepanova, A., Simstich, J. Downcore variations in species composition of ostracods and mollusks in core BP00-07/5 and some paleoenvironmental implementations. In: Progr. and Abstr. 4th Intern. Workshop SIRRO. Moscow: GEOKHI. 2003, 41p.

Tarasov, G.A. Peculiarities of distribution of bottom sediments main types. In: Biogeocoenoses of the West Arctic Glacial Shelves. Apatity: KSC RAS, 1996, pp. 66–80 (in Russian).

Tarasov, G.A., Pogodina, I.A., Khasankaev, V.B. et al. Sedimentation processes at glacial shelves. Apatity: KSC RAS, 2000, 473p. (in Russian).

Taylor, J., Dowdeswell, J.A., Siegert, M.J. Late Weichselian depositional processes, fluxes, and sediment volumes on the margins of the Norwegian Sea (62–75°N). Mar. Geol. 2002a, Vol. 188, pp. 61–77.

Taylor, J., Tranter, M., Munhoven, G. Carbon cycling and burial in the glacially influenced Polar North Atlantic. Paleoceanography. 2002b. Vol. 17. doi: 10.1029/2001PA000644.

The Nordic Seas. Ed. B.G. Hurdle. New York: Springer-Verlag, 1986, 777p.

Thiede, J., Tiedemann, R. The alternative natural climate change – Do we have to expect new glacial? In: Climate of the 21 st Century: Changes and Risks. Eds. J.L. Lozán, H. Graßl, P. Hupfer. Hamburg: Wissenschaftliche Auswertungen, 2001, pp. 190–195.

Trifonov, V.G. Neotectonics of Eurasia. Moscow: Nauchny mir, 1999, 252p. (in Russian).

Troitsky, S.L. Basic regularities of the fauna composition changes in cross-sections of marine intermoraine layers of the Ust'-Yenisei Depression and Low Pechora Depression. In: Paleogeography of the North Siberia Quaternary. Novosibirsk: SB Acad. Sci. USSR, 1964, pp. 48–64 (in Russian).

Troitsky, S.L. General review of Siberia marine Pleistocene. In: Problems of Siberia Quaternary Geology. Moscow: Nauka, 1969, pp. 51–62 (in Russian).

Troitsky, S.L. Quaternary deposits and relief of coastal plain zones of the Yenisei Gulf and adjacent Byrranga Mountaines. Moscow: Nauka, 1966, 115p. (in Russian).

Tütken, T., Eisenhauer, A., Wigand, B., Hansen, B.T. Glacial – interglacial cycles in Sr and Nd isotops composition of Arctic marine sediment triggered by the Svalbard/Barents Sea ice sheet. Mar. Geol., 2002, Vol. 182, pp. 351–372.

Tveranger, J., Astakhov, V., Mangerud, J. The margin of the last Barents-Kara ice sheet at Markhide, Northern Russia. Quatern. Res., 1995, Vol. 44, pp. 328–340.

Ul', G.F. The Kara Sea bottom and coast rock. Leningrad: NIIGA, 1936, 136p. (in Russian).

Untersteiner, N. Structure and dynamics of the Arctic Ocean ice cover. Geol. Soc. Am., 1990, Vol. 1, pp. 37–51.

Urban, B. Eemian and Early Weichselian vegetation and uranium/thorium and thermoluminescence age determinations from section in Northeast-Lower-Saxony, Germany. In: The Eemian: Local Sequences, Global Perspectives. Abstr. SEQS Symp. Kerkrade. 1998, pp. 86.

Valpeter, A.P. Ancient coast lines in the shelf and coastal zones of the marginal seas of the USSR East. In: Shelf structure of the USSR seas as a base for evaluation of engineer-geological conditions. Riga, 1984, pp. 13–26 (in Russian).

Vasil'chuk, Yu.K., Esipov, A.D., Oprunenko, S.F. et al. New data on oxygen isotope content in syngenetic repeteadly-vein ice of the Late Pleistocene of Kolyma River lower part. Repts Acad. Sci. USSR, 1985, Vol. 281, N 4, pp. 904–908 (in Russian).

Vedernikov, V.I., Demidov, A.B., Sud'bin, A.I. Primary production and chlorophille in the Kara Sea in September of 1993. Oceanology, 1994, N 5, pp. 693–703 (in Russian).

Veinsberg, I.G. Main stages of the Late Quaternary history of the marine coastal zones development of the East Siberian, Okhotsk, Aral, and Baltic Seas. In: Shelf Structure of the USSR Seas as a Base for Evaluation of Engineer-Geological Conditions. Riga, 1984, pp. 7–13 (in Russian).

Veinsberg, I.G., Stele, V.Ya., Savvaitova, A.S. et al. Late Quaternary history of the Pechora Sea coastal zones development. In: Correlation of Paleogeographic Events in the System Materic-Shelf-Ocean. Moscow: MSU, 1995, pp. 106–112 (in Russian).

Velichko, A.A., Andreev, A.A., Klimanov, V.A. Dinamics of vegetation and climate of the North Eurasia in the late glacial time and Holocene. In: Short-Period and Abrupt Environment-Climatic Changes in the Last 15 000 yrs. Moscow: IG RAS, 1994, pp. 4–60 (in Russian).

Vogelsang, E. Paläo-Ozeanographie des Europäshen Nordmeeres an Hand stabilen Kohlenstoff–Sauerstoffisotope. Ber. Sonderforsch., 1990, Vol. 313, (23), pp. 1–136.

Vogt, C. Zeitliche und räumliche Verteilung von Mineralvergesellschaftungen in spätquartären Sedimenten des Arktischen Ozeans und ihre Nützlichkeit als Klimaindikatoren während der Glazial/Interglazial-wechsel. Ber. Polarforsch., 1997, N 251, 354 s.

Vogt, C., Knies, J., Spielhagen, R.F., Stein, R. Detailed mineralogical evidence for two nearly identical glacial/deglacial cycles and Atlantic water advection to the Arctic Ocean during the last 90 000 years. Glob. Planet. Change., 2001, Vol. 31 (1–4), pp. 23–44.

Vogt, P.R. Seafloor topography, sediments, and paleoenvironments. In: The Nordic Seas. Ed. B.G. Hurdle. Berlin: Springer, 1986, pp. 237–412.

Vogt, P.R., Crane, K., Sundvor, E. Deep Pleistocene iceberg plowmarks on the Yermak Plateau: Sidescan and 3,5 kHz evidence for thick calving ice fronts and a possible marine ice sheet in the Arctic Ocean. Geology, 1994, Vol. 22, pp. 403–406.

Vöelker, A.H.L. Zur Deutung der Dansgaard-Oeschger Ereignisse in ultra-hochauflösenden Sedimentprofilen aus dem Europäischen Nordmeer. Ph. D. Dissertation. Kiel University, 1999, 155 s.

Volynets, O.N. Geochemical types, petrology, and genesis of the late Cenozoic volcanic rock from the Kurile-Kamchatka island-arc system. Intern. Geol. Rev., 1994, Vol. 36 (4), pp. 373–403.

Vorren, T.O., Laberg, J.S., Blaume, F. et al. The Norwegian-Greenland Sea continental margins: Morphology and late Quaternary sedimentary processes and environment. Quatern. Sci. Rev., 1998, Vol. 17, pp. 273–302.

Wahsner, M., Ivanov, G., Tarasov, G. Marine geological investigation of surface sediments in the Franz Josef Land area and the St. Anna Trough. Ber. Polarforsch., 1996, N 212, pp. 45–64.

Wahsner, M., Müller, C., Ivanov, G. et al. Clay mineral distributions in surface sediments from the Eurasian Arctic Ocean and the Eurasian continental margin as indicator for source areas and transport pathways of sediments: A synthesis. Boreas., 1999, Vol. 28, pp. 215–233.

Wallrabe-Adams, H.-J., Lackschewitz, K.S. Chemical composition, distribution, and origin of silic volcanic ash layers in the Greenland – Iceland – Norwegian Sea: Explosive volcanism from 10 to 300 ka as recorded in deep-sea sediments. Mar. Geol., 2003, Vol. 193, pp. 273–293.

Wang, R., Chen, R. *Cycladophora davisiana* (Radiolarian) in the Bering Sea during the late Quaternary: A stratigraphic tool and proxy of the glacial Subarctic Pacific Intermediate Water. Sci. in China. Ser. D. Earth Sci., 2005, Vol. 48, (10), pp. 1698–1707.

Weinelt, M., Vogelsang, E., Kucera, M. et al. Variability of North Atlantic heat transfer during MIS 2. Paleoceanography, 2003, Vol. 18 (3), doi: 1071 (16.1–16.18).

Willard, D.A. Pliocene-Pleistocene pollen assemblages from the Yermak Plateau, Arctic Ocean: Sites 910 and 911. In: Proc. ODP. Sci. Res., 1996, Vol. 151, pp. 297–305.

Wolf, E. European Project for Ice Coring in Antarctica (EPICA). PAGES News., 2006, Vol. 14 (1), pp. 31–33.

Wollenburg J.E., Kuhnt, W., Mackensen, A. Changes in Arctic Ocean paleoproductivity and hydrography during the last 145 kyr: The benthic foraminiferal record. Paleoceanography, 2001, Vol. 16, pp. 65–77.

Yudovich, Ya.E., Ketris, M.P. Fundamentals of lithochemistry. St. Petersburg: Nauka, 2000, 479p. (in Russian).

Zarkhidze, V.S., Krasnov, I.I., Spiridonov, M.A. et al. Main cross-sections of the Lower Pechora and their significance for understanding of latest stages of geological development of the north-eastern European part of the USSR. In: The Arctic Ocean and its coastal area in Cenozoic. Leningrad: Hydrometeoizdat, 1970, pp. 336–339 (in Russian).

Zeeberg, J.J., Forman, S.L., Polyak, L. Glacier extent in a Novaya Zemlya fjord during the ≪Little Ice Age≫ inferred from glaciomarine sediment records. Pol. Res., 2003, Vol. 22, pp. 385–394.

Zhuze, A.P., Poretzky, V.S. Diatom interglaciation sediments along Volga River. Repts Sov. Sect. AICHPE., Vol. I, 1937, pp. 178–205 (in Russian).

Znamenskaya, O.M., Cheremisinova, E.A. Distribution of Mga Interglaciation sea and main features of its paleogeography. In: Questions of Quaternary deposits stratigraphy of the north-eastern European part of the USSR. Leningrad: Gostoptechizdat, 1962, pp. 131–140 (in Russian).

Zubakov, V.A. Global climatic events of the Pleistocene. Leningrad: Hydrometeoizdat, 1986, 288p. (in Russian).

Index

A

Abrasion, 57, 101, 116, 177, 179, 198, 199, 203, 209, 216, 225, 243, 262, 264, 271, 289, 290, 291, 349
Abrupt climate change, 8
Absorption, 183, 209, 257
Advance, 37, 81, 109, 111, 121, 172, 293, 342, 343, 344
Advection, 78, 84, 85, 87, 95, 101, 102, 103, 104, 107, 171, 178, 237, 239, 254, 255, 298, 344, 351, 352, 353, 354
Age model, 68, 98–100, 149, 154, 161, 263, 294, 298
Alpha Ridge, 113, 122, 174
Aluminium, 201, 208, 261, 262, 270, 275, 281, 282, 322
Amerasian Basin, 113, 117, 121, 122
AMS ^{14}C, 68, 136, 138, 149, 173, 212, 224, 232, 241, 249, 256, 257, 274, 295, 309, 311, 341
Amundsen Basin, 113, 122, 138, 144, 174, 176, 341
Amur River, 306
Arctic Ocean, 3–9, 61, 108, 113–176, 177, 237, 239, 253, 254, 256, 272, 289, 291, 301, 310, 337, 341, 344, 349, 350–353, 354
Arctic water, 4, 15, 16, 17, 48, 61, 62, 63, 116, 169, 188, 189, 337
Atlantic Ocean, 113, 116, 351
Atlantic water, 62, 63, 64, 79, 84, 85, 86, 87, 88, 90, 101, 103, 104, 107, 116, 118, 119, 129, 135, 165, 171, 178, 188, 189, 193, 194, 222, 226, 238, 253, 254, 269, 295, 296, 337, 344, 351, 352, 353
Atmosphere, 57, 109, 127, 188, 327, 353, 355
Atmospheric pressure, 188

B

Barents Sea, 4, 28, 36, 43, 45, 57, 61, 62, 65, 74, 83, 84, 86, 101, 119, 121, 123, 127, 129, 132, 133, 135, 144, 164, 165, 172, 177–178, 189, 193, 194, 210–224, 228, 232, 234, 236, 246, 255, 299, 337, 342, 344, 345, 347, 348
Bathymetry, 290, 349
Bear Island, 4, 74, 177, 186, 342
Bear Trough, 178
Beaufort Gyre, 117, 118, 119, 154, 166, 172
Benthic foraminifers, 90, 91, 93, 94, 96, 97, 98, 100, 102, 103, 104, 109, 119, 130, 132, 133, 134, 136, 138, 165, 170, 173, 193, 197, 201, 203, 204, 206, 209, 210, 212, 213, 215, 222, 230, 231, 232, 237, 238, 239, 249, 253, 257, 260, 265, 266, 269, 270, 286, 299, 308, 310, 327
Bering Sea, 289, 301, 302, 304, 305, 306, 307–310, 329
Bering Strait, 113, 116, 289, 291, 301, 306, 310
Biogenic opal, 98, 176, 179, 203, 206, 207, 281, 291, 292, 299, 307, 311, 314, 316, 321, 325
Biosphere, 355
Biostratigraphy, 68, 136
Bivalves, 22, 133, 155, 156, 163, 193, 204, 208, 212, 226, 227, 230, 232, 233, 243, 249, 252, 254, 257, 259, 265, 274, 275, 281, 282, 286, 295, 308
Black ore minerals, 119, 120, 132, 151, 155, 156, 157, 160, 161, 165, 183, 190, 193, 201, 203, 204, 207, 208, 209, 228, 253, 260, 261, 267, 290, 308
Bore hole, 27, 241, 247, 339
Bottom current, 67, 104, 126, 134, 181, 183, 199, 208, 209, 213, 215, 216, 219, 226, 234, 239, 255, 270

Bottom sediment, 48, 79, 83, 87, 88, 89, 100, 104, 105, 108, 124, 127, 134, 164, 165, 170, 171, 172, 194, 195, 204, 206, 209, 226, 235, 236, 256, 257, 263, 272, 296, 319, 321, 329, 341, 343, 344, 351, 355
Bottom topography, 62, 129, 198, 199, 207, 208
Bottom water, 22, 24, 65, 69, 87, 90, 102, 103, 108, 117, 122, 129, 130, 134, 137, 188, 207, 226, 232, 237, 238, 242, 244, 264, 270, 338
Brines, 118, 186, 207, 224, 226, 238, 239
Bulk density, 91, 94, 95, 96, 97, 98, 154, 245, 257, 259, 260, 264, 265, 266, 270, 275, 310

C

Calcium, 208, 275, 276, 314, 315, 316, 317, 322
Calendar age, 49, 68, 100, 215, 227, 233, 257, 273
Canadian Basin, 122, 172
Carbonate nannoplankton, 90, 97
Carbonates, 64, 78, 79, 80, 83, 84, 87, 96, 98, 100, 103, 105, 109, 111, 125, 126, 154, 156, 157, 159, 160, 162, 168, 173, 179, 183, 189, 222, 230, 236, 239, 246, 251, 254, 287, 308, 314, 318, 354
Carbon isotopes, 65, 173, 212, 222, 231, 232, 238, 269, 286, 287
Channel, 81, 88, 179, 198, 199, 200, 206, 265, 270, 275, 281, 284, 287, 288, 289, 290, 293, 296
Chemostratigraphy, 136
Chlorite, 65, 90, 91, 93, 96, 97, 98, 101, 121, 122, 123, 125, 154, 157, 170, 186, 190, 192, 193, 194, 201, 203, 207, 219, 260, 266, 267, 290, 291, 307, 322, 323
Chukchi Sea, 47, 121, 123, 144, 150, 289, 291, 292, 298–299, 306, 344, 354
Circulation, 61, 62, 88, 90, 101, 103, 117, 118, 119, 121, 127, 169, 173, 179, 188, 189, 199, 256, 301, 302, 305, 306, 307, 310, 322, 325, 327, 329, 351, 353, 354, 355
Clastics, 43, 156, 307, 308, 309
Clay, 298–300, 322–324
Clay minerals, 82, 90, 91, 93, 95, 96, 98, 99, 101, 104, 107, 120, 121, 122, 123, 124, 129, 151, 152, 155, 156, 160, 165, 170, 186, 188, 189, 190, 192, 194, 197, 199, 201, 203, 204, 206, 207, 208, 219–221, 226, 253, 254, 257, 259, 260, 262, 265, 266, 268, 291, 307, 308, 316, 322–324, 327
Climate, 297–298, 333–355
Climate change, 8, 107, 238, 245, 327, 341, 354
Climatostratigraphy, 27, 33
Clinopyroxene, 119, 154, 157, 160, 165, 166, 168, 185, 186, 190, 191, 193, 201, 203, 204, 206, 207, 208, 219, 228, 235, 253, 255, 261, 267, 269, 270, 287, 290
Coast, 3–6
Coast line, 12
Coccoliths, 109, 151, 153
Cold epoch, 128, 147, 168, 354, 355
Color reflection, 314
Continental margin, 61, 65, 68, 74, 75, 76, 77, 79, 80, 81, 83, 84, 85, 86, 88, 90, 107–112, 113, 120, 121, 122, 129, 133, 138, 144, 145, 146, 149, 168, 170, 297, 321, 329, 349, 354
Continental rise, 61, 109, 113, 124, 138, 144
Continental slope, 48, 61, 81, 82, 87, 90, 102, 105, 107, 109, 113, 118, 129, 130, 132, 133, 134, 135, 138, 147, 163, 164, 170, 174, 177, 178, 291, 293, 294, 295, 296, 297–298, 348, 349
Contour currents, 87
Contourites, 109
Copper, 207, 209, 261, 286, 287
Correlation, 3, 5, 6–9, 26, 32, 33, 36, 97, 101, 105, 106, 121, 147, 154, 158, 160, 161, 163, 165, 172, 181, 182, 183, 208, 213, 214, 216, 221, 257, 269, 275, 277, 278, 279, 282, 283, 285, 311, 316, 317, 322, 323, 324, 352
Cryosoles, 354
Crystal lattice, 261, 262

D

Debris-flow, 105, 107
Deep, 307–310
Deep water, 22, 24, 33, 61, 62, 65, 69, 87, 90, 108, 113, 116, 118, 119, 122, 126, 129, 133, 138, 147, 150, 172, 186, 199, 210, 215, 291, 293, 298, 301, 305, 307–310, 345, 349, 350, 354
Deglaciation, 28, 29, 37, 38, 39, 57–58, 83, 99, 101, 103, 104, 109, 172, 213, 214, 223, 228, 232, 235, 236, 238, 239, 241, 246, 253, 255, 271, 296, 341, 342, 345, 348, 349–350
Diagenesis, 102, 236
Diamicton, 96, 98, 105, 107, 119, 210, 213, 216, 223, 225, 241, 242, 243, 245, 246, 247, 249, 251, 253, 254
Diatom clay, 312, 313, 314, 316, 327

Diatom ooze, 304, 307, 311, 312, 313, 314, 316, 317, 321, 322, 325, 328
Diatoms, 11, 17, 90, 91, 136, 195, 197, 201, 203, 206, 207, 256, 259, 260, 270, 291, 296, 298–299, 301, 304, 307, 308, 310, 311, 312, 313, 314, 316, 317, 321, 322, 324, 327, 328, 355
Dirty ice, 115, 129
Drilling, 16, 43, 68, 153, 305, 327

E
East Greenland Current, 61, 86, 87, 108, 112, 127, 129, 351, 354
East Siberian Sea, 53, 57, 121, 123, 289, 290–291, 292, 297, 298
Environment, 3, 4, 8, 9, 15, 24, 89, 90, 113–135, 164, 168, 172, 177–210, 224–226, 232, 234, 235, 241, 246, 253, 262, 268, 289–299, 301–307, 329, 334, 339, 342, 344, 352
Epidote, 25, 119, 154, 157, 160, 165, 166, 168, 169, 183, 184, 190, 191, 193, 201, 203, 204, 206, 207, 208, 219, 228, 235, 239, 253, 261, 262, 267, 269, 270, 290, 308
Erosion, 22, 31, 54, 91, 101, 128, 132, 164, 180, 185, 189, 193, 194, 198, 199, 200, 207, 208, 209, 215, 216, 219, 241, 243, 246, 247, 255, 264, 265, 295, 324, 329
Estuary, 45, 195, 198, 257, 262, 264, 271
Eurasian Basin, 113, 122, 172

F
Facies, 11, 14, 17, 24, 25, 33, 39, 65, 88, 124–135, 162, 193, 194–210, 213, 214, 215, 238, 246, 258, 262, 264, 271, 275, 282, 286, 349–350
Fan, 57, 61, 74, 83, 88, 144, 293, 296, 348
Feldspars, 40, 65, 93, 98, 132, 151, 152, 155, 156, 160, 170, 183, 186, 187, 189, 190, 193, 201, 203, 204, 206, 207, 208, 251, 254, 259, 261, 262, 265, 281, 291, 307, 308, 317
Ferruginous film, 25, 151, 152, 153, 155, 156, 157, 158, 160, 161, 165, 262
Fjord, 61, 62, 74, 75, 108, 109, 115, 177, 178, 219, 221, 344, 345, 346
Fram Strait, 61, 65, 75, 79, 80, 83, 84, 85, 87, 113, 115, 116, 118, 127, 128, 129, 138, 143, 145, 164, 170, 173, 178, 189, 237, 295, 337, 338
Franz Josef Land, 165, 168, 211, 255
Franz Victoria Trough, 177, 178, 186, 211, 213, 237, 238, 254, 269, 344
Fresh water, 62, 116, 138, 199, 209, 238, 260, 264, 270, 296, 345

G
Gakkel Ridge, 48, 113, 122, 138, 143, 144, 297, 349
Garnet, 25, 119, 120, 154, 157, 160, 161, 165, 168, 170, 183, 184, 186, 190, 191, 193, 203, 204, 207, 208, 219, 228, 235, 238, 290
Geochemical association, 292–295, 297–298
Geochemistry, 197, 257, 287, 296
Geoecology, 3
Glacial activity, 43–47, 333–355
Glacial continental margin, 146, 354
Glaciation, 4, 5–6, 9, 15, 27, 29, 31, 33, 36, 38, 39, 41, 54, 57–58, 69, 79, 83, 85, 87, 99, 100, 101, 103, 104, 107, 109, 111, 146, 147, 153, 172, 188, 193, 210, 214, 223, 228, 232, 235, 236, 238, 239, 241, 246, 292, 293, 322, 337, 339, 345–350, 355
Glacier, 4, 11, 29, 33, 38, 39, 41, 45, 57, 85, 86, 87, 102, 107, 112, 115, 125, 133, 147, 164, 166, 169, 172, 216, 219, 222, 223, 235, 246, 255, 324, 325, 342, 343, 344, 345, 346, 348, 349, 350, 354
Glacioisostasy, 84, 239
Grab samples, 197
Grain size, 65, 91, 92, 103, 104, 115, 126, 137, 154, 155, 159, 172, 181, 182, 183, 189, 193, 195, 197, 200, 201, 202, 204, 209, 211, 215–218, 226, 234, 235, 236, 239, 242, 243, 244, 245, 248, 249, 250, 251, 254, 256, 257, 258, 260, 262, 265–266, 270, 273, 275, 282, 289, 307, 308, 311, 314, 316
Gravel, 17, 24, 33, 54, 58, 96, 97, 98, 109, 119, 125, 128, 130, 131, 132, 135, 156, 160, 171, 180, 189, 200, 211, 213, 243, 249, 250, 251, 252, 291, 302, 314, 317, 324, 343
Gravity corer, 69, 148, 211, 268
Greenland Sea, 61, 62, 68, 74, 76, 79, 80, 83, 84, 85, 108, 134, 337
Gyre, 61, 117, 118, 119, 129, 154, 166, 172, 301, 306, 307, 324, 325, 327, 328

H
Heavy minerals, 119, 120, 127, 151, 157, 158, 161, 165, 166, 168, 169, 170, 172, 183, 184, 185, 186, 190, 197, 201, 203, 204, 206, 207, 208, 219, 221, 226, 228, 229, 253, 254, 255, 257, 259, 260, 261, 265, 266, 267, 269, 270, 290, 308

Heinrich event, 81, 137, 152, 153, 166, 351, 352
Hematite, 228
High productivity event, 85, 87, 97
Holocene, 43–45, 47–55, 57–58, 65, 68, 74, 83, 92, 105, 109, 111, 124–135, 138, 148–149, 153, 164, 171, 173, 174, 178, 179, 182, 198, 199, 200, 212, 224–240, 241, 245, 246, 254, 255, 256–288, 292–297, 307, 310, 321, 325, 337, 342, 344, 349, 353, 354
Holocene Optimum, 109, 344
Hornblende, 183, 184, 186, 190, 193, 203, 204, 208, 219, 228, 236, 261, 308
Hydrogen index (HI), 173, 174, 175, 176, 253, 294, 298
Hydrographic network, 199, 265

I

Iceberg, 81, 82, 86, 88, 89, 91, 96, 98, 99, 100, 101, 102, 103, 104, 105, 106, 107, 108, 111, 115, 120, 124, 126, 127, 133, 134, 137, 145, 153, 163, 165, 168, 170, 177, 179, 188, 193, 223, 253, 254, 310, 319, 329, 345, 354, 355
Ice rafted debris (IRD), 64, 65, 66, 88, 90, 91, 100, 101, 103, 105, 109, 110, 116, 119, 126, 128, 133, 137, 149, 152, 153, 154, 156, 162, 164, 166, 168, 169, 172, 200, 208, 209, 216, 226, 250, 253, 311, 312, 316, 325, 339, 343, 351, 353
Ice shed, 87
Ice sheet, 5, 36, 39, 78, 79, 80, 82, 83, 84, 86, 87, 101, 102, 103, 104, 108, 111, 127, 137, 138, 146, 172, 210, 222, 235, 239, 246, 253, 255, 273, 289, 293, 295, 298, 301, 327, 339, 345, 350, 351, 354
Ikaite, 282, 288, 296
Illite, 65, 90, 91, 95, 96, 98, 101, 121, 123, 170, 186, 190, 192, 193, 194, 201, 203, 219, 243, 253, 254, 260, 266, 269, 290, 291, 307, 322, 323, 324, 327
Indigirka River, 290
Inner shelf, 48, 82, 197, 199, 200, 204, 206, 207, 209, 291, 295, 296, 297
Interglaciation, 5, 16, 54, 79, 84, 87, 126, 193, 333, 334, 335, 337
Intermediate water, 62, 103, 178, 199, 226, 306, 351
Interstadial, 3, 4, 5, 6, 8, 15, 26, 33, 78, 85, 87, 110, 137, 162, 169, 170, 298, 337, 340, 354, 355
Iron, 119, 153, 154, 155, 156, 157, 160, 161, 163, 165, 169, 190, 193, 206, 208, 215, 228, 236, 261, 262, 267, 269

K

Kamchatka, 305, 306, 307, 314, 321, 322, 324, 325, 327, 329
Kamchatka Current, 306, 329
Kaolinite, 65, 91, 93, 96, 97, 98, 101, 102, 103, 121, 122, 123, 126, 170, 186, 190, 192, 193, 194, 201, 203, 204, 207, 219, 221, 253, 254, 260, 266, 291, 307, 322, 323
Kara Sea, 28, 114, 121, 122, 123, 144, 150, 165, 168, 172, 178–179, 185, 186, 194, 196, 199, 208, 209, 210, 237, 246, 256, 264, 265, 269, 270, 271, 272, 273, 280, 281, 288, 289, 295, 297, 344, 345
Kastenlot, 148
K-feldspar, 170, 203, 206
Khatanga River, 39, 168, 290
Kolguev Island, 226, 237, 238
Kolyma River, 119, 290, 291
Kuril Islands, 305, 307, 314, 321, 322, 324, 325, 327, 329

L

Laptev Sea, 9, 39, 47, 48, 114, 117, 121, 123, 138, 143, 144, 150, 165, 166, 168, 170, 234, 289–290, 291, 292–297, 344, 348, 349
Lead, 207, 262, 286, 287
Lena River, 119, 289, 295, 296
Light minerals, 187, 193, 197, 201, 203, 206, 291, 308, 310
Lithology, 64, 91, 129, 148–149, 150–153, 155–157, 159, 162–165, 180, 194, 197, 227, 250, 252, 256, 259, 265–266, 272, 274, 282, 286, 287, 301, 302, 304, 312
Lomonosov Ridge, 113, 122, 138, 143, 144, 145, 153, 166, 173, 174, 337, 341

M

Magnesium, 186, 208, 282, 322
Magnetic susceptibility, 110, 201, 206, 256, 265, 268, 269, 310, 311, 314, 316
Magnetostratigraphy, 136
Manganese, 119, 136, 137, 195, 199, 206, 208, 209, 215, 261, 275, 282, 286, 298, 299
Manganese nodules, 199, 206
Marine isotope stage, 69, 149, 151, 169, 307–310
Mass accumulation rate, 65, 68, 98–100, 101, 102, 103, 109, 112, 161, 181, 262–264, 268, 270, 271, 280, 283, 294, 295, 298, 301, 304, 307, 310, 326, 355
Melt water, 43, 232, 239, 255, 337, 341
Mendeleev Ridge, 113, 122, 143, 144
Mid-ocean ridge, 61

Mineral assemblage, 96, 97, 98, 129, 158, 161, 165, 166, 168, 172, 187, 191, 192, 257, 260
Mineralogy, 165–173, 201, 247–256, 322–324
Mixing zone, 198, 199, 200, 201, 203, 204, 207, 209, 243, 262, 263, 264, 271, 272, 275, 280, 282, 288
Mollusks, 14, 15, 17, 18, 22, 24, 25, 31, 36, 37, 39, 41, 47, 52, 53, 54, 96, 97, 98, 100, 119, 133, 134, 156, 215, 227, 230, 233, 243, 257, 259, 269, 273, 274, 276, 281, 282
Moraine, 16, 22, 25, 26, 27, 28, 29, 31, 32, 33, 36, 37, 38, 39, 40, 54, 58, 81, 105, 107, 109, 185, 188, 199, 200, 210, 213, 215, 223, 225, 247, 252, 253, 255, 339, 343
Multicorer, 128, 130, 173

N

Nansen Basin, 113, 122, 124, 127, 129, 138, 143, 177, 178
Nepheloid layer, 103, 134, 200, 207, 253, 256, 270
Non-glacial continental margin, 149
Nordic seas, 61, 62, 63, 64, 66, 67, 68, 69, 75, 76, 77, 78, 79, 80, 81, 82, 86, 107, 146, 296, 351, 354
Northern Island of Novaya Zemlya, 115, 165, 177, 178, 188, 234
Norwegian Current, 61, 62, 88, 90, 112, 178
Norwegian Sea, 4, 65, 68, 79, 80, 83, 84, 85, 88, 90, 98, 100, 101, 127, 129, 177, 337, 338, 339, 347
Novaya Zemlya, 28, 115, 123, 144, 165, 177, 178, 184, 185, 186, 188, 189, 190, 193, 194, 211, 219, 223, 224–240, 246, 253, 254, 344

O

Ob River, 31, 32, 119, 188, 195, 198, 201, 203, 204, 206, 272–288
Ob-Yenisei Shoal, 179, 195, 197, 199, 200, 204, 209, 286, 287, 288, 297
Oscillation, 8, 102, 103, 118
Ostracods, 119, 133, 136, 138, 172, 208, 222, 269, 344
Outer shelf, 48, 197, 200, 206, 207, 208, 209, 286, 291, 297
Outflow, 75, 129
Oxygen isotopes, 5, 6, 36, 65, 68, 70, 85, 88, 107, 110, 137, 138, 139, 149, 161, 162, 171, 172, 232, 237, 238, 239, 253, 270, 287, 294, 299, 309, 310, 311, 327, 328, 335, 338, 339

P

Pacific Ocean, 113, 117, 291, 301, 306, 316, 321, 329
Pack ice, 188, 193, 255, 325, 354
Paleoceanography, 127
Paleocirculation, 88, 327
Paleoenvironment, 210
Paleoproductivity, 100, 110, 111, 165, 224, 236–237, 239, 311, 317, 325, 326, 329, 354
Paleotemperature, 57, 98
Pathway, 65, 101, 127, 165, 183
Pechora River, 22, 29, 45, 57, 119, 239, 246, 348
Pechora Sea, 177–181, 183–186, 210, 224, 226, 228, 235, 236, 239, 241–247, 342, 348
Permafrost, 239, 246, 264, 288, 289, 339, 349, 350, 353
Phosphorus, 208, 209, 275, 282
Plagioclases, 170, 183, 186, 189, 203, 262, 275, 291, 308, 317
Planktonic foraminifers, 65, 88, 90, 91, 93–98, 100, 103, 133–135, 152–154, 156, 158, 165, 212, 213, 222, 230, 237, 238, 254, 299, 301, 351
Polar Drift, *see* Transpolar Drift
Polar Water, 62, 63, 64, 86, 118, 339, 351, 352
Pollen, 16, 47, 333, 334, 342
Potassium, 170, 183, 189, 201, 203, 262, 281, 282, 291, 308, 317
Primary production, 65, 103, 104, 119, 129, 134, 135, 165, 171, 178, 179, 223, 226, 235, 236, 239, 243, 255–256, 289, 290, 291, 298, 301, 306, 310, 325, 355
Provenance, 96, 101, 127, 204, 210, 255, 311, 327
Pyroxene, 120, 165, 169, 183, 190, 267, 308

Q

Quartz, 40, 65, 93, 98, 119, 132, 151, 152, 155–156, 160, 170, 183, 186, 187, 189, 190, 193, 201, 203, 204, 206–209, 248, 251–255, 259, 260, 265, 268, 280, 281, 284, 285, 291, 307, 308
Quaternary, 94, 96, 98, 161, 169, 224, 245–248, 324, 328, 329, 341, 353

R

Radiocarbon age, 28, 154, 215, 225, 234, 243, 257, 263, 273, 287
Radiolarians, 119, 138, 311, 321, 325
Rare earth elements, 268, 307, 319–322
Redox condition, 209
Regression, 28, 33, 43, 339, 349

Retreat, 33, 81, 103, 104, 109, 111, 172, 235, 239, 344
Riverine discharge, 219, 281
River valley, 16, 36, 54, 199, 201, 208, 209, 265, 271, 272, 281, 334, 348
Rock, 132–133, 151–160, 165, 166, 180, 187, 188–190, 193, 194, 198–199, 201, 204, 207, 208–210, 215, 225, 259–263, 302, 308, 310, 311, 317, 321, 349, 350, 353
Rubidium, 201, 275, 281, 282, 283, 314, 316, 317

S

St. Anna Trough, 123, 144, 147, 177, 178, 179, 186–194, 197, 200, 237, 249, 253
Salinity, 11, 22, 24, 37, 62, 63, 90, 116, 118, 128, 129, 130, 179, 188, 189, 197, 198–200, 209, 213, 214, 232, 237, 262, 264, 306, 351
Sand, 54, 93, 94, 97, 98, 107, 119, 125, 128, 130–132, 134, 135, 152, 153, 159, 179–181, 183, 189, 193, 199, 201, 207, 208, 209, 227, 247, 250–254, 257, 259–261, 302, 312, 313, 343
Scandinavia, 4, 11, 61, 65, 85, 107, 123, 133, 134, 178, 340, 345
Sea basin, 14, 120
Sea ice, 63, 113–124, 125–129, 133–135, 164–166, 168–170, 289–291, 297, 301, 304, 306, 309, 317, 325, 328, 329, 337–339, 345, 349, 350–355
Sea level, 4, 15, 16, 17, 24, 29, 43, 45, 47, 53, 57, 85, 109, 164, 172, 234, 246, 247, 268, 271, 289, 293, 296, 325, 345, 347–350
Sea of Okhotsk, 301–302, 305–307, 310, 324–327, 329
Sea surface temperature (SST), 92, 95, 104, 291
Sea water, 82, 101, 115, 128, 164, 178, 197–199, 201, 209, 214, 243, 253, 255, 262, 264, 271, 272, 280, 287, 288
Sedimentary basin, 209, 255, 256
Sedimentary components, 101, 104
Sedimentary material, 84, 87, 88, 90, 102, 103, 104, 105, 109, 115, 116, 125, 129, 133, 164, 165, 166, 168, 169, 172, 177, 179, 183, 184, 186, 188, 189, 198, 199, 200, 208, 209, 216, 221, 232, 234, 239, 270, 289, 290, 291, 322, 349
Sedimentation rate, 49, 68–82, 86, 98–107, 111, 136–139, 144–151, 155, 158, 161–163, 171, 195, 197, 203, 204, 206, 207, 209, 262–264, 268–270, 280, 283, 287, 294, 296–298, 309, 312, 316, 349, 350, 353–355
Sediment core, 69, 89, 91, 92, 94, 95, 96, 99, 102, 106, 107–108, 110, 111, 112, 127, 143, 149, 152, 154–161, 184, 211–222, 224, 225, 227, 229, 231, 233, 247, 249, 250, 252, 257, 259, 261, 263, 265, 266, 267, 269, 272, 273, 274, 276, 277–279, 280, 281–282, 283, 284, 286–287, 294, 295, 299, 304, 305, 309, 312, 315, 318, 319, 320, 323
Severnaya Zemlya, 47, 48, 115, 154, 165, 169, 172, 178–179, 195, 289, 292, 294, 339
Shelf, 100–105, 107–109, 199, 206–209, 298–299
Silica, 133, 201, 203, 206, 207
Silt, 91, 130, 152, 155, 159, 162, 163, 180, 250, 259, 274, 282, 302, 343
Smear-slide, 308, 310
Smectite, 65, 67, 91, 93, 95–98, 101–102, 104, 121–123, 170, 186, 190, 192–194, 201, 203, 204, 206, 207, 219, 221, 254, 260, 261, 266, 267, 290, 291, 322–324, 327
Sodium, 206, 248, 275
Source province, 101, 195, 255, 310, 319, 324, 327
Southern Island of Novaya Zemlya, 177, 224, 235, 238, 239, 246
South Novaya Zemlya Trough, 225, 240
Spitsbergen, 4, 61, 65, 74, 75, 82–87, 115, 118, 121, 123, 124, 125, 127–130, 132–135, 165, 177, 178, 338, 343–346, 352
Spreading, 61
Stadial, 6, 8, 15, 32, 33, 78, 172, 340
Staurolite, 154, 157, 160, 208
Strontium, 206, 208, 262, 275, 276, 282, 283, 314, 317, 318
Subarctic seas, 331–355
Subsidence, 271, 324
Sulphur, 205
Surface water, 61, 63, 65, 90, 97, 98, 102, 103, 104, 108, 116–118, 129, 135, 172, 174, 176, 179, 188–189, 198–200, 206, 207, 238, 262, 264, 270, 301, 305, 306, 327, 337, 339, 341, 342, 351
Surge, 58, 81, 101, 105, 107, 345, 355
Svalbard, 87, 107

T

Taz River, 195, 198, 200, 201
Terrestrial record, 3, 47, 138, 198, 210, 235, 246, 253, 255, 349
Terrigenous-mineralogical province, 185, 219, 228

Terrigenous sedimentation, 311, 354
Thermohaline circulation, 61, 101, 169, 256, 302, 354, 355
Till, 22, 24, 26, 58, 249, 343
Total organic carbon, 173, 210
Transform fault, 295, 297
Transgression, 15–29
Transpolar Drift, 87, 117, 119, 128, 145, 154, 165, 166, 168, 170, 172, 188, 290
Transportation, 154, 165, 169, 180, 189, 322
Turbidites, 88, 89, 105–107, 109, 119, 144, 153, 158–164, 166, 214, 223, 345, 347, 354

U
Underwater slump, 297

V
Vaigach Island, 185
Valday, 4, 8, 16, 22, 26–28, 33, 35, 36, 210, 212–213, 224–225, 234–235, 241, 243, 244, 246–247, 287, 337, 339, 351, 352
Volcanic activity, 83
Volcanic ash, 105, 308, 312, 313–314, 316–317, 323, 324
Volcanic glass, 151, 156, 158, 267, 308
Vøring Plateau, 4, 61, 75, 88–90, 98, 99, 101–107, 338
Voronin Trough, 123, 154, 178, 186, 195

W
Warm epoch, 85, 147, 355
Warming, 4–5, 8, 27, 36, 41, 53, 95, 97, 103–104, 110, 168, 235, 239, 254, 264, 269, 334, 337, 339, 341–344, 350, 352
Water content, 93–98, 213, 245, 257, 259, 260, 264–266, 275, 307, 308
Water current, 87, 127, 135
Water depth, 48, 61, 106, 111, 113, 130, 149, 154, 158, 174, 178, 185, 186, 190, 198, 210, 213, 224, 225, 257, 272, 281, 289–291, 294, 298, 304, 305, 307, 310
Watershed, 198, 288
Weichselian, 78, 80, 81, 84, 101, 109, 138, 210, 255, 287, 292
West Spitsbergen Current, 61, 86, 87, 118, 127, 129, 178, 352
Wind, 322
Wood, 53, 54, 58, 198, 201, 203, 259, 269, 334

Y
Yamal Peninsula, 33
Yana River, 57, 168
Yenisei Gulf, 198, 201
Yenisei River, 32, 119, 198, 201, 206, 208, 209, 256–257, 262, 264, 265, 268–272, 287
Younger Dryas, 43, 47, 48, 94, 98, 100, 101, 104, 166, 225, 239, 309, 341–342, 347, 349

Z
Zircon, 25, 183, 184, 186, 219, 228, 290

GPSR Compliance

The European Union's (EU) General Product Safety Regulation (GPSR) is a set of rules that requires consumer products to be safe and our obligations to ensure this.

If you have any concerns about our products, you can contact us on ProductSafety@springernature.com

In case Publisher is established outside the EU, the EU authorized representative is:

Springer Nature Customer Service Center GmbH
Europaplatz 3
69115 Heidelberg, Germany

Batch number: 10755244

Printed by Printforce, the Netherlands